T0180690

Advances in Intelligent Systems and Computing

Volume 435

Series editor

Janusz Kacprzyk, Polish Academy of Sciences, Warsaw, Poland
e-mail: kacprzyk@ibspan.waw.pl

About this Series

The series "Advances in Intelligent Systems and Computing" contains publications on theory, applications, and design methods of Intelligent Systems and Intelligent Computing. Virtually all disciplines such as engineering, natural sciences, computer and information science, ICT, economics, business, e-commerce, environment, healthcare, life science are covered. The list of topics spans all the areas of modern intelligent systems and computing.

The publications within "Advances in Intelligent Systems and Computing" are primarily textbooks and proceedings of important conferences, symposia and congresses. They cover significant recent developments in the field, both of a foundational and applicable character. An important characteristic feature of the series is the short publication time and world-wide distribution. This permits a rapid and broad dissemination of research results.

Advisory Board

Chairman

Nikhil R. Pal, Indian Statistical Institute, Kolkata, India
e-mail: nikhil@isical.ac.in

Members

Rafael Bello, Universidad Central "Marta Abreu" de Las Villas, Santa Clara, Cuba
e-mail: rbellop@uclv.edu.cu

Emilio S. Corchado, University of Salamanca, Salamanca, Spain
e-mail: escorchado@usal.es

Hani Hagras, University of Essex, Colchester, UK
e-mail: hani@essex.ac.uk

László T. Kóczy, Széchenyi István University, Győr, Hungary
e-mail: koczy@sze.hu

Vladik Kreinovich, University of Texas at El Paso, El Paso, USA
e-mail: vladik@utep.edu

Chin-Teng Lin, National Chiao Tung University, Hsinchu, Taiwan
e-mail: ctlin@mail.nctu.edu.tw

Jie Lu, University of Technology, Sydney, Australia
e-mail: Jie.Lu@uts.edu.au

Patricia Melin, Tijuana Institute of Technology, Tijuana, Mexico
e-mail: epmelin@hafsamx.org

Nadia Nedjah, State University of Rio de Janeiro, Rio de Janeiro, Brazil
e-mail: nadia@eng.uerj.br

Ngoc Thanh Nguyen, Wroclaw University of Technology, Wroclaw, Poland
e-mail: Ngoc-Thanh.Nguyen@pwr.edu.pl

Jun Wang, The Chinese University of Hong Kong, Shatin, Hong Kong
e-mail: jwang@mae.cuhk.edu.hk

More information about this series at http://www.springer.com/series/11156

Suresh Chandra Satapathy
Jyotsna Kumar Mandal · Siba K. Udgata
Vikrant Bhateja
Editors

Information Systems Design and Intelligent Applications

Proceedings of Third International Conference INDIA 2016, Volume 3

Springer

Editors
Suresh Chandra Satapathy
Department of Computer Science and Engineering
Anil Neerukonda Institute of Technology and Sciences
Visakhapatnam
India

Jyotsna Kumar Mandal
Kalyani University
Nadia, West Bengal
India

Siba K. Udgata
University of Hyderabad
Hyderabad
India

Vikrant Bhateja
Department of Electronics and Communication Engineering
Shri Ramswaroop Memorial Group of Professional Colleges
Lucknow, Uttar Pradesh
India

ISSN 2194-5357 ISSN 2194-5365 (electronic)
Advances in Intelligent Systems and Computing
ISBN 978-81-322-2756-4 ISBN 978-81-322-2757-1 (eBook)
DOI 10.1007/978-81-322-2757-1

Library of Congress Control Number: 2015960416

© Springer India 2016
This work is subject to copyright. All rights are reserved by the Publisher, whether the whole or part of the material is concerned, specifically the rights of translation, reprinting, reuse of illustrations, recitation, broadcasting, reproduction on microfilms or in any other physical way, and transmission or information storage and retrieval, electronic adaptation, computer software, or by similar or dissimilar methodology now known or hereafter developed.
The use of general descriptive names, registered names, trademarks, service marks, etc. in this publication does not imply, even in the absence of a specific statement, that such names are exempt from the relevant protective laws and regulations and therefore free for general use.
The publisher, the authors and the editors are safe to assume that the advice and information in this book are believed to be true and accurate at the date of publication. Neither the publisher nor the authors or the editors give a warranty, express or implied, with respect to the material contained herein or for any errors or omissions that may have been made.

Printed on acid-free paper

This Springer imprint is published by SpringerNature
The registered company is Springer (India) Pvt. Ltd.

Preface

The papers in this volume were presented at the INDIA 2016: Third International Conference on Information System Design and Intelligent Applications. This conference was organized by the Department of CSE of Anil Neerukonda Institute of Technology and Sciences (ANITS) and ANITS CSI Student Branch with technical support of CSI, Division-V (Education and Research) during 8–9 January 2016. The conference was hosted in the ANITS campus. The objective of this international conference was to provide opportunities for researchers, academicians, industry personas and students to interact and exchange ideas, experience and expertise in the current trends and strategies for Information and Intelligent Techniques. Research submissions in various advanced technology areas were received and after a rigorous peer-review process with the help of programme committee members and external reviewers, 215 papers in three separate volumes (Volume I: 75, Volume II: 75, Volume III: 65) were accepted with an acceptance ratio of 0.38. The conference featured seven special sessions in various cutting edge technologies, which were conducted by eminent professors. Many distinguished personalities like Dr. Ashok Deshpande, Founding Chair: Berkeley Initiative in Soft Computing (BISC)—UC Berkeley CA; Guest Faculty, University of California Berkeley; Visiting Professor, University of New South Wales Canberra and Indian Institute of Technology Bombay, Mumbai, India, Dr. Parag Kulkarni, Pune; Dr. Aynur Ünal, Strategic Adviser and Visiting Full Professor, Department of Mechanical Engineering, IIT Guwahati; Dr. Goutam Sanyal, NIT, Durgapur; Dr. Naeem Hannoon, Universiti Teknologi MARA, Shah Alam, Malaysia; Dr. Rajib Mall, Indian Institute of Technology Kharagpur, India; Dr. B. Majhi, NIT-Rourkela; Dr. Vipin Tyagi, Jaypee University of Engineering and Technology, Guna; Prof. Bipin V. Mehta, President CSI; Dr. Durgesh Kumar Mishra, Chairman, Div-IV, CSI; Dr. Manas Kumar Sanyal, University of Kalyani; Prof. Amit Joshi, Sabar Institute, Gujarat; Dr. J.V.R. Murthy, JNTU, Kakinada; Dr. P.V.G.D. Prasad Reddy, CoE, Andhra University; Dr. K. Srujan Raju, CMR Technical Campus, Hyderabad; Dr. Swagatam Das, ISI Kolkata; Dr. B.K. Panigrahi, IIT Delhi; Dr. V. Suma, Dayananda Sagar Institute, Bangalore; Dr. P.S. Avadhani,

Vice-Principal, CoE(A), Andhra University and Chairman of CSI, Vizag Chapter, and many more graced the occasion as distinguished speaker, session chairs, panelist for panel discussions, etc., during the conference days.

Our sincere thanks to Dr. Neerukonda B.R. Prasad, Chairman, Shri V. Thapovardhan, Secretary and Correspondent, Dr. R. Govardhan Rao, Director (Admin) and Prof. V.S.R.K. Prasad, Principal of ANITS for their excellent support and encouragement to organize this conference of such magnitude.

Thanks are due to all special session chairs, track managers and distinguished reviewers for their timely technical support. Our entire organizing committee, staff of CSE department and student volunteers deserve a big pat for their tireless efforts to make the event a grand success. Special thanks to our Programme Chairs for carrying out an immaculate job. We place our special thanks here to our publication chairs, who did a great job to make the conference widely visible.

Lastly, our heartfelt thanks to all authors without whom the conference would never have happened. Their technical contributions made our proceedings rich and praiseworthy. We hope that readers will find the chapters useful and interesting.

Our sincere thanks to all sponsors, press, print and electronic media for their excellent coverage of the conference.

November 2015

Suresh Chandra Satapathy
Jyotsna Kumar Mandal
Siba K. Udgata
Vikrant Bhateja

Organizing Committee

Chief Patrons

Dr. Neerukonda B.R. Prasad, Chairman, ANITS
Shri V. Thapovardhan, Secretary and Correspondent, ANITS, Visakhapatnam

Patrons

Prof. V.S.R.K. Prasad, Principal, ANITS, Visakhapatnam
Prof. R. Govardhan Rao, Director-Admin, ANITS, Visakhapatnam

Honorary Chairs

Dr. Bipin V. Mehta, President CSI, India
Dr. Anirban Basu, Vice-President, CSI, India

Advisory Committee

Prof. P.S. Avadhani, Chairman, CSI Vizag Chapter, Vice Principal, AU College of Engineering
Shri D.N. Rao, Vice Chairman and Chairman (Elect), CSI Vizag Chapter, Director (Operations), RINL, Vizag Steel Plant
Shri Y. Madhusudana Rao, Secretary, CSI Vizag Chapter, AGM (IT), Vizag Steel Plant
Shri Y. Satyanarayana, Treasurer, CSI Vizag Chapter, AGM (IT), Vizag Steel Plant

Organizing Chair

Dr. Suresh Chandra Satapathy, ANITS, Visakhapatnam

Organizing Members

All faculty and staff of Department of CSE, ANITS
Students Volunteers of ANITS CSI Student Branch

Program Chair

Dr. Manas Kumar Sanayal, University of Kalyani, West Bengal
Prof. Pritee Parwekar, ANITS

Publication Chair

Prof. Vikrant Bhateja, SRMGPC, Lucknow

Publication Co-chair

Mr. Amit Joshi, CSI Udaipur Chapter

Publicity Committee

Chair: Dr. K. Srujan Raju, CMR Technical Campus, Hyderabad
Co-chair: Dr. Venu Madhav Kuthadi,
Department of Applied Information Systems
Faculty of Management
University of Johannesburg
Auckland Park, Johannesburg, RSA

Special Session Chairs

Dr. Mahesh Chandra, BIT Mesra, India, Dr. Asutosh Kar, BITS, Hyderabad: "Modern Adaptive Filtering Algorithms and Applications for Biomedical Signal Processing Designs"
Dr. Vipin Tyagi, JIIT, Guna: "Cyber Security and Digital Forensics"
Dr. Anuja Arora, Dr. Parmeet, Dr. Shikha Mehta, JIIT, Noida-62: "Recent Trends in Data Intensive Computing and Applications"
Dr. Suma, Dayananda Sagar Institute, Bangalore: "Software Engineering and its Applications"
Hari Mohan Pandey, Ankit Chaudhary: "Patricia Ryser-Welch, Jagdish Raheja", "Hybrid Intelligence and Applications"
Hardeep Singh, Punjab: "ICT, IT Security & Prospective in Science, Engineering & Management"
Dr. Divakar Yadav, Dr. Vimal Kumar, JIIT, Noida-62: "Recent Trends in Information Retrieval"

Track Managers

Track #1: Image Processing, Machine Learning and Pattern Recognition—Dr. Steven L. Fernandez
Track #2: Data Engineering—Dr. Sireesha Rodda
Track #3: Software Engineering—Dr. Kavita Choudhary
Track #4: Intelligent Signal Processing and Soft Computing—Dr. Sayan Chakraborty

Technical Review Committee

Akhil Jose Aei, Vimaljyothi Engineering College (VJEC), Kannur, Kerala, India.
Alvaro Suárez Sarmiento, University of Las Palmas de Gran Canaria.
Aarti Singh, MMICTBM, M.M. University, Mullana, India.
Agnieszka Boltuc, University of Bialystok, Poland.
Anandi Giri, YMT college of Management, Navi Mumbai, India.
Anil Gulabrao Khairnar, North Maharashtra University, Jalgaon, India.
Anita Kumari, Lovely Professional University, Jalandhar, Punjab
Anita M. Thengade, MIT COE Pune, India.
Arvind Pandey, MMMUT, Gorakhpur (U.P.), India.
Banani Saha, University of Calcutta, India.
Bharathi Malakreddy, JNTU Hyderabad, India.
Bineet Kumar Joshi, ICFAI University, Dehradun, India.
Chhayarani Ram Kinkar, ICFAI, Hyderabad, India.
Chirag Arora, KIET, Ghaziabad (U.P.), India.
C. Lakshmi Devasena, IFHE University, Hyderabad, India.
Charan S.G., Alcatel-Lucent India Limited, Bangalore, India
Dac-Nhuong Le, VNU University, Hanoi, Vietnam.
Emmanuel C. Manasseh, Tanzania Communications Regulatory Authority (TCRA)
Fernando Bobillo Ortega, University of Zaragoza, Spain.
Frede Blaabjerg, Department of Energy Technology, Aalborg University, Denmark.
Foued Melakessou, University of Luxembourg, Luxembourg
G.S. Chandra Prasad, Matrusri Engineering College, Saidabad, Hyderabad
Gustavo Fernandez, Austrian Institute of Technology, Vienna, Austria
Igor N. Belyh, St. Petersburg State Polytechnical University
Jignesh G. Bhatt, Dharmsinh Desai University, Gujarat, India.
Jyoti Yadav, Netaji Subhas Institute of Technology, New Delhi, India.
K. Kalimathu, SRM University, India.
Kamlesh Verma, IRDE, DRDO, Dehradun, India.
Karim Hashim Kraidi, The University of Al-Mustansiriya, Baghdad, Iraq
Krishnendu Guha, University of Calcutta, India

Lalitha RVS, Sri Sai Aditya Institute of Science and Technology, India.
M. Fiorini, Poland
Mahdin Mahboob, Stony Brook University.
Mahmood Ali Mirza, DMS SVH College of Engineering (A.P.), India
Manimegalai C.T., SRM University, India.
Mp Vasudha, Jain University Bangalore, India.
Nikhil Bhargava, CSI ADM, Ericsson, India.
Nilanjan Dey, BCET, Durgapur, India.
Pritee Parweker, ANITS, Visakhapatnam
Sireesha Rodda, GITAM, Visakhapatnam
Parama Bagchi, India
Ch Seshadri Rao, ANITS, Visakhapatnam
Pramod Kumar Jha, Centre for Advanced Systems (CAS), India.
Pradeep Kumar Singh, Amity University, Noida, U.P. India.
Ramesh Sunder Nayak, Canara Engineering College, India.
R.K. Chauhan, MMMUT, Gorakhpur (U.P.), India
Rajiv Srivastava, Scholar tech Education, India.
Ranjan Tripathi, SRMGPC, Lucknow (U.P.), India.
S. Brinda, St. Joseph's Degree and PG College, Hyderabad, India.
Sabitha G., SRM University, India.
Suesh Limkar, AISSMS IOIT, Pune
Y.V.S.M. Murthy, NIT, Surathkal
B.N. Biswal, BEC, Bhubaneswar
Mihir Mohanty, SOA, Bhubaneswar
S. Sethi, IGIT, Sarang
Sangeetha M., SRM University, India.
Satyasai Jagannath Nanda, Malaviya National Institute of Technology Jaipur, India.
Saurabh Kumar Pandey, National Institute of Technology, Hamirpur (H.P.), India.
Sergio Valcarcel, Technical University of Madrid, Spain.
Shanthi Makka, JRE School of Engineering, Gr. Noida (U.P.), India.
Shilpa Bahl, KIIT, Gurgaon, India.
Sourav De, University Institute of Technology, BU, India.
Sourav Samanta, University Institute of Technology, BU, India.
Suvojit Acharjee, NIT, Agartala, India.
Sumit Soman, C-DAC, Noida (U.P.), India.
Usha Batra, ITM University, Gurgaon, India.
Vimal Mishra, MMMUT, Gorakhpur (U.P.), India.
Wan Khairunizam Wan Ahmad, AICOS Research Lab, School of Mechatronic, UniMAP.
Yadlapati Srinivasa Kishore Babu, JNTUK, Vizianagaram, India.

Contents

About the Editors

Dr. Suresh Chandra Satapathy is currently working as Professor and Head, Department of Computer Science and Engineering, Anil Neerukonda Institute of Technology and Sciences (ANITS), Visakhapatnam, Andhra Pradesh, India. He obtained his Ph.D. in Computer Science Engineering from JNTUH, Hyderabad and his Master's degree in Computer Science and Engineering from National Institute of Technology (NIT), Rourkela, Odisha. He has more than 27 years of teaching and research experience. His research interests include machine learning, data mining, swarm intelligence studies and their applications to engineering. He has more than 98 publications to his credit in various reputed international journals and conference proceedings. He has edited many volumes from Springer AISC and LNCS in the past and he is also the editorial board member of a few international journals. He is a senior member of IEEE and Life Member of Computer society of India. Currently, he is the National Chairman of Division-V (Education and Research) of Computer Society of India.

Dr. Jyotsna Kumar Mandal has an M.Sc. in Physics from Jadavpur University in 1986, M.Tech. in Computer Science from University of Calcutta. He was awarded the Ph.D. in Computer Science & Engineering by Jadavpur University in 2000. Presently, he is working as Professor of Computer Science & Engineering and former Dean, Faculty of Engineering, Technology and Management, Kalyani University, Kalyani, Nadia, West Bengal for two consecutive terms. He started his career as lecturer at NERIST, Arunachal Pradesh in September, 1988. He has teaching and research experience of 28 years. His areas of research include coding theory, data and network security, remote sensing and GIS-based applications, data compression, error correction, visual cryptography, steganography, security in MANET, wireless networks and unify computing. He has produced 11 Ph.D. degrees of which three have been submitted (2015) and eight are ongoing. He has supervised 3 M.Phil. and 30 M.Tech. theses. He is life member of Computer Society of India since 1992, CRSI since 2009, ACM since 2012, IEEE since 2013 and Fellow member of IETE since 2012, Executive member of CSI Kolkata Chapter. He has delivered invited lectures and acted as programme chair of many international conferences and also edited nine volumes of proceedings from Springer AISC series, CSI 2012 from McGraw-Hill, CIMTA 2013 from Procedia Technology, Elsevier. He is reviewer of various international journals and conferences. He has over 355 articles and 5 books published to his credit.

Dr. Siba K. Udgata is a Professor of School of Computer and Information Sciences, University of Hyderabad, India. He is presently heading Centre for Modelling, Simulation and Design (CMSD), a high-performance computing facility at University of Hyderabad. He obtained his Master's followed by Ph.D. in Computer Science (mobile computing and wireless communication). His main research interests include wireless communication, mobile computing, wireless sensor networks and intelligent algorithms. He was a United Nations Fellow and worked in the United Nations University/International Institute for Software Technology (UNU/IIST), Macau, as research fellow in the year 2001. Dr. Udgata is working as principal investigator in many Government of India funded research projects, mainly for development of wireless sensor network applications and application of swarm intelligence techniques. He has published extensively in refereed international journals and conferences in India and abroad. He was also on the editorial board of many Springer LNCS/LNAI and Springer AISC Proceedings.

Prof. Vikrant Bhateja is Associate Professor, Department of Electronics and Communication Engineering, Shri Ramswaroop Memorial Group of Professional Colleges (SRMGPC), Lucknow, and also the Head (Academics & Quality Control) in the same college. His areas of research include digital image and video processing, computer vision, medical imaging, machine learning, pattern analysis and recognition, neural networks, soft computing and bio-inspired computing techniques. He has more than 90 quality publications in various international journals and conference proceedings. Professor Vikrant has been on TPC and chaired various sessions from the above domain in international conferences of IEEE and Springer. He has been the track chair and served in the core-technical/editorial teams for international conferences: FICTA 2014, CSI 2014 and INDIA 2015 under Springer-ASIC Series and INDIACom-2015, ICACCI-2015 under IEEE. He is associate editor in International Journal of Convergence Computing (IJConvC) and also serves on the editorial board of International Journal of Image Mining (IJIM) under Inderscience Publishers. At present, he is guest editor for two special issues floated in International Journal of Rough Sets and Data Analysis (IJRSDA) and International Journal of System Dynamics Applications (IJSDA) under IGI Global publications.

A New Private Security Policy Approach for DDoS Attack Defense in NGNs

Dac-Nhuong Le, Vo Nhan Van and Trinh Thi Thuy Giang

Abstract Nowadays, the Distributed Denial of Service (DDoS) attack is still one of the most common and devastating security threats to the internet. This problem is progressing quickly, and it is becoming more and more difficult to grasp a global view of the problem. In this paper, we propose a new defense method used the bandwidth in second that a server can use for UDP packets is set as a parameter for controlling a DDoS attack by using the number of UDP packets available. It is registered in the private security policy as a parameter for detecting a flood attack. The efficiency of our proposed method was also proved in the experiments with NS2. DDoS attack is controlled effectively by the private security policy the bandwidth of the regular traffic would be maintained.

Keywords Distributed denial of service · UDP flood attack · Private security policy · Next generation network · NS2

1 Introduction

The DDoS is developing quickly and become is progressing quickly, and it is becoming more and more difficult to grasp a global view of the problem. The attackers have been abused communication protocols such as character generator, NTP and DNS. These are all based on the UDP which indirectly allows attackers to

D.-N. Le (✉)
Hai Phong University, Hai Phong, Vietnam
e-mail: Nhuongld@hus.edu.vn

V.N. Van
Duy Tan University, Da Nang, Vietnam
e-mail: vovannhan@duytan.edu.vn

T.T.T. Giang
Hanoi University of Science, Vietnam National University, Hanoi, Vietnam
e-mail: trinhthuygiang@hus.edu.vn

© Springer India 2016
S.C. Satapathy et al. (eds.), *Information Systems Design and Intelligent Applications*, Advances in Intelligent Systems and Computing 435,
DOI 10.1007/978-81-322-2757-1_1

disguise their identities through the fake address so they are not immediately determine the source of an attack. Attackers send small request packets to intermediary victim servers, and those servers in turn respond to the attackers intended target. The availability of these vulnerable protocols, which are often enabled by default in server software, make the Internet a ready-to-use botnet of potential victim devices that can be exploited by malicious actors to launch huge attacks [1, 2].

In this paper, we propose an approach for guaranteed QoS to normal users under DDoS flood attack based on bandwidth dynamic assignment in order to sustain the server. The rest of this paper is organized as follows: Sect. 2 introduces state of art DDoS attacks defense. Section 3 presents our method suggested to bandwidth dynamic controlled UDP DDoS flood attack. Section 4 analysis experiment results. Finally, Sect. 5 concludes the paper.

2 Related Works

The authors in [3] evaluated the efficiency of access control list against DDoS attacks used the rules affected the collateral damage. The unique technique to detect DDoS attacks on networks called Distributed Change Point Detection using change aggregation tree is presented in [4]. An evaluated the impact of a UDP flood attack on a network using a testbed method and negative selection algorithm in [5, 6]. The author in [7] presented an approach to conduct DDoS experiments based on TCP and UDP flood attacks used access control lists and rate limiting. An UDP flood attack defense called Protocol Share Based Traffic Rate Analysis introduced in [8]. In [9], the authors conducted an experiment on DDoS attacks using simulation software called Bonesi.

A UDP flood attack in DDoS attacks is a method causing host based denial of service. It occurs when an attacker crafts numerous UDP packets to random destination ports on a victim computer. The victim computer, on receipt of the UDP packet, would respond with appropriate ICMP packets if one of these ports is closed. Numerous packet responses would slow down the system or crash it [6].

3 Defending UDP Attack Based on Private Security Policy

In the experiments, we assumes that the target network is the NGN, the ISP provide their services over the NGN which is a private extension and the edge routers can be controlled. The NGN topology has two types of routers: Entrance-Side Edge Routers (*EnRouter*) and Exit-Slide Edge Router (*ExRouter*). The abnormal IP packets are detected by the carrier and ISP side. Individual information is used and determined by a *Private Security Policy* (PSP) that a user had registered before the normal IP packets are never discarded as a result of being false recognized (Fig. 1).

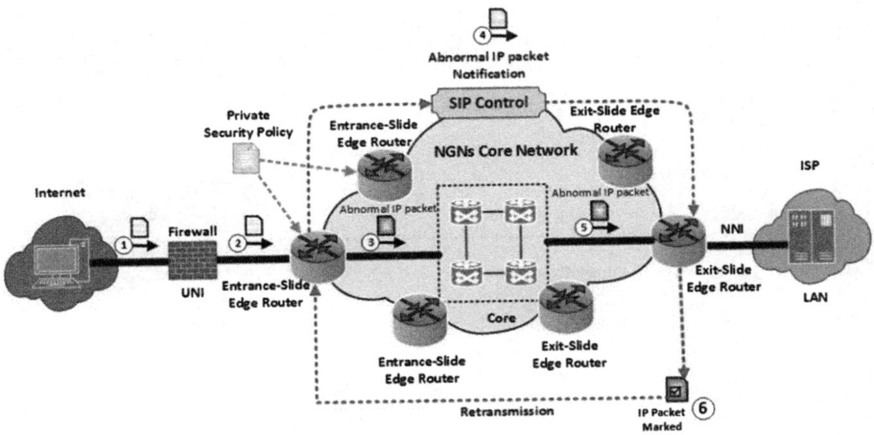

Fig. 1 UDP flooding DDoS attack defense method using private policy

DDoS attack using UDP accelerates the load in the network. It is hard to counter because UDP packets sent from network are limited by the LAN, and enough countermeasures have not been developed. Hence, our method targets DDoS attacks using UDP. UDP flood attack is detected by judging whether it exceeds a certain number of UDP packets (*available bandwidth*) in the unit time output from the NGN side. The number is determined according to the private security policy on the LAN side and registered in the PSP before the DDoS attack detection starts. The UDP DDoS attack can be controlled down to the minimum bandwidth available for forwarding UDP packets from the NGN to a LAN, which is registered similarly. Therefor, the attacks is controlled by delaying UDP packets judged to be DoS attack packets. Our process used private policy has four phases is shown in Fig. 2.

Phase 1 *Initialization private security policy at EnRouter*. The bandwidth in a time unit that a server can use for UDP packets is set as a parameter for controlling a DDoS attack by using the number of UDP packets available bandwidth in time unit (*seconds*). It is registered in the private security policy as a parameter for detecting a UDP-based DDoS attack. The PSP includes destination IP address, DoS attack detection threshold and UDP communication bandwidth.

Phase 2 *Detection abnormal IP packet at EnRouters*. The number of UDP packets arriving at a server is always observed at EnRouters. The attack is judged to occur when the number of UDP packets (*for one second*) exceeds the values registered in the PSP such as: the packet type, destination IP address, and destination port number are examined. When the measured number of UDP packets in time unit equals to the number registered in the PSP or more are detected, they are recognized as abnormal packets. So, those packets are judged to be the attack packets. Destination IP address "*IP*", destination port number "*Port*", and average packet size are read from packets judged to be the attack packets (*bytes*). The

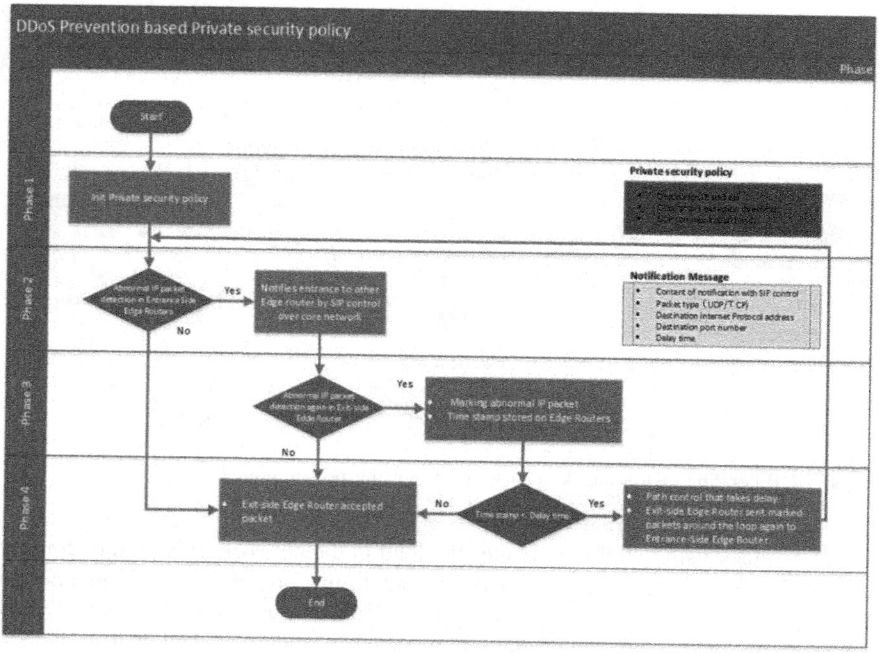

Fig. 2 DDoS prevention based on private security policy

bandwidth of UDP packets must be controlled and limited, then we calculate the packet delay time. After the attack is detected at ExRouter, all EnRouters are notified of the attack by SIP control. This SIP is assumed to be our private extension. The SIP based notification contains information about the type of attack packets and their destination IP address and destination port number and the required delay. This information is used for marking IP packets. The delay time is used by the path control, which imposes this delay.

Phase 3 *Marked abnormal IP packets.* The EnRouter checks all IP packets as they pass through. It checks for IP packet information corresponding to the packet type, destination IP address, and destination port number of attack packets reported by ExRouter by SIP control and writes the extraction time (*as a timestamp*) in the extension header field of the IP packet. If IP packet being MTU size, the fragmentation of IP packets will be done before the timestamp is written in these IP packets. All IP packet are examined and they having destination port number "*Port*" and packet type "*UDP*" and destination IP address "*IP*" are regarded as attack packets. The information extraction time is written into the IP packets in order to mark them.

Phase 4 *Controlled the abnormal IP packets at ExRouter.* All marked IP packets are forwarded by the EnRouter around a loop so that they return later. When an IP packet is received, the marked timestamp is compared with the current time to ensure that the required delay had been added. The path control for delaying IP packets marked at EnRouter by retransmitted along a route with a loop added that returns to the EnRouter. When a marked IP packet is received at EnRouter, the Edge Router judges whether the current time minus the marking time is greater than the required delay time. Then, IP packets that have been delayed long enough are transmitted to the destination IP address through the NGN core router, while ones with insufficient delay are sent around the loop again. So, IP packets transmitted from the ExRouter to the destination IP address "*IP*" and destination port number "*Port*" are transmitted at a reduced bitrates.

Thus, our method decreases the number of IP packets per time unit by applying a path control that imposes a delay. In this way, the attacks can be controlled. Note that the point here is to delay them. Packets that are clearly malicious are discarded. However, genuine or borderline packets are delayed so that the user can decide. As a result, the influence of DDoS packets on other packets user traffic becomes extremely small.

4 Experiment and Results

4.1 Network Topology and NS2 Implementation

The network topology for the simulation is shown in Fig. 3. Network architecture consists HTTP Server DNS1 Server, DNS2 Server, NTF Server, regular Users, DDoS Attack connected by NGN network. In NGN network, the routers are divided into 3 categories Core Router, EnRouters, ExRouters. We assumed both of them send UDP, TCP packets via NGN from the Internet to the servers. The bandwidth

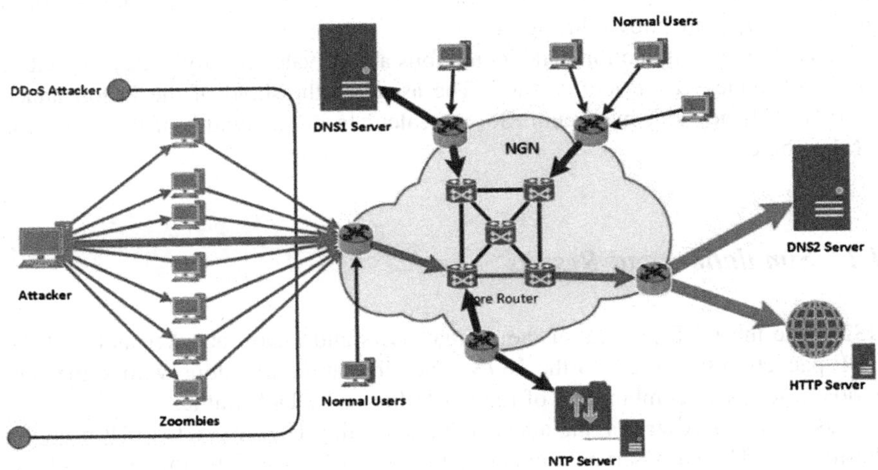

Fig. 3 Network topology for simulation

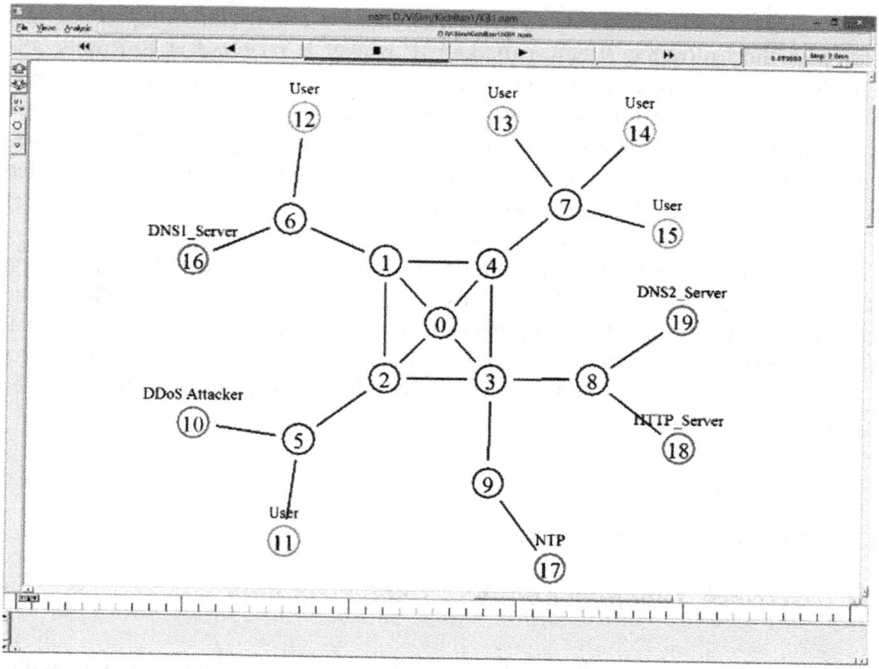

Fig. 4 Network structure simulation in NS2

of the access line between NGN and servers was 10 Mbps, and that of both the Internet and NGN was much broader 10 Mbps.

We evaluated the effect of our defense technique using Network Simulator version 2.3.3 (NS2) [10]. The network topology implemented include 20 nodes $(n_0, n_1, \ldots, n_{19})$ are shown in Fig. 4.

Table 1 shows simulation detail conditions and private security policy to control the DDoS attacks using our method. The assumed threshold of the DDoS attack is 10 % of bandwidth maximum. So, we calculate to bandwidth of the control is 1 Mbps here.

4.2 Simulation and Results

NS2 made the UDP packets of the regular users and DDoS attacker and sent the UDP packets respectively to the NGN. The simulation was done with some scenarios in terms of combination of regular traffic and DDoS traffic.

Case study 1 After 5 s, the attacker starts sending UDP packets to Http Server during 30 s. The Http Server is congested temporarily caused by the DDoS attack. Therefore, all requests from user send to the server will be rejected. We started set

Table 1 Object's parameter detail for simulation in NS2

Classification	NS-2 object	Parameter	Values
Core router	n0, n1, n2, n3, n4		
EnRouters	n5, n6, n7, n9		
ExRouter	n8		
DDoS attacker	n10	Protocol	UDP
		Size of packet	64 Byte
		Transmission pattern	Constant bit rate
		Transmission bandwidth	10 Mbps
		Port	53
Users	n11, n12	Protocol	TCP
		Size of packet	64/512/1024/1500 Bytes
		Transmission pattern	Constant Bit Rate
	n13, n14, n15	Transmission bandwidth	1.3 Mbps
		Transmission bandwidth	5 Mbps
		Port	80
DNS1 server	n16	Protocol: UDP	Bandwidth maximum: 10 Mbps
		Port UDP: 53	Transmission bandwidth: 5 Mbps
		Transmission pattern	Constant bit rate
NTP server	n17	Protocol: UDP	Bandwidth maximum: 10 Mbps
		Port UDP: 123	Transmission bandwidth: 5 Mbps
		Transmission pattern	Constant bit rate
HTTP server	n18	Protocol: UDP/TCP	Bandwidth maximum: 10 Mbps
		Port TCP: 80	Port UDP: 123
		normal udp bandwidth	0.175 Mbps
		DDoS attack detection	Threshold 10 % = 1 Mbps
DNS2 server	n19	Protocol: UDP	Bandwidth maximum: 10 Mbps
		Port UDP: 53	Normal UDP bandwidth: 0.35 Mbps
		Port UDP: 123	Normal UDP bandwidth: 0.175 Mbps
		DDoS attack detection	Threshold 10 % = 1 Mbps

up our private security policy in the EnRouters and ExRouters. When our control method had been executed, the communication bandwidth of regular users was secured and these users could communicate without any influence of the attack. It is also confirmed that traffic does flow into the internal NGN, because the attack is controlled effectively by the EnRouter is illustrated in Fig. 5.

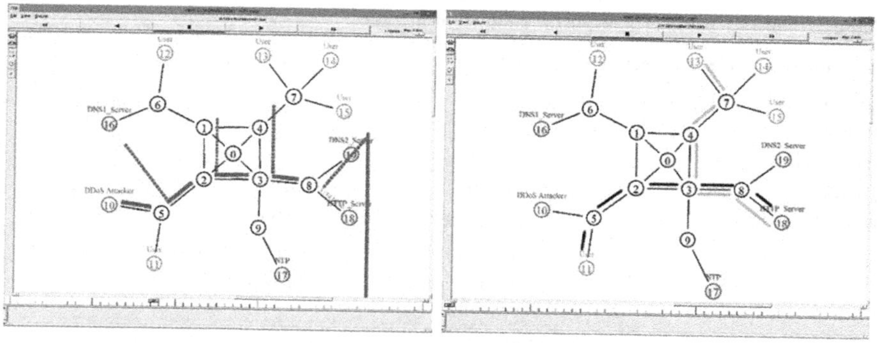

Fig. 5 Congestion was temporarily caused by the DDoS attack in HTTP Server and DDoS attack is controlled effectively by the private security policy in case study 1

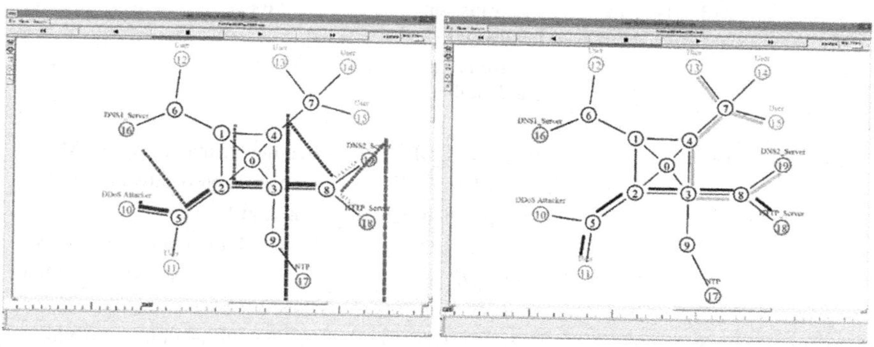

Fig. 6 Congestion was temporarily caused by the DDoS attack in HTTP server and DDoS attack is controlled effectively by the private security policy in case study 2

Case study 2 Similar, after 5 s, the attacker starts sending UDP packets to HTTP Server and DNS2 Server during 30 s. Both Http Server and DNS2 Server are congested temporarily caused by the DDoS attack. We had been executed our proposed method and the DDoS attack is controlled effectively is illustrated in Fig. 6.

To evaluate the effectiveness of the proposed method, we compared the UDP bandwidth of servers. The results shows that from 0 to 5 s, traffic ratio of both users and DDoS traffic were less than 5 Mbps around and no bandwidth degradation occurred on the regular traffic. Starting at 5th seconds, the DDoS attack occupying almost full bandwidth of the access link ($\approx$10 Mbps) and our private security policy had been executed. After 5 s the DoS attack started, our countermeasures worked effectively, the bandwidth of the DDoS traffic was decrease by the proposed countermeasures so that the bandwidth of the regular traffic would be maintained.

5 Conclusion and Future Works

In this paper, we propose a UDP DDoS attack defense method using private security policy. The efficiency of our proposed method was also proved in the experiment with NS2. DDoS attack is controlled effectively by the private security policy the bandwidth of the regular traffic would be maintained. Our next goal is guaranteed QoS to normal users under DDoS flood attack in NGNs and suggest an intelligent Intrusion Detection System and evaluate the impact of a UDP flood attack on various defense mechanisms (Figs. 7 and 8).

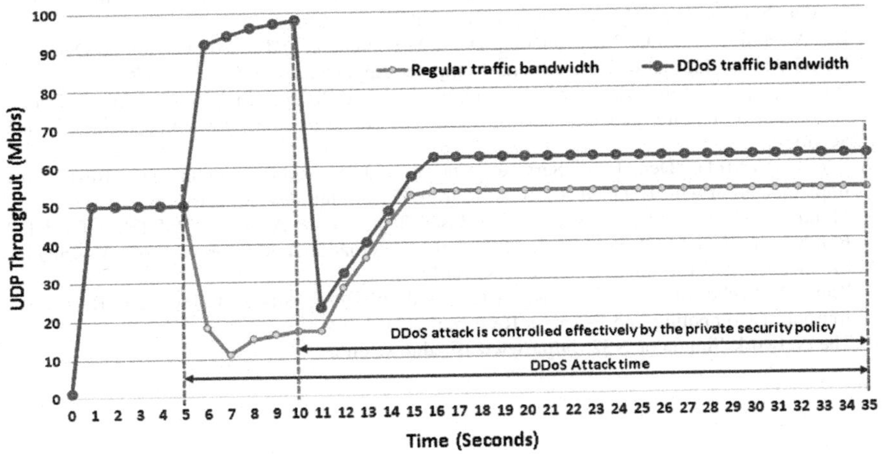

Fig. 7 Evaluated UDP throughput before and during attack in case study 1

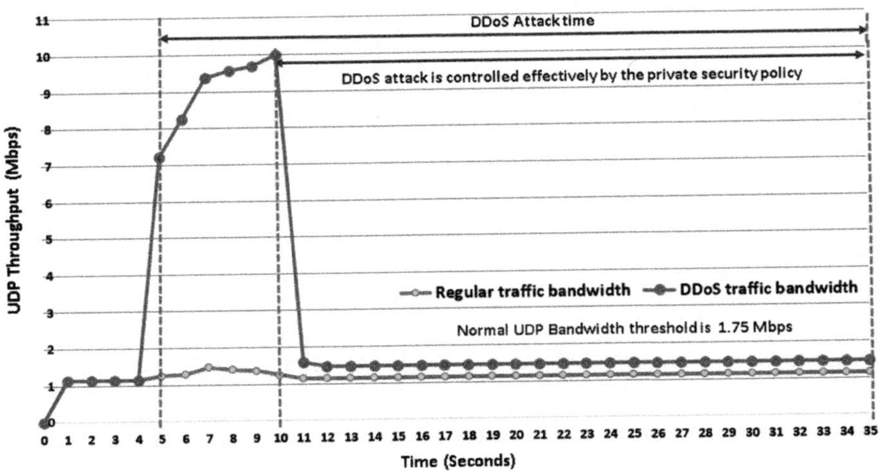

Fig. 8 Evaluated UDP throughput before and during attack in case study 2

References

1. Deepika Mahajan and Monika Sachdeva (2013). DDoS Attack Prevention and Mitigation Techniques—A Review. International Journal of Computer Applications 67(19):21–24. Published by Foundation of Computer Science, USA.
2. Arun Raj Kumar et al (2009), Distributed Denial-of-Service Threat in Collaborative Environment A Survey on DDoS Attack Tools and Traceback Mechanisms, IEEE International Advance Computing Conference (IACC 2009) Patiala, India.
3. Pack, G., Yoon, J., Collins, E., Estan, C. (2006). On Filtering of DDoS Attacks Based on Source Address Prefixes. The Securecomm and Workshops, Baltimore. doi:10.1109/SECCOMW.2006.359537.
4. David, D. (2007). Prevent IP spoofing with the Cisco IOS. Retrieved from http://www.techrepublic.com/article/prevent-ip-spoofing-with-the-cisco-ios.
5. Lu, W. et al (2009). One-Way Queuing Delay Measurement and Its Application on Detecting DDoS Attack. Journal of Network and Computer Applications, 32(2), 367–376.
6. Rui, X., Li, M., Ling, Z. (2009). Defending against UDP Flooding by Negative Selection Algorithm Based on Eigenvalue Sets. International Conference on Information Assurance and Security.
7. Rao, S. (2011). Denial of Service Attacks and Mitigation Techniques: Real Time Implementation with Detailed Analysis. Retrieved from http://www.sans.org.
8. Mohd, Z. et al (2011). Protocol Share Based Traffic Rate Analysis (PSBTRA) for UDP Bandwidth Attack. Paper presented at the International Conference on Informatics Engineering and Information Science, Kuala Lumpur.
9. Kaur, D., Sachdeva, M., Kumar, K. (2012). Study of DDoS Attacks Using DETER Testbed. Journal of Computing and Business Research, 3(2), 1–13.
10. The Network Simulator NS2, http://www.isi.edu/nsnam/ns.

An Effective Approach for Providing Diverse and Serendipitous Recommendations

Ivy Jain and Hitesh Hasija

Abstract Over the years, recommendation systems successfully suggest relevant items to its users using one of its popular methods of collaborative filtering. But, the current state of recommender system fails to suggest relevant items that are unknown (novel) and surprising (serendipitous) for its users. Therefore, we proposed a new approach that takes as input the positive ratings of the user, positive ratings of the similar users and negative ratings of the dissimilar users to construct a hybrid system capable of providing all possible information about its users. The major contribution of this paper is to diversify the suggestions of items, a user is provided with. The result obtained shows that as compared to general collaborative filtering, our algorithm achieves better catalogue coverage. The novelty and serendipity results also proved the success of the proposed algorithm.

Keywords Recommender systems · Collaborative filtering · Weighted catalog coverage · Novelty · Serendipity

1 Introduction

Recommendation systems or recommender systems are used very frequently now days to provide ratings or preferences to a given item like movie, restaurants, etc. [1]. Recommender systems are basically classified into three types [2] and they are:

I. Jain (✉)
Vivekananda Institute of Professional Studies, Delhi, India
e-mail: ivy.jain@yahoo.in

H. Hasija
Delhi Technological University, Delhi, India
e-mail: hitoo.hasija@gmail.com

© Springer India 2016

S.C. Satapathy et al. (eds.), *Information Systems Design and Intelligent Applications*, Advances in Intelligent Systems and Computing 435,
DOI 10.1007/978-81-322-2757-1_2

1. Content filtering approach
2. Collaborative filtering approach
3. Hybrid filtering approach

In collaborative filtering approach [3], the recommendations are provided by determining similar kinds of users and then analysing their preferences, considering that similar kinds of users have same likings and same ratings for different items. Last.fm is an example of it. Collaborative recommender is further classified into model and memory based approaches [4]. In memory based approach the rating is based on aggregate of similar user ratings for the same item while in model base approach, ratings are used to train a model which is further used to provide ratings.

In content filtering approach [5], past ratings provided by a user are used to provide new recommendations with various algorithms like Bayesian classifier, decision trees, genetic algorithm etc. [6–8]. For example, based on the previous news browsing habits of a user, it could be determined whether particular news would be liked by user or not. In hybrid based approach [9–13], the characteristics of both content and collaborative are combined into a single model to overcome the shortcomings of both simultaneously. Netflix is the best example of it.

This paper consists of 5 sections, Sect. 2 consists of background details, methodology used to solve the problem is described in Sect. 3, results and comparison with previous approaches has been covered in Sect. 4, finally conclusion and future work are described in Sect. 5.

2 Background

Collaborative filtering works on the opinion of people that have interest like you as seen in the past, to predict the items that may interest you now. While providing recommendation, in collaborative filtering approach similar users are determined by various approaches like k-nearest neighbour classifier in such a way that the thinking of all these users is same. Hence, the recommendations provided by these users would also be similar. As one of the approaches to determine similar thinking users in collaborative filtering is Pearson's correlation coefficient. So for our work, we would be using Pearson's correlation coefficient [14, 15] to find the similarity of a user with other users. Pearson's correlation coefficient is calculated by the formula:

$$\rho = \frac{\sum (x - \bar{x})(y - \bar{y})}{\sqrt{\sum (x - \bar{x})(x - \bar{x})}\sqrt{\sum (y - \bar{y})(y - \bar{y})}} \tag{1}$$

$$\bar{x} = \frac{1}{n}\sum_{i=1}^{n} xi \tag{2}$$

As stated in Eq. (1), Pearson's correlation coefficient signifies the linear relationship between the two variables. As stated in Eq. (2), x (bar) is the calculation of sample mean. According to Cohen [16], the value of Pearson's correlation coefficient ranges from −1 to +1. The zero value of coefficient signifies that, there is no dependency between the two variables. The negative value implies inverse relationship i.e. as the value of one variable increases the value of another variable decreases. The positive value implies positive relationship that is, as the value of one variable increases so as the value of another variable. Pearson's correlation coefficient is symmetric in nature. It is also invariant to the scaling of two variables. Thus, it can also be used for determining correlation between two different variables like the height of a player and his goal taking capability in a basketball match. While determining the value of Pearson's correlation coefficient between two different variables, their units are not changed. Because, the method of determining this coefficient is such that it do not depends on the units of those variables. But, collaborative filtering suffers from some major problems as stated in [17]. Sparse data problem deals with the few or none ratings provided by user for some of the input data, and there are various approaches as justified by [18–20] to deal with it, in different situations. Hence, opting different approaches for different problem domains is a cumbersome task [21].

3 Methodology Used to Solve the Problem

If the value of Pearson's correlation coefficient is determined between two users by comparing their recommendations for two different movies, then that value will provide us the information that whether these two users have similar liking or not. If a high positive value which is greater than 0.5 is obtained for any two users then they could be considered as friends of each other. Similarly, friends of friends could be determined. Now, if the value obtained is less than −0.5 then, those two users would be considered as enemies of each other. But, enemies of enemies have similar likings as that of its friends, just like the opposite of opposite would be equivalent to the similar one. In this way, first of all friends as well as friends of friends are determined and then enemies of enemies are determined. In a nutshell, it will provide us with those users whose thinking's are similar. This approach is just like an alternative to the various approaches used in collaborative filtering. Finally, five recommendations are obtained based on friends of a particular user i.e. based on collaborative filtering approach. Then, using positive ratings of similar users i.e. likes of likes for a user, three novel recommendations are obtained. At the end, taking negative ratings of dissimilar users i.e. taking dislikes of dislikes for a user, two serendipitous recommendations are obtained.

4 Results and Comparison

A. Study of the system

To evaluate the performance of our system we created browser based software and conducted a user study for 100 users and 200 movies in IT laboratory. The group selected for study included IT professional and the movies selected were from "movie lens" dataset of varying genres. The screenshot of the system is shown in Fig. 1.

The system does not provide any recommendation till the user provide ratings of at least 5 movies. To help user's in selecting a movie to watch, this system suggests some movies based on their popularity following a non personalized recommender approach. After the user provided some ratings, the system suggested some normal recommendations based on the popular collaborative filtering, some novel recommendations based on the positive ratings of similar user and some serendipitous recommendations based on the negative ratings of the dissimilar users.

The user then evaluates the recommendations provided by giving a rating on scale 1–5. The user is also asked a question- "are you familiar with the movie" to know the novelty content provided by the system. The user is also provided with an explanation facility, using a question mark beside the novel and serendipitous recommendation in the software. This explanation facility is to let the user

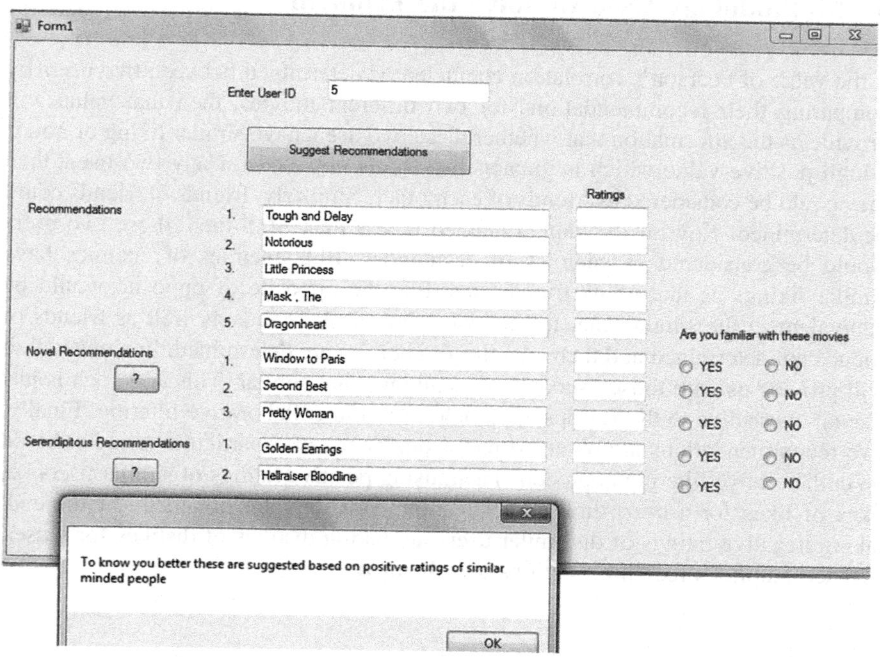

Fig. 1 Snapshot of the system designed on proposed approach

understand the reason behind separating the recommendations in three different categories and help them evaluate the alternative recommendations better without affecting the interest, user have in the recommender system.

B. Performance

For the evaluation of system performance, we calculated the metrics based on novelty and serendipity factor only. As, the system has also been provided with recommendations from the collaborative filtering approach the accuracy of the system is not affected and need no attention. The following measures are used for evaluation.

4.1 Precision for less popular items—Precision in general, refers to the number of movies rated positive (above 3) by a user to the total number of movies recommended. We calculated precision for movies recommended as novel and serendipitous. As a result, even low value achieved for such precision would be quite beneficial for the system.

4.2 Weighted catalogue coverage—it is the proportion of useful movies that is recommended by the system for a particular user. It is given by Eq. (3)

$$Weighted\_catalogue\_coverage = \frac{(S_r) \cap (S_s)}{S_s} \tag{3}$$

where S_r the set of movies is recommended to a user and S_s is the set of movies considered useful for that user.

4.3 Serendipity—to measure serendipity we used concept similar in. It is based on the distance applied to movie labels (genre). The comparison between movies is given by Eq. (4)

$$Dist(i,j) = 1 - \frac{(G_i) \cap \left(G_j\right)}{(G_i) \cup \left(G_j\right)} \tag{4}$$

where G_i is the set of genres describing positively rated serendipitous movie and G_j is the set of genres describing positively rated movie by collaborative filtering.Then to measure surprise of movie i we take,

$$\mathrm{S_{count}(i) = min\, dist(i, j)} \tag{5}$$

In addition, to analyze the novelty of items suggested we consider the answer to the question—"Are you familiar with the movie" as given by the users.

C. Results

The system was operational for a period of 4 weeks and the ratings and feedback from the users were stored in the system. Precision value of about 60 % was achieved for less popular items. The figures for the precision value obtained are low

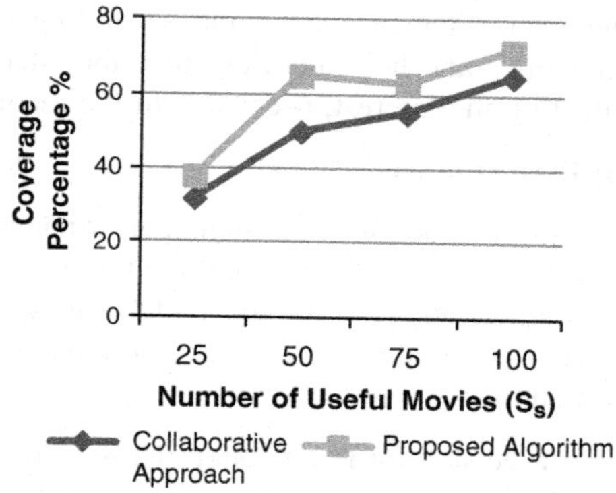

Fig. 2 Graph representing catalogue coverage percentage

but considering it only for the less popular items was significant. For novelty we found that users were not familiar with the movies suggested under the novelty and serendipity section and due to the explanation provided to them, they were interested to try them (Fig. 2).

The major achievement of the system was in case of the weighted catalogue coverage. The result in figure shows that our algorithm was able to provide a better coverage of the movies that was useful to a user. Thus, the proposed algorithm was able to suggest movies rated positive by a user and not included in the list provided by collaborative filtering.

Finally, for label-based surprise metric our algorithm was able to provide good results with an average S_{count} of all items rated positive in serendipitous list = 0.653.

5 Conclusion and Future Work

In summary, we proposed an effective recommender system that focussed on providing diverse recommendations to its users that are new and surprising in addition to being relevant. In our method, a system has been designed using positive rating of item given by users (collaborative) to find relevant item, then using the positive rating of the similar users and negative rating of the dissimilar users to suggest the novel and serendipitous recommendations respectively. The system suggests serendipitous movies as a separate list so that even if these movies do not match with user interest, then user should not stop using the recommender system as a whole. In addition to this, user is provided with an explanation of as to why the list is separated thus helping the user to evaluate these movies better.

Result showed that the method helps the system to know user choices better and recommend some diverse movies completely unknown to its users. The precision values calculated in the result section are low but considering it for unknown items is quite high. However, the overall precision for the complete system remains unaffected. The weighted catalogue coverage for a particular user is found to be better than collaborative filtering as a user is provided with variety of options that matches the user interest. Based on calculation of serendipitous factor, it was showed that a movie rated positive by a user from the serendipitous list is far from the profile of movies in the normal recommendation list. This helps the users to understand the system better.

As a part of future work we would like to study the trade off between coverage obtained and the overall accuracy of the system. We would also like to validate the performance of the proposed algorithm by obtaining user opinions in live study to understand the reason for trying out any recommendation that is completely unknown to them.

References

1. Francesco Ricci and Lior Rokach and Bracha Shapira, Introduction to Recommender Systems Handbook, Recommender Systems Handbook, Springer, (2011), pp. 1–35.
2. Gediminas Adomavicious et al, "Incorporating Contexual information in recommender systems using a multidimensional approach", ACM transactions on Information Systems, vol. 23, Issue1, (2005).
3. Hosein Jafarkarimi; A.T.H. Sim and R. Saadatdoost A Naïve Recommendation Model for Large Databases, International Journal of Information and Education Technology, June (2012).
4. Juan F. Huete, J.M. Fernandez-Luna, Luis M. de Campos, Miguel A. Rueda-Morales, "using past-prediction accuracy in recommender systems", Journal of Information Sciences 199(12), pp. 78–92, (2012).
5. Prem Melville and Vikas Sindhwani, Recommender Systems, Encyclopedia of Machine Learning, (2010).
6. R. Agrawal, T. Imielinski, and A. Swami, Mining association rules between sets of items in large databases. In Proceedings of the ACM SIGMOD Int'l Conf. Management of Data, pp. 207–216, (1993).
7. J. S. Breese, D. Heckerman, and C. Kadie, Empirical analysis of predictive algorithms for collaborative filtering. In Proceedings of the 14th Annual Conference on Uncertainty in Artificial in Intelligence, pp. 43–52, (1998).
8. J.-S. Kim, Y.-A Do, J.-W. Ryu, and M.-W. Kim. A collaborative recommendation system using neural networks for increment of performance. Journal of Korean Brain Society, 1 (2):234–244, (2001).
9. R. J. Mooney and L. Roy (1999). Content-based book recommendation using learning for text categorization. In Workshop Recom. Sys.: Algo. and Evaluation.
10. Walter Carrer-Neto, Maria Luisa Hernandez-Alcaraz, Rafael Valencia-Garcia, Francisco Garcia-Sanchez, "Social knowledge-based recommendation system. Application to the movies domain", Journal of Expert Systems with Applications 39(12), pp. 10990–11000, (2012).
11. Duen-Ren Liu, Chin-Hui Lai, Wang-Jung Lee, "A hybrid of sequential rules and collaborative filtering for product recommendation", Journal of Information Sciences 179(09), pp. 3505–3519, (2009).
12. Lingling Zhang, Caifeng Hu, Quan Chen, Yibing Chen, Young Shi, "Domain Knowledge Based Personalized Recommendation Model and Its Application in Cross-selling", Journal of Procedia Computer Science 9(12), pp. 1314–1323, (2012).
13. Tian Chen, Liang He, "Collaborative Filtering based on Demographic Attribute Vector", In Proceedings of International conference on future Computer and Communication, IEEE, (2009).

14. N. Tintarev and J. Masthoff, "The effectiveness of personalized movie explanations: An experiment using commercial meta-data," in Proc. International Conference on Adaptive Hypermedia, (2008), pp. 204–213.
15. Juan Pablo Timpanaro, Isabelle Chrisment, Olivier Festor, "Group-Based Characterisation for the I2P Anonymous File-Sharing Environment".
16. Jacob Cohen. Statistical Power Analysis for the Behavioral Sciences (2nd Edition). Routledge Academic, 2 edition, January (1988).
17. Sanghack Lee and Jihoon Yang and Sung-Yong Park, Discovery of Hidden Similarity on Collaborative Filtering to Overcome Sparsity Problem, Discovery Science, (2007).
18. George Lekakos, George M. Giaglis, Improving the prediction accuracy of recommendation algorithms: Approaches anchored on human factors, Interacting with Computers 18 (2006) 410–431.
19. Tae Hyup Roh, Kyong Joo Oh, Ingoo Han, The collaborative filtering recommendation based on SOM cluster-indexing CBR, Expert System with Application 25(2003) 413–423.
20. Shih, Y.-Y., & Liu, D.-R., Product recommendation approaches: Collaborative filtering via customer lifetime value and customer demands, Expert Systems with Applications (2007),
21. DanEr CHEN, The Collaborative Filtering Recommendation Algorithm Based on BP Neural Networks, (2009) International Symposium on Intelligent Ubiquitous Computing and Education.

Envelope Fluctuation Reduction for WiMAX MIMO-OFDM Signals Using Adaptive Network Fuzzy Inference Systems

Khushboo Pachori, Amit Mishra, Rahul Pachauri and Narendra Singh

Abstract In this article, the envelope fluctuation i.e., peak-to-average power ratio (PAPR) reduction technique is developed and analyzed using an Adaptive-Network based Fuzzy Inference System (ANFIS) for multiple input multiple output combined with orthogonal frequency division multiplexing (MIMO-OFDM) system under fading environment. The proposed method involves the training of ANFIS structure by the MIMO-OFDM signals with low PAPR obtained from the active gradient project (AGP) method, and then combined with partial transmit sequence (PTS) PAPR reduction technique. This method approximately reaches the PAPR reduction as the active partial sequence (APS) method, with significantly less computational complexity and convergence time. The results depict that proposed scheme performs better other conventional than that of the other conventional schemes.

Keywords Multiple input multiple output · Peak-to-average power ratio · Adaptive-network fuzzy inference system · Active partial sequence

---

K. Pachori (✉)
Department of Electronics and Communication, Jaypee University of Engineering and Technology, Guna 473226, MP, India
e-mail: khushboopachori@gmail.com

A. Mishra
Department of Electronics and Communication, Thapar University, Patiala 147004, Punjab, India
e-mail: amitutk@gmail.com

R. Pachauri · N. Singh
Jaypee University of Engineering and Technology, Guna 473226, MP, India
e-mail: pachauri.123@gmail.com

N. Singh
e-mail: narendra.singh@juet.ac.in

© Springer India 2016
S.C. Satapathy et al. (eds.), *Information Systems Design and Intelligent Applications*, Advances in Intelligent Systems and Computing 435,
DOI 10.1007/978-81-322-2757-1_3

1 Introduction

The great demand for higher data transmission rate over fading channels environment, requires combination of two models, i.e., multiple input multiple output and orthogonal frequency division multiplexing (MIMO-OFDM). It offers numerous benefits, such as high spectral efficiency, multipath delay spread tolerance, and diversity. The MIMO-OFDM has been widely adopted in several application areas viz., digital audio broadcasting (DAB), HIPERLAN/2, wireless local area network (WLAN), WiMAX, 3G Long Term Revolution (LTE) and 4G telecommunication systems [1–3]. In spite of numerous advantages, it still suffers from a major drawback i.e., high instantaneous peak-to-average power ratio (PAPR). Therefore, methods to reduce the PAPR of multicarrier signals [4–10] have been proposed.

Amongst the existing method, the partial transmit sequence (PTS) [11] and active gradient project (AGP) [12] are interesting techniques due to better PAPR performance and less computational complexity, respectively. The combinational approach of the above mentioned schemes [13], which known as active partial sequence (APS). The APS scheme shows the superior achievement than that of other conventional techniques, without any degradation in bit error rate (BER) and power spectrum (PSD). The computational complexity of APS method is further reduced by using adaptive network fuzzy inference system (ANFIS) [14] approach.

The Adaptive Network Fuzzy Inference (ANFIS) [10] is used to solve problems with unknown physical description, and provide some fuzzy (heuristic) rules that synthesize the behavior of system model which is widely employed in various practical applications of signal processing such as signal identification, adaptive channel equalization and inverse modeling.

The organization of the paper is as follows: In Sect. 2, Adaptive Network Fuzzy Inference System and Proposed Method are discussed. The computational complexity of the proposed method is formulated in Sect. 3 while the analysis of simulation results are provided in Sect. 4 followed by the conclusions in Sect. 5.

2 System Model

Let us consider, the MIMO-OFDM ($N_T \times N_R$) system with N_T number of transmit and N_R receive antennas, respectively. The input data $X^{(t)}$ stream is then mapped into N_c number of orthogonal symbols.

$$X_{N_T}^{(t)} = X_{N_T}^{k}, k = 0, 1, \ldots, N_c \tag{1}$$

The time domain signal $x_i = [x_i(0), \ldots, x_i(LN_c - 1)]$ could be obtained via IFFT operations as

$$x_i(n) = \frac{1}{\sqrt{N_c}} \sum_{k=0}^{LN_c-1} X_i(k) e^{\frac{j2\pi kn}{LN_c}} \tag{2}$$

where $i = 1, \ldots, N_T$ and $n = 0, 1, \ldots, LN_c - 1$. The oversampling factor L is an integer.

Generally, the PAPR computed from L times oversampled time domain MIMO-OFDM signal samples $x_i(n)$ is given as

$$PAPR = \frac{1}{E[|x^i(n)|^2]} \max_{0 \le n \le LN_c - 1} |x^i(n)|^2 \tag{3}$$

where $E[.]$ represents the expectation.

The complementary cumulative distribution function (CCDF) defines the PAPR performance as the probability of the transmitted signal exceeding a given PAPR threshold η, and evaluated as

$$CCDF = 1 - Pr\{PAPR \le \eta\} = 1 - (1 - e^{\eta})^{N_c} \tag{4}$$

2.1 Adaptive Network Fuzzy Inference System

The combination of Artificial Neural Networks (ANN) and Fuzzy Inference System (FIS) offers a viable alternative to solve realistic problem. To facilitate the learning of [15] inference system, a [16] hybrid learning algorithm is used. In the backward pass of the hybrid learning algorithm, the error rate propagates backward and the parameters are updated using by the gradient descent method whilst in the forward pass, functional signals move forward and the premise parameters are found by using least squares estimate. The combined network architecture is therefore known as ANFIS, which offers a scheme with fuzzy modeling to learn the input-output mapping from a given data set.

The implementation of ANFIS structure in this work is already discussed in [14], where η denotes Real or Imaginary part of Net_η depending on the model (as shown in Fig. 1).

2.2 Philosophy and Implementation Algorithm of Proposed Method

In this paper, a data set of signals with low PAPR obtained by AGP method is used to train the ANFIS structure as shown in Fig. 1. Due to compatibility of ANFIS

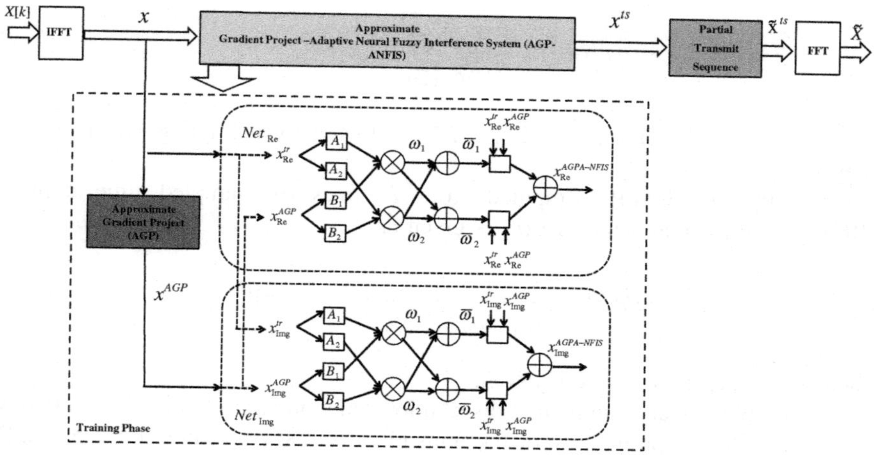

Fig. 1 Block diagram of proposed scheme using ANFIS structure

with integer data, the time domain training signals are first decomposed into their real and imaginary parts. The steps of training algorithm are given below.

- Input the time domain original data x to the AGP algorithm and get $\mathbf{x}^{AGP}$ signal with reduced PAPR.
- Divide input data into two sets, namely, ($\mathbf{x}^{tr}$, $\mathbf{x}^{AGP,tr}$) the training sets, and the test set ($\mathbf{x}^{ts}$, $\mathbf{x}^{AGP,ts}$).
- Decompose the original data and AGP algorithm output data into real and imaginary parts as $\mathbf{x}^{tr}$ ($\mathbf{x}^{tr}_{Re}$, $\mathbf{x}^{tr}_{Im}$) and $\mathbf{x}^{AGP,tr}$ ($\mathbf{x}^{AGP,tr}_{Re}$, $\mathbf{x}^{AGP,tr}_{Im}$), respectively.
- Compute $\mathbf{x}^{AGP-ANFIS}_{Re}$ and $\mathbf{x}^{AGP-ANFIS}_{Im}$ through training of two separate ANFIS networks $\mathbf{Net}_{Re}$ and $\mathbf{Net}_{Im}$ by [($\mathbf{x}^{tr}_{Re}$, $\mathbf{x}^{tr}_{Im}$), $\mathbf{x}^{AGP,tr}_{Re}$] and [($\mathbf{x}^{tr}_{Re}$, x^{tr}_{Im}), $\mathbf{x}^{AGP,tr}_{Im}$] to both the networks simultaneously.
- To validate the networks $\mathbf{Net}_{Re}$ and $\mathbf{Net}_{Im}$, test with the values of $\mathbf{x}^{ts}$(test set).
- Use x_{ts} (test set) as input to PTS method and partition the data block $x^{v}_{ts}[n]$ into V sub blocks as follows

$$x_{ts}[n] = x^{v}_{ts}[n] = [x^{1}, x^{2}, \ldots, x^{(V)}]^{T}$$

where x^{v}_{ts} is the sub block that are consecutively located and are of equal size.
- Generate the phase vector b^{v}, and optimize the value for minimum PAPR.

$$x_{ts}[n] = \sum_{v=1}^{V} b^{v} x^{v}_{ts}[n]$$

where $b^{v} = e^{j\phi v}$, $v = 1, 2, \ldots, V$.

- The corresponding time domain signal with the lowest PAPR vector can be expressed as

$$\tilde{x}_{ts}[n] = \sum_{v=1}^{V} \tilde{b}^v x_{ts}^v[n]$$

where

$$[\tilde{b}^1, \ldots, \tilde{b}^V] = arg \min_{\tilde{b}^1,\ldots,\tilde{b}^V} \left(\max_{n=0,1,\ldots,N-1} \left| \sum_{v=1}^{V} b^v x_{ts}^v[n] \right| \right)$$

- Obtain $\tilde{X}$ by applying FFT to $\tilde{x}_{ts}$. Finally, prepare side information (SI) for receiver and send $\tilde{X}$ to SFBC block of MIMO-OFDM transmitter.
  # Once the ANFIS structure is trained and tested (offline), it becomes ready to work with unknown MIMO-OFDM signals and do not need to be retrained until the number of sub-carriers is changed.

3 Complexity Analysis

In Table 1, the computation complexity of various conventional methods and proposed method is summarized. It can be observed that the proposed model need to perform only one I(FFT) operations, so, the computational complexity of proposed method in terms of integer multiplication ($18N + V$) and additions ($16N + (M + 4V + 2)$) per MIMO-OFDM symbol with N, V and $M = 4^{V-1}$, where number of sub-carriers, sub-blocks and candidate sequence, respectively.

Table 1 Computational complexity comparison of conventional and proposed (WiMAX) method

Operations	Design examples				
	Worldwide interoperability for microwave access (WiMAX) IEEE 802.16e				
	Approximate gradient project (AGP) method	Partial transmit sequence (PTS) method	Selective mapping method	Active partial sequence (APS) method [13]	Proposed method
Complex multiplications	5822	3058	3458	5826	–
Complex additions	11,644	125,359	6115	11,726	–
Integer multiplications	–	–	–	–	**3604**
Integer additions	–	–	–	–	**3282**

In general, any integer addition and multiplication operations are two and four times simpler than their complex counterparts respectively [10]. Therefore, the computational complexity of proposed scheme is very less as compared with APS [13] and other conventional methods, such as PTS [11], AGP [12] and SLM [5].

4 Results

In this paper, 50,000 (approx.) randomly generated QPSK modulated symbols have been generated to obtain the low PAPR signals through APS method for IEEE 802.16e. This simulation has been carried out in MATLAB (R2011a) environment. The training phase simulation of ANFIS have been carried out with 70 % data for training and 30 % data for testing.

The WiMAX mandates 192 data N_d, 8 pilot data N_{pilot} and 56 null subcarriers N_{null} subcarriers $N = 200$. The operating frequency for IEEE 802.16e is 2500 MHz. The MIMO-OFDM system uses 2 transmit antennas and 2 receive antenna (2 × 2), the number of iterations N_{iter} and number of sub-blocks V, for proposed scheme are initialized as 5 and 4, respectively.

Figure 2 shows the complementary cumulative distribution function (CCDF) curves of various schemes, and gives clear observation that the proposed method (≈8.2 dB for WiMax at CCDF of 1×10^{-3}) is 3.2 dB (approx.) lower than MIMO-OFDM (original) scheme (≈11.2 dB for WiMAX standard at CCDF of 1×10^{-3}). The BER plots for different schemes in WiMAX MIMO-OFDM standards are shown in Fig. 3. All the schemes have been simulated in the presence of Rayleigh fading environment.

Figure 4 shows the power spectral density curves of proposed and other schemes. It can be observed that the proposed scheme maintains out-of-band

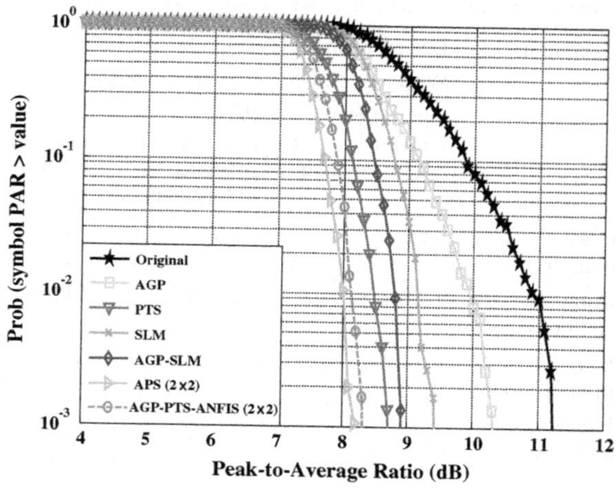

Fig. 2 Comparison of CCDF for WiMAX standard

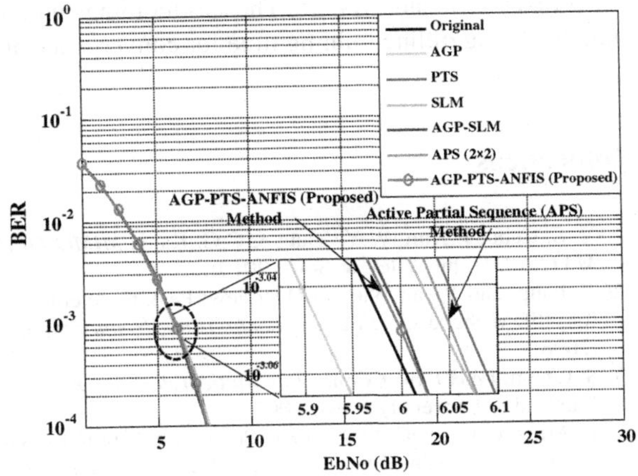

Fig. 3 BER comparison for WiMAX standard

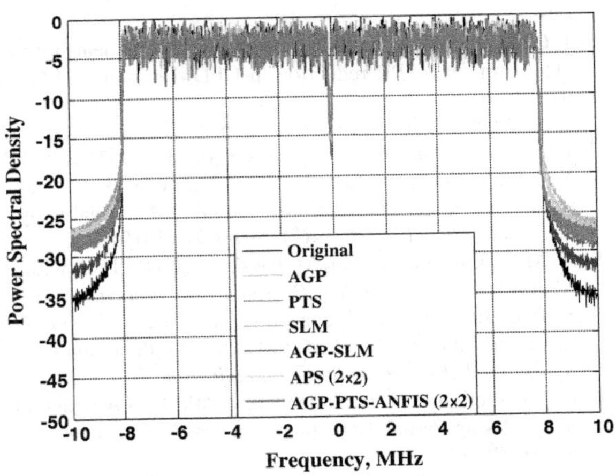

Fig. 4 Power spectrum density comparison for WiMAX standard

radiations and in-band ripples for WiMAX MIMO-OFDM standards as compared to other conventional methods.

5 Conclusion

In this paper, modeling of APS [13] scheme, a combinational method for PAPR reduction in MIMO-OFDM signals, is trained using ANFIS structure. The proposed method shows 3.2 dB reduction in PAPR as compared to standard simulation (original), maintains BER performance with no data rate loss. Moreover, the proposed scheme significantly reduces the computational (≈60 %) complexity of additions and time complexity due to feed-forward structure of adaptive network based fuzzy inference system. The shortcomings inherent in ANFIS based modeling include, a long offline training time for convergence and fixed setting of parameters.

References

1. Y. S. Ceho, J. Kim, W.Y. Yang & C.G. Kang, *MIMO-OFDM Wireless communications with MATLAB*, (John Wiley & Sons, 2010).
2. T. Jiang, and Yiyan Wu, An Overview: Peak-to-Average Power Ratio Reduction Techniques for OFDM Signals," *IEEE Transactions on Broadcasting*, 0.4em vol. 54, no. 2, pp. 257–268, (2008).
3. S. Litsyn, *Peak Power Control in Multicarrier Communications*, 1em plus 0.5em minus 0.4em Cambridge University Press, 2007.
4. J. Armstrong, "Peak-to-Average Power Reduction for OFDM by Repeated Clipping and Frequency Domain Filtering," *IEEE Electronics Letter*, 38(8), 246–247 (2002).
5. R. W. Bäuml, R.F.H. Fischer & J.B. Huber, "Reducing the peak-to-average power ratio of multicarrier modulation by selected mapping," *IEEE Electronics Letters*, 32(22), 2056–2057 (1996).
6. J. C. Chen & C. P. Li, Tone reservation using near-optimal peak reduction tone set selection algorithm for PAPR reduction in OFDM systems, *IEEE Signal Process. Lett.*, 17(11), 933–936 (2010).
7. Amit Mishra, Rajiv Saxena & Manish Patidar, "AGP-NCS Scheme for PAPR Reduction," *Wireless Personal Communications*, doi:10.1007/s11277-015-2275-8, (2015).
8. Y. Jabrane, V. P. G. Jimenez, A. G. Armeda, B. E. Said, & A. A. Ouahman, "Reduction of power envelope fluctuations in OFDM signals by using neural networks," *IEEE Communication Letters*, 14(7), 599–601 (2010).
9. A. Mishra, R. Saxena, and M. Patidar, "OFDM Link with a better Performance using Artificial Neural Network," *Wireless Personal Communication*, 0.4em vol. 77, pp. 1477–1487, (2014).
10. V. P. G. Jimenez, Y. Jabrane, A. G. Armeda, B. E. Said, & A. A. Ouahman, "Reduction of the Envelope Fluctuations of Multi-Carrier Modulations using Adaptive Neural Fuzzy Inference Systems," *IEEE Trans. on Communication*, 59(1), 19–25 (2011).
11. L. J. Cimini, and N. R. Sollenberger, "Peak-to-Average Power Ratio Reduction of an OFDM Signal Using Partial Transmit Sequences," *IEEE Communication Letters*, 0.4em vol. 4, no. 3, pp. 86–88, (2000).
12. B. S. Krongold, and D. L. Jones, "PAR reduction in OFDM via Active Constellation Extension," *IEEE Transaction on Broadcasting*, 0.4em vol. 49, no. 3, pp. 258–268, (2003).
13. K. Pachori, A. Mishra, "PAPR Reduction in MIMO-OFDM by Using Active Partial Sequence," *Circuits Syst Signal Process.*, doi:10.1007/s00034-015-0039-z, (2015).
14. J. -S. R. Jang, "Adaptive-network-based fuzzy inference system," *IEEE Trans. Commun.*, vol. 43, no. 6, pp. 2111–2117, (1995).
15. H. Takagi, and M. Sugeno, "Derivation of fuzzy control rules from human operator's control actions," *Proc. IFAC Symp. Fuzzy Inform., Knowledge Representation and Decision Analysis.*, pp. 55–60, (1983).
16. J. -S. Roger Jang, "Fuzzy modeling using generalized neural networks and Kalman filter algorithm, " *Proc. Ninth Nat. Conf. Artificial Intell. (AAAI-91)*, pp. 762–767, (1991).

Modeling and Performance Analysis of Free Space Quantum Key Distribution

Minal Lopes and Nisha Sarwade

Abstract With the technological development, the demand for secure communication is growing exponentially. Global secure communication have become crucial with increasing number of internet applications. Quantum Cryptography (QC) or Quantum Key Distribution (QKD) in that regime, promises an unconditional security based on laws of quantum principles. Free space QKD allows longer communication distances with practical secure key rates to aid secure global key distribution via satellites. This is encouraging many research groups to conduct QKD experimentation. But it is observed that such experiments are very complex and expensive. This paper thus attempts to establish a model for analysis of free space QKD through simulation. The model will be verified against experimental results available from different literature. It can be seen that the simulation approach stands effective for performance analysis of such complex systems. The developed model, test parameters like quantum bit error rate and secret key rate against mean photon number of laser pulses and quantum channel loss, and proves to fit well with the experimental results.

Keywords Free space quantum cryptography · Quantum key distribution · Free space optical transmission · QC · QKD

M. Lopes (✉) · N. Sarwade
Department of Electronics Engineering, Veermata Jijabai Institute of Technology, Mumbai 400019, India
e-mail: minal.lopes@sfitengg.org

N. Sarwade
e-mail: nishasarvade@vjti.org.in

© Springer India 2016

S.C. Satapathy et al. (eds.), *Information Systems Design and Intelligent Applications*, Advances in Intelligent Systems and Computing 435,
DOI 10.1007/978-81-322-2757-1_4

1 Introduction

In past twenty decades Quantum Cryptography have been studied, explored and experimented rigorously. It is considered as an optimal solution for the inherent problems (eavesdropping and dependency on mathematical complexity) endured by modern cryptography. Quantum cryptography promises unconditional security by exploiting the quantum principles, such as principle of uncertainty, quantum entanglement and no-cloning principle. A simple QC communication is shown in Fig. 1.

Quantum cryptography demand two separate channels (quantum channel and classical channel) for its operation. A quantum channel is for exchange of single photons, which the legitimate parties use for generation of a common key through some post processing techniques communicated over classical channel. Eve represents an illegitimate entity, equipped with advanced resources and is assumed to have complete access to both the communication channels.

Many QKD experiments have exploited, wired (fiber optic) and wireless (free space optics, FSO) possibilities of the essential quantum channel. The obvious limitation of wired medium (communication distance) put forth the demand for more and more free space experiments. Due to this many research groups have contributed towards experimentation in free space quantum key distribution. A handful of pioneering experiments are been identified which include Buttler [1], Rarity [2], Kurtsifier [3] and Hughes [4]. These experiments rather promised the practical transmission distances with free space QKD. In recent years QKD also have found an interesting application in global key distribution via satellites. Literature [5–7] have reported interesting results that points towards the feasibility of satellite based quantum communication.

From these experiments it is evident that laboratory test beds or experimental set ups are dynamic, complex and costlier due to the stringent requirements such as single photon sources, detectors, transmitter and receiver telescope alignments and the atmospheric conditions. Modeling free space QKD transmission stands

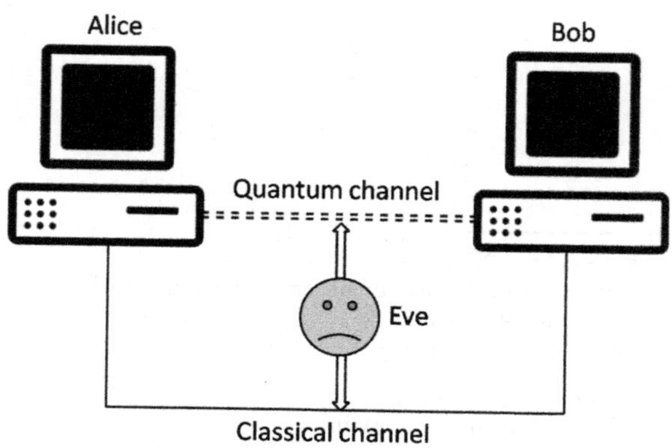

Fig. 1 A simple quantum cryptographic communication

beneficial for analyzing the effect of various parameters on the key generation rates. QKD protocols have ambiguous categorization depending upon the type of protocols (discrete or continuous variable), type of photon sources (prepare and measure or Entanglement) and the attack mitigation strategies (decoy state, single photon sources). This paper specifically discusses the discrete, prepare and measure BB84 prototype QKD transmission with Poisson distributed laser sources.

The organization of this paper is as follows: In Sect. 2 the effects of atmospheric disturbances on free space optical transmission and modeling free space QKD is discussed in detail. Section 3 will present a comprehensive review of selective literature, reporting free space QKD and enumerate the parameters essential for modeling. Section 4 will evaluate the developed model against experimental results discussed in Sect. 2. Section 5 will conclude and explore the future scope for free space quantum communication.

2 Free Space Quantum Cryptography

This section first analyzes effects of atmospheric disturbances on optical transmission. A mathematical model for free space QKD is built subsequently.

2.1 FSO Link

Free space optics (FSO) was considered as an alternate in quantum key distribution to cater to the need for longer communication distances. A typical FSO link (refer Fig. 2) consist of a transmitter with laser source, modulator and a transmitting telescope unit. The optical beam traverse through the free space where statistical fluctuations in the atmosphere plays an important role. The receiver module contains photo-detectors and demodulator assembly.

The propagation of light through free space is affected mainly by atmospheric turbulence, path loss factor and the beam wandering errors. Atmospheric turbulence causes intensity fluctuation of the laser beam. These fluctuation are assumed to be log-normally distributed in the regime of weak turbulence [8, 9]. Thus the probability distribution of fluctuating atmospheric transmission coefficient is given as,

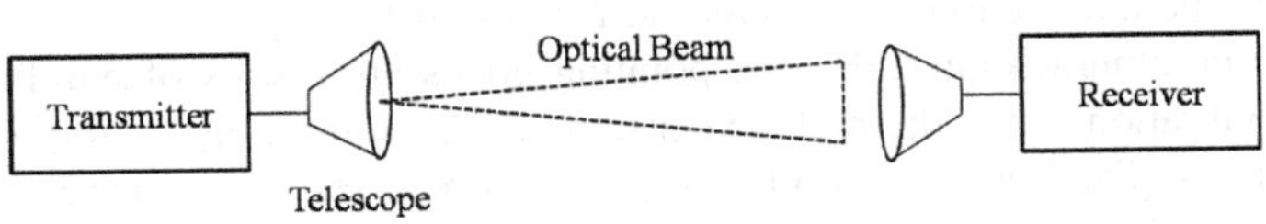

Fig. 2 Typical FSO link

$$f(\eta_{atm}) = \frac{1}{2\eta_{atm}\sqrt{2\pi\sigma_x^2}} \exp\left(\frac{(\ln \eta_{atm} + 2\sigma_x^2)^2}{8\sigma_x^2}\right) \quad (1)$$

where,

(a) The log amplitude variance σ_x^2 is given as,

$$\sigma_x^2 \approx \frac{\sigma_r^2}{4} \quad (2)$$

Here, σ_r^2 is the intensity scintillation-index and is defined by Rytov's variance [10] as,

$$\sigma_r^2 = 1.23\, C_N^2 K^{7/6} L^{11/6} \quad (3)$$

where C_N^2 is the structure constant of refractive index fluctuations.
$K = \frac{2\pi}{\lambda}$ is the optical wave number with wavelength λ.
and L is the propagation path length between transmitter and receiver.

(b) η_{atm} is the atmospheric transmittance given as,

$$\eta_{atm} = \frac{2P_R}{\pi W^2} \quad (4)$$

where, W is beam width and P_R is a received power given by the equation [11],

$$P_R = \frac{\pi^2 P_T D_T^2 D_R^2}{16 L^2 \lambda^2} \quad (5)$$

Here, P_T—Transmission power.
D_T—Diameter of Transmitter optical antenna.
and D_R—Diameter of Receiver optical antenna.

Path Loss Factor represents the attenuation (absorption) of laser power through the stable free space and is described by the exponential Beer-Lambert law as,

$$Path\,Loss = \exp(-\rho L) \quad (6)$$

where, ρ is the attenuation coefficient and L is the path length.

In FSO communication links, the pointing errors are results of building sway, building vibration and thermal expansion of buildings used for mounting trans-receivers. The fraction of collected power at receiver in such case is given as [12],

$$h_p(r; L) \approx A_0 \exp\left(-\frac{2r^2}{w_{z_{eq}}^2}\right) \quad (7)$$

where r is instantaneous radial displacement between the beam centroid and the detector center. L is a path length.

A_0 is the fraction of collected power at $r = 0$ given as $A_0 = [erf(v)]^2$ and $w_{z_{eq}}^2$ is the equivalent beam-width given as

$$w_{z_{eq}}^2 = w_z^2 \frac{\sqrt{\pi} erf(v)}{2v\exp(-v^2)}$$

here v is the ratio between aperture radius and beam width given as,

$$v = \sqrt{\frac{\pi}{2}} \frac{a}{w_z}.$$

The error function $\mathrm{erf}(x) = \frac{2}{\sqrt{\pi}} \int_0^x e^{-t^2} dt$ and beam-width $w_z = \theta z$, where θ is the transmit divergence angle describing the increase in beam radius with distance from the transmitter.

The mathematical analysis discussed above gives clear idea about the selection of locale for testing the quantum communication. It is thus evident that most of the free space QKD test beds are located indoor or at a very weak atmospheric turbulence places. Also to keep the QKD experimentation simple it is recommended to directly indicate the total attenuation caused by free space in decibels. This definitely do not limit the scope of quantum communication, but provides a good base to explore free space QKD to its full strength.

2.2 *Free Space QKD Modeling*

For any QKD system, there will be two communications,

- quantum communication and
- classical communication

In quantum communication, Alice transmits photons to Bob using a Poisson distributed laser source (weak coherent pulses, WCP). This probability of transmitting a pulse with i photons is given as,

$$p(\mu, i) = \frac{e^{-\mu} \mu^i}{i!} \quad (8)$$

where, μ is the mean photon number (MPN).

At receiver Bob's detector setup detects the signal per laser pulse (P_{signal}) from Alice. Bob's detector usually have its own limitations and count even due to false photon count, known as dark counts (P_{dark}). Thus Bob's total detection probability P_D can be given as,

$$P_D = P_{signal} + P_{dark} \tag{9}$$

The above equation assumes that P_{signal} and P_{dark} are two mutually exclusive events. The value of P_{dark} is a measured value obtained from experimentation. The value of P_{signal} can be calculated as,

$$P_{signal} = \sum_{i=1}^{\infty} \eta_i p(\mu, i) \tag{10}$$

where η_i is the total efficiency of detecting a pulse with i of photons given by a binomial distribution as,

$$\eta_i = \sum_{k=1}^{i} C_i^k (\eta)^k (1-\eta)^{i-k} = 1 - (1-\eta)^i \tag{11}$$

From Eqs. (8), (10) and (11), P_{signal} can be calculated as,

$$P_{signal} = \sum_{i=1}^{\infty} \frac{\mu^i \left[1 - (1-\eta)^i\right]}{i!} e^{-\mu} = 1 - e^{-\eta\mu} \tag{12}$$

In Eq. (12), η represents the actual photon detection efficiency of Bob's detector module which is contributed by the free space transmission efficiency, receiver module' s efficiency and photon detector efficiency. Therefore,

$$\eta = \eta_{FSO} \times \eta_{Bob} \times \eta_D \tag{13}$$

where, η_{Bob} is internal transmission efficiency of Bob's receiver module (specified through experiments) and detector efficiency η_D is a characteristic efficiency of single photon detectors. The free space transmission efficiency η_{FSO} is given as,

$$\eta_{FSO} = 10^{-\gamma/10} \tag{14}$$

where, γ is the total atmospheric attenuation in dB. In general γ is the addition of losses introduced by atmospheric turbulence, path loss and pointing errors discussed in Sect. 2.1.

P_D and P_{signal} contribute to quantum bit error rate (QBER) which is an important performance parameter in QKD analysis. Thus,

$$QBER = \frac{P_{error}}{P_D} = \frac{\frac{1}{2}P_{dark} + e_{error}P_{signal}}{1 - e^{-\eta\mu} + P_{dark}} \quad (15)$$

The term $\frac{1}{2}P_{dark}$ represents random occurrence of dark counts and e_{error} is the probability of error due to imperfect trans-receiver alignment. At the end of this quantum communication, a sifted key rate can be calculated as [13],

$$R_{sifted} = c \times P_D \times \exp(-c \times P_D \times t_d) \quad (16)$$

where,
c system clock frequency and
t_d receiver's (detection system) dead time

The value of QBER further decides the session continuation over classical channel for post processing and secure key generation. The error correction and privacy amplification (post processing) steps ensures secure and unambiguous key generation against eavesdropper. Thus these steps certainly accounts for a loss of fraction of bits η_{post}. The value of η_{post} is given as [14],

$$\eta_{post} = \max\left\{f1\left[1 - \log_2\left(1 + 4\left(\frac{QBER}{f1}\right) - 4\left(\frac{QBER}{f1}\right)^2\right)\right] - f(QBER)H_2(QBER),\ 0\right\} \quad (17)$$

Here, $f_1 = \frac{P_{signal}}{P_D}$ is a rate of photon detection. $H_2(QBER)$ is a conditional entropy indicating error correction and is given as,

$$H_2(QBER) = -QBER\log_2(QBER) - (1 - QBER)\log_2(1 - QBER) \quad (18)$$

and $f(QBER)$ is a correction factor used for non ideal error correction procedures.
Lastly, the Secure (Secret) Key Rate, R can be given as,

$$R = c \times \eta_I \times P_D \times \eta_{post} \quad (19)$$

where, η_I is the protocol's Intrinsic Efficiency. (e.g. BB84 has intrinsic efficiency of 0.5. SARG04 has intrinsic efficiency of 0.25.)

As stated earlier, the model developed above, caters specifically to the BB84 prototype QKD protocols. The performance parameters such as QBER, sifted key rate and secure key rate can be calculated by using Eqs. (15), (16) and (19) respectively.

3 Test Data for Free Space QKD Simulation

Free space QKD experiments are reviewed and discussed over the past decade with an emphasis on technological challenges in quantum communication [15]. The experiments reported in the literature includes, those conducted along horizontal propagation paths of varying distances, with communication paths from ground-to-aircraft, ground-to-space, and demonstrations in the laboratory. The available data mainly characterize propagation distances, transmission speeds, quantum key distribution (QKD) protocols, and quantum bit error rates (QBER). This section discusses the experimental results of three specific QKD literature cited as Bienfang [16], Tobias [17] and Kim [18], designated after the first author and the year of experimentation. The test data obtained from these experiments is tabulated (refer Table 1) and used for verification of a model developed in Sect. 2.2.

Bienfang demonstrates the exchange of sifted quantum cryptographic key over a 730 m free-space link between two buildings, at the rate up to 1 Mbps. A fairly constant QBER of 1.1 % is reported. The experiment claims a unique clock recovery and synchronization mechanism with 1.25 Gbps system clock rate. It includes, a free-space 1550 nm optical link that operates continuously between the buildings, which is used as a classical channel for QKD protocol. A quantum channel runs parallel to the classical channel and operates at 845 nm. Typical quantum channel source used is, 10 GHz vertical-cavity surface-emitting lasers (VCSELs).

Tobias, reports experimental implementation of a BB84 protocol type quantum key distribution over a 144 km free-space link using weak coherent laser pulses. This link is established between the Canary islands of La Palma and Tenerife. The 144 km path length is much longer than from the lower earth orbit (LEO) satellites to a ground station and serves as a realistic test bed for future quantum communication to space. The experiment also optimize the link transmission with bidirectional active telescope tracking, which claims a secure key rate of 12.5 kbit.

Kim, implements a BB84 quantum key distribution protocol over a free-space optical path on an optical table. The sifted key generation rate of 23.6 kbits per second have been demonstrated at the average photon number per pulse $\mu = 0.16$. The experiment also test the long-distance capability of QKD system by adding optical losses to the quantum channel. The additional channel losses were introduced, by inserting a set of uncoated glass plates at normal angles to the path of the photon and $\mu = 0.242$. The section ahead verify the free space QKD model using the set of parameters from Table 1. It is essential to mention here, that the simulation assumes default values for $e_{error} = 0.01$ and $f(QBER) = 1.22$ [14].

Table 1 Test data for free space QKD model verification

Parameters	Bienfang [16]	Tobias [17]	Kim [18]
Transmission distance	730 m (between two buildings)	144 km (between canary islands)	17 m (table top experiment)
Wavelength 'λ'	845 nm	850 nm	780 nm
Mean photon number 'μ'	0.1	0.3	0.16 and 0.242
Clock rate 'c'	10 GHz	10 MHz	1 MHz
Detector efficiency 'η_D'	0.48	0.25	0.41
Channel loss 'γ'	−5 dB	−28 dB	−0.18 dB
Dark count probability $Pdark = No.of\ detectors \times DCR \times t_w$	Dark count rate $(DCR) = 1$ kHz, timing window $(t_w) = 350$ ps, no. of detectors = 2	Dark count rate $(DCR) = 1$ kHz, timing window $(t_w) = 5.9$ ns, no. of detectors = 2	Dark count rate $(DCR) = 300$ Hz, timing window $(t_w) = 125$ ns, no. of detectors = 2
Detection system dead time 't_d'	1.6×10^{-9}	–	–

4 Verification of Free Space QKD Model

This section discusses the results from selected free space QKD experiments and reproduce them through simulation using MATLAB software. Figures 3 and 4, represents effect of mean photon number on QBER and Sifted Key Rate. These results can be compared with Fig. 3 of Bienfang. As discussed in Sect. 3, the simulation produces a fairly constant QBER of 1 % and the sifted key rate approaching 1 Mbps.

Figure 5, reproduces the results of Fig. 1 from Tobias. The graph represents dependence of key generation rate on quantum channel loss. It suggests the

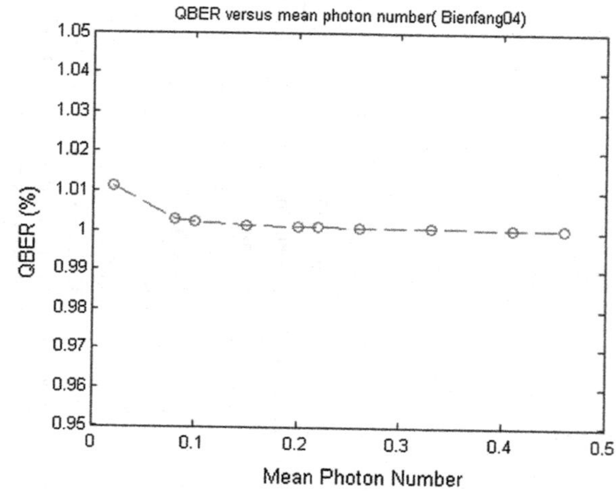

Fig. 3 QBER versus mean photon number (Bienfang)

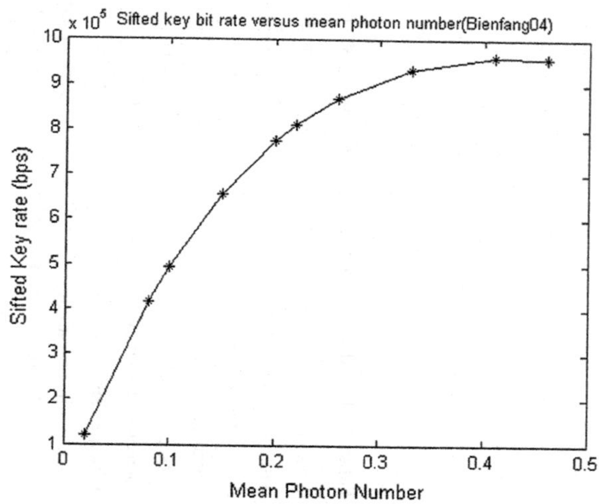

Fig. 4 SKR versus mean photon number (Bienfang)

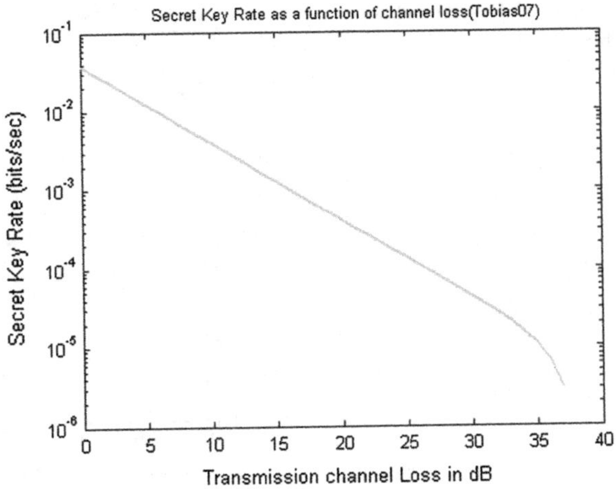

Fig. 5 SKR versus channel loss (Tobias)

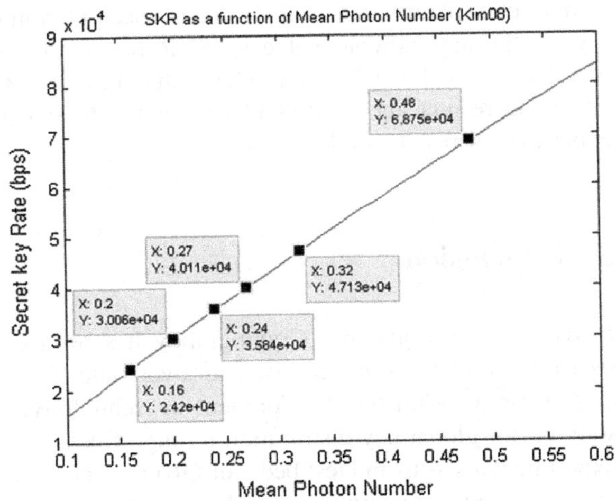

Fig. 6 SKR versus mean photon number (Kim)

practical key generation rates up to 35 dB channel loss. It is evident from Fig. 5 that the simulation results are in good agreement with the experimental outcomes. Figure 6, simulates the secret key rate versus mean photon number, for a set of parameters from Kim. The data points specifically marked in the graph are indicating a match with the results from Table 1 of Kim. Figure 7 shows simulations for effect of channel loss on secret key rate for Kim.

Kim reports addition of channel loss by inserting an uncoated glass plates within the photon's free space path. Although, for simulation purpose the channel loss value is considered in dB and is increased gradually from 0.01 to 4 dB. The secret key rate values are then matched with values from Kim [18], to find the channel loss introduced due to insertion of each glass plate. This simulated results can be verified

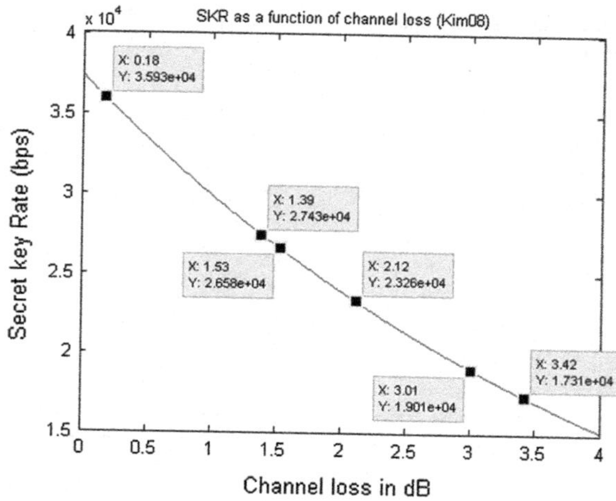

Fig. 7 SKR versus channel loss (Kim)

from a set of parameters for the value of mean photon number equal to 0.242. From Fig. 6, the approximate value of SKR for $\mu = 0.242$ is 35.8 kHz, which is in coordination with the 35.9 kHz SKR value for 0.18 dB channel loss given by Fig. 7. From the results above, it is clear that the model developed in Sect. 2 fits the experimental results well.

5 Conclusion

In recent years, Quantum Cryptography has been explored intensively by many research groups, as it promises to overcome the basic drawbacks of modern cryptography. With the development in technology, QC is now even seen as the solution for global key distribution through free space optical communication. The experimental set-up and test beds for QKD systems are dynamic and complex. Thus simulation and modeling approach serves as a great benefit for analysis of such systems. In this paper, a mathematical model for effect of atmospheric turbulence, path loss and pointing errors on the performance of free space QKD is developed. This model is verified against the parameters from some chosen experiments reported in literature.

Free space QKD can definitely prove promising for longer communication distances against its wired counterpart. Thus a comprehensive analysis of effect of atmospheric interference on free space optical communication is presented. But it is important to note here that many QKD experiments do not provide minute parametric analysis to avoid complications, rather it is preferred to address the overall atmospheric attenuation in decibels as reported in Table 1. From the results discussed in Sect. 4, it is evident that the developed free space QKD model is accurate. It can also be used as an elementary framework for simulation of other QKD protocols. The trivial differences in the performance parameters are observed due to unavailability of some parameters for which the default values are been considered.

The constraints of this model is that it is valid only for prepare and measure type of QKD protocols. Where as many experiments reports free space QKD with entangled photon sources and decoy states to enhance the transmission distance. Thus it is necessary to model these QKD system variants. The model developed in this paper allows easy extensions to suit any protocol type and can be used with minor modifications to accommodate various QKD systems.

References

1. W. Buttler, R. Hughes, P. Kwiat, S. Lamoreaux, G. Luther, G. Morgan, J. Nordholt, C. Peterson, and C. Simmons, "Practical free-space quantum key distribution over 1 km," Physical Review Letters, vol. 81, no. 15, p. 3283, 1998.
2. J. Rarity, P. Gorman, and P. Tapster, "Secure key exchange over 1.9 km free-space range using quantum cryptography," Electronics letters, vol. 37, no. 8, pp. 512–514, 2001.
3. C. Kurtsiefer, P. Zarda, M. Halder, P. M. Gorman, P. R. Tapster, J. Rarity, and H. Weinfurter, "Long-distance free-space quantum cryptography," in Photonics Asia 2002. International Society for Optics and Photonics, 2002, pp. 25–31.
4. R. J. Hughes, J. E. Nordholt, D. Derkacs, and C. G. Peterson, "Practical free-space quantum key distribution over 10 km in daylight and at night," New journal of physics, vol. 4, no. 1, p. 43, 2002.
5. R. Ursin, F. Tiefenbacher, T. Schmitt-Manderbach, H. Weier, T. Scheidl, M. Lindenthal, B. Blauensteiner, T. Jennewein, J. Perdigues, P. Trojek et al., "Entanglement-based quantum communication over 144 km," Nature physics, vol. 3, no. 7, pp. 481–486, 2007.
6. E. Meyer-Scott, Z. Yan, A. MacDonald, J.-P. Bourgoin, H. Hübel, and T. Jennewein, "How to implement decoy-state quantum key distribution for a satellite uplink with 50-db channel loss," Physical Review A, vol. 84, no. 6, p. 062326, 2011.
7. S. Nauerth, F. Moll, M. Rau, C. Fuchs, J. Horwath, S. Frick, and H. Weinfurter, "Air-toground quantum communication," Nature Photonics, vol. 7, no. 5, pp. 382–386, 2013.
8. M. T. Rahman, S. Iqbal, and M. M. Islam, "Modeling and performance analysis of free space optical communication system," in Informatics, Electronics Vision (ICIEV), 2012 International Conference on. IEEE, 2012, pp. 211–218.
9. C. Erven, B. Heim, E. Meyer-Scott, J. Bourgoin, R. Laflamme, G. Weihs, and T. Jennewein, "Studying free-space transmission statistics and improving free-space quantum key distribution in the turbulent atmosphere," New Journal of Physics, vol. 14, no. 12, p. 123018, 2012.
10. H. Henniger and O.Wilfert, "An introduction to free-space optical communications," Radioengineering, vol. 19, no. 2, pp. 203–212, 2010.
11. J. Franz and V. K. Jain, Optical Communications: Components and Systems: Analysis-design-optimization-application. CRC press, 2000.
12. F. Yang, J. Cheng, T. Tsiftsis et al., "Free-space optical communication with nonzero boresight pointing errors," Communications, IEEE Transactions on, vol. 62, no. 2, pp. 713–725, 2014.
13. H. Shibata, T. Honjo, and K. Shimizu, "Quantum key distribution over a 72 db channel loss using ultralow dark count superconducting single-photon detectors," Optics letters, vol. 39, no. 17, pp. 5078–5081, 2014.
14. N. Lütkenhaus, "Security against individual attacks for realistic quantum key distribution," Physical Review A, vol. 61, no. 5, p. 052304, 2000.

15. A. Tunick, T. Moore, K. Deacon, and R. Meyers, "Review of representative free-space quantum communications experiments," in SPIE Optical Engineering + Applications. International Society for Optics and Photonics, 2010, pp. 781 512–781 512.
16. J. Bienfang, A. Gross, A. Mink, B. Hershman, A. Nakassis, X. Tang, R. Lu, D. Su, C. Clark, C. Williams et al., "Quantum key distribution with 1.25 gbps clock synchronization," Optics Express, vol. 12, no. 9, pp. 2011–2016, 2004.
17. T. Schmitt-Manderbach, H. Weier, M. Fürst, R. Ursin, F. Tiefenbacher, T. Scheidl, J. Perdigues, Z. Sodnik, C. Kurtsiefer, J. G. Rarity et al., "Experimental demonstration of free-space decoy-state quantum key distribution over 144 km," Physical Review Letters, vol. 98, no. 1, p. 010504, 2007.
18. Y.-S. Kim, Y.-C. Jeong, and Y.-H. Kim, "Implementation of polarization-coded free-space bb84 quantum key distribution," Laser Physics, vol. 18, no. 6, pp. 810–814, 2008.

Design of a Low-Delay-Write Model of a TMCAM

N.S. Ranjan, Soumitra Pal and Aminul Islam

Abstract In this paper, a novel version of Ternary Magnetic-Content-Addressable Memory (TMCAM) is proposed for a low-delay-write operation. This is attained from the connections of circuit and majorly due to the exceptional operational features of CAM integrated with MTJ. While the previous TMCAM required each one of the MTJ to be written separately, this model attempts to nullify the problem. Consequently, a reduction in delay by almost twice is obtained in comparison to the previous TMCAM with a 22 nm CMOS technology used for simulation purposes. This can be effectively employed in adaptive biomedical signal processing applications where writing is often and hence, delay cannot be compromised.

Keywords Biomedical applications · Magnetic tunnel junction (MTJ) · Content addressable memory · Ternary magnetic-content-addressable memory (TMCAM)

N.S. Ranjan (✉) · A. Islam
Electronics and Communication Engineering, Birla Institute of Technology, Mesra, Ranchi 835215, Jharkhand, India
e-mail: be1062313@bitmesra.ac.in

A. Islam
e-mail: aminulislam@bitmesra.ac.in

S. Pal
Applied Electronics and Instrumentation Engineering, C. V. Raman College of Engineering, Bidya Nagar, Mahura, Janla, Bhubaneswar 752054, India
e-mail: soumitra10028.13@bitmesra.ac.in

© Springer India 2016
S.C. Satapathy et al. (eds.), *Information Systems Design and Intelligent Applications*, Advances in Intelligent Systems and Computing 435,
DOI 10.1007/978-81-322-2757-1_5

1 Introduction

In a conventional searching system, the time required to search a given set of data is directly proportional to the size of the data. Unlike this, Content-Addressable Memory (CAM) has the advantage of parallel-searching the dataset in a single-clock period. This advantage, has brought revolution, by its immense amount of applications in image coding [1], network routers [2], pattern recognition [3] and many more in the field of signal processing and networking.

Though it is advantageous, the hardware cost of CAM becomes expensive, as its network includes many transistors for the reasons that it consists individual cells for comparison and 10 transistors per cell. So, to curb this, a special device known as MTJ (Magnetic tunnel junction) has been employed which alone can replace 10 transistors in each CAM cell [4].

The ternary CAM (TCAM) [5], magnetic CAM (MCAM) [6] and ternary magnetic CAM (TMCAM) employ the MTJ to utilize the advantages of non-volatile memory, low power dissipation, less area and high speed searching. The TMCAMs are better in terms of power and speed compared to the previous MCAM and TCAM systems [7]. The TMCAM consists of two MTJs per cell, each of them controlled by two control signals. A bit is stored as the resistance value in the MTJs, i.e., if the value of resistance is high (antiparallel), then the value stored is 1, and if the value of resistance is low (parallel), then the value stored is 0. For ternary state both the MTJs are written with either Rp or Rap. Writing is done using spin-transfer torque writing [8, 9], that occurs in the direction of the passing current itself and the time taken for writing either 1 (Rap) or 0 (Rp), is almost the same [10]. The reported TMCAM has a drawback of writing each MTJ individually, which cannot be compromised in systems where writing is predominant, and hence, demand high speed. This paper modifies the reported TMCAM for higher writing speeds, which can prove to be twice as fast as the previous ones.

2 Basics of MTJ

An MTJ model is shown in Fig. 1. It consists of two layers—pinned layer and free layer; both of them separated by an oxide layer such as MgO [11, 12], which acts as a tunneling barrier. The resistance MTJ changes with the relative alignment of magnetization of free layer with respect to pinned layer [13, 14]. When the alignment of the magnetization of the free layer is same as that of the pinned layer then the device is said to be in parallel state (Rp) and if it is opposite, then it is said to be in antiparallel parallel state (Rap). On a scale of the magnitude of resistance, the parallel state has low resistance whereas the antiparallel has high resistance. The TMR ratio is given by, (Rap − Rp)/Rp and greater the ratio value, greater is the sense margin for reading the bit value stored in it. In the following section, a brief description of the operation of a TMCAM is mentioned.

Fig. 1 A basic MTJ structure and its states

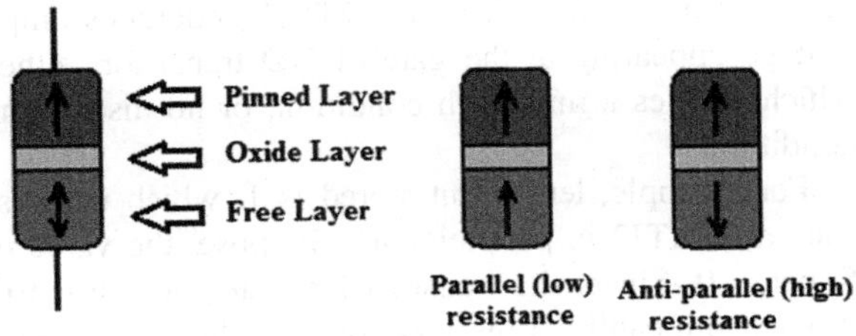

3 Basics of TMCAM

Each TMCAM cell consists of two MTJs connected to transmission gates, transmission gate 1 and 2 (TG1 and TG2) with control signals C1, C1_bar and C2, C2_bar respectively, as shown in Fig. 2, which in turn are connected to the bit lines S and S_bar. The junction formed by the MTJs is connected to a buffer, which amplifies the signal, and also to a writing transistor (V_W).

During the search operation, the writing transistor (Vw) is switched off and both the transmission gates are turned ON. The match-line (ML) is pre-charged. The bit line S is made 1 or 0, depending on the bit to be searched. The resultant voltage at

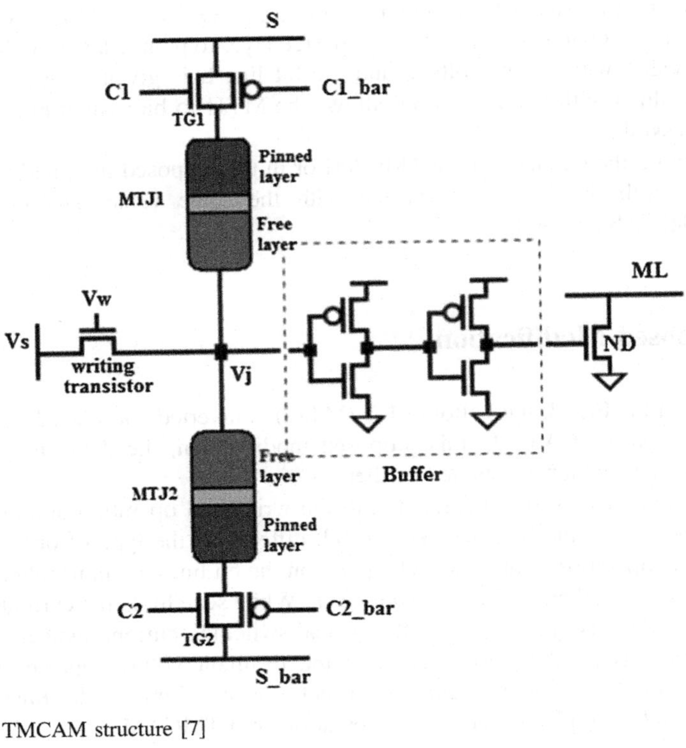

Fig. 2 A TMCAM structure [7]

the junction of the MTJ1 and MTJ2 undergoes amplification by the buffer. The voltage appearing at the gate of ND transistor, either discharges the match-line, which implies a mismatch condition, or no discharging occurs, implying a match condition.

For example, let the bit stored is 1, which implies the MTJ1 is in antiparallel state and MTJ2 in parallel state. Suppose, the value on the search line S is 1 and S_bar = 0. Since, the transmission gates are turned ON, the 'high' voltage on S drops significantly at the junction (Vj) due to the high resistance of MTJ1. On the other hand, as MTJ2 has a low resistance, the voltage '0' at S_bar is easily carried to the junction (Vj). Hence, the resultant voltage at Vj is almost null. This null value fails in switching ON the transistor ND, and hence the match-line (ML) doesn't discharge, implying a match condition.

During write operation, the writing transistor (Vw) is turned ON. If MTJ1 is to be written, then the transmission gate 1 (TG1) is turned ON and the transmission gate 2 (TG2) is maintained in OFF state. Depending on the value to be written on the MTJ1, the bit line S is made high/low and correspondingly the voltage at Vs is made low/high. If MTJ1 has to be written from parallel to antiparallel state then the switching current has to flow from the pinned layer of the MTJ to the free layer and if it has to be written from antiparallel to parallel then the switching current has to flow from free layer to pinned layer.

Suppose, MTJ1 has to be written from parallel state to antiparallel. To achieve this, the writing transistor (Vw) and TG1 are turned ON and TG2 is in OFF state. To allow the switching current to flow from free layer to pinned layer, the voltage at Vs is provided with 'high' voltage and the bit line S is given 'low'. Given this condition, along with a clock gating, allows the MTJ1 to be written to antiparallel state successfully.

For writing the ternary state in TMCAM or in the proposed modified version of TMCAM, both the MTJs are written with the same value, i.e., either both anti-parallel or both parallel.

4 Proposed Modification

As shown in the Fig. 3, the bottom MTJ (MTJ2) is inverted and placed, in contrast to the reported TMCAM. In this proposed modification, the delay in writing is reduced by half, which is detailed further.

The bit line drivers are integrated with the write/read operation and hence, the values appearing at the bit lines, S1 and S2, differ with the type of operation i.e., read or write operation. During search operation the bit lines are in negation to each other i.e., if S1 = 1 then S2 = 0 and vice versa. While searching, the current flowing through the MTJs is quite less than the critical switching current, hence, the MTJs only function as parallel resistances and the resultant voltage appearing at the junction is amplified and it eventually detects the matching or the mismatching condition, and is similar to the search operation of a TMCAM.

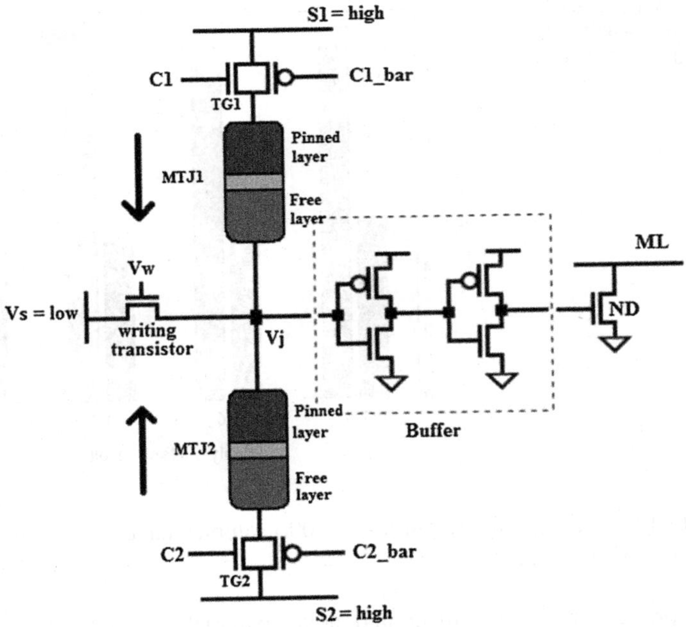

Fig. 3 The TMCAM structure with the proposed modification

During writing operation, the search driver integrated with the write enable operation, make the bit lines, S1 and S2, take up same values, i.e., either S1 = S2 = 0 or S1 = S2 = 1.

For example, if MTJ1 is to be programmed to antiparallel state from parallel and MTJ 2, to parallel from antiparallel, then the writing transistor is turned on and the bit lines S1 and S2 are made high. Correspondingly, Vs is made low. Current flows as shown in Fig. 3, and resultant states are written. For writing them again to their previous states, the bit lines are made low and Vs is provided with a high voltage.

5 Results and Simulation

The issue of write-delay has been dealt in this paper and sound results have also been obtained.

A. Simulation

The PTM (Predictive Technology Mode) model of 22 nm CMOS technology has been used in the simulation of this design. An MTJ of TMR value 2 with switching currents being, I_{ptoap} = 90 uA and I_{aptop} = 100 uA. The voltage on the bit lines for search operation is 1 V while for write operation is varied from

Fig. 4 The comparison of the write delay of the TMCAM and the modified TMCAM circuits

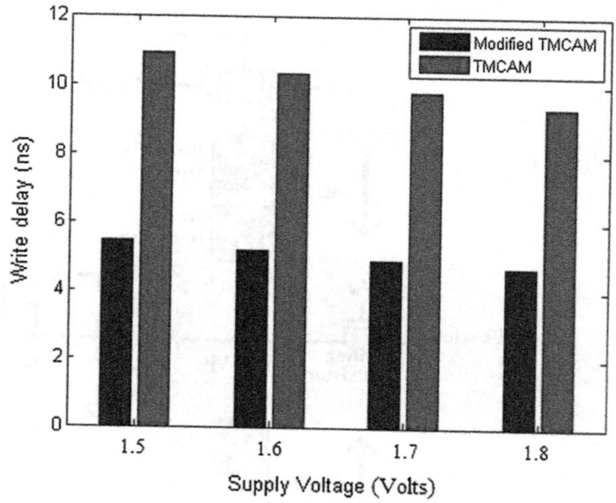

1.5 V to 1.8 V. For accuracy purposes, 1000 samples have been aggregated to obtain a refined result.

B. Result

The improvement in the power and delay parameters during the searching operation are negligible, as is evident from the design that the MTJs act as parallel resistances as the read current is significantly lower than the switching current.

As the bit lines in the new circuit are feasible in holding same or different values, the improvement in write-delay is quite significant. At four different voltage the time delay of the two designs, with TMR = 2, are compared in Fig. 4. It's clear that the time-delay has been reduced to half of that of the previous design. As the improvement is due to the design, flexibility of bit line operation and the symmetrical operation of MTJ, the results are the same for higher TMRs.

6 Conclusion

The proposed modification of TMCAM has very low write-delay and hence, would make it ideal in systems where the writing is predominant and writing speed cannot be compromised. Since, TMCAM is a very important part and is used in several systems, the demands it has are innumerable. Over the proposed advantage, the characteristic features of TMCAM like robustness, searching speed, power consumption and reliability are retained.

References

1. S. Panchanathan and M. Goldberg, "A content-addressable memory architecture for image coding using vector quantization," *IEEE Trans. Signal Process.*, vol. 39, no. 9, pp. 2066–2078, Sep. 1991.
2. T.-B. Pei and C. Zukowski, "VLSI implementation of routing tables: Tries and CAMs," in *Proc. IEEE INFOCOM*, vol. 2. Apr. 1991, pp. 515–524.
3. T. Ikenaga and T. Ogura, "A fully parallel 1-MbCAMLSI for real-timepixel-parallel image processing," *IEEE J. Solid State Circuits*, vol. 35, pp. 536–544, Apr. 2000.
4. W. Wang and Z. Jiang, "Magnetic content addressable memory," *IEEE Trans. Magn.*, vol. 43, no. 6, pp. 2355–2357, Jun. 2007.
5. S. Choi, K. Sohn, and H.-J. Yoo, "A 0.7-fJ/bit/search 2.2-ns search time hybrid-type TCAM architecture," *IEEE J. Solid-State Circuits*, vol. 40, no. 1, pp. 254–260, Jan. 2005.
6. W. Wang, "Magnetic content addressable memory design for wide array structure," *IEEE Trans. Magn.*, vol. 47, no. 10, pp. 3864–3867, Oct. 2011.
7. Mohit Kumar Gupta and Mohd Hasan, "Robust High Speed Ternary Magnetic Content Addressable Memory," *IEEE Trans. on Electron Devices,* vol. 62, no. 4, pp. 1163–1169, April 2015.
8. J. C. Slonczewski, "Current-driven excitation of magnetic multilayers," *J. Magn. Magn. Mater.*, vol. 159, no. 1/2, pp. L1–L7, Jun. 1996.
9. M. Hosomi, H. Yamagishi, T. Yamamoto, K. Bessho, Y. Higo, K. Yamane, H. Yamada, M. Shoji, H. Hachinoa, C. Fukumoto, H. Nagao, and H. Kano, "A novel nonvolatile memory with spin torque transfer magnetization switching: Spin-RAM," in *Int. Electron Devices Meeting (IEDM) Tech. Dig.*, 2005, pp. 459–462.
10. Takayuki Kawahara, Riichiro Takemura, Katsuya Miura, Jun Hayakawa, Shoji Ikeda, Young Min Lee, Ryutaro Sasaki, Yasushi Goto, Kenchi Ito, Toshiyasu Meguro, Fumihiro Matsukura, Hiromasa Takahashi, Hideyuki Matsuoka, and Hideo Ohno, "2 Mb SPRAM (SPin-Transfer Torque RAM) With Bit-by-Bit Bi-Directional Current Write and Parallelizing-Direction Current Read," *IEEE Journal of Solid-state circuits*, vol. 43, no. 1, pp. 109–120, Jan. 2008.
11. S. Yuasa, T. Nagahama, A. Fukushima, Y. Suzuki, and K. Ando, "Giant room-temperature magnetoresistance in single-crystal Fe/MgO/Fe magnetic tunnel junctions," *Nat. Mater.*, vol. 3, p. 868, 2004.
12. S. S. Parkin, C. Kaiser, A. Panchula, P. M. Rice, B. Hughes, M. Samant, and S.-H. Yang, "Giant tunneling magnetoresistance at room temperature with MgO (100) tunnel barriers," *Nat. Mater.*, vol. 3, p. 862, 2004.
13. T. Miyazaki and N. Tezuka, "Giant magnetic tunneling effect in Fe/Al2O3/Fe junction," *J. Magn. Magn. Mater.*, vol. 139, no. 3, pp. L231–L234, Jan. 1995.
14. T. Miyazaki and N. Tezuka, "Giant magnetic tunneling effect in Fe/Al2O3/Fe junction," *J. Magn. Magn. Mater.*, vol. 139, no. 3, pp. L231–L234, Jan. 1995.

SLA Based e-Learning Service Provisioning in Cloud

Mridul Paul and Ajanta Das

Abstract Cloud services allow individuals and businesses to use software and hardware that are managed by third parties at remote locations. The cloud computing model allows access to information and computing resources from anywhere through only the successful network connection. Cloud computing provides Infrastructure, Software and Platform as-a-service to its users with huge data storage, networks, processing power, and applications. Compared to traditional computing, its resources and services are available and scalable according to consumers need. Hence, e-Learning has been an interesting area where use of cloud can be leveraged, to reach online education across the globe. Provisioning of specific services is always depending on relevant Service Level Agreements that need to be agreed by both parties: provider and consumer. The objective of this paper is to provision e-Learning service on cloud. In order to accomplish this goal, negotiation of corresponding SLA parameters specific to the e-Learning service is necessary.

Keywords Cloud computing · Service level agreement (SLA) · e-Learning

1 Introduction

With increasing popularity and usage of the Internet and World Wide Web (WWW), researchers started experimenting with techniques to utilize infrastructure as extension to their existing designs. The designs leaned towards Service Oriented Architecture (SOA) and leveraged internet for certain common services provided by

M. Paul (✉) · A. Das
Department of Computer Science & Engineering, Birla Institute of Technology, Mesra, Deemed University, Kolkata Campus, Kolkata, India
e-mail: mridulpaul2000@yahoo.com

A. Das
e-mail: ajantadas@bitmesra.ac.in

© Springer India 2016
S.C. Satapathy et al. (eds.), *Information Systems Design and Intelligent Applications*, Advances in Intelligent Systems and Computing 435,
DOI 10.1007/978-81-322-2757-1_6

third parties with concept around computing services firmed up. Organizations such as Microsoft, Amazon and Google [7] started investing in building scalable platforms and opened gates for the public computing community usage over internet. Thus a Cloud movement started providing services around infrastructure, platform and software reducing total cost of ownership with high availability [2].

Cloud computing uses a similar architecture described in Service Oriented Architecture (SOA). Service consumers have flexibility to choose services in cloud according to their needs. The standards related to services consumption is maintained through Service Level Agreements (SLAs). SLA, by definition [1], is a negotiated document that includes a description about agreed service, service level parameters and guarantees. The main idea of having SLAs for any service in cloud is to provide a clear definition about service terms like performance, availability and billing.

Traditional e-Learning systems came into existence from the need of delivering learning content anytime to the users with no limitation of their age. However, traditional e-Learning systems faced challenges when the number of users multiplied or course contents increased. The tight coupling of modules within the system made it difficult to implement any changes or enhancements. It became even more tedious to integrate with other external systems. Thus the cost for support and maintenance of such systems increased over time.

In this scenario, provisioning of e-Learning as a service in cloud is a novel approach towards providing reliable, cost effective solution. The hardware infrastructure, in cloud computing, can easily be scaled up allowing e-Learning system to easily address scalability issues. The functionalities of e-Learning system can be implemented as services on cloud based on design standards that enforce loose coupling. This enables relatively easier to make any changes to the e-Learning system. The user expectations can be addressed through negotiating SLA contract. The importance of SLAs in cloud is outlined in the review on provisioning of services [14]. The objective of the paper is to provision e-Learning services in cloud. The services are defined in SLA. The SLA parameters pertaining to the user requirements can also be tracked through metrics captured when the services are consumed by the users.

The remainder of this paper is organized as follows: Sect. 2 presents related work. A brief overview on cloud and e-Learning is described in Sect. 3. Section 4 presents e-Learning SLAs and Sect. 5 proposes an e-Learning service in cloud. Section 6 details out the test case results and discussion of e-Learning service, followed by the Sect. 7, conclusion of the paper.

2 Related Work

Several researches have been conducted to optimize underlying cloud resources to host and manage services. A noted work from Bonvin et al. [3, 4] proposed that adaptive mechanism of service provisioning has higher benefits than static mechanism. In static service provisioning, the number of virtual machines (VMs) used by web application is fixed and does not change with time. In adaptive provisioning, the economic fitness of components is derived. The economic fitness is used for deciding whether that component stays in the cloud or be moved from one VM to another or be destroyed. However specific SLAs are not specified for the web application and hence metrics are captured for measuring performance.

Iqbal et al. [8] proposed algorithm that guarantees on minimizing response time for multi-tier cloud based web applications with maximum resource utilization. The algorithm detected bottlenecks by capturing details of CPU, memory, and I/O resource usage of each tier through real time monitoring and processing of each tier's log files. On detection of any performance bottleneck, corrective action is taken. Lodde et al. [10] stressed on service response time as a parameter for measuring and controlling service provisioning. A history of requests and associated response time is maintained where services are hosted. The underlying framework maintains a history of requests and associated response time for services hosted and calculates optimal resources (number of VM instances) required to comply with SLAs.

From above research papers, the consumer's view of SLA is not taken into consideration. SLA related to services can vary from user to user. A service consumer and provider first need to agree on the parameters which define expectations for the service. The service provider has to then capture these parameters and finalize metrics to monitor those SLA parameters at regular interval. Hence, this paper aims to establish a legal document, SLA, specifically non-functional SLAs and provide metrics to monitor these SLA parameters continuously related to e-Learning services.

3 Overview of Cloud and e-Learning

Cloud computing as defined by the National Institute of Standards and Technology's as model [13] that is universal, easy, on-demand access to shared pool of computing resources that can instantly and automatically provisioned. It has five basic and essential characteristics—*Resource Pooling* enables cloud providers to serve multiple consumers in a multi-tenant model, *Broad Network Access* enables consumers to access computing capabilities any type of internet enabled devices, *On-Demand Self-service* provides flexibility to choose computing capabilities, such as server time and network storage, *Rapid Elasticity* to provision computing capabilities on request for demand and *Measured Service* for consumers use services that can be calibrated and measured on parameters which are transparent to both provider and consumer.

Therefore cloud computing provides Infrastructure, Software and Platform as-a-service to its end users with huge data storage space, networks, processing power, and applications. Traditional e-Learning systems came into existence from the need of delivering learning content from and to anywhere. Learners could interact with their trainers, access materials and notes from any location. In order to remove the limitations for traditional e-Learning, provisioning e-Learning service in cloud is necessary. To achieve this goal, negotiating service level parameters between consumer and provider become essential.

4 Service Level Agreements

Service Level Agreement (SLA) is a key to service provisioning in cloud. It is a legal format that lays down rules for delivering services as well as framework for charging these services. Both service provider and consumer need to consider SLAs to define, create and operate services. SLAs have following important characteristics:

- describe service so that consumer can understand functionalities of the service
- articulate the performance levels of the service
- provide performance parameters for measuring and monitoring the services
- impose penalties in case the service does not meet requirements.

The expectations of consumer with respect to the cloud services can be categorized under two main subsets—functional and non-functional requirements. Functional requirements are those which are pertaining to the activity or functionality of the services. In context to e-Learning services, a simple functionality can be uploading and downloading of learning materials for a particular course. Another can be of tracking and reporting progress of the learner. Non-functional requirements or parameters, such as the performance, availability, security and billing of the service related to the consumers' expectation. SLA document for e-Learning comprises of the details on parties involved, service level parameters and objectives of the service [9]. SLA parameters are specified by metrics. Metrics can be simple or aggregated measure of variables associated with service level parameters. A metric description also includes the measurement and mathematical formula. The document consists of three sections—first section contains description of parties that are involved, their roles (provider, customer, third parties), the second section describes service level parameters are measurable indicators that provide insight into the service being offered. The last section has service level objectives that define the obligations or guaranteed condition of a service in a given period. Action guarantees represent promises of parties. It sends a notification in case the guarantees are not met.

In this paper, the non-functional requirements of SLA is defined and measured towards e-Learning service. The non-functional requirements considered here are *Availability* and *Billing*. In order to define SLAs, specific parameters are chosen and the values for these parameters are captured on the cloud infrastructure. Finally, these parameters are reported in the form of metric. Table 1 presents the corresponding metrics associated with the proposed SLA parameters for e-Learning service. The service level parameters related to these requirements in context of e-Learning are mentioned in the following:

Table 1 SLA parameters and associated metrics

SLA parameters	Metrics
Service availability	Uptime % = (Uptime $_{actual}$/Uptime $_{scheduled}$) * 100
	Uptime $_{actual}$ = no_actual_up_hrs
	Uptime $_{scheduled}$ = no_sched_up_hrs
Service outage	Outage % = (Downtime $_{scheduled}$/Total hours in a month) * 100
	Downtime $_{scheduled}$ = no_sched_down_hrs
Service backup	Backup $_{duration}$ = no_backup_duration
	Backup $_{Interval}$ = no_backup_frequency
Service billing	Price $_{usage}$ = Service usage time * Usage price per hour

- **Service Availability**—The service provider is responsible for availability of e-Learning service in cloud. However, it cannot be held liable for problems in network connection from consumers' machine to cloud. Service Availability does not include scheduled outages that are planned in advance by the service providers. The metric for Service Availability is Uptime percentage (*Uptime %*) which tracks ratio of actual uptime over scheduled uptime. Details of this metric is defined in Table 1.
- **Service Outage**—A Service Outage is defined as a period in which the consumers are unable to access the e-Learning services in cloud. The outages in consideration are planned downtimes that are needed for maintenance (upgrade and repairs) for the services. Hence, Service Outage duration need to be mentioned in SLA. The metric to track this parameter is Outage percentage (*Outage %*) which is the ratio of scheduled outage time over total time in a month. Details of this metrics are defined in Table 1.
- **Service Backup**—This parameter is related to *Availability*. Backups are procedures that keep a copy of user data (both static and transactional) in another secondary location, so that in case of unforeseen data loss event, the service provider can restore data from that secondary location. The associated metrics are Backup $_{duration}$ and Backup $_{Interval}$ to capture duration and frequency of the backups to be organized.
- **Service Billing**—As cloud computing is based on pay-as-you-go pricing model. The service consumers are charged based on the subscription model or course of the learning material they chose and duration of their usage. Price $_{usage}$ is the metric considered for this parameter.

5 e-Learning Services in Cloud

This section first briefly compares the traditional e-Learning and e-Learning system in computational Grid. Then it elaborates the improvements made in the existing e-Learning system to make it executable in cloud infrastructure.

Traditional e-Learning system [16] has five distinct layers. Layers at the highest level manage user profiles and identify them for smooth access to e-Learning services. *Learning Management* layer handles backend management of curriculum, resources, instructors and learners through self-contained learning objects (LOs); two layers at lowest level manage internal hardware resources, data storage, access and sharing. In our previous researches [12, 15], we proposed multi-agent based e-Learning framework with the capability of handling multimedia related documents and usage of huge data storage available in Grid.

However Grid computing [6] has some improvement areas that can be addressed by cloud computing. In Grid computing, the service provider has to manage the Grid infrastructure along with the application. In the cloud, the SaaS, PaaS and IaaS architecture models [5] allow service providers to focus on the services, while the responsibility of infrastructure resides with the cloud provider. In Grid computing, infrastructure is distributed across several locations with multiple parties through different ownership. This increases both administrative and communication cost to manage resources. In cloud computing, infrastructure is provided as a service. Hence, the service provider can communicate with cloud provider of that domain and substantially reduce both administrative and communication costs.

This paper proposes an improved layered architecture in cloud keeping all the functionalities same as in e-Learning system in Grid. The proposed architecture comprises of three layers—*e-Learning Portal*, *User Management* and *Learning Management*. The interactions among these layers are briefly described as follows:

- *e-Learning Portal (Layer 1)*—This layer serves as first interaction point for users to e-Learning services. The portal allows user to access relevant parts of e-Learning services via standard web browser. The users log in through the portal with appropriate credentials authenticated by User Management module.
- *User Management (Layer 2)*—This layer provides functionalities such as SLA management, access rights and role assignment. Each user has to go through the process of SLA creation before accessing any service. SLAs and related metrics are stored in the database. During service usage, a set of *Interacting Agents* collect SLA metrics from other modules and store information in the database. *Metric Reporter* then reads this information from the database and presents reports SLA status to the user. In case the SLAs do not meet the defined criteria, *Action Controller* sends appropriate messages to the service provider for corrective actions.
- *Learning Management (Layer 3)*—This layer manages learning courses, processes and contents. The courses designed for user needs are maintained here and functionalities such as course bookings, preparation and search, are provided. Learning processes such as tracking learners' progress, group discussions, email services, bulletin boards, assignments evaluation are defined in this module.

We choose Google's PaaS product, Google App Engine (GAE) [11], for implementing our e-Learning services. PaaS provides necessary server softwares, operating system, storage and computing infrastructure to establish robust and scalable e-Learning services. A Case Study on e-Learning Service is described in the next section.

6 Test Case Results and Discussion

This section describes the results from a case study implementation of e-Learning services on GAE. SLAs and related parameters are first captured, and then tracked through metrics. For our evaluation, we developed course management service that displays available courses to the user. The *Interacting Agent* collects the availability status of this service and passes information to the *Parameter Collector* for storage. *Metric Reporter* measures the SLA parameters based on acceptable range which is defined. The range for *Service Uptime* has lower limit as 99.5 % and upper limit as 100 %. *Service Outage* range is from 0.00 to 0.85 %. However, *Uptime %* and *Outage %* are independent metrics. Provisioning of e-Learning service is considered as test case for defining and evaluating SLA parameters.

This test case implementation provides a portal which acts as interface to e-Learning services. The users login in the portal as Service Consumer or Service Provider. Before accessing any service, the portal directs users to a negotiate SLA. Various courses are uploaded by the providers and consumers can download those courses through SLA management. Figure 1 shows the relevant SLA parameters, which consumers need to, fill and submit in the portal.

- *Availability*—This SLA parameter enables users to provide start and end time for services to be available. It also prompts for schedule downtime in hours when the system will be unavailable. In this case, the values considered for

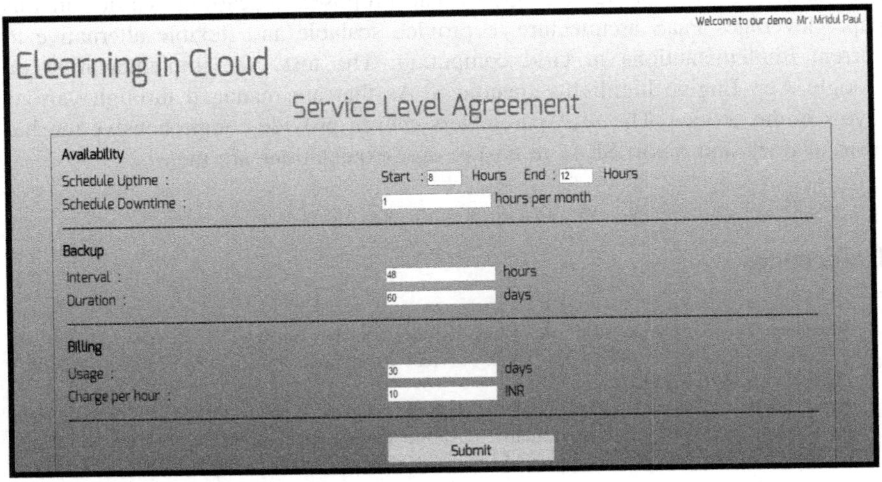

Fig. 1 SLA submission in e-Learning portal

availability is 0800 h (start time) and 1200 h (end time). Therefore service availability will be measured for 4 h of duration. The schedule downtime has been considered 1 h per month.

- *Backup*—The users are required to enter time interval for backups to be taken according to their choice. The value considered in this case is 48 h or 2 days. The users also need to provide duration of backups which signifies the length of time the backups will contain history data. The unit for this parameter is days. We have considered a value of 60 days for backup duration.
- *Billing*—Billing SLA has two parameters, usage duration of use of service (in days) and price to be charged per hour for the services. The usage value here is 30 days and the second parameter is pre-selected as 10 INR.

The evaluation of this test case is monitored for 5 days. The Metric Collector captured values for Uptime % (Service Availability), Outage % (Service Outage) and Billing for service usage during that period. While the Uptime % for this test case is 100 %, Outage % is 0 %. Both these values are found to be within stipulated range. In this test case, current Billing for 5 days of usage is 200 INR.

7 Conclusion

The main intention of this paper is to draw up research agenda for exploring cloud computing in the field of e-Learning. We presented important aspects of cloud computing and e-Learning system. As cloud implementations based on SOA architecture, functional and non-functional SLAs have to be thoroughly reviewed in order to move traditional e-Learning system to cloud. Besides, identification of SLA parameters and related metrics to measure SLAs are extremely critical in maintaining Quality of Service (QoS). The proposed e-Learning service in this paper leverages PaaS architecture to provide scalable and flexible alternative to current implementations in Grid computing. The test case implementation on Google App Engine highlights specific SLAs that are managed through various layers in the service. The SLA parameters chosen provide comprehensive mechanism to track and report SLAs to ensure user expectations are met.

References

1. Andrieux A., K. Czajkowski, A. Dan, K. Keahey, H. Ludwig, T. Nakata, J. Pruyne, J. Rofrano, S. Tuecke, Ming Xu "Web services agreement specification (WSAgreement)", 2006, Open Grid Forum.
2. Armbrust M., A. Fox, R. Griffith, A.D. Joseph, R. Katz, A. Konwinski, G. Lee, D. Patterson, A. Rabkin, I. Stoica, and M. Zaharia, "Above the Clouds: A Berkeley View of Cloud Computing", Advanced Computing Machines, Volume 53 Issue 4, April 2010, pp. 50–58.

3. Bonvin N., T.G. Papaioannou and K. Aberer, "An economic approach for scalable and highly-available distributed applications" IEEE 3rd International Conference on Cloud Computing, 2010, pp. 498–505.
4. Bonvin N., T. G. Papaioannou and K. Aberer, "Autonomic SLA-driven Provisioning for Cloud Applications", 11th IEEE/ACM International Symposium on Cluster, Cloud and Grid Computing, 2011, pp. 434–443.
5. Furht B., A. Escalante, "Handbook of Cloud Computing", Springer, 2010, pp 3–8.
6. Foster I., C. Kesselman, J. M. Nick and S. Tuecke, "Grid Computing: Making of Global Infrastructure a Reality", Wiley 2003, pp. 217–249.
7. Garfinkel S.L., "An Evaluation of Amazon's Grid Computing Services: EC2, S3 and SQS", Center for Research on Computation and Society, Harvard University, Technical Report, 2007.
8. Iqbal W., Matthew N. Dailey, D. Carrera, "SLA-Driven Dynamic Resource Management for Multi-tier Web Applications in a Cloud", 10th IEEE/ACM International Conference on Cluster, Cloud and Grid Computing, 2010, pp. 37–46.
9. Keller A, H. Ludwig, "Defining and Monitoring Service Level Agreements for dynamic e-Business", Proceedings of the 16th System Administration Conference, Nov 2002.
10. Lodde A., A. Schlechter, P. Bauler, F. Feltz, "SLA-Driven Resource Provisioning in the Cloud", 1st International Symposium on Network Cloud Computing and Applications, 2011, pp. 28–35.
11. Malawski M., M. Kuźniar, M. Wójcik, P. Bubak, M., "How to Use Google App Engine for Free Computing", Internet Computing, IEEE, 2013, Volume: 17, Issue: 1, pp. 50–59.
12. Mitra S., A. Das and S. Roy, "Development of E-Learning System in Grid Environment", Information Systems Design and Intelligent Applications, January 2015, Volume 2, pp: 133–141.
13. NIST Cloud Computing Program http://www.nist.gov/itl/cloud/index.cfm (2006).
14. Paul M., A. Das, "A Review on Provisioning of Services in Cloud Computing", International Journal of Science and Research, November 2014, Volume 3 Issue 11, pp. 2692–2698.
15. Roy S., A. De Sarkar and N. Mukherjee, "An Agent Based E-Learning Framework for Grid Environment", E-Learning Paradigms and Applications, Studies in Computational Intelligence, 2014, Volume 528, pp 121–144.
16. Shen Z., Y. Shi, G. Xu. Shi, "A learning resource metadata management system based on LOM specification", Proceedings of the 7th International Conference on Computer Supported Cooperative Work in Design, Sept 2002, pp. 452–457.

Estimating the Similarities of G7 Countries Using Economic Parameters

Swati Hira, Anita Bai and P.S. Deshpande

Abstract The contribution of this paper is to estimate the equality and differences between and within G7 countries: Canada, France, Germany, Italy, Japan, United Kingdom and United States. We used five parameters (GDP per capita, Employment, Population, General government revenue, General government total expenditure, Gross national savings) which widely correlate economic growth in the G7 countries. The means of the seven countries are identically equal is considered as a null hypothesis for each five parameters. We are using One-way analysis of variance statistical technique. Furthermore, the complete data set is evaluated to test the equivalence of the means between the G7 countries and each of a seven countries. The results show significant gaps between the group of G7 countries as well as selected parameters.

Keywords GDP per capita · Employment · Population · General government revenue · General government total expenditure · Economic growth

1 Introduction

G7 countries—common terms, which are indicated by seven countries—Canada, France, Germany, Italy, Japan, United Kingdom and United States to monitor developments in the world economy and assess economic policies. A country's economic health can usually be measured by looking at that country's economic growth. Most of the economist as Vinkler [1] Ekaterina and Sergey [2] and Narayan

S. Hira (✉) · A. Bai · P.S. Deshpande
Visvesvaraya National Institute of Technology, Nagpur, India
e-mail: dt11cse076@cse.vnit.ac.in

A. Bai
e-mail: anita.bai@students.vnit.ac.in

P.S. Deshpande
e-mail: psdeshpande@cse.vnit.ac.in

© Springer India 2016
S.C. Satapathy et al. (eds.), *Information Systems Design and Intelligent Applications*, Advances in Intelligent Systems and Computing 435,
DOI 10.1007/978-81-322-2757-1_7

[3] measures the growth of country on the basis of GDP per capita. Economy growth is also related with other economic parameters (Inflation, General government net lending/borrowing, Gross national savings, General government revenue, General government total expenditure, Population and Employment etc.) and non-economic parameters (Education, Health, Agriculture and environment etc).

From the previous researches we analyze that different parameters are used for various countries, For example the author Koulakiotis et al. [4] presented the economy growth on European countries on the basis of economic parameters by using a GARCH model to analytically investigate the casual relationship between GDP and inflation and also to observe the level of volatility for the case of inflation.

Results from Ekaterina and Sergey [2] indicate the socio-economic parameters of development in BRIC to describe GDP growth, income per capita, currency movements and also analyzed the rate of unemployment, employment in the informal sector using the latest demographic projections and a model of capital accumulation and productivity growth.

We also observed that Narayan [3] used GDP per capita, Koulakiotis et al. [4] used Inflation, GDP for European Countries using GARCH model, Du [5] used environmental pollution, economic growth and population parameters to find interrelation between them in economic zones in China using Environmental Kuznets Curve (EKC). EKC indicates the reverse U relation between the environmental quality and per capita income to identify whether economic growth is a method to resolve the environmental issue or a factor that causes the environmental issue. Dao [6] observed population and economic growth in developing countries using the least-squares estimation technique in a multivariate linear regression. Hull [7] examined the Relationship between economic growth, employment and poverty reduction for low income countries. Yu [8] examined the role of employment in the process of investment and economic growth for 18 cities in Henan province by using hypothesis testing. Taha et al. [9] describes case study of Malaysia on the basis of stock market, tax revenue and economic growth using Granger causality [10] and VECM framework. Chen [11] use linear Granger causality to investigate the relationship between government expenditure and economic growth in China. Tao [12] analyze the efficiency of current fiscal expenditure for economic growth and agriculture in rural China, and attain the efficiencies of the different expenditure. Yang and Zheng [13] estimate the long-term and short-term elasticity of the local fiscal expenditure on the economic growth, and then study the interaction between these two factors for Fujian province in China.

From above literature we observed that various economic parameters are used to indicate the economic growth of country but some of them which mainly describes the economic status of country are GDP per capita, Employment, Population, General government revenue, Gross national savings and General government total expenditure. At the same time we also observed that not significant research has been done to indicate the gaps between the G7 countries by using above parameters to facilitate economic cooperation among the world's and discuss the actions on economic and commercial matters and works to aid the economies of other nations. So we decided to use above five parameters for G7 countries in our research.

Table 1 Sample dataset

Country/series-specific notes	1980	1981	1982	1983	1984	1985
GDP per capita	9,629.89	9,227.53	8,746.22	8,281.16	7,818.35	8,292.22
Employment	25.086	24.43	23.951	23.775	24.285	24.593
Population	56.33	56.357	56.291	56.316	56.409	56.554
General government revenue	39.58	41.688	42.775	41.508	41.425	41.199
General government total expenditure	42.742	45.953	45.355	44.789	44.957	43.956
Gross national savings	49.57	31.676	52.795	38.535	42.435	38.149

2 Data Set

Yearly data on above five parameters were obtained from the IMF Finance data [14] over the period 1980–2013 for the G7 countries Canada, France, Germany, Italy, Japan, United Kingdom and United States. The data has been transformed to normalized form. Table 1, describes the IMF sample dataset use for further processing.

3 Methodology: Anova and Z-Statistics

We have adopted the One way Analysis of variance approach in order to examine the equality of means between G7 countries and within a country for each parameter in the period from 1980 to 2013.

To evaluate the results, we calculate the mean and variance for each parameter and defining the terms as follows:

x_{ij}	values of observations (years) i for country j; j = 1, 2,…, k,
n_j	size of jth country,
$\bar{x}_j$	sample mean of jth country,
$\bar{\bar{x}}$	grand mean,
k	number of country (k = 7),
$n_T = \sum n_j$	Total country size,
s_j^2	sample arithmetic variance for parameter j,
σ_b^2	between column variance (e.g. between Canada and France),
σ_w^2	within column variance (e.g. within Canada)

The arithmetic mean formula for country j as follows:

$$\bar{x}_j = \sum_i \frac{x_{ij}}{n_j}, \quad \text{where i} = 1, 2, \ldots, n_j \tag{1}$$

which indicates

$$\sum_i x_{ij} = n_j \bar{x}_j \tag{2}$$

The arithmetic variance formula for country j as follows:

$$s_j^2 = \frac{\sum_i \left(x_{ij} - \bar{x}_j\right)^2}{\left(n_j - 1\right)}, \quad \text{where i} = 1, 2, \ldots, n_j \tag{3}$$

$\bar{\bar{x}}$, is the sum of all year's data divided by the total number of years n_T,

$$\bar{\bar{\mathrm{x}}} = \sum_{\mathrm{j}} \sum_{\mathrm{i}} \mathrm{x_{ij}} / \mathrm{n_T} \tag{4}$$

Analysis of variance is calculated according to Levin and Rubin [15] as *follows:* Total sum of squares SST of a parameter is calculated as sum of variation between different countries and variation within countries.

$$\mathrm{SST} = \mathrm{SSB} + \mathrm{SSW} \tag{5}$$

where

$$SSB = \sum n_j (\bar{x}_j - \bar{\bar{x}})^2 \tag{6}$$

and

$$SSW = \sum_j (n_j - 1) s_j^2 \tag{7}$$

Estimation of between column (country) variance (MSB):

$$\sigma_b^2 = \frac{\sum n_j \left(\bar{x}_j - \bar{\bar{x}}\right)^2}{k - 1} \tag{8}$$

$$\text{or} \quad MSB = \frac{SSB}{k - 1} \tag{9}$$

Estimation of within column (country) variance (MSW):

$$\sigma_w^2 = \sum_j \left(\frac{n_j - 1}{n_T - k}\right) s_j^2 \tag{10}$$

$$\text{or} \quad MSW = \frac{SSW}{\mathrm{n_T - k}} \tag{11}$$

The F-test statistics is use to test equality of the means in G7 countries for a parameter as follows:

$$F = MSB/MSW \tag{12}$$

Equations (9) and (11) are used to calculate MSB and MSW. We are comparing F with its critical value F_{crit} (α, k − 1, $\mathrm{n_T}$ − k) for significance, where α = 0.05.

If the hypothesis of similarity of means is rejected, a query occurs as to which means are distinct. The statistics methods use to resolve this query are known as multiple paired comparison procedures. We calculated it by using two tailed significance level estimates (z-test) Levin and Rubin [15].

We calculated hypothesis at α = 0.05 level that there is no difference between parameters of G7 countries versus one Country.

The Hypothesis is

$H_0 : \mu_1 = \mu_2$, Null Hypothesis, there is no differences exists $H_1 : \mu_1 \neq \mu_2$, Alternative Hypothesis, differences exists

The z-test Statistics are

$$Z = \frac{(\bar{x}_a - \bar{x}_i) - (\mu_1 - \mu_2)_{H_0}}{\sigma_{x_a - x_j}} \tag{13}$$

where,

j = 1, 2,…, 7 denotes G7 countries,

x_a, denotes means of all G7 countries,

$\sigma_{x_a - x_j}$, Standard error of the difference between two means,

and $\sigma_{x_a - x_j}$ is calculated as,

$$\sigma_{x_a - x_j} = \sqrt{\frac{\sigma_1^2}{n_1} + \frac{\sigma_2^2}{n_2}} \tag{14}$$

where σ_1 and σ_2 are calculated from Eq. (3) and

σ_1^2 denotes variance of G7 counties

σ_2^2 denotes variance of one country.

4 Results

Equations (6), (7) and (12) are used to calculate Table 2 describes the ANOVA results representing SSB, SSW and TSS which shows their percentage of contribution for each parameter. The differences between the G7 countries is much smaller than the within countries differences.

This is notably observable when examining Gross domestic product per capita, where only 2.78 % of the difference is explained by group of countries and 97 % of the differences exist within the country. Population displays the maximum difference between countries at approximately 17 %. Similarly Employment, General government revenue, General government total expenditure and Gross national savings shows 8, 11.82, 11 and 12.60 % difference exist between countries respectively.

The F-statistic is calculated using Eq. (12), we observed that the F-test statistic exceeds the significance level of F_{crit} (0.05) except Gross domestic product per capita for the degree of freedom (df) in all cases.

The results are shown in Table 3 calculated using Eq. (13), describes the results of testing for similarity of means between the G7 countries for the five parameters. Each country is significantly different from the group of all G7 countries.

We observed that GDP per capita is the only parameter where we did not get any exception. The only exception is country Japan in relation to employment parameter. In population parameter we got exception in three countries as Italy, Japan and United Kingdom. Similarly we got the exceptions for other parameters. This is strong indication of each country identity in these five parameters.

Table 2 Anova results

Parameters	SSB	%TSS	SSW	%TSS	F-statistics	F-critical
Gross domestic product per capita	0.630	2.78	22.04	97.21	1.101	2.137
Employment	2.007	8.05	22.91	91.94	3.372	2.137
Population	4.643	16.92	22.79	83.07	7.842	2.137
General government revenue	3.718	11.82	27.73	88.17	5.162	2.137
General government total expenditure	3.230	11.00	26.11	88.99	4.762	2.137
Gross national savings	3.670	12.60	25.45	87.39	5.967	2.137

Table 3 Multiple paired comparisons

Comparison	Z-stastics	Decision
Gross domestic product per capita		
All countries versus Canada	−1.13763	Accept hypothesis of equal means
All countries versus France	−0.10141	Accept hypothesis of equal means
All countries versus Germany	0.189697	Accept hypothesis of equal means
All countries versus Italy	−0.08686	Accept hypothesis of equal means
All countries versus Japan	1.305578	Accept hypothesis of equal means
All countries versus UK	−0.51883	Accept hypothesis of equal means
All countries versus US	0.362444	Accept hypothesis of equal means
Employment		
All countries versus Canada	−0.31877	Accept hypothesis of equal means
All countries versus France	−1.1457	Accept hypothesis of equal means
All countries versus Germany	−1.59073	Accept hypothesis of equal means
All countries versus Italy	−0.61265	Accept hypothesis of equal means
All countries versus Japan	2.46643	Fail to accept the hypothesis of equal means
All countries versus UK	0.201679	Accept hypothesis of equal means
All countries versus US	1.05261	Accept hypothesis of equal means
Population		
All countries versus Canada	−0.00068	Accept hypothesis of equal means
All countries versus France	−0.1702	Accept hypothesis of equal means
All countries versus Germany	1.753566	Accept hypothesis of equal means
All countries versus Italy	−2.82057	Fail to accept the hypothesis of equal means
All countries versus Japan	3.100156	Fail to accept the hypothesis of equal means
All countries versus UK	−2.23987	Fail to accept the hypothesis of equal means
All countries versus US	0.09982	Accept hypothesis of equal means
General government revenue		
All countries versus Canada	1.436764	Accept hypothesis of equal means
All countries versus France	0.550711	Accept hypothesis of equal means
All countries versus Germany	2.57576	Fail to accept the hypothesis of equal means
All countries versus Italy	3.68759	Fail to accept the hypothesis of equal means
All countries versus Japan	3.229094	Fail to accept the hypothesis of equal means
All countries versus UK	−1.33073	Accept hypothesis of equal means
All countries versus US	−0.96342	Accept hypothesis of equal means
General government total expenditure		
All countries versus Canada	0.66326	Accept hypothesis of equal means
All countries versus France	4.322691	Fail to accept the hypothesis of equal means
All countries versus Germany	2.854144	Fail to accept the hypothesis of equal means
All countries versus Italy	4.475404	Fail to accept the hypothesis of equal means
All countries versus Japan	0.79585	Accept hypothesis of equal means
All countries versus UK	3.858912	Fail to accept the hypothesis of equal means
All countries versus US	−0.49286	Accept hypothesis of equal means

5 Conclusion

The results of our approach indicate that substantial gap between the G7 countries as well as within countries. Regardless of the strengths of globalization throughout the globe, there exists economic inequality between nations and within countries. This paper used the time series data provided by IMF and perform essential calculations for ANOVA and z significance test. The main contribution of our paper it to test the similarity of means among G7 countries (Canada, France, Germany, Italy, Japan, United Kingdom and United States) for five parameters (GDP per capita, Employment, Population, General government revenue, General government total expenditure). Further calculations were performed for multiple paired comparisons and testing for similarity between the group of G7 countries and one country for each parameter. We used above five parameters in our paper to determine the uniqueness of each country. From our results we observed that hypothesis is only followed by GDP per capita, i.e. there exists some type of similarity between countries and also at the same time we observed that all the other parameters were not following the same behaviour as GDP per capita, i.e. parameter was not having the same characteristic for G7 countries. So in our research we provide evidence that each country is having its own identity with respect to its parameters.

References

1. Vinkler, P.: Correlation between the structure of scientific research, scientometric indicators and GDP in EU and non-EU countries. Scientometrics 74(2). (2008) 237–254. doi:10.1007/s11192-008-0215-z.
2. Ekaterina, D., Sergey, K.: The Chinese Economy and the Other BRIC Countries: The Comparative Analysis. IEEE International Conference on Management Science & Engineering, Moscow, Russia, September. (2009). doi:10.1109/ICMSE.2009.5318207.
3. Narayan, P.K.: Common Trends and Common Cycles in Per Capita GDP: The Case of the G7 Countries, 1870–2001. International Advances in Economic Research 14(3). (2008) 280–290. doi:10.1007/s11294-008-9162-y.
4. Koulakiotis, A., Lyroudi, K., Papasyriopoulos, N.: Inflation, GDP and Causality for European Countries. International Advances in Economic Research 18(1). (2012) 53–62. doi:10.1007/s11294-011-9340-1.
5. Du, W.: Interrelations between Environmental Pollution,Economic Growth and Population Distribution in Gradient Economic Zones in China. IEEE International Conference on Remote Sensing Environment and Transportation Engineering (RSETE). (2011) doi:10.1109/RSETE.2011.5965738.
6. Dao, M.Q.: Population and economic growth in developing countries. International Journal of Academic Research in Business & Social Sciences 2(1). (2012).
7. Hull, K.: Understanding the relationship between economic growth, employment and poverty reduction. Organization for Economic Cooperation and Devlopment (OECD) Promoting pro-poor growth: employment. OECD. (2009) 69–94.

8. Yu, Z.: The Role of employment in the process of investment and economic growth. IEEE 2nd International Conference on Artificial Intelligence, Management Science and Electronic Commerce (AIMSEC). (2011) 832–834. doi:10.1109/AIMSEC.2011.6010482.
9. Taha, R., Colombage, S. RN., Maslyuk, S.: Stock Market, Tax Revenue and Economic Growth: A Case-Study of Malaysia. IEEE International Conference on Management Science & Engineering, Melbourne, Australia. (2010) 1084–1090. doi:10.1109/ICMSE.2010.5719932.
10. Engle, RF., Granger, CWJ.: Co-integration and error correction: Representation, estimation and testing. Econometrica 55(2). (1987) 251–276.
11. Chen, Z.: Government Expenditure and Economic Growth in China. IEEE International Conference on Computer Application and System Modeling (ICCASM 2010). (2010) doi:10.1109/ICCASM.2010.5622654.
12. Tao, Y.: Efficiency Analysis on Fiscal Expenditure Based on Economic Growth in Rural China. IEEE International Conference on Management Science & Engineering, September. (2009)1019–1023, Moscow, Russia. doi:10.1109/ICMSE.2009.5318195.
13. Yang, G., Zheng, H.: The Research on relationship of Local Fiscal Expenditure for S&T and Economic Growth- Based on Cointegration and VA R Model Analysis. International Conference on Management and service Science (MASS). (2011) 1–4.
14. IMF: International Monetary Fund. UN conference in Bretton Woods. http://www.imf.org/external/data.htm. Accessed July (1944).
15. Levin, RI., Rubin, DS.: Statistics for Management. Pearson Prentice Hall, University of North Carolona. (2006).

Offline Malayalam Character Recognition: A Comparative Study Using Multiple Classifier Combination Techniques

Anitha Mary M.O. Chacko and K.S. Anil Kumar

Abstract Malayalam character recognition has gained immense popularity in the past few years. The intrinsic challenges present in this domain along with the large character set of Malayalam further complicate the recognition process. Here we present a comparative evaluation of different multiple classifier combination techniques for the offline recognition of Malayalam characters. We have extracted three different features from the preprocessed character images—Density features, Run-length count and Projection profiles. These features are fed as input to three different neural networks and finally the results of these three networks were combined and evaluated using six different classifier combination methods: Max Rule, Sum Rule, Product Rule, Borda Count Rule, Majority Voting and Weighted Majority voting schemes. The best recognition accuracy of 97.67 % was attained using the Weighted Majority scheme considering top 3 results.

Keywords Pattern recognition · Multiple classifier system · Neural networks · Character recognition · Classifier combination

1 Introduction

Offline Character Recognition is the task of recognizing and converting handwritten text into a machine editable format. Automatic Recognition of handwritten text is one of the most active research areas in the domain of pattern recognition. The inherent challenges present in this domain along with the wide variety writing styles by different writers at different times further hinders the recognition process.

A.M.M.O. Chacko (✉) · K.S. Anil Kumar
Department of Computer Science, Sree Ayyappa College,
Eramallikkara, Chengannur 689109, Kerala, India
e-mail: anithamarychacko@gmail.com

K.S. Anil Kumar
e-mail: ksanilksitm@gmail.com

© Springer India 2016

S.C. Satapathy et al. (eds.), *Information Systems Design and Intelligent Applications*, Advances in Intelligent Systems and Computing 435,
DOI 10.1007/978-81-322-2757-1_8

The problem gets even more complicated for South Indian languages due to its extremely large character set formed by loops and curves which makes the development of a complete OCR system for these languages a highly challenging task [1]. The high similarity in the writing style of different characters also complicates the recognition process. Among the Indian languages, the research on Malayalam scripts has gained high attention in the recent years due to the initiatives of the Government of India, Ministry of Communications and Information Technology and Technology Development of Indian Languages (TDIL).

In this paper, we investigate the results of combining multiple classifiers for the recognition of handwritten Malayalam characters. The extracted features are based on the density features, run-length count and projection profiles. These features are fed as input to three feedforward neural networks. The final results are obtained by combining the results of individual classifiers using six different combination schemes: Max Rule, Sum Rule, Product Rule, Borda Count Rule, Majority Voting and Weighted Majority voting rule.

This paper is structured as follows: Sect. 2 presents the related works. Section 3 introduces the architecture of the proposed system. Section 4 presents the experimental results and finally, Sect. 5 concludes the paper.

2 Related Works

In the literature, several techniques can be traced for the recognition of offline handwritten characters. An excellent survey on OCR research in South Indian Scripts is presented in [2, 3]. A detailed survey on handwritten Malayalam character recognition is presented in [4].

Rajashekararadhya et al. [5] proposed an offline handwritten numeral recognition system for the four South Indian languages—Malayalam, Tamil, Kannada and Telugu. The proposed system used zone centroid and image centroid based features. Here, Nearest Neighbour and Feedforward neural networks were used as classifiers. Sangame et al. [6] proposed a system for unconstrained handwritten Kannada vowels recognition based upon Hu's seven invariant moments. A Euclidian distance criterion and K-NN classifier were used for classification. Ragha and Sasikumar [7] proposed a recognition method based on extraction of moment features from Gabor directional images of handwritten Kannada characters. An MLP trained with backpropagation was used for classification purposes. Aradhya et al. [8] proposed a technique for Kannada character recognition based on Fourier Transform and Principal Component Analysis using a Probabilistic Neural Network. Bhattacharya et al. [9] proposed a two-stage recognition scheme for handwritten Tamil characters. Sutha and Ramaraj [10] proposed an approach to recognize handwritten Tamil characters using Fourier descriptor based features and multi-layer perceptron (MLP) classifier. Hewavitharana and Fernando [11] proposed a two-stage classification approach which is a hybrid of structural and statistical techniques.

The effect of multiple classifier combination techniques for Malayalam Character Recognition is a less explored area. In [12], a classifier based weighted majority scheme was proposed for the recognition of Malayalam Characters using Chain Code histogram and Fourier Descriptors. A multiple classifier system using gradient and density features were proposed in [13]. These works proves that the multiple classifier combination schemes offers immense potential for Malayalam character recognition which if properly explored can aid the development of a highly accurate recognition system.

3 Proposed System

The proposed system consists of mainly 4 stages: Preprocessing, Feature Extraction, Classification and Post Processing. The scanned image is first preprocessed and then subjected to feature extraction. Here, we have extracted three features-density features, run-length count and projection profiles. The outputs of these networks are combined using different combination techniques and finally the characters are mapped to their corresponding Unicode characters in the post processing stage. Figure 1 shows the architecture of the proposed system.

3.1 Preprocessing

The objective of this phase is to remove distortions from scanned images that occur due to the poor quality of the document or scanner. We have applied the same sequence of preprocessing steps as in [14].

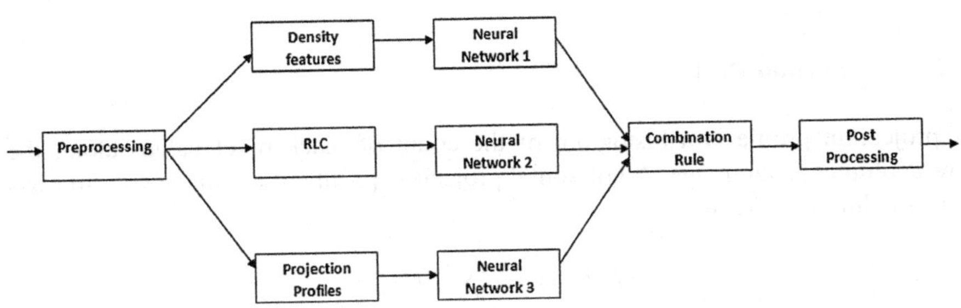

Fig. 1 Architecture of the proposed system

3.2 *Feature Extraction*

Feature extraction is the most important phase in character recognition that determines the success rate of the underlying OCR system. In a comparative study of different feature extraction techniques for Malayalam, the best recognition rate with minimal feature size have been obtained using Density features. So, here we have Profiles.

3.2.1 Density Features

Density of a zone is defined as the ratio of number of foreground pixels of each zone divided by the total number of pixels in each zone. For extracting the density features, the preprocessed character images are divided into 4 × 4 zones. The 16 density features obtained from the 16 zones forms the input to the first neural network.

$$\mathrm{d(i)} = \frac{\text{Number of foreground pixels in zone i}}{\text{Total number of pixels in zone i}} \tag{1}$$

3.2.2 Run Length Count

Run length count is the count of successive 1's obtained in each row/column of the image. The points of transitions between 0 and 1 mark the edges of the characters. From the 16 zones of each character, we calculate the horizontal and vertical RLC features. The horizontal RLC is the count of continuous 1's encountered in a left to right scan of the image. The vertical RLC is the count of continuous 1's encountered in a top to bottom scan of the image. Thus the input to the second neural network consists of 32 RLC features—16 horizontal RLC features and 16 vertical RLC features.

3.2.3 Projection Profiles

A projection profile is a histogram of the count of black pixel values along the rows/columns in an image. A horizontal projection profile of an image with m rows and n columns is defined as:

$$HPP(i) = \sum_{1 \le j \le n} f(i,j)$$

The vertical projection profile which is the count of black pixels along each column of the image is defined as:

$$VPP(i) = \sum_{1 \leq i \leq m} f(i,j)$$

The sum of vertical and horizontal projection profiles are computed for each of the 16 zones. Thus the input to the third neural network consists of 32 projection features—16 HPPs and 16 VPPs.

3.3 *Classification*

In this stage, the characters are classified into corresponding class labels. We have used three feedforward neural networks for the density features, RLC features and projection profiles respectively. The results of these three classifiers were combined and evaluated using six combination rules.

4 Classifier Combination Methods

Combining classifiers increase the accuracy of individual classifiers as the misclassifications made by different classifiers trained with different features may not overlap. The challenge involved here is to find a suitable combination method which can significantly improve the result of base classifiers. In this paper, we have evaluated the performance of our system using the following six combination strategies. d^{com} denotes the decision of the combined classifier and δ_{ik} denotes the activation value given by the kth classifier for the ith class.

4.1 *Max Rule*

This rule selects the class with maximum activation value among the three individual classifiers. The final combined decision d^{com} is:

$$d^{com} = \max_i(\delta_{ik})$$

4.2 *Sum Rule*

This rule sums up the activation values given by each of the classifiers for each of the classes and selects the class with the maximum value as the final output

$$\mathrm{d^{com}} = \max_{\mathrm{i}}\left(\sum_{\mathrm{k}=1,2,3} \delta_{\mathrm{ik}}\right)$$

4.3 Product Rule

This rule multiplies the activation values given by each of the classifiers for each of the classes and selects the class with the maximum product value as the final output.

$$d^{com} = \max_i\left(\prod_{k=1,2,3} \delta_{ik}\right)$$

4.4 Borda Count Rule

The borda count for a class is the number of classes that are ranked below it by each classifier. In this method, each of the classifiers sorts the output classes according to descending order of their borda count values. The class with the highest borda count value is selected as the final output class.

4.5 Majority Voting Rule

This method selects the final output class as that class which receives the maximum number of votes/classifications.

4.6 Weighted Majority Voting Rule

The weighted majority scheme for combining the results of individual classifiers was proposed in [14]. Here, each classifier is assigned weights based on their individual success rates. The final combined decision is:

$$\mathrm{d^{com}} = \max_{\mathrm{i}} \sum_{\mathrm{k}=1,2,3} \mathrm{w_k} * \delta_{\mathrm{ik}} \quad 1 \le \mathrm{i} \le \mathrm{N_C}$$

where

$$w_k = \frac{p_k}{\sum_{k=1,2,3} p_k}$$

p_k denotes the success rate of kth classifier.

5 Experimental Results

The proposed system was tested using a database consisting of 990 samples of Malayalam characters collected from 30 people belonging to different age groups and professions. The database consists of 33 selected characters of Malayalam—8 isolated vowels and 25 consonants.

The results of the experiment are summarized in Tables 1 and 2. Table 1 shows the accuracies that we have obtained for the individual classifiers trained with the density, RLC and projection profile features. Table 2 summarizes the results obtained by combining the individual classifiers. It is evident from the table that all the combination schemes attain better results than the single classifier.

The best recognition accuracy of 97.67 % was obtained using the Weighted Majority Voting scheme considering the top 3 recognition results. The combined classifier gave an overall accuracy of 92.53 % considering only the top 1 result. The combination system using the Sum rule also attains a high accuracy of 92.12 %. The lowest accuracy was reported by the Borda count combination method. The performance results are also summarized in Fig. 2.

Table 1 Individual classifier results

Feature	Size	Accuracy (%)
Density	16	85.35
RLC	32	75.25
Projection profiles	32	72.22

Table 2 Classifier combination results

Feature	Accuracy (%)
Max rule	91.92
Sum rule	92.12
Product rule	88.48
Borda count rule	87.27
Majority voting	88.68
Weighted (Top 1)	92.53
Weighted (Top 2)	96.26
Weighted (Top 3)	97.67

Fig. 2 Performance results

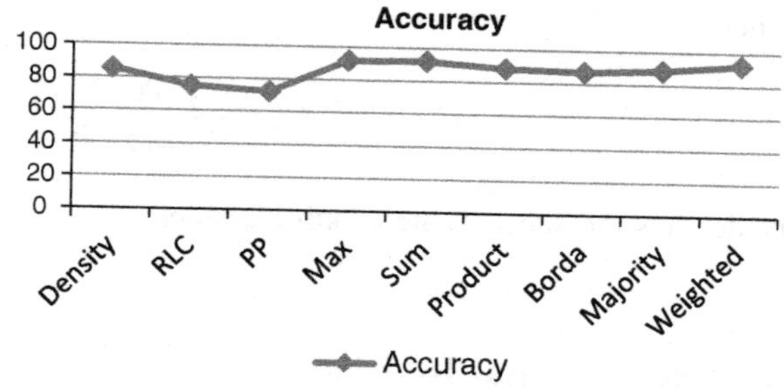

6 Conclusion

In this paper, we have presented a comparative evaluation of different multiple classifier combination techniques for the offline recognition of Malayalam characters. We have extracted three different feature sets—Density features, Run-length count and Projection profiles. These features were fed as input to three different neural networks and finally the results of these three networks were combined using six different classifier combination methods: Max Rule, Sum Rule, Product Rule, Borda Count Rule, Majority Voting and Weighted Majority voting schemes. The best recognition accuracy of 97.67 % was attained using the Weighted Majority scheme considering top 3 results. From the experiment, we have concluded that most of the misclassifications were due to similar shaped character pairs. So, the future works aim to reduce these errors in the post processing stage and thus develop a 100 % accurate OCR system.

References

1. R. Plamondan, S.N. Srihari, Online and offline handwriting recognition: A comprehensive survey", IEEE Trans. On PAMI, Vol22(1), pp 63–84 (2000).
2. Jomy John, Pramod K. V, Kannan Balakrishnan, Handwritten Character Recognition of South Indian Scripts: A Review", National Conference on Indian Language Computing, Kochi, Feb 19–20 (2011).
3. Abdul Rahiman M., Rajasree M. S., A Detailed Study and Analysis of OCR Research in South Indian Scripts". In: Proceedings of International Conference on Advances in Recent Technologies in Communication and Computing, IEEE, pp 31–38 (2009).
4. Anitha Mary M.O. Chacko and Dhanya P.M., Handwritten Character Recognition in Malayalam Scripts -A Review", Int. Journal of Artificial Intelligence & Applications (IJAIA), Vol. 5, No. 1, January 2014, pp 79–89 (2014).
5. Rajasekararadhya SV, Ranjan PV, Efficient zone based feature extraction algorithm for handwritten numeral recognition of popular south Indian scripts", J Tech Appl Inform Technol 7(1): pp 1171–1180 (2009).
6. Sangame S.K., Ramteke R.J., Rajkumar Benne, Recognition of isolated handwritten Kannada vowels", Adv. Computat. Res. Vol.1, Issue 2, 5255 (2009).

7. L R Ragha, M Sasikumar, Using Moments Features from Gabor Directional Images for Kannada Handwriting Character Recognition. In: Proceedings of International Conference and Workshop on Emerging Trends in Technology (ICWET) TCET, Mumbai, India (2010).
8. Aradhya V. N. M, Niranjan S. K. and Kumar, G. H. Probabilistic neural network based approach for handwritten character recognition", Int. J. Comp. Comput. Technol. (Special Issue) 1, 913 (2010).
9. Bhattacharya U., Ghosh S. K., and Parui S. K. A two-stage recognition scheme for handwritten Tamil characters", In: Proceedings of the 9th International Conference on Document Analysis and Recognition (ICDAR07), pp 511–515 (2007).
10. Sutha, J. and Ramaraj, N. Neural network based offline Tamil handwritten character recognition system.", In: Proceedings of ICCIMA07, pp 446–450 (2007).
11. Hewavitharana S. and Fernando H. C. A two-stage classification approach to Tamil handwriting recognition", In: Proceedings of the Tamil Internet Conference, pp 118–124 (2002).
12. Anitha Mary M.O. Chacko and Dhanya P.M., "Combining Classifiers for Offline Malayalam Character Recognition", Emerging ICT for Bridging the Future - Proceedings of the 49th Annual Convention of the Computer Society of India CSI Volume 2, Advances in Intelligent Systems and Computing Volume 338, pp 19–26 (2015).
13. Anitha Mary M.O. Chacko and Dhanya P.M., "Multiple Classifier System for Offline Malayalam Character Recognition", International Conference on Information and Communication Technologies (ICICT-2014), Procedia Computer Science 46 pp 86–92 . (2015).
14. Anitha Mary M.O. Chacko and Dhanya P.M., "A Comparative Study of Different Feature Extraction Techniques for Offline Malayalam Character Recognition", International Conference on Computational Intelligence in Data Mining (ICCIDM-2014), Springer (2014).

A Comparative Study on Load Balancing Algorithms for SIP Servers

Abdullah Akbar, S. Mahaboob Basha and Syed Abdul Sattar

Abstract The widespread acceptance and usage of smartphones deployed with high end operating systems have made Voice over IP applications extremely popular and prevalent globally. A large set of users amongst these use a plethora of Internet based applications after configuring them on their devices. SIP Proxy servers are predominantly used in these VOIP networks for the routing challenges arising from the requirement of supporting millions of VOIP concurrent/subsequent calls and also increasing the QoS (Quality of Service) of the routed calls. For intelligent load balancing, call dispatchers are used to achieve high throughput and minimum response times by balancing the calls amongst SIP proxy servers. Several load balancing algorithms are used like round robin, Call-Join-Shortest-Queue (CJSQ), Transaction-Join-Shortest-Queue (TJSQ) and Transaction-Least-Work-Left (TLWL). In this paper, we present a comparative analysis of load balancing algorithms for SIP servers with respect to call response time and server throughput performance.

Keywords Session initiation protocol (SIP) · Dispatcher · Load balancing · High availability · Fault tolerance · Algorithm · Performance · Server

A. Akbar (✉)
Jawaharlal Nehru Technological University, Hyderabad, India
e-mail: akbar.jntuphd@gmail.com

S. Mahaboob Basha
Al Habeeb College of Engineering & Technology, Chevella, India
e-mail: smbasha.phd@gmail.com

S.A. Sattar
Royal Institute of Technology & Science, Chevella, India
e-mail: syedabdulsattar1965@gmail.com

© Springer India 2016
S.C. Satapathy et al. (eds.), *Information Systems Design and Intelligent Applications*, Advances in Intelligent Systems and Computing 435,
DOI 10.1007/978-81-322-2757-1_9

1 Introduction

SIP or Session Initiation Protocol is a standard protocol used to for the control and maintenance of multimedia sessions in VOIP networks and is defined in RFC 2543 [1]. 3GPP (3rd Generation Partnership Project) has also accepted and adopted SIP for the management of media sessions in IMS (Internet Protocol Multimedia Systems). RFC 2543 explains that SIP is relevant in the application layer of OSI interconnection model for initiation, establishment, adjustment and termination of multimedia sessions. H.323 has been made defunct by SIP and has been widely replaced in VOIP software product lines. The ever-growing demand of enabling millions of VOIP clients and their calls have made leading ISPs use a cluster of servers. Dispatchers are required for logical distribution of incoming calls to SIP proxy servers. These important call dispatchers employ the use of various load balancing algorithms for intelligent decisions.

SIP uses addresses to locate callers and intended recipients/callee. An elementary SIP call flow is shown in Fig. 1. Location services are used to get the SIP address of the callee and a call is initiated by sending an INVITE request to the nearest proxy server. Users typically register addresses with SIP location servers.

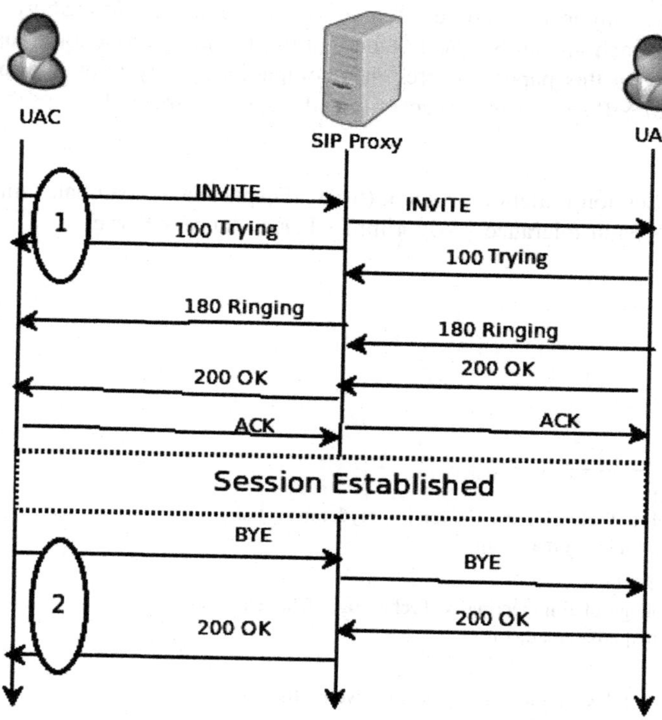

Fig. 1 An elementary SIP call flow

The proxy server forwards the INVITE request to the callee and responds with a 100 TRYING message. Once the callee phone starts ringing, it sends a 180 RINGING message to proxy server which retransmits it to the caller. If both the user agent server and client agree, the response is 200 OK. Then the proxy server is by passed for the media communication. SIP distinguishes clearly between call and transaction. SIP also makes it compulsory for all response messages to have the same value for parameters like Call-ID, CSEQ, TO, and FROM fields for their respective requests. This helps in the request response matching. SIP does not specify any particular transport protocol. For the transport layer, SIP can be used in combination with TCP (Transmission Control Protocol), Real-time Transport Protocol (RTP), UDP (User Datagram Protocol) or SCTP (Stream Control Transmission Protocol). A SIP invitation is considered successful when INVITE is responded with ACK. But ACK is not a part of the SIP transaction. Disconnection of a SIP call is done by either the caller or the callee sending the BYE request. 200 OK is required after the BYE request to conclude the SIP transaction flow. In case of proxy server failure, the redirect server forwards the INVITE requests.

This paper is about load balancing algorithms used in SIP network. Section 2 covers related work done in SIP server load balancing. In Sect. 3 we provide details of different load balancing algorithms. Section 4 gives implementation details and Sect. 5 gives experimental test bed infrastructure description to evaluate load balancing algorithms. Brief results evaluating INVITE response time, BYE response time, and throughput/good put are discussed in Sect. 6 and conclusion is drawn in Sect. 7.

2 Related Work

Solutions for web server load balancing and fail over mechanisms are evaluated by Ciardo et al. [2]. SIP differs from HTTP essentially with respect to session establishment and management. SIP call dispatcher is required to store the session information for correct proxy server call routing [3]. In web servers, high availability is achieved by IP address takeover and MAC address takeover schemes. Hong et al. divided the SIP overload algorithms in two main categories: Load reducing and/or Load balancing approach [4]. Load reducing is quite blunt and just rejects the incoming requests for new calls when an overload is detected. On the other hand, load balancing aims at equal distribution and redistribution of the SIP traffic among the servers. Again, load balancing methods are categorized as either static or dynamic. Cisco has used static load balancing based on DNS SRV in its architecture as well as Naming Authority Pointer records also. The decision taken by Load dispatcher for the server selection is on the basis of priority and weight pertaining to these records. Kundan Singh et al. employed web server redundancy techniques for SIP calls management. Their work has evaluated several load balancing methodologies relying on DNS, SIP header parameters, unused servers with exclusive IP addresses and NAT on a two tier architecture [5]. Their evaluative study also took into consideration fail over mechanisms using DNS, clients, database replication and IP address takeover. Hongbo et al. have proposed a set of three dynamic algorithms for SIP server farms [6]. These algorithms make request assignments based with session data awareness. Transaction Least Work Left (TLWL) betters the performance of the other two algorithms in terms throughput and response time [7].

3 Load Balancing Algorithms

In this particular section we explain the load balancing algorithms used for selecting SIP server by the Load balancer. UAC's send INVITE, BYE request to the load balancer, load balancing algorithm selects a best SIP server to forward the request. The servers transmit responses like 100 trying, 180 ringing to the load balancer, then the load balancer forwards to UAC as shown in Fig. 2.

Load balancing algorithms are classified depending on where they have implemented such as Receiver Initiated, Sender Initiated. Static load balancing algorithms takes decision on predefined policy whereas dynamic load balancing algorithm takes decision on present CPU load, Memory Utilization, Response time parameters of SIP servers.

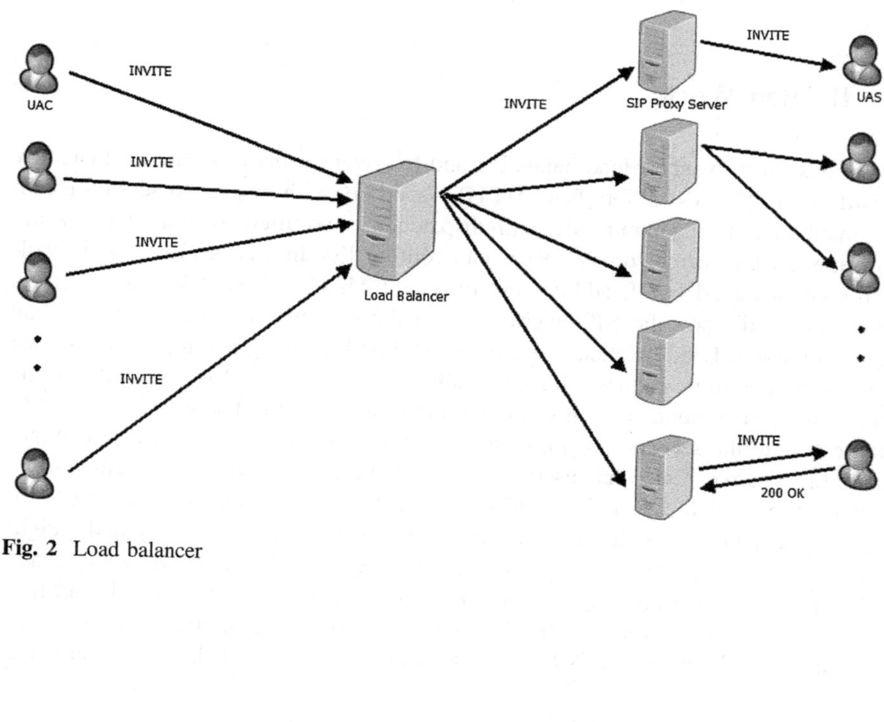

Fig. 2 Load balancer

3.1 Round Robin

Round Robin algorithm is the default algorithm implemented in most of practical Load balancers like Kamailio and Mobicents servers. It randomly select the SIP server for every fixed round-trip time. The algorithm does not consider the current load experienced SIP server and takes decision unconsciously so it has poor performance.

3.2 Call Join Shortest Queue (CJSQ)

Call Joint Shortest Queue algorithm keep tracks the number of calls dispatched to each SIP server by maintaining counters for each server. Load balancer routes the new requests to the server which have least number of calls. Counter is incremented when a new INVITE request is forwarded to the SIP server. The algorithm take decision on number of calls allocated to each SIP server. But it is not adequate because SIP protocol distinguishes between a call and transaction. Processing activity of the INVITE request requires more resources compared to the processing of the BYE request.

3.3 Transaction Join Shortest Queue (TJSQ)

Transaction Join Shortest Queue algorithm distinguish between calls and transactions. A SIP transaction is different from a call, a SIP call start by sending INVITE request and ends with BYE request. SIP transaction is the one in which INVITE request was sent and corresponding 200 OK was received. Load balancer which use TJSQ algorithm dispatches the new INVITE request to a server with minimum transactions. TJSQ algorithm keeps track of completion of call so that it has a finer grain control of loads to the server. Transaction Join Shortest Queue gets better performance than CJSQ because calls have variable durations. The number of calls routed to a server is not an accurate measure of load experienced by each SIP server.

3.4 Transaction Least Work Left (TLWL)

Transaction Least Work Left algorithm takes decision on least work (Load) to be done to route the new call to the SIP server. Here the load is calculated by giving relative weights to INVITE and BYE transactions. Since INVITE transaction consumes more resources compared to BYE transaction. INVITE needs database lookup whereas BYE transaction involves release of resources. Cost ratio of 1.75:1 was assigned for INVITE and BYE respectively.

4 Implementation

We had implemented load balancing algorithms by modifying dispatcher module of Kamailio [8]. Kamailio happens to be an excellent open-source SIP router present in many VOIP networks. Kamailio can be configured as a Proxy server, Registrar server, Back-to-Back User Agent client/server, Presence Server as well as Load Balancer. It supports both IP Version 4 and IP Version 6 addresses. Kamailio is a communication protocol which supports TCP, UDP, TLS and SCTP protocols. Kamailio can support thousands of calls per second even on low-budget hardware. I&I Internet AG one of leading ISP in Germany used Kamailio to provide service to millions of VOIP users. Kamailio has been written in C language. Apart from C it supports scripting languages such as LUA, Perl, java and python. Due to its modular design, Kamailio modules can be selectively loaded by changing the configuration file. Kamailio supports NAT traversal and database support. The core module of Kamailio provides basic functionality such as Transport Management, memory management and shared memory. Remaining features such as Presence, PSTN routing, accounting, load balancing, Security are implemented as separate modules.

In our implementation we has used dispatcher module for Load balancing of SIP requests. Dispatcher module of Kamailio provides stateless load balancer to dispatch SIP traffic. Many algorithms like round-robin, hash-over SIP attributes, Call load distribution and weight based load balancing already implemented in this module. We had added new load balancing algorithms such as Call Join Shortest Queue (CJSQ), Transaction Join Shortest Queue (TJSQ), and Transaction Least Work Left (TLWL) to the *ds_select_dst* function. Destination addresses of proxy servers are given as a list file.

5 Experimental Test Bed

For evaluation of the load balancing algorithms, we have designed an experimental test bed infrastructure. The system design was depicted in Fig. 3. We have used experimental test bed infrastructure consisting of three systems. The hardware configuration used in the test bed are listed in the Table 1. Standalone components of the test bed infrastructure have been described below.

5.1 Load Generator

Load generation for our experiments was done using SIPp, a popular industry used open source platform tool for load testing. The version of SIPp was 3.3. SIPp allows

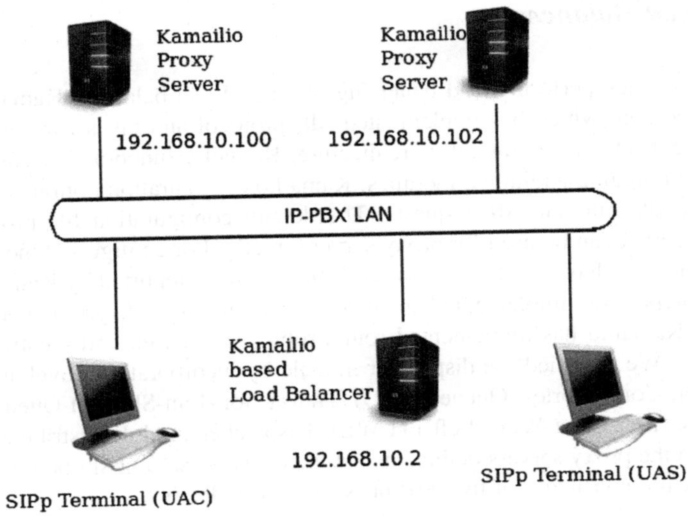

Fig. 3 SIP load balancing test bed

Table 1 Hardware test-bed characteristics

PC 1	
CPU	Intel Pentium D 3.0 GHz
RAM	1 GB
OS	Ubuntu 13.10
Role	UAC
PC 2	
CPU	Intel core i3
RAM	4 GB
OS	Ubuntu 13.10
Role	Upstream Proxy, UAS
PC 3	
CPU	Intel core i5
RAM	4 GB
OS	Ubuntu 13.10
Role	Downstream Proxy

for emulation of UAC or UAS depending on the XML configuration provided. Our UAC was emulated with our XML SIPp configuration and we used the SIPp features to record the statistics like response time. SIPp was also configured to increase the call rate by 5 cps for every 5 s till the maximum limit of 10,000 calls.

5.2 *Load Balancer*

The entity which performs load balancing is called Load balancer. Kamailio has a kernel type core which has implemented SIP protocol and takes care of memory management. Due to its modular architecture, Kamailio functionality can be configured by plugging additional modules. Kamailio configuration options provides a way to handle different SIP requests. The default configuration file provided by Kamailio server can be used for proxy server directly. For configuring the server as load balancer we have used a custom configuration file supported by Kamailio. The load balancing was implemented in dispatch module. The dispatch module provided by Kamailio has implemented round-robin, load based and identifier based algorithms. We modified the dispatcher module by incorporating novel algorithms such as Call-Join-Shortest-Queue (CJSQ), Transaction-Join-Shortest-Queue (TJSQ) and Transaction-Least-Work-Left (TLWL). Dispatcher module distribute the call requests to the proxy servers defined as list file or database. For simplicity we have used a list file consisting of five SIP proxy server addresses.

5.3 *SIP Proxy Server*

PJSIP is one of open source SIP stack written in C++. Its design has a small footprint and can be portable to any platform. To implement SIP Proxy server we developed an application based on PJSIP library which fulfils basic server functionality. We run SIP Proxy on five different ports on the same machine to simulate five SIP proxy servers.

5.4 *User Application Server*

For call termination, UAS was configured by instantiating the PJSIP stack platform. The UAS was used to send 200 OK for every request so that the SIP transaction is closed. This resulted in successful transactions for INVITE and BYE messages.

6 Results

Keeping in mind the end goal to quantify INVITE response time and BYE response time we have utilized a tweaked XML situation record in SIPp. At whatever point INVITE solicitation comes timer 1 began and it is ceased while the related 200 OK

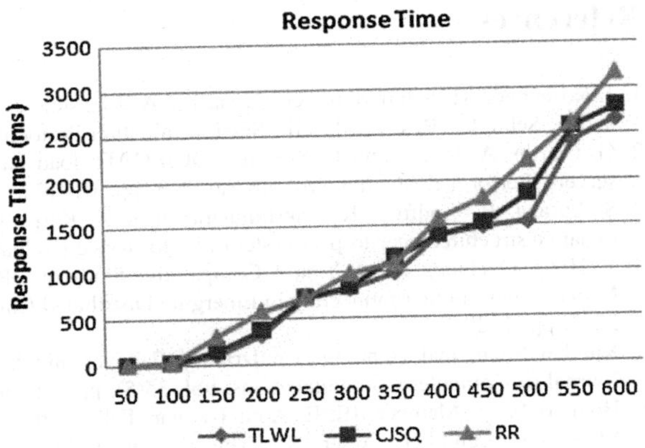

Fig. 4 Response time

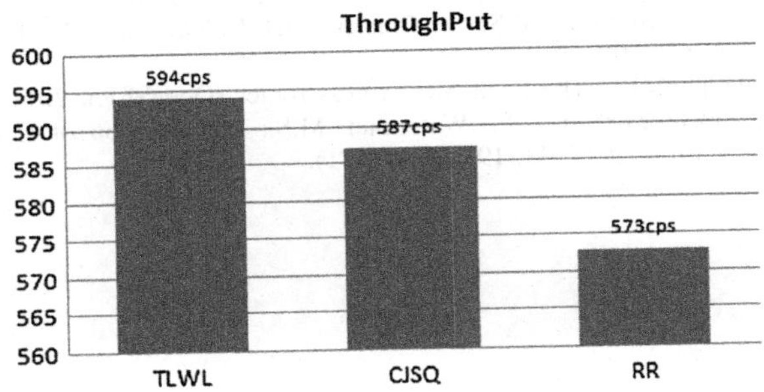

Fig. 5 Throughput

was received correspondingly timer 2 measured the BYE reaction time. We have collected the statistics employing—trace stat and—trace rtt options of SIPp. We had increased the SIP calls by bit by bit increasing the call rate until 600 cps was reached. In Fig. 4 response time for Round Robin, CJSQ, TLWL algorithm depicted. Transaction Least Work Left algorithm has minimum response time. In Fig. 5 throughput for Round Robin, CJSQ, TLWL algorithm depicted.

7 Conclusion

We have compared novel load balancing methods for SIP proxy servers in terms of throughput and response time. The algorithms are implemented using Kamailio server which is a famous open source SIP Proxy server. The results have proven that Session Aware Request Assignment based algorithms dramatically increase the good put/throughput and decrease the response time.

References

1. J. Rosenberg, H. Schulzrinne, G. Camarillo, A. Johnston, J. Peterson, R. Sparks, M. Handley, and E. Schooler. RFC 3261: SIP: Session Initiation Protocol. Technical report, IETF, 2002.
2. G. Ciardo, A. Riska, and E. Smirni, EQUILOAD: load balancing policy for clustered Web servers, Perform Evaluating, vol. 46, no. 2–3, pp.101–124, 2001.
3. S. Surana, B. Godfrey, K. Lakshminarayanan, R. Karp, and I. Stoica. Load balancing in dynamic structured peer-to-peer systems. Performance Evaluation, 63(3):217–240, March 2006.
4. Y. Hong, C. Huang, and J. Yan, A Comparative Study of SIP Overload Control Algorithms, in Network and Traffic Engineering in Emerging Distributed Computing Applications, IGI Global, 2012, pp. 1–20.
5. Kundan Singh and Henning Schulzrinne" Failover and Load Sharing in SIP Telephony", Journal of Computer Communications, vol. 35(5), pp. 927–942, 2007.
6. Hongbo Jiang, Member, IEEE, Arun Iyengar, Fellow, IEEE, Erich Nahum, Member, IEEE, Wolfgang Segmuller, Asser N. Tantawi, Senior Member, IEEE, Member, ACM, and Charles P. Wright, Design, Implementation, and Performance of a Load Balancer for SIP Server Clusters, 2012.
7. Hongbo Jiang; Iyengar, A.; Nahum, E.; Segmuller, W.; Tantawi, A.; Wright, C.P., "Load Balancing for SIP Server Clusters," INFOCOM 2009, IEEE, vol., no., pp. 2286, 2294, 19–25 April 2009.
8. OpenSIPS, The open SIP express router (OpenSER), 2011 [Online]. Available: http://www.openser.org.th, T.F., Waterman, M.S.: Identification of Common Molecular Subsequences. J. Mol. Biol. 147, 195–197 (1981).

Design of Wireless Sensor Network Based Embedded Systems for Aquaculture Environment Monitoring

Sai Krishna Vaddadi and Shashikant Sadistap

Abstract This paper proposes an process/environmental parameters monitoring framework based on Wireless Sensor Networks (WSNs) using embedded systems. The developed system will have many applications particularly in the area of aquaculture, tea plantations, vineyards, precision agriculture, green houses monitoring etc. The complexity of the system increases for applications involving aquaculture. The developed systems are tailored for sensing in wide farmland's without (or) very little human supervision. The fully designed and developed system consists of sensors for monitoring the parameters, sensor node with wireless network topology, gateway for collecting the information from the sensor node and transmitting the information to the central control unit. Data base, real-time monitoring and visualization facilities, controlling of sensor nodes and sensors are provided at the central control unit. Sensor nodes, sensor assembly and gateways should with stand demanding environmental conditions like rain, dust, heat, moisture, relative humidity etc. With these systems deployed in large farm's bodies the scientist's will have more and more field data with very little human intervention. The acquired data can be used to provide quantitative indications about the farm's status, thus easy for analysis and decision making. In this paper we are focusing our discussion about WSNs in the field of aquaculture.

Keywords Wireless sensor networks (WSNs) · Aquaculture · Distributed monitoring systems · Communication protocol

---

S.K. Vaddadi (✉)
Academy of Scientific and Innovative Research, Pilani 333031, India
e-mail: saishrustee@gmail.com

S.K. Vaddadi · S. Sadistap
AEG, CSIR-CEERI, Pilani 333031, India

© Springer India 2016
S.C. Satapathy et al. (eds.), *Information Systems Design and Intelligent Applications*, Advances in Intelligent Systems and Computing 435,
DOI 10.1007/978-81-322-2757-1_10

1 Introduction and Motivation

If we consider the case of Indian fisheries and aquaculture industry, it is an important sector providing occupation, revenue generation and nutritional values to the public. Presently India is second largest in the area of aquaculture and have third position in global scenario in terms of production of fishes (4.4 %) of global fish production [1]. The future development of aquaculture in India depends on adoption of new and innovative production technologies, management and utilization of less water resources and proper market tie-ups etc. [1]. Finding new areas and producing high quality and reliable products will improve the value of Indian fisheries in global scenario. In the developed countries aquaculture is far more advanced with use of sophisticated and state of the art technologies, while in the developing countries like India we are still practicing traditional forms of aquaculture [2]. The developed countries are achieving their production with less human resources, where as in India most of the process are still human dependent.

Fish farms usually spread in large areas are not easily accessible, and are not exactly suitable for electronic products due power restrictions, communications and harsh environmental conditions. The other issues involving in an aquatic environment are possible water infiltrations, algae deposits on the sensor, communication inefficiency due to RF reflections and a contained elevation of the antenna with respect to water surface. The water quality information, such as rate of dissolved oxygen generation, pH, conductivity, temperature and other dissolved chemicals can be used to predict the health of the fish, production of the fish and make decision about application of food (or) pesticides. Therefore, how to gather such comprehensive information accurately and reliably is the core of the aquaculture monitoring.

Distributed environmental monitoring with wireless sensor networks (WSNs) is one of the most challenging research activities faced by the embedded system community in the last decade [3–5]. WSNs, formed by several tiny sensor nodes, are useful and important for monitoring physical phenomenon with large resolutions in space and time. The sensor node is capable of sensing the parameters, process and wirelessly communicate the information. These sensor nodes are useful in the applications like precision agriculture, aquaculture, farming, green house, industrial control.

The aim of this paper is to present an aquatic monitoring framework based on WSNs that is scalable, adaptive with respect to topological changes in the network. The proposed framework addresses all aspects related to environmental monitoring: sensing, local and remote transmission, data storage, and visualization. We are discussing about WSNs used to measure the water quality of fish farms and the water parameters are dissolved oxygen (DO), pH, temperature and conductivity.

2 Relevant WSNs Applications

Deployments of WSNs rarely address adaptability to environmental changes and energy-aware issues; such a limitation affects lifetime and Quality of Service of the monitoring system [6]. The work of [7] presents a star-based topology for seabirds habitat monitoring (the gateway collects data from the sensor nodes and forwards them to a remote control station for further processing). The work in [8] proposes a system for monitoring volcanic eruptions; battery based solutions are used for the gateway and the three sensor units.

Several general purpose sensor node platforms like Mica [9] and Telos [10] have been developed in the last decade which were widely tested for several applications in different parts of the world. These sensor nodes, also called as motes, were designed to support an efficient, low power and light weight operating system called TinyOS [11, 12]. The core operating system in TinyOS is as small as 400 Bytes while many applications are expected to consume only 16 kB of memory. The nesC language [13] has been used to implement the TinyOS and several protocols and applications run on the TinyOS. This approach requires large amount of development time and it is an open source application, there is a possibility of support issues in future. In Agrisens project [14], the deployment of wireless sensor networks has been carried out for precision agriculture in Sula vineyard. Similarly, monitoring of high-quality wine production in Sicilian winery using Wireless Sensor Networks has been presented in [15]. In Sensor Scope [16], the application specific sensor box design approach was developed which resulted in a cost effective solution for environmental monitoring for better reliability in the harsh environmental conditions of Switzerland.

The available general purpose sensor network platforms [9, 10] requires several modifications for being suitable for the agriculture applications. While, the other platforms that are presently designed for different applications are very expensive to be feasible for facilitating marginal farmers in the developing nations.

3 System Hardware Aspects

In this section, we discuss about the hardware of the proposed WSNs for aquaculture environment monitoring. The hardware is divided into sensor node, sensors, mechanical aspect and transceiver.

3.1 Sensor Node

The sensor node of the WSN consists of four modules: control unit, signal acquisition and conditioning, wireless communication and data storage. To full-fill the requirements of the sensor node an 8-bit RISC architecture based PIC18F4520 [17]

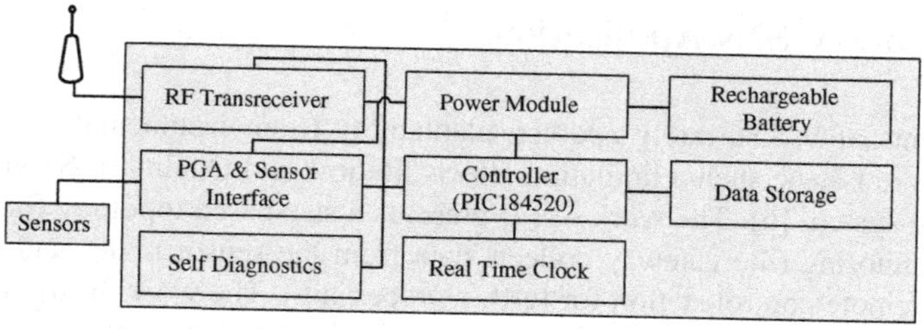

Fig. 1 Architecture of the sensor node embedded system

microcontroller is used. The microcontroller monitors and controls analog-to-digital conversion, real time clock, memory storage, data validation and wireless transmission. The same hardware is used for both the sensor node and the gateway module. The only difference between the sensor node and gate way is their mounting. Sensor nodes are typically floating bodies, so their wireless antenna will be below the ground level height. In order to communicate for longer distances, we need to keep the antenna above the ground level for line of sight between various Zigbee modules. So for gateway module, the system will be mounted above the ground level. Basic building blocks of the hardware used for the sensor node is as shown in Fig. 1.

External memory and real time clock (RTC) are connected to microcontroller through I^2C peripheral interface. Serial-peripheral-interface (SPI) based programmable gain amplifier (PGA) is used for increasing the number of analog to digital channels. Using this PGA, 8 analog channels are interfaced and only one A/D of the microcontroller is used. Since, the microcontroller only supports either I^2C or SPI communication, we need to build our own SPI software library. RTC is used for maintaining the date and time, external memory of (1 M bits) is used for storing the parameters information.

3.2 Sensors and Mechanical Aspect

Sensor Interfaced with the sensor node were dissolved oxygen, pH, temperature, conductivity and relative humidity of the atmosphere. All sensors used in this WSN gives 4–20 mA output, the reason for selecting the current output is for minimizing the voltage drop losses due to long cables (or) wires. Protection mechanism and interface circuits have been designed to convert the current signal from the sensor to voltage form and protect the system from high voltages. The range of various sensors are: dissolved oxygen [18] measuring range 0–200 %, conductivity [19] measuring range of 0–2000 mS, pH [20] measuring range of 0–14, temperature measuring range 0–100 °C and relative humidity measuring range 0–100 %.

The mechanical aspect of the system is very important for systems which will be used in aquatic environment. IP 67 waterproof PVC box is used for this purpose, where IP means Ingress Protection. 67 indicates the degree of protection against dust and water.

4 System Firmware and Networking Protocol Aspects

4.1 Sensing and Storage

A timer interrupt periodically wakes up the sensor node, which acquires the data of all sensors, validates and stores them. After completion of the activity the sensor node go to sleep mode. The sensing, storage and transceiver scheme is shown Fig. 2. The sensor node will wake from the sleep mode in one of the two conditions, timer interrupt or receiving any data through RF. The validated data is accumulated and stored in memory in a predefined format with a date and time stamping.

The memory is divided into two sections one is short memory section and the other is long term memory. In short term section the data of each individual channel is accumulated for a user defined duration along with number of samples accumulated. Data packet is build and is stored in the short term memory location. Similar mechanism is used for long term, where the time duration (sample for every 5 min) is longer than short term and stored in long term memory location. In short term memory only recent (last 4 h) information is available and in long term last 27 days of information is available. The advantage having an internal memory on the sensor node, it is not always necessary to send the data once acquired, the gateway can schedule the communication with the sensor node.

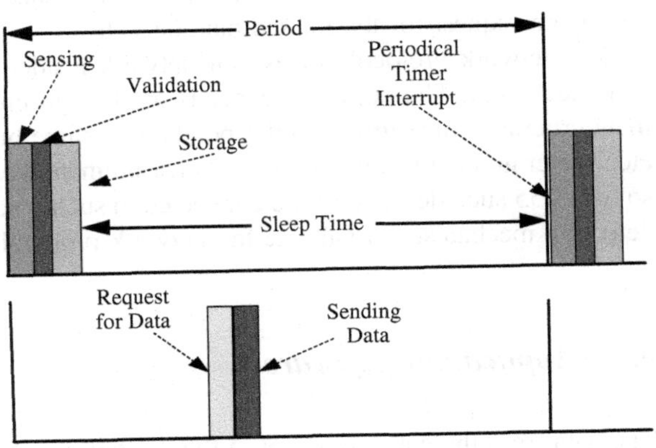

Fig. 2 Timing diagram of sensing, storage and transceiver

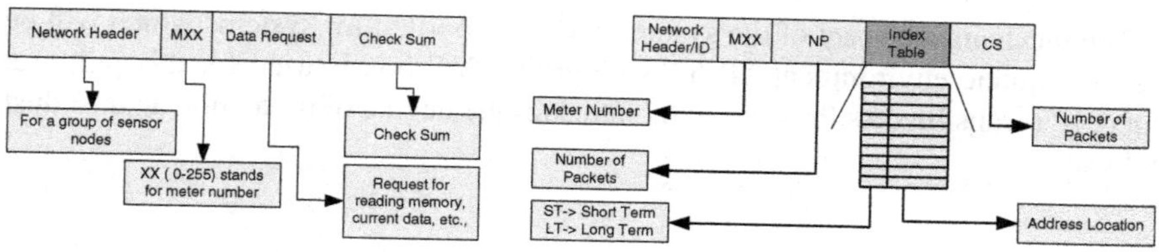

Fig. 3 Command to sensor node and response from sensor node

4.2 *Network Protocol*

Each sensor node in the network possesses a wireless device for communication with the remote station. A predefined network protocol is necessary to maintain a harmonious communication between the measurement system and remote station. Measurement system acts as a slave in the network where as the remote station acts a master. The request for the data from the gateway to sensor node is shown in Fig. 3.

The response of the sensor node to a query from gate way is shown in the Fig. 3. It consists of number of packets will be transmitted and the index table containing data from which memory location they will be transmitted along with their addresses. Each sensor node in the network will have a network header other than the sensor node number. Using this we can setup different clusters of sensor nodes for different applications. After establishing the communication between sensor node and gateway, whenever there is a request for data packet, the sensor node will send an index table before the start of the data transmission. The index table indicates how many packets of information is about to be transferred. Each bit in the index table indicates one data packet. After successful reception of data packet at the gateway, the gateway will update its index table. After receiving all the data, the gateway will check against the index table. If the index table matches with the data packets received, then the communication is successful, otherwise the gateway will again request for the missed data packets.

The network protocol consists of network header, meter number, data request and check sum. The network header is used to group devices of similar measurement criteria or measuring similar ponds with same breeding capacity or same fish etc., Meter number is unique for each measurement device and it is an 8-bit number, so only 255 such devices can be connected in such a network. Checksum is an error detection mechanism to validate the network protocol.

4.3 *Application Software*

The data from the gateway is not in the desired user format (hexa-decimal format), an application software is developed for maintaining the data base, real-time monitoring and visualization facilities, controlling of sensor nodes and sensors. The application software converts the data from hexa-decimal format to decimal/floating

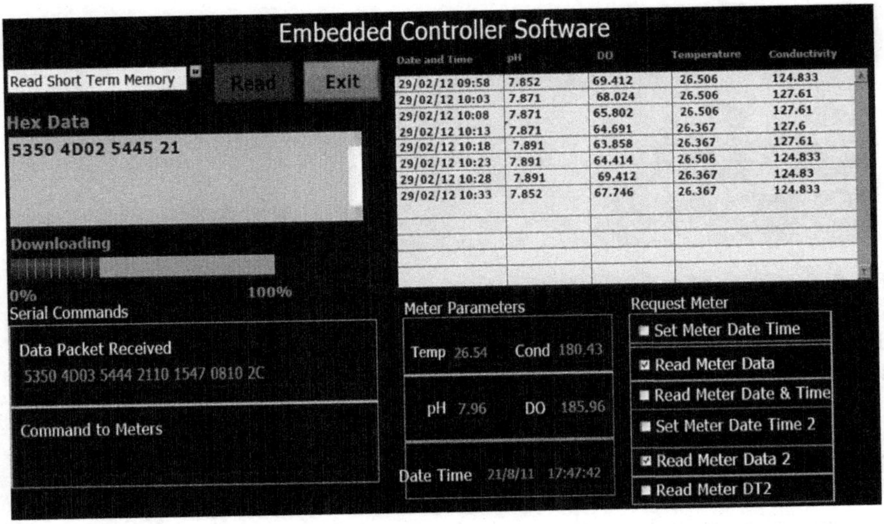

Fig. 4 GUI of the application software

format. The application software is developed using Lab VIEW. The GUI of the application software is shown in Fig. 4.

The data from the sensor nodes are stored in a text file in a predefined format. Presently the information that is available is limited, if the number of sensor nodes are increased, then we need to have a data base like oracle etc. The application software has features like read/write/erase memory, set/read date time, read current data of the sensor node. Only one device can be communicated with the central unit at a time. Individual actions like aerator operation, feeding etc., can also be controlled.

5 Deployment and Future Work

Three sensor nodes and application software for control unit are developed and deployed for aquaculture environment monitoring. Other than environment monitoring, few sensor nodes do other activities like feeding, movement etc.

The developed systems are installed at ICAR-CIFA Bhubaneswar ponds. The developed systems are shown in Fig. 5. During the initial communication testing, we didn't implemented the index table. Due to this there is loss of data packets and we unable to trace what are the data packets that are lost. It is becoming difficult to find out which data packets are to be retransmitted. The initial mechanical systems (boxes) are developed using aluminum and after a duration of 6 months, it is observed that most of components (screws, nuts, bolts etc.,) are corroded. Due to higher moisture and relative humidity levels the PCB's are started corroding and salts are forming at junctions and terminals. New PCB with acrylic coating is

Fig. 5 Simply sensor node, auto feeding system and smart boat

developed and installed. Index table based communication between control unit and sensor node and IP67 mechanical boxes are used.

It is observed that algae is forming around the sensors, due to this, the measurement errors are increasing. So it proposed, time to time maintenance is required for calibration and cleaning purpose of the sensors.

Presently the sensor nodes are battery operated, there is a large scope for implementing solar-energy-harvesting methods. Also proper mechanism for sensors is necessary to protect the sensors from algae and other materials present in water. Soft computing algorithms can also be implemented to find the stress factor in fish based on the measured parameters, so that preventive action can be taken. Few parameters in the aquaculture shows correlation, based on this information, algorithms can be developed to predict whether all the sensors are working correctly or not.

6 Conclusion

Wireless sensor network based aquatic environment monitoring is developed and deployed at ICAR-CIFA Bhubaneswar. Earlier, to monitor pond parameters like dissolved oxygen, pH a scientific personnel need to be sent from the lab for data gathering, it is a risky method due to unfavorable pond conditions. Also this process is time consuming and costly. With these senor nodes, the user can gather the information about pond with minimum human intervention.

The WSNs are successfully tested and deployed in aquatic environment, so these nodes can be implemented in other less demanding situations like precision agriculture, food storage, tea plantations and manufacturing vineyards and wine manufacturing etc.

Acknowledgments The author thanks DBT, New Delhi for approving and funding this "Smart Pond Management System" project. Also author thanks CSIR-CEERI Director Dr. Chandra Shekar for allowing us to work in this area and publish this research. Thanks are due to Dr. S.S. Sadistap, Head, AEG for providing information, motivation and logistics. Thanks are also due to AEG team members who helped us during testing, assembly and installation. Also special thanks to Dr P C Das, Principal Scientist, ICAR-CIFA, Bhubaneswar for providing support during testing and installation.

References

1. The State of Fisheries and Aquaculture 2010 by The State of World Fisheries and Aquaculture (SOFIA).
2. Prospects for coastal aquaculture in India by k. alagarswami, Central Marine Fisheries Research Institute, Cochin, in CMFRI Bulletin 20.
3. D. Estrin, L. Girod, L. Pottie, and M. Srivastavam, "Instrumenting the world with wireless sensor networks, "in Proc. IEEE Int. Conf. Acoust., Speech, Signal Process., 2001, vol. 4, pp. 2033–2036.
4. I.F. Aakyildiz, W. Su, Y. Sankarasubramaniam, and E. Cayirci, "Wireless sensor networks: A survey," Comput. Networks, vol.38, no.4, pp.393–422, Mar. 15, 2002.
5. G. J. Pottie and W. J. Kaiser, "Wireless integrated network sensors," Commun. ACM, vol. 43, no. 5, pp. 51–58, May 2000.
6. C. Alippi, R. Camplani, C. Galperti, and M. Roveri, "A robust, adaptive, solar-powered WSN framework for aquatic environmental monitoring," IEEE Sensors J., vol. 11, no. 1, pp. 45–55, Jan. 2011.
7. E. Biagioni and K. Bridges, "The application of remote sensor technology to assist the recovery of rare and endangered species," Int. J. High Perform. Computing Applic., vol. 16, no. 3, pp. 315–324, 2002.
8. P. Zhang, C. M. Sadler, S. A. Lyon, and M. Martonosi, "Hardware design experiences in ZebraNet," in Proc. 2nd Int. Conf. Embedded Networked Sensor Syst., New York, 2004, pp. 227–238.
9. J. L. Hill and D. E. Culler, "Mica: A wireless platform for deeply embedded networks," IEEE Micro, vol. 22, no. 6, pp. 12–24, Nov. 2002. [Online]. Available: http://dx.doi.org/10.1109/MM.2002.1134340.
10. J. Polastre, R. Szewczyk, and D. E. Culler, "Telos: enabling ultra-low power wireless research." in IPSN, 2005, pp. 364–369.
11. D. Gay, P. Levis, and D. E. Culler, "Software design patterns for tinyos," in LCTES, 2005, pp. 40–49.
12. P. Levis, S. Madden, J. Polastre, R. Szewczyk, A. Woo, D. Gay, J. Hill, M. Welsh, E. Brewer, and D. Culler, "Tinyos: An operating system for sensor networks," in Ambient Intelligence. Springer Verlag, 2004.
13. D. Gay, M. Welsh, P. Levis, E. Brewer, R.V. Behren, and D. Culler, "The nesc language: A holistic approach to networked embedded systems," In Proceedings of Programming Language Design and Implementation (PLDI), 2003, pp. 1–11.
14. S. Neelamegam, C. P. R. G. Naveen, M. Padmawar, U. B. Desai, S. N. Merchant, and V. Kulkarni, "Agrisens: Wireless sensor network for agriculture - a sula vineyard case study," in Proc. 1st International Workshop on Wireless Sensor Network Deployments, Santa Fe, New Mexico, USA, June 2007.
15. G. Anastasi, O. Farruggia, G. L. Re, and M. Ortolani, "Monitoring high quality wine production using wireless sensor networks," in HICSS'09, 2009, pp. 1–7.
16. F. Ingelrest, G. Barrenetxea, G. Schaefer, M. Vetterli, O. Couach, and M. Parlange, "Sensorscope: Application-specific sensor network for environmental monitoring," ACM Trans. Sen. Netw., vol. 6, no. 2, pp. 17:1–17:32, Mar. 2010.
17. PIC 18F4520 Data Sheet Microcontrollers with 10-Bit A/D and nanoWatt Technology.
18. 35151-10 - Alpha DO 500 2-Wire Dissolved Oxygen Transmitter by EUTECH Instruments.
19. ECCONCTP0500 - Alpha COND 500 2-Wire Conductivity Transmitter by EUTECH Instruments.
20. 56717-20 - pH 500 pH/ORP 2-wire LCD Transmitter with display by EUTECH Instruments.

Touch-to-Learn: A Tangible Learning System for Hard-of Hearing Children

Mitali Sinha, Suman Deb and Sonia Nandi

Abstract Children's learning styles can be categorized mainly into visual, auditory, tactile and kinesthetic learning. The lack of auditory learning capability deprive the hard-of-hearing children from indulging into traditional learning environment. To facilitate the learning of the stone-deaf children we propose a "Touch-to-learn" system, a manifestation of a tangible learning system for the elementary learners. A prototype of the system is presented focusing on children's dental health and proper eating style which can be extended in other learning areas. The system effectively make use of the technological development embedding technology into learning. It also bridges the gap between the physical and digital interaction by introducing the technologies like RFID and Wii Remote providing a tangible learning environment and thus accelerating the learning process of hard-of-hearing children.

Keywords Tangible learning · RFID · Wii remote

1 Introduction

Children learn intuitively using the basic sense of sight, sound, touch and movement. They construct a mental model of the world by exploring their surroundings [1]. However a section of hard-of-hearing children experience various difficulties in the traditional learning approaches. Hard of hearing children often faces problem to follow the instructors. They may miss some words or sentences while the instructor

M. Sinha (✉) · S. Deb · S. Nandi
Computer Science and Engineering Department, NIT, Agartala, India
e-mail: mitalisinha93@gmail.com

S. Deb
e-mail: sumandebcs@gmail.com

S. Nandi
e-mail: sonianandi90@gmail.com

© Springer India 2016
S.C. Satapathy et al. (eds.), *Information Systems Design and Intelligent Applications*, Advances in Intelligent Systems and Computing 435,
DOI 10.1007/978-81-322-2757-1_11

moves to the next concept. Embedding modern technology into learning system can provide a solution to fill up this gap in learning.

Studies [2] are carried out to cope with the difficulties faced by deaf and hard-of-hearing children. Various suggestions like explanation with pictures and animation, using concrete physical objects, visual prompts were proposed that were found to be effective [3–5]. Many researches [6] have been carried out in proving the fact that concrete and hands-on experiences produces better results rather than abstract representation in developing thinking and learning capabilities among children. The cognitive load of thinking among children is reduced by using Tangible User Interfaces. It provides a connection between the physical (tactile) and digital (visual) learning environment.

The advanced and low cost technologies like RFID and Wii Remote can be explored effectively to provide a solution to the difficulty in learning process. These technologies provide a tangible method of learning for children in a play full environment. Our proposed "Touch-To-Learn" system uses the RFID and Wii Remote for providing an environment where the hard of hearing children can intuitively learn using their sense of touch and sight.

2 Technology

2.1 RFID

RFID (Radio Frequency Identification) Technology has been extensively used in various educational interactive applications [7, 8]. RFID is the wireless use of electromagnetic field to transfer data for the purposes of automatically identifying and tracking tags attached to objects [9]. It mainly consists of two parts: Data-collecting readers and data-providing transponders or tags. A tag consists of an antenna for receiving or transmitting information and an integrated circuit for storing information. A tag reader is generally coupled to computational devices and whenever any tagged object is brought to its proximity the communication between the tag and the tag reader takes place. The readers transmits signals on the system's operating frequency in order to identify a tag within its range. The tag respond to the signal and transmits the information stored in it (Figs. 1 and 2).

2.2 Wii Remote

The Wii Remote, informally known as Wiimote is basically a remote controller for a gaming console named Wii, developed by a Japanese company Nintendo Co.ltd. The motion sensing capability of Wii Remote distinguishes it from other traditional Remotes. The key feature of Wiimote is its ability to digitally manipulate items providing gesture recognition and pointing of objects on screen. The three-axes

Fig. 1 RFID tag [10]

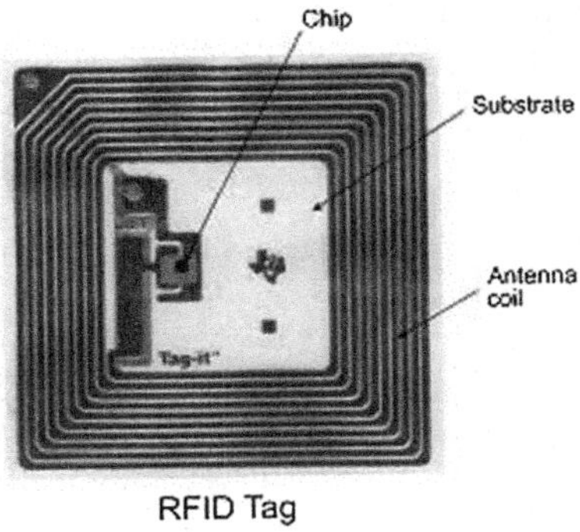

Fig. 2 RFID communication system [11]

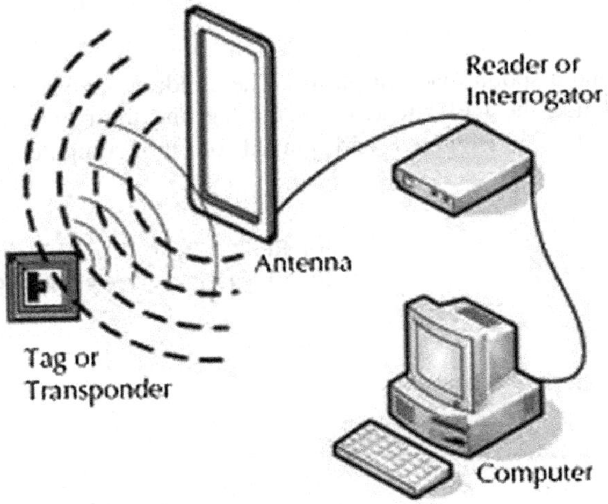

accelerometer of Wiimote helps in gesture recognition and its IR sensing capability determine where the Wiimote is pointing on the screen. The communication process of Wii Remote to its original gaming console Wii is accomplished by Bluetooth Wireless Technology which is based on Bluetooth Human Interface Device (HID) protocol. Due to its simple connection technology the Wii Remote can be easily connected to a Bluetooth enabled computer. Therefore the information from the Wiimote camera and the accelerometer can be easily directed to the computer for further interpretation (Fig. 3).

3 Proposed Design

Touch-to-learn system was initially proposed for deaf or hard-of-hearing children to provide them a proper health education with an aim of its further application. A set of RFID tagged artifacts were presented to the children that included a set of a brush

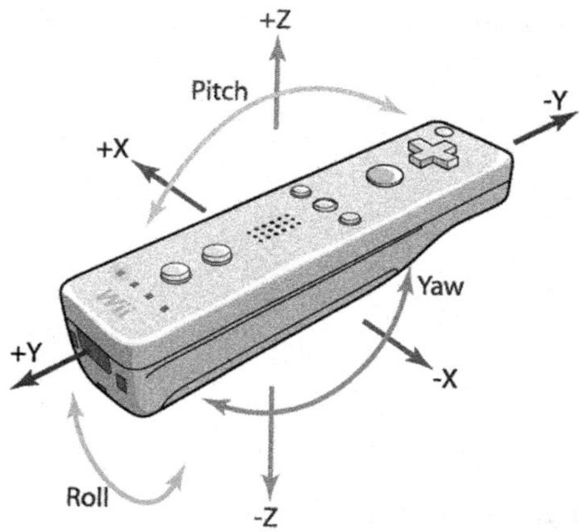

Fig. 3 6 DOF of Wii remote [12]

and a spoon made of plastic. The children were asked to lift a tangible artifact and touch the RFID reader. As the tag and the reader starts to communicate a short video is triggered and played on the computer screen. The video provides

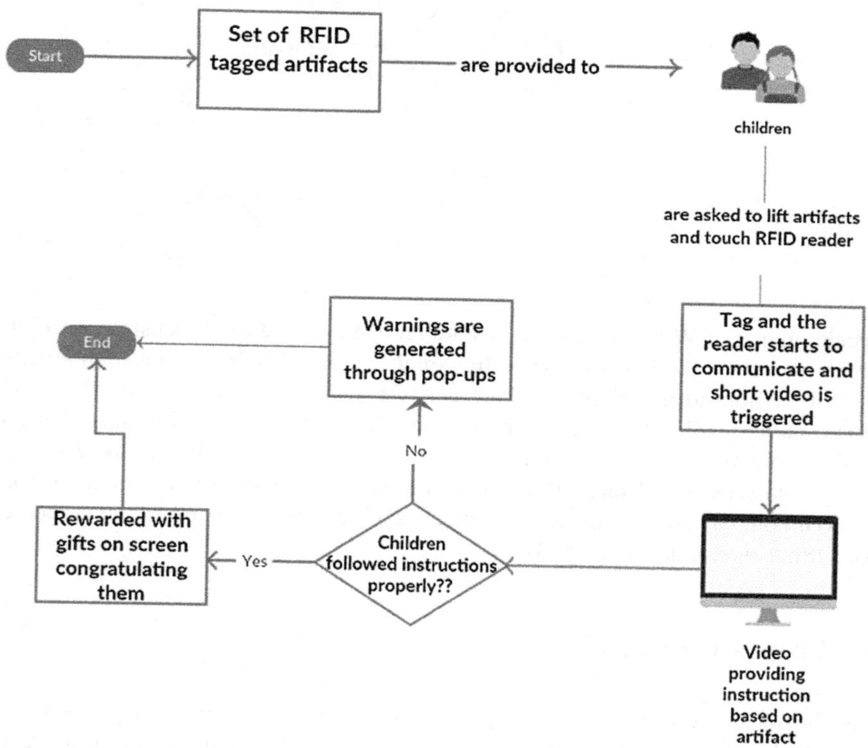

Fig. 4 Schematic of methodology

Fig. 5 A child holding the Wiimote and manipulating the brush on screen

instruction to proper brushing technique or good eating habit depending upon the artifact chosen. It presented simplified text and pictorial or animated content for easy understanding of the hard-of hearing children. Followed to the short instructional video an interface open up where the artifacts are simulated on screen using a Wii Remote. If the children properly followed the instruction, they were rewarded with gifts on screen congratulating them on their success. On the other hand, if they failed to do the same, warning were generated through pop-ups (Figs. 4 and 5).

4 Conclusion

Our proposed system presented tangible application of two modern technologies, RFID and Wii Remote into learning especially for hard-of-hearing children. The usance of tangible devices proved to be effective in learning where children learned intuitively in a play full environment. The system was especially designed for the hard-of-hearing children taking into consideration their limitations. With the use of proper artifacts and simulating them on-screen the system was able to enhance the understandability among children.

4.1 Declaration

Authors are responsible for the inclusion of human images in the paper. Proper permission has been taken from the child's guardian regarding publishing the image as a part of the experiment.

References

1. Hengeveld, B., Voort, R., Balkom, H., Hummels, C, & Moor, J. (2007). Designing for diversity: Developing complex adaptive tangible products. Paper presented at the 1st International Conference on Tangible and Embedded Interaction (TEl), Baton Rouge, Louisiana.

2. Y. Wang, "Inquiry-based science instruction and performance literacy for students who are deaf or hard of hearing," American Annals of the Deaf, 156(3), pp. 239–254, 2011. W. (eds.) Euro-Par 2006. LNCS, vol. 4128, pp. 1148–1158. Springer, Heidelberg (2006).
3. T.J. Diebold and M.B. Waldron, "Designing instructional formats: The effects of verbal and pictorial components on hearing impaired students' comprehension of science concepts," American Annals of the *Deaf*, *133*(1), pp. 30–35, 1988.
4. H. Lang, and D. Steely, "Web-based science instruction for deaf students: What research says to the teacher," Instructional Science, *31*, pp. 277–298, 2003.
5. Jeng Hong Ho, S. ZhiYing Zhou, Dong Wei, and Alfred Low, " Investigating the effects of educational Game with Wii Remote on outcomes of learning", Transactions on Edutainment III, LNCS 5904, pp. 240–252, 2009.
6. Marshall, P. (2007). Do tangible interfaces enhance learning? Paper presented at the 1" International Conference on Tangible and Embedded Interaction (TEI), Baton Rouge, Louisiana.
7. S. His and H. Fait, "RFID enhances visitors' museum experience at the Exploratorium," Communications of the ACM, 48(9), pp. 60–65, 2005.
8. I. Whitehouse and M. Ragus, "E-learning using radio frequency identification device scoping study," Australian Government – Department of Education, Science, and Training.
9. https://en.wikipedia.org/wiki/Radio-frequency_identification.
10. http://www.google.co.in/imgres?imgurl=http://www.osculator.net/doc/_media/faq:prywiimote.gif.
11. http://www.google.co.in/imgres?imgurl=http://endtimestruth.com/wpcontent/uploads/2014/01/RFID-chip-and-antenna-3.png&imgrefurl=http://endtimestruth.com/mark-of-thebeast/rfid.
12. http://www.google.co.in/imgres?imgurl=http://www.osculator.net/doc/_media/faq:prywiimote.gif.

Creating Low Cost Multi-gesture Device Control by Using Depth Sensing

Sonia Nandi, Suman Deb and Mitali Sinha

Abstract In real life, most of the works are done in a gradient manner of interaction. That means binary result do not come of every actions. Holding, touching and reacting requires the knowledge of weight, surface, friction etc. and depending on that user applies force and hold on the object. In this paper we tried to introduce similar kind of sensation for interacting with an object. The goal is to build a natural way to interact with different devices. Natural interaction involves gradient control of limbs which can be named as 'Gradient gestures interact of things'. It has been long practice for gesture control devices like TV, electric appliances control, air condition control, etc. but by introduction of gradient gesture interact, switching on or off will be far beyond binary controls. It can control intermediate different segments of operation which can be termed as gradient control. For example, the intensity of light, the speed of fan, air condition temperature can be controlled by the proximate distance or closeness from the device. The Gradient Gestures interact of things is mostly important for gaming and interaction where depending on people's proximate distance from each other we can introduce new gaming rules which can be used in more natural way. It can be used for entertainment purpose also.

Keywords Depth sensor · Gradient gestures interact of things · Kinect · Skeleton tracking · Gradient feeling

---

S. Nandi (✉) · S. Deb · M. Sinha
National Institute of Technology, Agartala, India
e-mail: sonianandi90@gmail.com

S. Deb
e-mail: sumandebcs@gmail.com

M. Sinha
e-mail: mitalisinha93@gmail.com

© Springer India 2016
S.C. Satapathy et al. (eds.), *Information Systems Design and Intelligent Applications*, Advances in Intelligent Systems and Computing 435,
DOI 10.1007/978-81-322-2757-1_12

1 Introduction

In this developing era, there are many gesture controlled devices that had been brought in practice like myo, leap motion, etc. but all these devices are very expensive. Whereas we can build a cost effective gesture controlled devices incorporating depth sensor which not only be used for minute identification of movements but also can be used for controlling large appliances like regulating doors in the buildings or gates, rail safety, etc.

There are multiple sensors like infrared, ultrasonic having the above capabilities but they have some limitations. Infrared sensors are incapable of providing accurate ranging as well as it cannot be used in bright sunlight. Even irregular movements also can cause triggering of that infrared sensor. Again in ultrasonic sensor, sound absorbing materials like sponge and the 'ghost echo' can cause problem. To combat all those things, we are proposing in this paper to introduce multiple gesture sensing using the concept of depth sensing. The idea of depth sensing not only makes it efficient but also can identify the relative proximity or relative difference between the object with the subject or with the application device to the user.

Depth sensing is again important for making a gradient feeling, the feeling that changes with gesture, time and position with respect to object. For example, if we are using binary switching for motion, it will either on or off but if we can make a gradient reaction along with movement of particular gesture, it will be something different other than binary switching. For example-Color gradient. In it, the color is dependent on the position of a particular region. It may be axial gradient where the color changes along the line or radial gradient where the edge of a circle has one color and gradually the color changes and gets a different color at the center [1] (Figs. 1 and 2).

Here, in this paper, we tried to incorporate gradient feeling of an object. We tried to describe how depth sensor can be introduced to bring a gradient sensation to the

Fig. 1 Axial gradient [2]

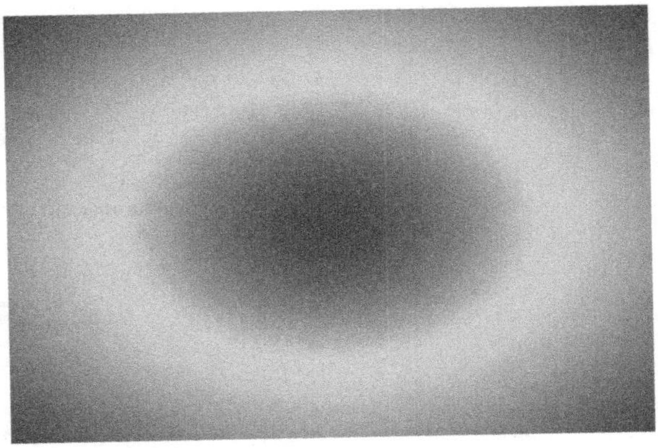

Fig. 2 Radial gradient [3]

machines. By different gestures in front of the sensor embedded appliances, we can make the appliance react (Figs. 3 and 4).

2 Depth Sensor

Depth sensing technology used in kinect uses IR camera and IR laser projector. The raw depth data can be extracted by stereo triangulation process [6]. Two images of a single point are required to get the depth information of that point. This technology can detect up to six users present within the sensor's viewing range and can provide the data of two skeletons in detail. It can track up to twenty joints by processing the raw depth information.

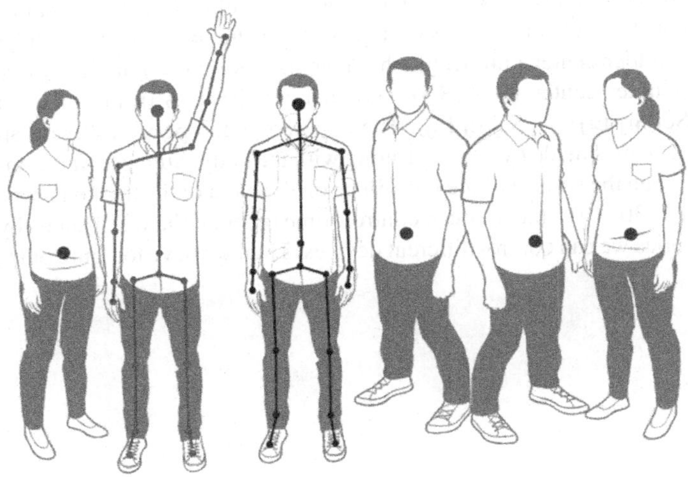

Fig. 3 Kinect recognizing multiple persons [4]

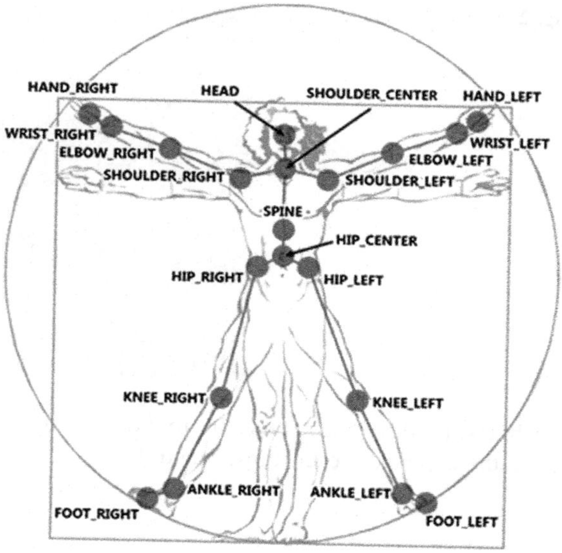

Fig. 4 Kinect tracking 20 joints [5]

3 Methodology

The idea in this paper is to bring the gradient feeling of an object. More and more the person is coming closer to an object, the device reacts accordingly. We don't have to adjust manually. For example, as the person is coming closure to the fan, the speed of the fan is getting changed. We can even select different devices just by moving our hand. To get the epitome of that feeling, first we have to track the skeleton of the human body. After that we have to point out the shoulder left (Slj) or shoulder right (Srj) joint of the skeleton. The next step is to get the angle made in the 3D space between shoulder center joint (Scj), shoulder right joint (Srj) and elbow right joint (Erj) or shoulder center joint (Scj), shoulder left joint (Slj) and elbow left joint (Elj) i.e. (Scj Srj Erj) or (ScjSlj Elj). The aim here is to operate different appliances based on the angles made by the hand along with the body. Suppose, if the angle made is 10°–30°, then the control of the fan will be on the hand of the tracked user and if the angle made is 30°–50°, the user can control some other appliance. That is, by making different angles we can control different devices. Then we have to figure out the elbow left (Elj) and elbow right joint (Erj) and then calculate the angle between shoulder left joint (Slj), elbow left joint (Elj) and wrist left joint (Wlj) or shoulder right joint (Srj), elbow right joint (Erj) and wrist right joint (Wrj) and by making different angles it will regulate the speed, volume or intensity of different appliances.

4 Representation

In the figure below, we have described in detail how we introduced the gradient feeling by making different angles with our hand standing in front of kinect. Suppose, we are controlling TV, by making angle a shown as the Fig. 6, we can increase the volume of TV to 10 %, angle b will increase the volume to 30 % and so on (Figs. 5, 6, 7, 8 and 9).

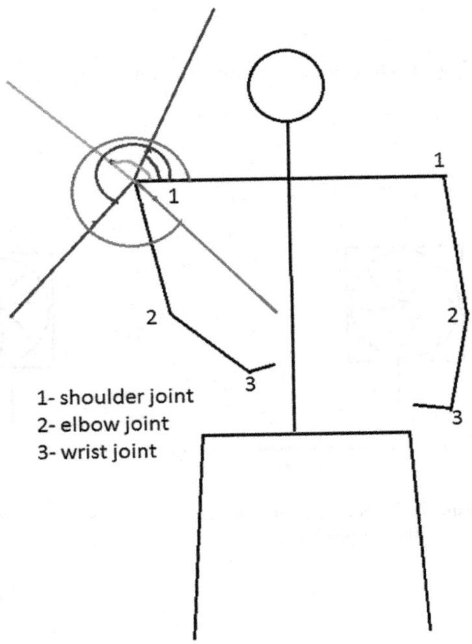

Fig. 5 Different angles for selecting different devices

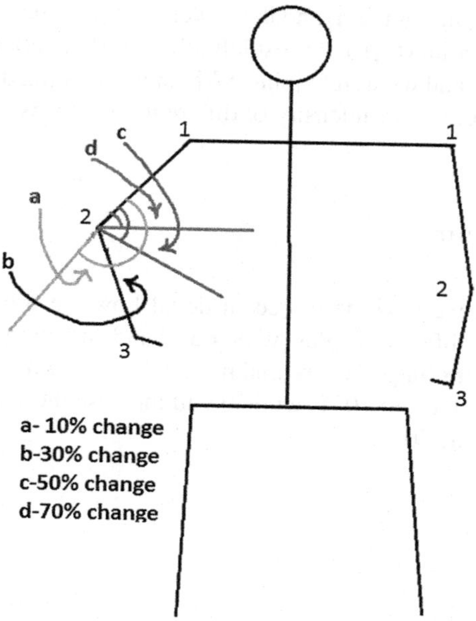

Fig. 6 Different angles for regulating a device

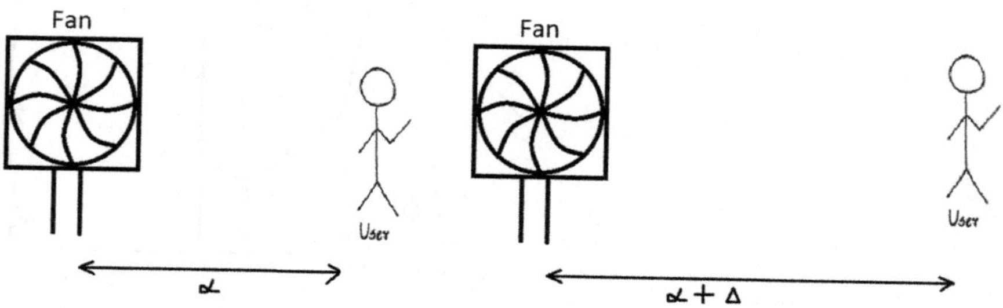

Fig. 7 Depending on the proximate distance between the user and fan, the speed of fan is controlled automatically

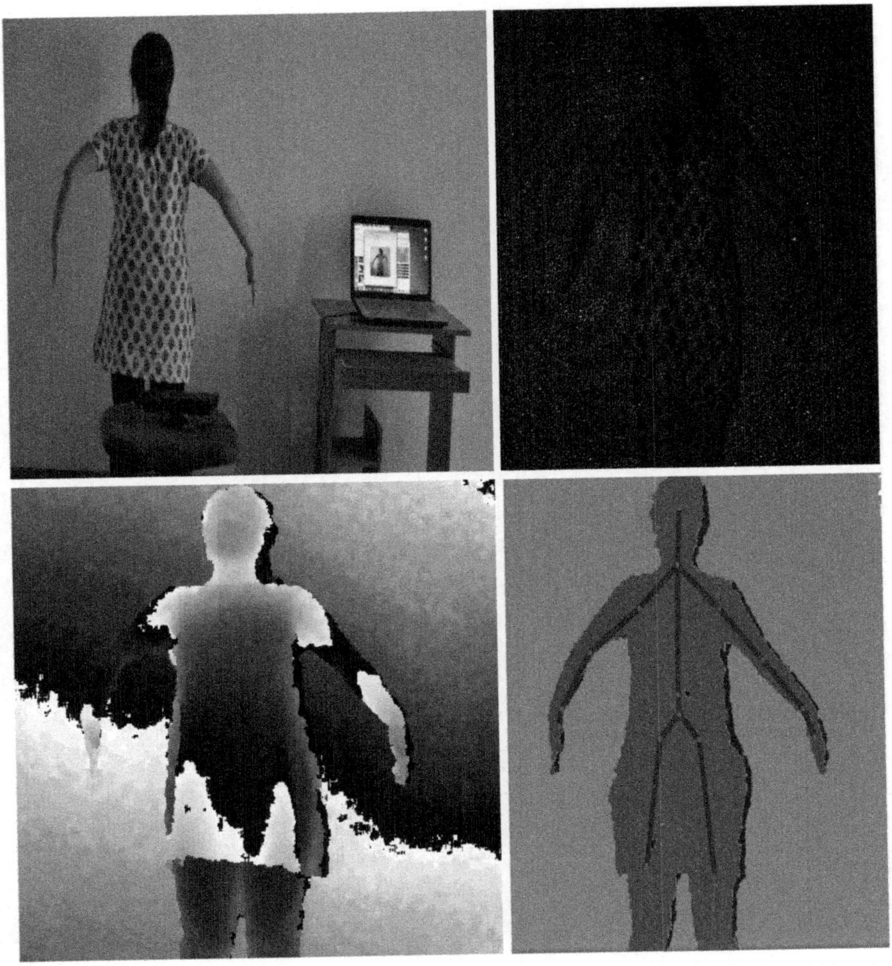

Fig. 8 *Upper left* CMOS view. *Upper right* IR view, *lower left* depth view, *lower right* skeletal view

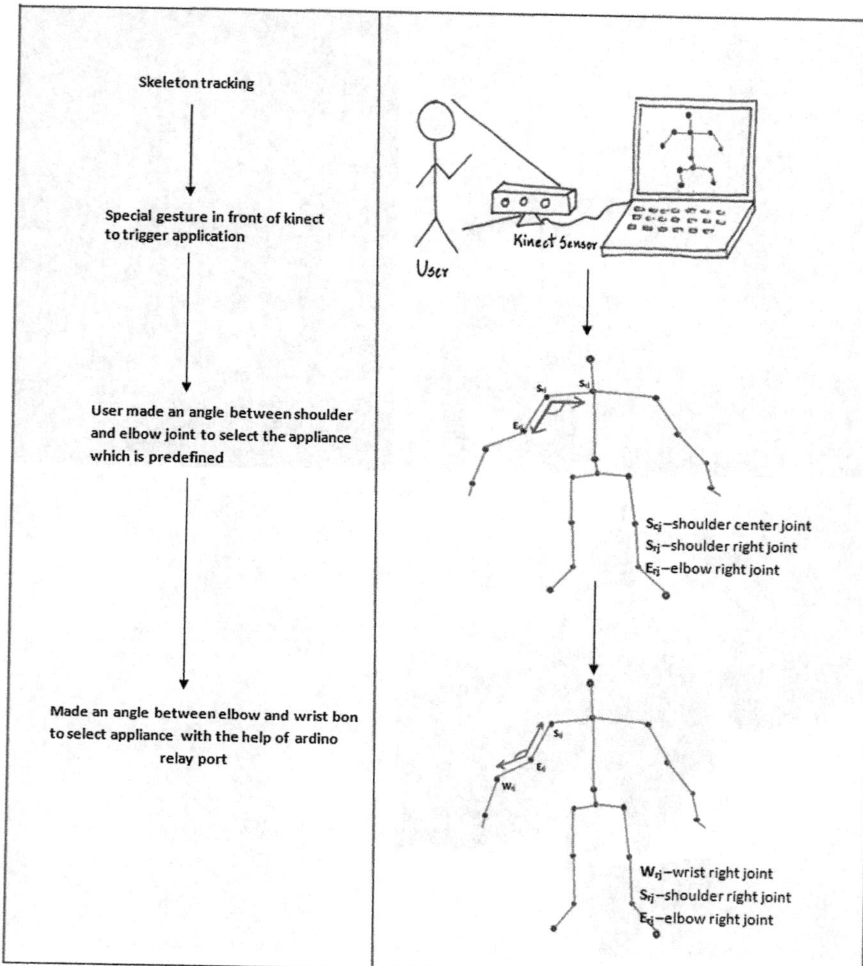

Fig. 9 Schematic of the methodology

5 Conclusion

Controlling a device in binary mode is a traditional manner in which we switch on or off and the device reacts. In this paper, we have successfully experimented and found gradient control using depth sensor making interactive appliance for daily use where people can put multilevel controlling of devices. The intensity control of light can be easily created an controlled by only human proximity movement from particular device. As it is been gradient control or multi steps control, so depending on the choice of the person and proximity distance, these devices can put in a very comfortable manner. Till up to the present work, the system is limited only for one person. In future we will be trying to incorporate more than one person. Moreover the same device can be created for challenging game environment, virtual reality environment and entertainment. The gradient control can be a prospect for even making for vehicle and hazardous position where thee presence of any people can be automatically sensed by the machine and as per the degree of closeness to danger the alarms or mitigation process can be initiated. The concept of gradient control in this paper tried to focus on depth sensing for implementing gradient control using optical depth sensor like kinect. Our future work endeavors for making the same gradient control systems for blind people where they can approximate the distance or some dynamic object towards them.

References

1. https://en.wikipedia.org/wiki/Color_gradient.
2. http://i.stack.imgur.com/3HYKK.jpg.
3. http://www.w3schools.com/css/css3_gradients.asp.
4. https://msdn.microsoft.com/en-us/library/hh973074.aspx.
5. http://www.contentmaster.com/kinect/kinect-sdk-skeleton-tracking/.
6. S. Rusinkiewicz, M. Levoy, "Efficient variants of the ICP algorithm". In Third International Conference on 3D Digital Imaging and Modeling, 2001.

Evaluation of Glomerular Filtration Rate by Single Frame Method Applying GATES Formula

A. Shiva, Palla Sri Harsha, Kumar T. Rajamani, Siva Subramanyam and Siva Sankar Sai

Abstract This paper aims to assess the utility of the single frame Method in the calculation of Glomerular Filtration Rate (GFR) by using GATES equation and compare it with GFR calculated by grouping frames. The DICOM image has number of frames in which the timely report of the activity of tracer is acquired accordingly at various instances of frames. Here we take a single frame from an image collection of 60 frames where the activity of the radio tracer is at maximum. The activity is expected to be maximized at the 40th frame which is a conjunction of filtration phase and excretion phase, which is proved by visual perception of the image. The GFR is calculated by using gates formula and the counts are obtained by semi-automatic segmentation tool called Fiji, ImageJ. This renal uptake study provides all the structural and functional information, in accordance to the traditional methods and proves to efficient when compared to the mathematical complexity.

Keywords FIJI · Gates equation · GFR · Renal uptake · Single frame

A. Shiva (✉) · P.S. Harsha · S.S. Sai
Sri Sathya Sai Institute of Higher Learning, Anantapur, India
e-mail: shiva.amruthavakkula@gmail.com

P.S. Harsha
e-mail: harshasri439@gmail.com

S.S. Sai
e-mail: sivasankarasai@sssihl.edu.in

K.T. Rajamani
Robert Bosch, Anantapur, India
e-mail: kumartr@gmail.com

S. Subramanyam
Sri Sathya Sai Institute of Higher Medical Sciences, Anantapur, India
e-mail: sivasubramaniyan.v@sssihms.org.in

© Springer India 2016
S.C. Satapathy et al. (eds.), *Information Systems Design and Intelligent Applications*, Advances in Intelligent Systems and Computing 435,
DOI 10.1007/978-81-322-2757-1_13

1 Introduction

The rate at which plasma is filtered (measured in mL/min) is known as Glomerular filtration rate (GFR) [1]. It is also defined as the amount of filtrate formed in all the renal corpuscles of both kidneys each minute. The GFR indicates the proper functioning of kidneys. In adults, the GFR averages 125 mL/min in males and 105 mL/min in females [2]. The renal scan and uptake provide the information about the structure and the functioning of the kidneys. Various methods are described to measure renal function in renal scintigraphy.

GFR is customarily assessed by Cockcroft-Gault (CG) method which includes measuring the concentrations of serum markers such as serum creatinine and blood urea nitrogen [3]. Although widely used, these endogenous markers are not ideal and occasionally do not perform well. Serum creatinine is not considered appropriate for estimation of renal function due to its tubular secretion and also its variability with body mass, age, gender and race. And the Scr does not increase until renal function decreases to 50 % of its normal value [4]. Its excretion rate changes with age, sex, health and lean body mass. The population variance of serum creatinine level is large making it a poor measure for comparison with a reference range [5]. Nuclear Imaging methods, which depend on renal uptake early after tracer injection reflects renal activity, can calculate renal function from imaging data alone without blood sampling [2, 6].

Technetium-99m is a metastable nuclear isotope of technetium-99 that is used in several medical diagnostic procedures, making it as the most commonly used radio isotope. It is a chelating agent and was introduced into renal nuclear medicine in 1970 [7] since it was known that chelating agents used in toxic metal poisoning are eliminated by glomerular filtration without any metabolic alteration. It is very well suited for medical diagnosis because it emits gamma rays of 140 keV and has a half-life of 6.0058 h. Diethylene Triamine Penta Acetic Acid (DTPA) is glomerular.

ltering agent which is used to obtain GFR value, a marker for Kidney functioning [3].

2 Materials and Methods

2.1 Acquisition and Imaging

A dual Head variable angle Gamma camera (SIEMENS, e-cam) was used with a "Low Energy All Purpose" parallel hole collimator placed in the front (anterior) detector. After the radio-pharmaceutical was administrated intravenously into an anticubital vein, the image acquisition was started with the presence of the aoritic line in the display and the data is acquired in frame mode [8]. Initially images were acquired at a rate of 2 s per frame for 60 s to assess the renal perfusion followed by

30 s per frame for the next 30 frames. The total scan of patient is for 18 min excluding pre-syringe, post syringe and anti-cubital [4].

The imaging is done as per diuretic renogram protocol with a 64 × 64 matrix size. The acquired image is a 16-bit image. Therefore the maximum intensity value of the pixel is 65,536. The counts is the measurement of Intensity values over the region of interest. Each patient data consists of sixty-three 4D unit 16-bit DICOM images, namely pre syringe, kidney, post syringe and the anticubital. A DICOM image consists of 63 frames in which the first frame is pre syringe, the frames from 1 to 60 are renal images, 61 frame is post syringe, 62 frame is anticubital [8, 9]. The procedure of imaging is as follows:

1. A full syringe loaded with Tc99m is imaged before injection for 1 min.
2. The renal region is scanned for 15 min, in which for 15 images are scanned for every 2 s and the next 15 images are scanned for every 30 s.
3. Empty syringe is imaged for next 1 min.
4. Anticubital is imaged for 1 min.

The above formed images are coupled together to form a multi frame image or a composite DICOM image. Each single frame has the activity over that period.

2.2 Uptake and Calculations

Traditionally the GFR is calculated by grouping the frames in the secretary phase, i.e. from frame 1 to frame 34 and the sum total of all counts are taken as 3 the reference counts of the desired region in the renal cavity [4]. In contrast in this paper, we only take frame 40 and calculate the counts for desired region in the renal cavity. The then obtained counts are substituted in the Gates Formula to measure the Renal Uptake and GFR. The Gates formula is most commonly used to compute the GFR in DTPA renogram. In Gates method the renal uptake percentage (RU) is computed first and then the RU value is substituted in GFR equation,

$$\%RU = \frac{\frac{RKcts - RKbkgd}{\exp(-\mu * x)} + \frac{LKcts - LKbkgd}{\exp(\mu * x)}}{Pre\ injection\ cts - Post\ injection\ cts} \tag{1}$$

The Renal uptake denotes the percent of the total uptake by the renal cavity. Typically it ranges from 1 to 20. The lesser the uptake the lesser the GFR, which denotes the abnormal condition. GFR Equation is derived by sectoring the renal cavity and defining the regions. The calculation of GFR is performed through Gates Equation [2] (Fig. 1)

$$GFR = RU \times 9.8127 - 6.82519 \tag{2}$$

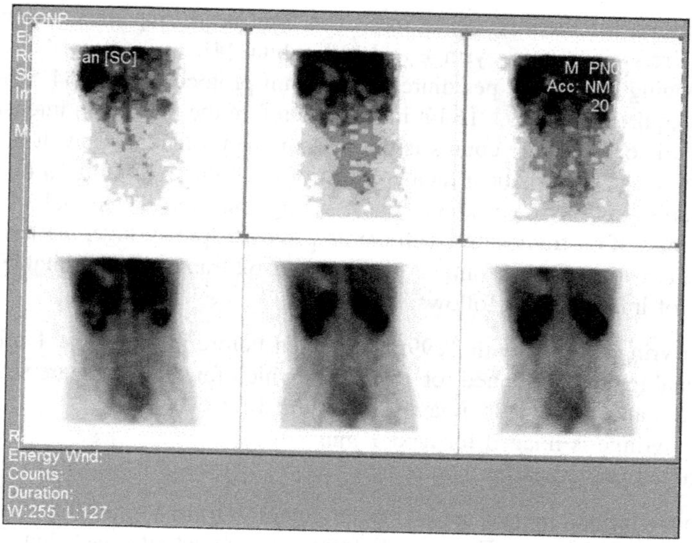

Fig. 1 Interactive tool at SSSIHMS

The 1 is the interactive GUI which is an assembly of SIEMENS Dual head gamma camera named ICON. The images presented in that window give the user to see the timely activity at various instances. The Fig. 2 is the analysis window where the graph represents activity-time.

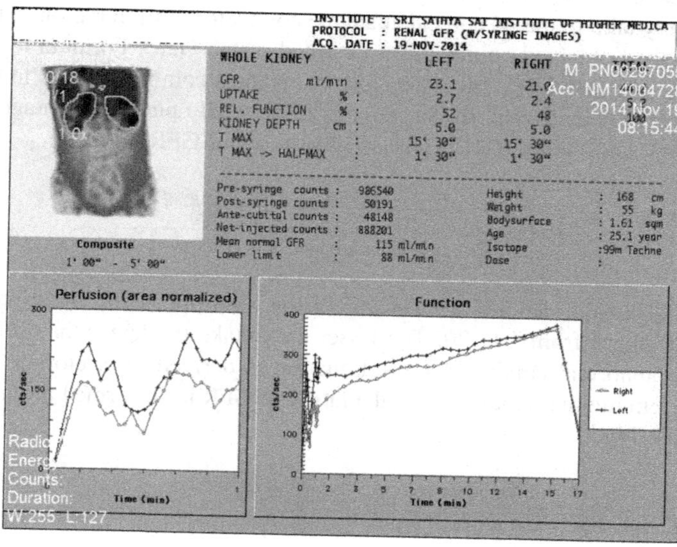

Fig. 2 Analysis window

2.3 *Interactive GUI Tool*

FIJI is an open source image processing software package based on ImageJ and one can perform various operations on an image such as Image enhancement, Registration, Morphological Processing, Segmentation etc. Fiji's main purpose is to provide a distribution of ImageJ with many bundled plug-ins. It is compatible with various operating systems such as Windows, Linux and MAC. There are various languages supported by Fiji framework such as JRuby, Clojure, Bean Shell and more. Apart from these Fiji has the ease of comfort in calculating the area in Region of Interest. The desired region is selected with a polygon ROI and the area is calculated by the following Formula

$$Number\ of\ pixels = \frac{Effective\ area\ of\ total\ pixels}{Effective\ area\ of\ a\ pixel} \tag{3}$$

This gives the number of pixels and their counts is the product of number of pixels and the mean intensity value.

Figure 3a represents the first frame in the entire image sequence. It determines the number of counts in a syringe loaded with radio tracer Fig. 3b Post syringe counts which is the empty syringe scanned after the syringe is injected. The Fig. 3c shows the time at which he uptake is maximum and it is the 40th frame in the frame sequence. The Fig. 3d shows the Fiji Tool and the results. From the images we can infer that the area of a pixel can be calculated and the number of pixels.

3 Results and Analysis

The study was done on 14 patients with various kidney disorders. The results are tabulated as shown in Tables 1. The significant difference between the obtained value and the utmost 2 proving to be a satisfactory result. However there may be discrepancies in the number of counts from the kidneys as well as the syringe counts. It may be due to difference in the region of interest. Table 1 summarizes the comparison of the values between the SIEMENS software and open source GUI Tool. The error percentage is taken as reference to expose the accuracy of the results. The Siemens assembly has ICON processing tool that is used to process the images whose values are taken as standard reference. Therefore the error is calculated based on the formula:

$$Error(\%) = \frac{|Siemens\ value - SingleFrameValue|}{Siemens\ Value} \tag{4}$$

From Eq. 4 the standard value can be stated the value obtained from Siemens Assembly. From the Table 1 the patient 3 and 4 showed no difference from the Siemens value. The patients 5, 6 and 12 showed more difference from the expected

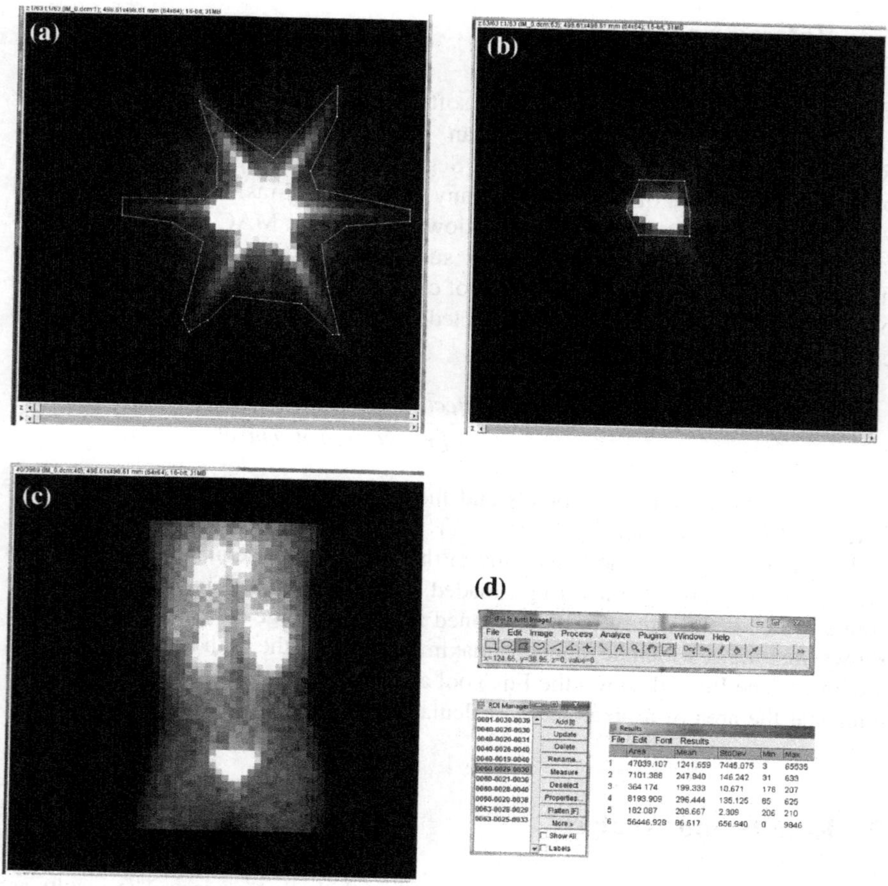

Fig. 3 **a** Full syringe. **b** Empty syringe. **c** Renal section. **d** Interactive tool

value. Statistical analysis will done after acquiring sufficient number of case studies. This would give more emphasis on the result to standardize it. Here the Siemens gamma camera assembly has processing tool called ICON. The results obtained from ICON are considered to be the standard to which the results obtained were compared.

Table 1 Comparison of renal uptake and glomerular filtration rate (GFR)

Patient ID	Renal uptake (Siemens)	Renal uptake (Single frame)	GFR (Siemens)	GFR (Single frame)	Error (%)
Patient 1	7.45	7.44	67.92	66.18	3
Patient 2	5.7	5.4	50.2	46.16	8
Patient 3	9.7	9.8	89.4	89.34	0
Patient 4	9.5	9.5	86.29	86.4	0
Patient 5	10.2	12.03	93.15	111.22	19
Patient 6	4.2	4.6	34.35	38.31	12
Patient 7	5.3	5.63	45.13	48.42	7
Patient 8	6.2	6.37	53.95	55.68	3
Patient 9	7.8	8.42	69.63	75.8	9
Patient 10	10.9	11.17	100.01	102.78	3
Patient 11	8.1	7.39	72.57	65.69	9
Patient 12	10.3	11.91	94.13	110.04	17
Patient 13	8.8	8.97	79.43	81.19	2
Patient 14	9	9.5	81.39	86.4	6

4 Conclusion

Computing the targeted frame may throw more light on the Renal studies. As the uptake in the 40th frame is comparatively approximate, it shows the result when grouping technique fails. Also 40th frame can be categorized as the frame in the transition from secretory to excretory phase whose study can yield many important revelations in the renal uptake studies. This tool becomes handy when the existing grouping technique fails. It also reduces the computation complexity when compared to the grouping technique. The Frame 40 is expected and selected based on the visual impression, which shows the fragility of the selection algorithm. Hence an improvised algorithm to select the appropriate frame where uptake is maximum gives much efficiency to this technique. This can be embedded into a simple portable devices to reduce the mathematical complexity.

References

1. Website: http://www.webmd.com/a-to-z-guides/kidney-scan, Healthwise for every health decision, September, 2014, Healthwise Incorporated.
2. Comparison of glomerular filtration rate by Gates method with Cockroft Gault equation in unilateral small kidney, The Egyptian Journal of Radiology and Nuclear Medicine (2013) 44, 651–655.
3. Pedone C, Corsonello A, Incalzi RA. Estimating renal function in older people: a comparison of three formulas. Age Ageing. 2006; 35:121–126.
4. Magdy M. Khalil: Basic Sciences of Nuclear Medicine. Springer verlag Heidelberg 2011.

5. Delpassand ES, Homayoon K, Madden T et al. Determination of glomerular filtration rate using a dual-detector gamma camera and the geometric mean of renal activity: correlation with the Tc-99m DTPA plasma clearance method. Clinical Nucl Med. 2000; 25(4):258–262.
6. Siddiqui AR. Localization of technetium-99m DTPA in neurofibroma. J Nucl Med 1986; 27: 143–144.
7. Glomerular Filtration Rate: Estimation from Fractional Renal Accumulation of 99mTc-DTPA. Review paper.
8. Sai Vignesh T, Subramaniyan V.S, Rajamani K.T, Sivasankar Sai, An interactive GUI tool for thyroid Uptake studies using Gamma Camera, ICACCI, September 2014.
9. Kuikka JT, Yang J, Kiiliinen H., Physical performance of the Siemens E.CAM gamma camera, Nuclear Medicine Communications, 1998 May; 19(5):457–62.

Design and Development of Cost Effective Wearable Glove for Automotive Industry

Manoharan Sangeetha, Gautham Raj Vijayaragavan, R.L. Raghav and K.P. Phani

Abstract In an automotive production unit considerable amount of time is spent on testing and observation. Therefore a need arises to address this issue with the help of any automation technique. In this paper, a cost effective wearable glove for monitoring automotive parameters such as temperature, flaw detection, Electro Motive Force (EMF) leakage and Direct Current (DC) measurement etc. To perform these measurements, various sensor units like temperature sensor for monitoring temperature, ultrasonic sensor for flaw detection are interfaced with the Arduino board and the monitored parameters are displayed in a graphical Liquid Crystal Display (LCD). The proposed design for wearable glove using Arduino Board enables time efficient continual monitoring of the parameters.

Keywords Arduino board · DC voltmeter · EMF detection · Temperature sensor · Ultrasonic sensor

1 Introduction

Employing wearable technology in professions where workers are exposed to dangers or hazards could help save their lives and protect health-care personnel [1].

A prototype wearable device for monitoring the breathing rate of a number of subjects was proposed by Brady et al. in [2]. Wearable technologies can collect data, track activities, and customize experiences to users' needs and desires [3]. Wearable technologies find its application in the 5G wireless Communication where the promising technology is Internet of Things (IoT). The integration of IoT with Wearable technology promises to have widespread societal influences in the coming years [4].

Manoharan Sangeetha (✉) · G.R. Vijayaragavan · R.L. Raghav · K.P. Phani
Department of ECE, SRM University, Kattankulathur, Chennai, India
e-mail: sangeetha.m@ktr.srmuniv.ac.in

© Springer India 2016
S.C. Satapathy et al. (eds.), *Information Systems Design and Intelligent Applications*, Advances in Intelligent Systems and Computing 435,
DOI 10.1007/978-81-322-2757-1_14

A method to design an Electronic Hand Glove which would help deaf and blind people to communicate easily is proposed in [5]. The proposed model uses 26 gestures of hand to communicate alphabets and 10 more gestures to communicate numbers. This would help the deaf person to communicate with others by typing text on LCD screen through hand gestures. The text is converted into speech so that the blind person could hear and communicate.

Wearable devices found a prominent application in the field of health care and medicine. There are a lot of electronic gloves researches undertaken for Automation and Medical Research purposes. In this paper, a method to design an electronic glove using Arduino Board for monitoring parameters related to an automotive industry.

In automotive industry, considerable amount of time is spent on testing and observation; this could be solved by using a wearable glove that is powered by an Arduino board to monitor the Temperature, to identify the Flaws, to detect the EMF leakage and also to serve as a DC voltmeter by measuring the voltage across a circuit. Using a compactly fitted glove to monitor the necessary parameters helps an individual to move around easily without carrying individual testing equipment for individual parameter measurements. Further continuous monitoring of accurate readings is possible in a time efficient manner using this glove.

The main aim of this work is to assist workers in an automotive industry where unnecessary time is being spent on monitoring and observation. The wearable glove provides ease of use and for monitoring and testing the parameters. The sensors interfaced with Arduino board measure and send accurate readings to graphic LCD which in turn displays the reading for the user to observe and record.

Hence an Electronic Glove is designed and developed to reduce the complexities in Testing and Monitoring services in an automotive Industry by reducing the amount of cost and time being deliberately spent in vain. The proposed design could assist the user with four different parameter measurements and monitoring and it is the first of its kind in the automotive testing Industry.

2 Block Diagram

An Arduino board is interfaced with a temperature sensor to monitor the temperature inside the testing area and an ultrasonic sensor is used to find flaws in an object and the corresponding temperature and distance reading is displayed on 16 * 2 graphics LCD. A voltage divider circuit and an EMF detection circuit are also interfaced with the Arduino board to serve as a DC voltmeter and EMF detector respectively. The DC voltmeter consists of two probes which can be placed across any electronic circuit and corresponding reading is displayed on graphic LCD. The proposed hand glove model is shown in Fig. 1

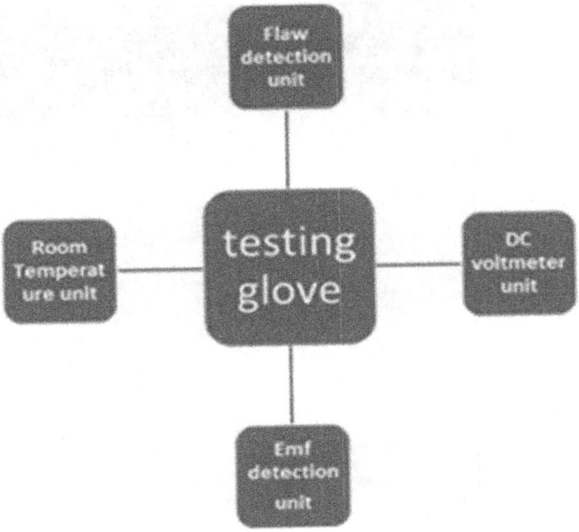

Fig. 1 Block diagram for the working of the interface

3 Block Diagram Description and Design Specification

Temperature monitoring unit, Flaw detection unit, EMF detection unit and DC voltmeter unit are all fitted compactly into a Wearable glove. The temperature sensor measures the room temperature and displays the readings in the Graphics LCD. The Ultrasonic sensor detects the presence of the flaw (if there is any) and displays the accurate range readings on graphics LCD. A buzzer is made to alarm if the readings measured are greater than 10 cm. This is done to ensure accurate readings are obtained every time. The DC voltmeter unit uses the voltage divider circuit to measure the voltage across any circuit. Two probes namely positive and negative placed across the circuit to measure its voltage ranging from 0 to 30 V. The EMF detection circuit triggers a LED to indicate the presence of EMF leakage. All these units are compactly fitted in a single glove in order to enable continuous monitoring of parameters in a time efficient and convenient way.

3.1 Arduino Board

The Arduino UNO is a microcontroller board based on the ATmega328 (Fig. 2).

3.2 Temperature Monitoring Unit

The Arduino is interfaced with temperature sensor at analog pin A1. The temperature proportional to output voltage is measured using the sensor and the

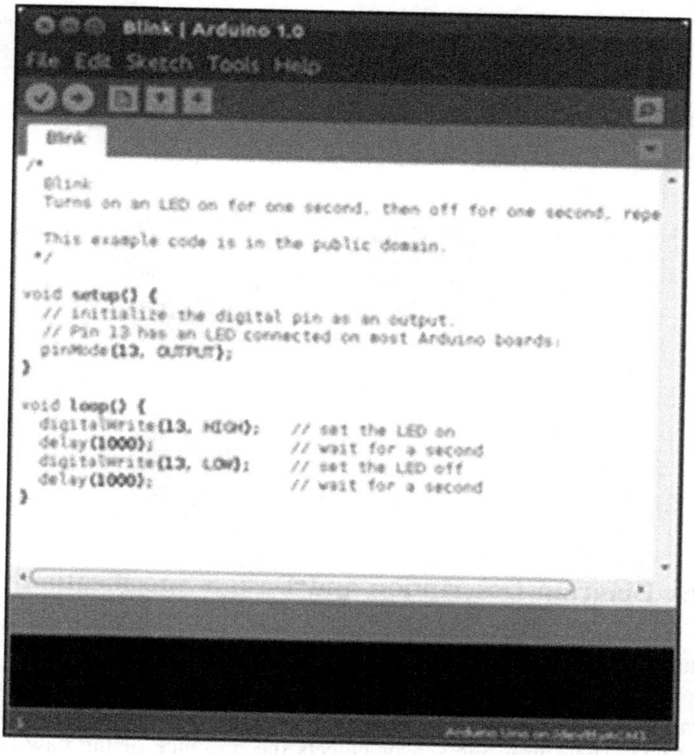

Fig. 2 Screenshot of the Arduino coding workspace

corresponding reading in Celsius is displayed on the 16 * 2 graphics LCD. Figure 3 shows the interfacing of the temperature sensor with the Arduino.

3.3 Ultrasonic Flaw Detection Unit

The flaw detection is achieved with the help of Ultrasonic sensor HC-SR04. Ultrasonic sensor module HC—SR04 has a non-contact measurement function, and provides accuracy up to even 3 mm. The range of this sensor varies from 2 to 400 cm. The sensor pins are interfaced in the following order: Vcc to pin 2, Trig to pin 8, Echo to pin 9, GND to pin 5 of the Arduino. The range values are recorded by the sensor, if it is found to be constant throughout the surface of an object, then it is concluded that there is no flaw present in the object, Otherwise if there is actually a flaw in the object, then deviations in the recorded values can be observed, thus proving that there exists a flaw in the object. The range values are displayed on the 16 * 2 graphic LCD continuously for the user to observe any deviation is observed.

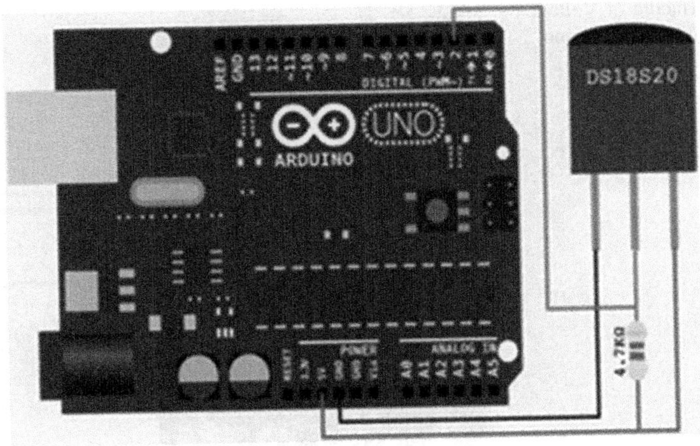

Fig. 3 Interfacing of LM35 with Arduino

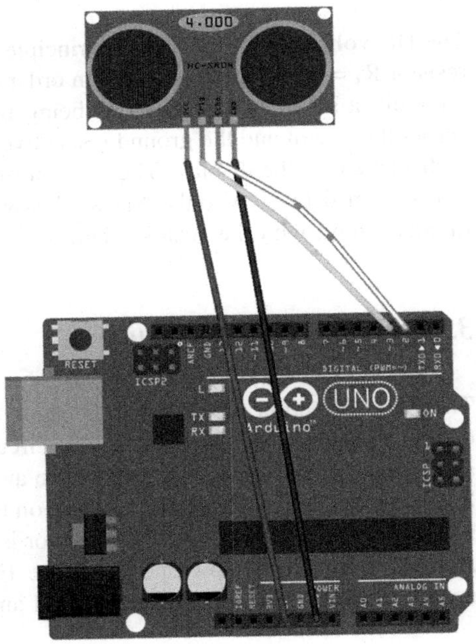

Fig. 4 Interfacing of HC-SR04 with Arduino

In order to get notified about the accurate readings a buzzer is made to sound if the range recorded is more than 10 cm, as accurate readings are observed when sensor is as close to the object under observation [6] (Fig. 4).

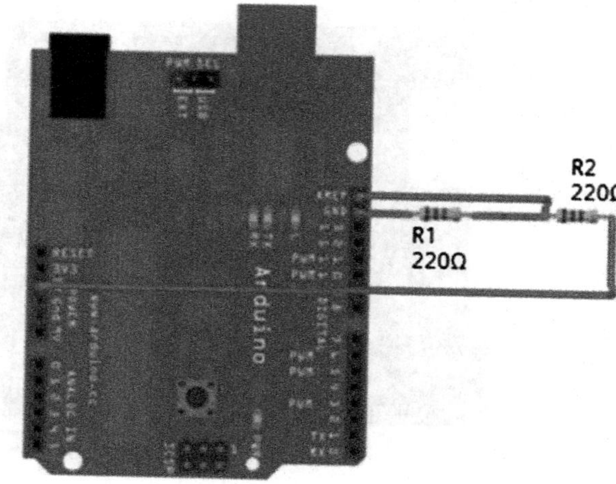

Fig. 5 Interfacing of Voltage divider circuit with Arduino

3.4 DC Voltmeter Unit

The DC voltmeter unit works on principle of voltage divider circuit. In the design, resistor $R_1 = R_2 = 220\ \Omega$ is taken. In order to find out the voltage across a device or a circuit, a set of two probes one being positive and other one negative is paced across the circuit and the ground respectively. The corresponding reading measured is displayed on the display. The DC voltmeter can be used to measure voltages of range from 0 to 30 V only. Shown below in Fig. 5 is the interfacing of voltage divider circuit with the Arduino board.

3.5 EMF Detection Unit

The EMF detection unit is used to check whether any EMF leakage occurs in a circuit or device. The EMF detection circuit basically consists of resistor wound across a copper wire and is connected to analog pin 5 of the Arduino and incase of EMF leakage, a LED is made to switch on to indicate the presence of EMF leakage happening in the circuit. A 3.3 M resistor is used in the EMF detection circuit with Arduino Interface. Below shown in the Fig. 6 is the interfacing between EMF detection circuit with the Arduino board and the LED indicator is also shown.

3.6 LCD Display

The Measured parameters are displayed in the LCD Display which is placed at the top of the glove. The LCD Display used over here is 16 * 2 Graphics LCD Display [7] (Fig. 7).

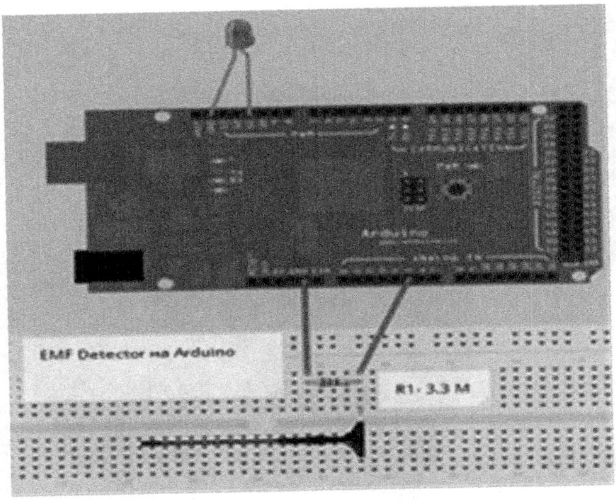

Fig. 6 Interfacing of EMF detection circuit with Arduino

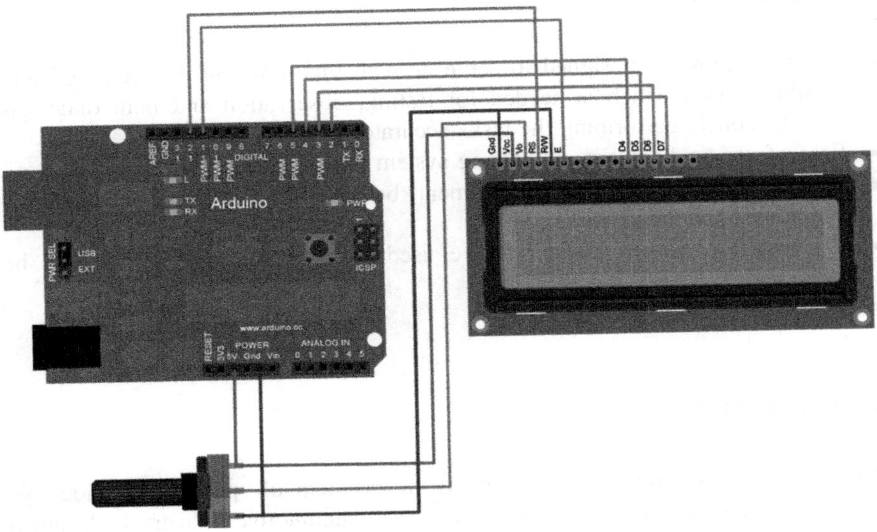

Fig. 7 Interfacing of LCD display with Arduino

4 Prototype Model

Below shown Fig. 8 is the prototype of the hand glove.

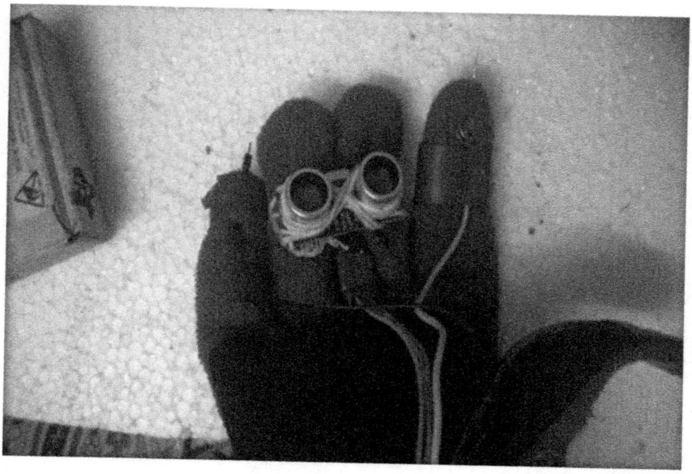

Fig. 8 Prototype model of the proposed system

5 Advantages

- Wearable glove is designed to serve as both time efficient and economically feasible device which is needed for testing, observation and fault diagnosis simultaneously performing the tasks accurately
- Ease of use as the entire monitoring system is compactly fitted in Glove
- First of its kind in testing environment, because of its multiple features in a single wearable hand glove
- Rechargeable Power source can be used i.e. 9 V batteries which can be recharged.

6 Conclusion

Wearable glove serving as a solution to the problem of spending considerable amount of time on testing and observation in an automotive industry is designed and developed. This glove is fitted with four units to measure temperature, identify flaws, detect EMF leakage and finally can also serve as digital voltmeter. All the four units are compactly fitted with the glove to enable continuous monitoring of parameters in a time efficient and convenient way. It is also economically feasible device that can be used for testing, observation and fault diagnosis in an automotive industry.

References

1. Rutherford, J.J, "Wearable Technology," IEEE Engineering in Medicine and Biology Magazine, vol. 29, no. 3, pp. 19–24 (2010).
2. Brady, S. Carson, B., O'Gorman, D. and Moyna, N., "Combining wireless with wearable technology for the development of on-body networks," International Workshop on Wearable and Implantable Body Sensor Networks (BSN) 2006, Cambridge, MA, pp. 36, (2006).
3. Charles McLellan, M2 M and the Internet of Things: A Guide ZDNET, http://www.zdnet.com/m2m-and-the-internet-of-things-7000008219 (2013).
4. David Evans, "The Future of Wearable Technology: Smaller, Cheaper, Faster, and Truly Personal Computing," (2013).
5. Gupta, D.; Singh, P.; Pandey, K.; Solanki, J., "Design and Development of a low cost electronic hand glove for deaf and blind," Proceedings of the 2nd International Conference on Computing for Sustainable Global Development (INDIACom), New Delhi, pp. 1607–1611, 11-13 (2015).
6. Smith, G.B., "Smart sensors for the automotive industry," IEE Colloquium on ASICs for Measurement Systems, (1992).
7. http://www.maximintegrated.com/en/products/analog/sensors-and-sensor-interface/DS18B20.html.
8. http://forum.arduino.cc/.
9. http://arduino.cc/en/Tutorial/LiquidCrystal.

Tampering Localization in Digital Image Using First Two Digit Probability Features

Archana V. Mire, Sanjay B. Dhok, Narendra J. Mistry and Prakash D. Porey

Abstract In this paper, we have used the first digit probability distribution to identify inconsistency present in the tampered JPEG image. Our empirical analysis shows that, first two digits probabilities get significantly affected by tampering operations. Thus, prima facie tampering can be efficiently localized using this smaller feature set, effectively reducing localization time. We trained SVM classifier using the first two digit probabilities of single and double compressed images, which can be used to locate tampering present in the double compressed image. Comparison of the proposed algorithm with other state of the art techniques shows very promising results.

1 Introduction

JPEG fingerprint is one of the important passive digital image forensic techniques having potential to unfold most of the tampering operations [1]. Mostly JPEG fingerprint based forgery detection techniques try to identify the absence of Aligned Double JPEG (ADJPEG) compression artifacts [1–3]. During JPEG compression, image is converted to YCbCr space and 8 × 8 block DCT transform is applied to each individual color channel. Each DCT coefficient in the 8 × 8 block gets quantized with the corresponding quantization step of 8 × 8 quantization table. All these quantized

A.V. Mire (✉) · N.J. Mistry · P.D. Porey
Sardar Vallabhbhai National Institute of Technology, Surat, India
e-mail: archam2002@yahoo.co.in

N.J. Mistry
e-mail: njm@coed.svnit.ac.in

P.D. Porey
e-mail: prakashporey@gmail.com

S.B. Dhok
Visvesvaraya National Institute of Technology, Nagpur, India
e-mail: sanjaydhok@gmail.com

© Springer India 2016
S.C. Satapathy et al. (eds.), *Information Systems Design and Intelligent Applications*, Advances in Intelligent Systems and Computing 435,
DOI 10.1007/978-81-322-2757-1_15

coefficients are collected in zigzag order for entropy coding [4]. Since coefficients present at the same position in different 8 × 8 blocks get quantized with the same quantization step, consistent artifact get introduced at each AC frequency position. It is difficult to create convincing tampering by aligning 8 × 8 DCT grid in tampered region. At least some portion of tampered region undergoes Non Aligned Double JPEG (NADJPEG) compression, where double compression artifacts are missing. Figure 1 shows an example of tampering where star segment from source image (a) (compressed at quality Q_1) is copied and pasted into destination image (b) (compressed at quality Q_2) to create composite double compressed image (c) (compressed at quality Q_3). Grids in Fig. 1 represent 8 × 8 DCT grid (shown with dark edges) used for compression. Star shaped copied from source image (a) shows significant shifting within different 8 × 8 grid in figure (c). Hence pasted star shape gets NADJPG compression, while smiley shape and all remaining background regions undergo ADJPEG compression. DCT coefficients of a star shape get quantize by quantization step q_{1_a} followed with q_2 while remaining background region get quantized with q_{1_b} followed with q_2. Based on this, tampered region can be identified as a region where aligned double compression artifacts are missing.

Hanny [5] used difference between test image and its recompressed versions to identify non aligned double compressed region. Zhouchen et al. [2] recognized the periodicity in the histogram of double quantized DCT coefficients for double compression detection. Tiziano et al. [3] used Expectation Maximization algorithm to find tampering probability map. Fan et al. [6] used light weight edges present inside 8 × 8 block due to non aligned double compression. Weiqi et al. [7] proposed block artificial characteristics matrix to identify double compressions. Dongdong et al. [8] proposed generalized Benford Model for double compression detection. Bin et al. [9] further extended this work to show that probability distribution of top 15–25 AC modes following the Generalized Benford's law well. Xiang et al. [10] and Irene et al. [11] used First Digit Probability Distribution (FDPD) of single compressed images and their double compressed counterparts to train SVM classifier. To enhance the location precision Feng et al. [12] combined the moments of characteristic function features with the FDPD features and formed 436-D vector to train and test the classifier.

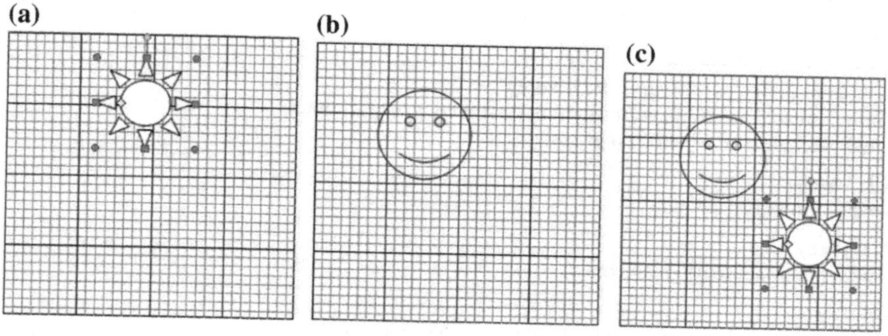

Fig. 1 Tampering showing aligned and non aligned double JPEG compression. **a** Source image. **b** Background image. **c** Composite image

Remaining of the paper is organized as follows. In Sect. 2 we discussed first digit probability distribution and showed that the first digit probability of digit ‘1’ and ‘2’ is sufficient to locate tampering present in the image. We proposed faster tampering localization algorithm in Sect. 3. In Sect. 4, we have discussed our experimental setup and results. Finally, paper is concluded with future work in Sect. 5.

2 First Digit Probability Distribution

Benford’s law [13] states that, in the set of naturally generated data, probability distribution of first digits d (d = 1, 2, …, 9) follow logarithmic nature as shown in Eq. (1).

$$p(d) = \log_{10}(1 + 1/d) \quad d = 1, 2...9 \tag{1}$$

where $p(d)$ stands for the probability of the first digit as d.

Although this law may not get satisfied in all domains, Dongdong et al. [8] demonstrated it for DCT coefficients of uncompressed images. They fitted single compressed images in a new logarithmic model using Eq. (2), called as Generalized Benford First Digit Probability Distribution (GBFDPD) Model. Through empirical analysis, they showed that double compressed images show significant divergence from it.

$$p(d) = N \log_{10}(1 + 1/(s + d^q)) \quad d = 1, 2...9 \tag{2}$$

where N is a normalization factor which makes $p(d)$ a probability distribution, s and q are model parameters specific to the quality of compression.

Irene et al. [11] used the probability of first digits ‘2’, ‘5’, ‘7’ to train the SVM classifier and claimed faster comparable performance.

2.1 *Impact of Double Compression on FDPD*

We analyzed the effect of compression on the various first digit probabilities for feature set reduction. We used 1338 uncompressed TIFF images from UCID database [14] and performed single compression followed with aligned and non aligned double compression. Each of the double compressed images was divided into 8 × 8 blocks and first nine AC frequencies were selected in zigzag order. Average FDPD of all nine frequencies for ADJPEG and NADJPEG compressed images is computed and plotted in Fig. 2. Due to space constraint we have plotted FDPD only at few quality set-ups, but we got similar results for other qualities too.

In Fig. 2, probability of the first digit ‘2’ shows a substantial difference in ADJPEG and NADJPEG images at all the qualities of compression. This makes the

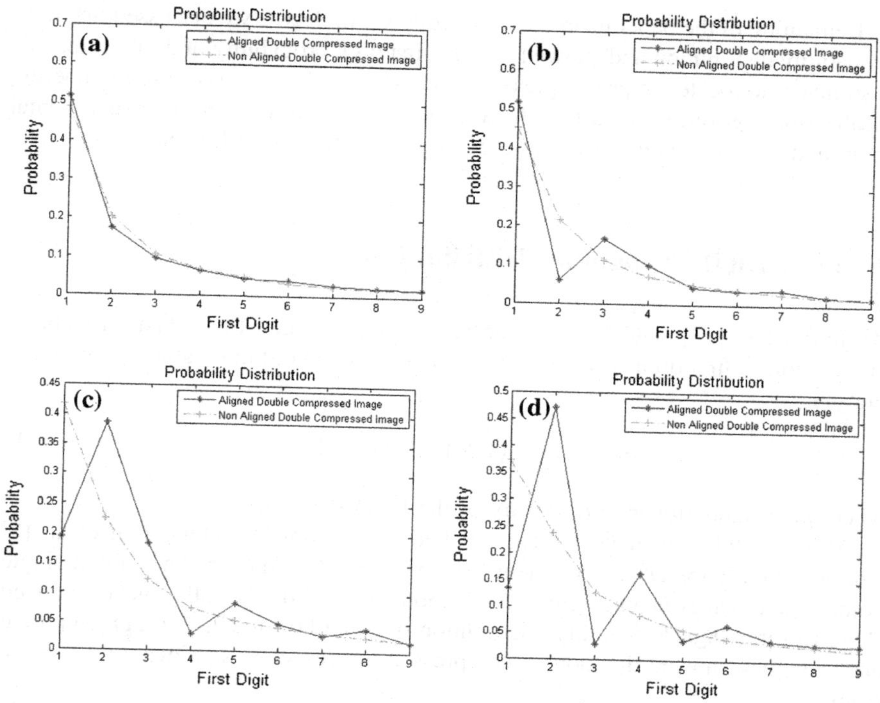

Fig. 2 Average FDPD of first 9 AC frequencies from UCID Database. **a** $Q_1 = 60$, $Q_2 = 65$. **b** $Q_1 = 60$, $Q_2 = 70$. **c** $Q_1 = 60$, $Q_2 = 75$. **d** $Q_1 = 60$, $Q_2 = 80$

probability of digit '2' as the first choice in feature selection. At lower quality difference, we can observe that probabilities of first digits '7', '8', '9' almost coincide for aligned and nonaligned double compressed images. Hence, it is unwisely to use it for feature set reduction. At higher quality, probability of first digit '3' and '4' show higher difference than digit '5'. Thus, using probabilities of first digit '1', '3' and '4' is a better option than using digit '5'. Since probability difference of digit '1' is greater than digit '3' and '4', we have used digit '1' probability. Thus, we used digit '1' and digit '2' probability from first 9 AC frequencies forming $9 \times 2 = 18$ dimensional feature vector.

3 Proposed Approach

We trained SVM classifier using 800 uncompressed images from UCID database [14]. Each image was single compressed and aligned double compressed to generate features for tampered and untampered images respectively. Each image was divided into 8×8 blocks to collect coefficients present at first 9 AC frequencies. For each of the first 9 AC frequencies, we computed FDPD using all first digits present at same

AC frequencies. Out of 9 first digit probabilities, we used probabilities of digit '1' and '2' to train the classifier. We trained different classifier with various sets of primary and secondary compression qualities varying from 60 to 90 in the step 5.

4 Performance Analysis

To evaluate the performance of the proposed Algorithm 1, we used MATLAB 2015 (64 bit version). For comparison, we maintained the same setup for Irene et al. [11] and Xiang et al. [10], except number of First Digit Probabilities used as a feature. We extracted DCT coefficients using Sallee, Matlab JPEG Toolbox [15]. For all the three approaches, SVM classifier was trained using features computed from 800 UCID images. For comparison, we evaluated these three approaches using two different setups, at various qualities of single and double compression using UCID images [14] and Dresden Images [16].

4.1 Feature Set Evaluation

We used random 400 raw images from Dresden database [16] for evaluation of the feature set selection. All these images were converted to TIFF format using UFRaw software suite for windows environment [17]. Each single compressed image was tampered by pasting random 128 × 128 DCT grid aligned blocks, where 8 × 8 grid of secondary compression coincides with the primary compression grid. Resultant tampered images were compressed at quality Q_2. While testing each image was divided into 64 × 64 non overlapping blocks and each block was classified as tampered or untampered using respective trained classifier for quality Q_1, Q_2. AUC values were computed using ROC plot for each tampered image and averaged over all 400 tampered images. In Fig. 3, each AUC value represents the average of all AUC values computed by varying secondary compression qualities with respect to each primary compression quality ($Q_1 = 60, 65 \ldots 85$, $Q_2 = Q_1 + 5, Q_1 + 10, \ldots 95$).

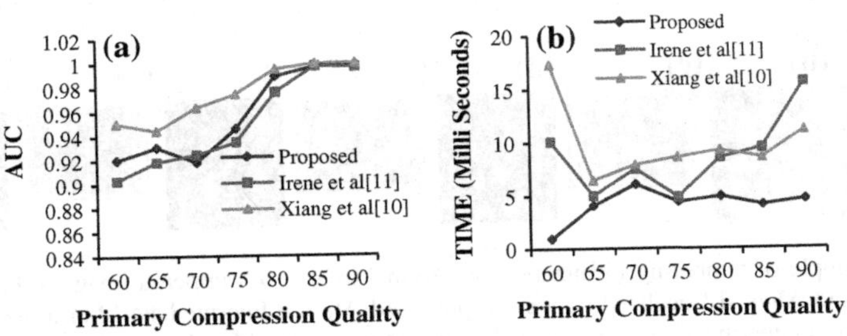

Fig. 3 Performance with Dresden images [16] at various primary compression qualities Q_1. **a** AUC curve. **b** Average testing time

```
Initialize map(size of test image I) = 0
Divide test image I into 40 × 40 blocks with 8 × 8 overlapping.
For each block B = 1:B_n
        Divide block B into 8 × 8 blocks B_DCT
        For i = 1:9 //AC frequencies in zigzag order from each block B_DCT
           Coeff_i = Coefficient present at frequency 'i'
           Find probability p(d) using Coeff_i,// d = 1:9
           F[(i − 1) * 2 + 1] = p(1), F[(i − 1) * 2 + 2] = p(2)
        end
class = Classify F using SVM classifier
 map(B) = map(B) + class
end-for
map = map >= 12
map = largest-connected_component(map)
Tampered_region = map
```

Algorithm 1. Tampering Localization

From Fig. 3a, one can easily conclude that performance of Xiang et al. [10] is best. Since it uses all the 9 first digit probability distribution, it is but obvious. Figure 3 also shows that the performance of our proposed algorithm is better than Irene et al. [11]. Since proposed technique uses first 2 digits, it is fastest among all three approaches, which is clearly visible in Fig. 3b. Proposed algorithm takes least testing time at all the compression qualities.

4.2 Tampering Localization Evaluation

Since tampered regions rarely follows uniform DCT block boundaries, overlapped block investigation is very necessary. For fine grain localization 8 × 8 overlapping is the best option. Being high resolution images, Dresden images need lots of time with 8 × 8 overlapping. Hence, we used 384 × 512 resolution images from UCID database [14]. Each image was JPEG compressed at primary compression quality Q_1 and random 100 × 100 regions is pasted to it. Resultant images were compressed at various secondary compression qualities $Q_2 > Q_1$. Algorithm 1 is used to localize such random tapering. While localizing tampering, each image was divided into 40 × 40 block with 8 × 8 overlapping and threshold = 12 used for selecting

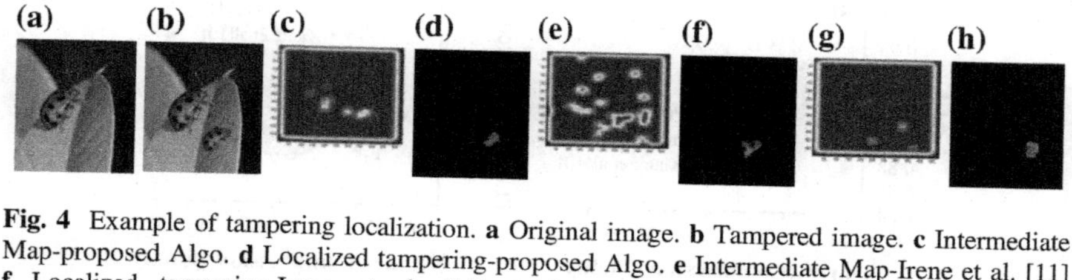

Fig. 4 Example of tampering localization. **a** Original image. **b** Tampered image. **c** Intermediate Map-proposed Algo. **d** Localized tampering-proposed Algo. **e** Intermediate Map-Irene et al. [11] **f** Localized tampering-Irene et al. [11]. **g** Intermediate Map-Xiang [10]. **h** Localized tampering-Xiang et al. [10]

Table 1 TPR at various values of accuracy score (Proposed/Irene et al. [11]/Xiang. et al. [10])

	$Q_1 = 60, Q_2 = 65$	$Q_1 = 60, Q_2 = 70$	$Q_1 = 60, Q_2 = 75$	$Q_1 = 60, Q_2 = 80$	$Q_1 = 60, Q_2 = 85$
0.1	75.14/77.02/62.71	93.03/91.53/71.37	81.73/75.71/75.14	87.57/76.46/86.06	90.15/73.05/89.59
0.2	63.47/71.37/44.63	89.83/87.76/57.44	76.65/67.80/67.42	83.99/68.93/82.49	86.80/65.99/85.87
0.3	51.22/64.78/37.66	86.82/83.99/50.47	73.63/62.90/60.45	79.66/61.96/80.23	83.83/59.29/83.27
4	38.79/56.8729.94	83.05/80.23/45.57	71.19/58.76/57.63	75.52/57.82/77.40	81.23/55.58/80.11
0.5	27.12/47.27/24.29	80.23/75.52/41.81	67.98/54.80/55.37	73.45/54.61/75.33	77.32/52.60/78.07
0.6	18.27/34.84/19.59	76.84/68.93/39.36	64.97/52.17/51.60	68.55/50.28/72.88	70.82/48.33/75.65
0.7	11.49/22.98/14.31	67.80/58.95/35.40	59.32/48.59/49.72	62.90/46.14/68.74	63.20/44.05/70.26
0.8	04.33/13.37/09.04	54.05/42.56/30.51	48.78/43.69/45.20	50.66/37.10/60.64	50.37/36.43/61.52
0.9	01.32/03.95/03.20	34.65/21.66/17.89	31.45/29.57/33.15	31.45/19.59/42.94	29.74/18.96/43.31
1	00.00/00.38/00.75	03.95/02.26/02.64	03.39/03.39/03.39	02.26/00.56/03.95	03.16/00.74/05.20

tampered pixels. UCID database consist of 1338 images, so remaining 538 images were tested using these trained classifiers (500 were used for training). For clarity Fig. 4a, b shows one of the original and its tampered version from CASIA V.2 tampered image database [18]. Since the compression quality of image was unknown and visual quality of the images was very high, we applied classifier trained with the higher quality of compression ($Q_1 = 85$, $Q_2 = 95$). Figure 4c, e, f show the localization map generated by applying an SVM classifier trained using all the three approaches. Figure 4d–f shows final localized tampered region after thresholding and selecting largest connected component.

For comparison, we computed ratio of actual tampered pixels identified as tampered (True positive-T_p) with total detected tampered pixels (All positive-A_p) and called it as *Accuracy Score* as shown in Eq. (3). We set up various threshold values for *Accuracy Score* from 0.1 to 1 and computed total number of output images giving *Accuracy Score* greater than or equal to threshold. We called these images as true positive images. For each value of the threshold, we computed ratio of true positive to total number of test images and called it as True Positive Rate (TPR) as given by Eq. (4). Table 1 shows TPR at various thresholds of Accuracy Score where performance of the proposed algorithm is better than Irene et al. [11]. Since Xiang et al. [10] uses all 9 digits probabilities it is better than our proposed algorithm at higher *Accuracy Score*. Still at $Q_1 = 60$, $Q_2 = 70$ and $Q_1 = 60$, $Q_2 = 75$, our proposed algorithm shows better localization results at higher *Accuracy Score*.

$$Accuracy\,Score = T_p / A_p \tag{3}$$

$$TPR = \frac{\sum(\text{Accuracy Score} \geq \text{Threshold})}{\text{Total no of Images}} \times 100 \tag{4}$$

5 Conclusion

We have proposed a lightweight algorithm which can automatically localize tampering present in the double compressed image. Our empirical analysis shows that FDPD feature size can be efficiently reduced by selecting first two digit probabilities. Performance analysis shows that the proposed algorithm works better than the other lightweight algorithm over a diverse range of quality factors. Since the algorithm uses 18 dimensional features, it has a major scope in prima facie tampering localization. Tampering followed with multiple compressions may destroy FDPD of tampered region, making its localization difficult. In future, we will try to make this algorithm robust against such anti forensic attacks.

References

1. Archana, V. M., Sanjay, B. D., Narendra., J. M. Prakash., D. P,: Factor Histogram based Forgery Localization in Double Compressed JPEG Images, Procedia Computer Science Journal, vol. 54(2015), pp. 690–696 (2015).
2. Zhouchen, L., Junfeng, H., Xiaoou, T., Chi-keung, T.: Fast, automatic and fine-grained tampered JPEG image detection via DCT coefficient analysis. Pattern Recognition, vol(42), pp. 2492–2501 (2009).
3. Tiziano, B., Alessandro, P.: Detection of Non-Aligned Double JPEG Compression Based on Integer Periodicity Maps. IEEE Transactions on Information Forensics and Security, vol 7(2), pp. 842–848 (2012).
4. Rafael, C. G., Richard, E. W.: Digital Image Processing", Prentice Hall, (2007).
5. Hanny, F.: Exposing digital forgeries from JPEG ghosts. IEEE Transactions on Information Forensics and Security, 2009, vol. 4(1), pp. 154–160, (2009).
6. Fan, Z., De Queiroz, R. L.: Identification of bitmap compression history: Jpeg detection and quantizer estimation. IEEE Transaction on Image Processing, vol 14(2), pp. 230–235 (2003).
7. Weiqi, L., Zhenhua, Q., Jiwu, H., Guoping, Q.: A novel method for detecting cropped and recompressed image block. IEEE International Conference on Acoustics, Speech, and Signal Processing (ICASSP '07), Vol. 2, pp. II-217-II-220 (2007).
8. Dongdong, F., Yun, Q. S., Wei, S.: A generalized Benford's law for JPEG coefficients and its applications in image forensics. SPIE Conference on Security, Steganography, and Watermarking of Multimedia Contents, (2007).
9. Bin, L., Shi, Y. Q., Jiwu, H.: Detecting doubly compressed JPEG images by using mode based first digit features. IEEE International Workshop on Multimedia Signal Processing Cairns, Queensland, Australia, pp. 730–735, (2008).
10. Xiang, H., Yu, Q. Z., Miao L., Frank, Y. S., Yun, Q. S.: Detection of the tampered region for JPEG images by using mode-based first digit features. EURASIP Journal on Advances in Signal Processing, 2012:190, (2012).
11. Irene, A., Rudy, B., Roberto C., Andrea, D. M.: Splicing Forgeries Localization through the Use of First Digit Features. proceedings of IEEE International Workshop on Information Forensics and Security (WIFS), (2014).
12. Feng, Z., Zhenhua, Y. U., Shenghong, L.: Detecting Double Compressed JPEG Images by Using Moment Features of Mode Based DCT Histograms. In International Conference on Multimedia Technology (ICMT), pp 1–4, (2010).
13. Benford,F.:The law of anomalous numbers. Proc. American Philosophical Society, vol. 78, pp. 551–572 (1938).
14. Gerald, S., Michal, S.: UCID - An Uncompressed Colour Image Database. Proc. SPIE 5307, Storage and Retrieval Methods and Applications for Multimedia 472 (2003).
15. P. Sallee, Matlab JPEG toolbox 1.4, [online], Available: http://dde.binghamton.edu/download/jpeg_toolbox.zip.
16. Thomas, G., Rainer, B.: The Dresden Image Database for benchmarking digital image forensics. Proceedings of the 2010 ACM Symposium on Applied Computing, ser. SAC '10. New York, NY, USA: ACM, 2010, pp. 1584–1590 (2010).
17. http://ufraw.sourceforge.net/Install.html.
18. CASIA Tampered Image Detection Evaluation Database http://forensics.idealtest.org:8080/index_v2.htm.

Classification of Bank Direct Marketing Data Using Subsets of Training Data

Debaditya Barman, Kamal Kumar Shaw, Anil Tudu and Nirmalya Chowdhury

Abstract Nowadays, most business organizations practice Direct Marketing. One of the promising application areas of this type of marketing practice is Banking and Financial Industry. A classification technique using subsets of training data has been proposed in this paper. We have used a real-world direct marketing campaign data for experimentation. This marketing campaign was a telemarketing campaign. The objective of our experiment is to forecast the probability of a term-deposit plan subscription. In our proposed method we have used customer segmentation process to group individual customers according to their demographic feature. We have used X-means clustering algorithm for customer segmentation process. We have extracted few appropriate collection of customers from the entire customer database using X-means cluster algorithm, on the basis of demographic feature of individual customers. We have tested our proposed method of training for classifier using three most widely used classifiers namely Naïve Bayes, Decision Tree and Support Vector Machine. It has been found that the result obtained using our proposed method for classification on the banking data is better compare to that reported in some previous work on the same data.

Keywords Direct marketing · Customer segmentation · X-means clustering

D. Barman (✉)
Department of Computer and System Sciences, Visva-Bharati, Santiniketan, India
e-mail: debadityabarman@gmail.com

K.K. Shaw
Tata Institute of Fundamental Research, Colaba, Mumbai, India
e-mail: Kamal.shaw@tifr.res.in

A. Tudu · N. Chowdhury
Department of Computer Science and Engineering, Jadavpur University, Kolkata, India
e-mail: anil.cst@gmail.com

N. Chowdhury
e-mail: nirmalya_chowdhury@yahoo.com

© Springer India 2016
S.C. Satapathy et al. (eds.), *Information Systems Design and Intelligent Applications*, Advances in Intelligent Systems and Computing 435,
DOI 10.1007/978-81-322-2757-1_16

1 Introduction

Direct Marketing is a special type of marketing technique where marketers target a specific group of consumers who may find an existing product or a new product interesting. In this type of marketing technique, the marketer tries to acquire new customers by using unsolicited direct communications. Existing as well as past customers are also approached by the marketers. During direct marketing campaign, marketers offers various exciting schemes to get customers to respond promptly.

There are three basic categories of Direct Marketing—Telemarketing, Email Direct Marketing and Direct Mail Marketing. In telemarketing a marketer tries to sell a product to consumers or inform about some new product via phone. In Email Direct Marketing, a marketer targets consumer through their email accounts. In Direct Mail Marketing a marketer sends the product brochure, product-catalogue, advertising material directly to consumers' address.

A well-performed direct marketing campaign can show approximate number clients who are interested in the product. This type of campaign certainly can deliver a positive return on investment and it is attracted to many marketers because of its directly measureable positive results. For example, if a marketer got 200 response against the 1000 solicitations that he or she has send to customers via mail during a marketing campaign, then we can say that the campaign has achieved 20 % direct responses. Direct marketing is always found to be measurement oriented. Various type of metrics (like '*Response rate*', '*Cost per lead/order*', '*ROI*', '*Churn*' etc.) are employed to measure up the performance of a marketing campaign.

One of the promising application areas of Direct Marketing practice is Banking and Financial Industry. Banks are spending a lot of money on Direct Marketing not only to acquire new customers but also to cross sell a product or service [1]. In this work, we have proposed a classification scheme using subsets of training data. We have used a dataset collected by a Portuguese bank [2]. In fact the Portuguese bank offered a term deposit to its customer, the bank then arranges a telephonic campaign on their said product to its existing and new customer. It also recorded the outcome of the campaign i.e. noting the customer who subscribed the term-deposit and those who did not.

The Portuguese bank created a database that contains a large number of attributes for each customer collected by telephonic calling. They had used their own contact-center to do the marketing campaigns. Mode of Direct Marketing is mostly telephonic. The campaign was conducted between May 2008 and November 2010. Consumers are offered a long term deposit plan. Various number of attributes of this telephonic conversion and an attribute (target attribute) denoting the results (subscribed to the term deposit plan or not) of such a campaign. There are sixteen attributes in the data set. These attributes can be grouped into two types: client demographic data and campaign related data. In the client demographic data, there are eight attributes, namely *age, job, marital status, education, credit status, bank balance, personal loan* and *housing loan*. In the campaign related data, there are eight attributes, namely *contact type, last communication day of the month, last communication month of the year, duration of the last communication with the client during previous marketing campaign, number of times client was contacted in the present marketing campaign, number of days passed by after last contact with the client, number of times client was contacted before present marketing campaign* and *outcome of the previous marketing campaign*. Descriptions and values of these attributes can be found in UCI Machine Learning Repository (http://mlr.cs.umass.edu/ml/datasets/Bank+Marketing). We have described the process of data collection and data pre-processing in Sect. 5.

As stated above our proposed classification scheme is based on using the subsets of training data. Since our database contains the customers of Portuguese bank, using subsets of training data implies segmentation of customers. By definition *Customer Segmentation* (CS) is "…the process of finding homogenous sub-groups within a heterogeneous aggregate market". It is also known as *Market Segmentation* (MS). Both CS and MS are used in direct marketing to identify potential consumer group. Elrod and Winer [3], defined MS as the technique to get a number of similar groups from a market. The principles of MS are very well-established in marketing theory and an important component of marketing strategy [4, 5]. A market can be segmented using four fundamental segmentation techniques: geographic, demographic, psychographic and behavioural [6]. In geographical segmentation we deal with attributes like—size of the country, weather, number of city, number of villages, population etc. The attributes in demographic segmentation are—age, sex, job-profile, education, salary, nationality, marital status etc. Psychographic segmentation deals with attributes like standard of living, position in the social hierarchy, food-habit etc. Finally, behavioural segmentation deals with attributes like purchase pattern, attitude towards newly launch product, rate of usage [3, 7] etc. In this work, we have used customer segmentation process to group individual customers according to their demographic information.

Note that Clustering techniques are not only employed to search for homogeneous groups of consumers [8] but also used in data driven segmentation [9]. In this work, we have used *X*-means clustering algorithm for customer segmentation purpose. We have discussed *X*-means algorithm in Sect. 3. Note that, only client demographic information has been used as an input to the clustering algorithm as a result we can obtain clusters of similar type of customers. Application of *X*-means algorithm on Portuguese banking data set results in four clusters of customers. In this way the entire customer database has been divided into four groups. We have applied our proposed method of classification using subsets of training data to forecast the probability of the term-deposit subscription. Three most widely used classifiers namely Naïve Bayes (NB), Decision Tree (DT) and Support Vector Machine (SVM) are used for experimentation with our proposed technique of training.

To measure the accuracy of our classifier we have used the area under the ROC (Receiver operating characteristics) curve or ROC-area or AUC because sometimes classification accuracy measured in simple percentage can be misleading. In general the best classifier will return an area of 1.0 and a random classifier or a bad classifier will return an area of 0.5.

Next section presents the description of the problem. In Sect. 3 we have discussed X-means algorithm. Our proposed method has been discussed in Sect. 4. We have provided experimental results in Sect. 4.1. Concluding comments has been incorporated in Sect. 5.

2 Statement of the Problem

Effects of mass marketing campaigns on population are continuously decreasing. Nowadays, Marketing managers prefer directed campaigns [2] with small number of carefully selected contacts. This type of campaigns can be improved using classification techniques.

To increase financial assets Banks are offering different long-term deposit plans with decent interest rates. Most of the times the Banker prefer direct marketing techniques to reach out to the clients. Thus, there is a need for efficient methodologies, so that in spite of contacting lesser number of customers we should have significantly high rate of success (clients subscribing the deposit).

We have performed our experiments on a real-world marketing campaign data from a Portuguese Banking Institution. Moro et al. [2] had preprocessed this data set and used it to forecast weather a client would subscribe to the term deposit plan. They had used various classification algorithms (NB, SVM and DT) to do the prediction task.

As we have discussed in the earlier section, there are sixteen attributes in the data set. These attributes can be grouped into two types: client demographic data and campaign related data. We have used this client demographic data to do the customer segmentation process to group the customers who are similar to each other. The whole concept behind the customer segmentation process has been explained in the previous section. Clustering techniques are very popular technique which successfully been applied to do the segmentation process. We have discussed about it in earlier section of this paper. By applying X-means clustering technique we have partitioned the whole database into appropriate groups. Each such group contains customer having similar features. Then several classifiers have been trained using these groups of data. The intention behind this approach is to train the classifiers separately for each group of customers so that they can provide better results.

In this paper, our focus is on applying our proposed classification technique on bank direct marketing dataset to help different financial institutions like banks, investment companies to determine product subscription analysis of customers. These organizations can be able to contact small number of consumers with greater probability of subscription by deploying our proposed technique on their large consumer databases.

3 Description of *X*-Means Algorithm

Most of the times, conventional clustering algorithms like *K*-means uses hit-and-miss type of approaches to determine the 'correct' number of clusters in a particular dataset. These approaches are very time and cost consuming in terms of determining what number of cluster constitutes 'correct' clustering [10]. *X*-Means [11] clustering algorithm is an extension of the very well-known *K*-means clustering algorithm. Unlike *K*-means, *X*-means do not need to know the number of Clusters (*K*) a priory [5]. *X*-means estimates the number of clusters very efficiently. This algorithm uses a heuristic to deliver the accurate number of centroids or clusters. It starts with a small set of centroids then gradually exploit if using more number of centroid deliver good performance.

In *X*-Means algorithm, a user need to specify the upper bound (K_{upper}) and the lower bound (K_{lower}) of the number of cluster (*K*). This algorithm uses a model selection criterion (i.e. BIC (Bayesian information criterion)) to determine the true *K* [12]. This algorithm starts with setting the value of *K* equal to lower bound specified by user and continues to add centroids as needed, until upper bound is obtained [13]. In our experiment the lower bound (K_{lower}) is one and the upper bound (Maximum number of cluster, K_{upper}) is 151. We have calculated the upper bound using the following formula suggested by Mardia et al. [14].

$$K_{upper} \approx \sqrt{\frac{n}{2}}, \textit{where n is the number of data points} \tag{1}$$

Please note that in our experiment $n = 45{,}211$. During this process, the set of centroids that achieves the best BIC score is recorded, and is treated as the output of the algorithm. We have used the BIC formula suggested by Pelleg and Moore [11] in our experiment.

4 Proposed Method

There exists numerous instances in the literature [15–17] that we can improve the out-come of a marketing campaigns by using various Classification algorithms with different capabilities. A forecasting model capable of labeling a data item into one of several predefined classes (e.g. in our case it is "yes" and "no") can be built using these classification algorithms.

In our experiments, we have decided to include these classifiers: decision tree, support vector machine, naïve bayes. This choice of classifiers was motivated by some preliminary research and methods used by the authors of a previous work on this dataset [2]. In this experiment Moro et al. [2] used the NB, DT and SVM classifiers on the dataset and software they used was *rminer* library for *R* tool. For our experiment we have used standard implementations of Support Vector Machine

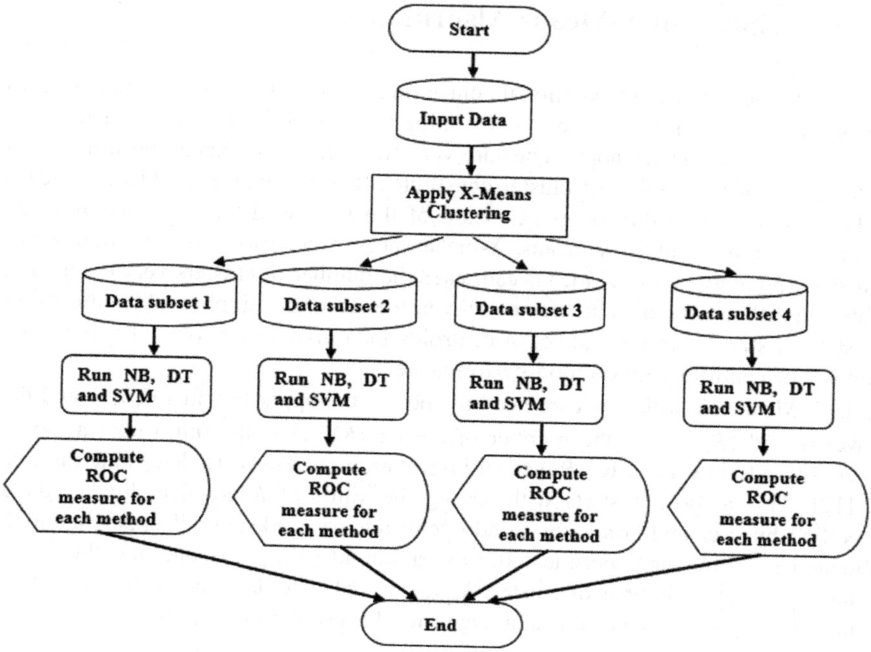

Fig. 1 Design of the classification system

(SVM) in rattle package of *R* tool. We have also used *weka* data mining software suite for NB and DT (C4.5) algorithms.

In our experiment, we have used customer segmentation process to group similar customers based on their demographic data. We have used X-means clustering algorithm for customer segmentation process. The whole customer database has been divided into four groups. Each such group further divided into 2:1 ratio for training and testing purpose respectively of the classifier models. For the prediction task we have applied several classifiers namely Naïve Bayes, DT and SVM on these four groups separately. We have used AUC to measure our proposed classifier's performance. We have compared our prediction results with that of a previous work on the same dataset (Fig. 1).

4.1 Experimental Result

As mentioned above, the dataset used for experimentation in this paper was collected by the authors of [2] from a Portuguese bank. It contains results of direct bank marketing campaigns. Specifically it includes seventeen campaigns of a Portuguese bank conducted between May 2008 and November 2010. These were

Table 1 Experiment results

Classification algorithm	Data set (subset number)	No. of runs of the classifier	AUC	Average AUC	AUC achieved in previous work [2]
Naïve Bayes	S 1	R-1	0.8920	0.8644	0.8700
		R-2	0.8927		
		R-3	0.8925		
	S 2	R-1	0.8706		
		R-2	0.8708		
		R-3	0.8703		
	S 3	R-1	0.8454		
		R-2	0.8459		
		R-3	0.8456		
	S 4	R-1	0.8487		
		R-2	0.8485		
		R-3	0.8488		
Decision tree (C4.5)	S 1	R-1	0.9001	0.9137	0.8680
		R-2	0.9004		
		R-3	0.9008		
	S 2	R-1	0.9154		
		R-2	0.9150		
		R-3	0.9156		
	S 3	R-1	0.9087		
		R-2	0.9086		
		R-3	0.9088		
	S 4	R-1	0.9301		
		R-2	0.9304		
		R-3	0.9309		
Support vector machine	S 1	R-1	0.9622	0.9456	0.9380
		R-2	0.9620		
		R-3	0.9625		
	S 2	R-1	0.9465		
		R-2	0.9465		
		R-3	0.9465		
	S 3	R-1	0.9374		
		R-2	0.9377		
		R-3	0.9374		
	S 4	R-1	0.9361		
		R-2	0.9360		
		R-3	0.9364		

always conducted over the phone and the customer was offered a long term-deposit scheme. Originally the dataset contained 79,354 contacts, out of which contacts with missing data or inconclusive results being discarded leading to a dataset with 45,211 instances with two possible outcomes—either the client signed for the long-term deposit or not. [2] obtain their best prediction results using twenty nine attributes. But the dataset) they had made public for research purposes have only sixteen attributes. So we have assumed that these sixteen attributes are important ones among the twenty nine attributes.

Among these sixteen attributes or input variables there are eight variables related to clients' personal information and eight variables related to information collected by telephonic campaign.

Table 1 presents the experimental results obtained by application of our proposed method of training with several classifiers on the Portuguese bank dataset. A comparison of the results obtained by our method and that of a previous work on the same data can also be found in Table 1. The entire dataset is initially divided into four groups using the *X*-means algorithm as stated above. For each run of the classifiers, each such group is again randomly divided in three parts, two of which is used for training and the third one is used for testing purpose. Each classifier is run three times independently for each such group and the results are given in fourth column in Table 1.

As stated above we have used AUC to measure the success rates of the classifiers using our proposed training method. The average of the results obtained for all the four groups for each classifier are shown in column five of Table 1. The results obtained by a previous work [2] on the same dataset using are shown in column six of Table 1. It can be seen that although the result using NB classifier is slightly poor than that of the previous work, but the results with other two classifiers using our proposed method of training are found to be better. It can also be noted that, in all the runs of NB classifier on Subset 1 and Subset 2, are better than that of the previous work [2]. We have used *rattle* package for SVM and *weka* package for DT (C4.5) and NB for building and analysis of the classifier model.

5 Conclusion

In this paper, a novel classifier training technique based on customer segmentation process has been proposed. We have applied our proposed method on a client database of a Portuguese banking institution. We have done customer segmentation for this database using client demographic features in an attempt to increase the performance of the classifiers. This technique can be useful to develop policies and to identify more customers with similar characteristics.

It can be concluded from the experimental results presented in Table 1 that overall performance of our proposed (second to the last column of Table 1) method is better compare to that of a previous work (last column of Table 1). Thus it seems that training with subsets of customers having similar features has improved the success rate of classifiers. Scope for further research work may include searching for more appropriate clustering algorithm for segmenting customer database.

References

1. Todé, C.: Banks increase direct marketing to customers, http://www.dmnews.com/banks-increase-direct-marketing-to-customers/article/129580/(2009) (Accessed on 26th December 2013).
2. Moro, S., Laureano, R., Cortez, P.: Using data mining for bank direct marketing: An application of the crisp-dm methodology, Proceedings of the European Simulation and Modelling Conference - ESM'2011, pp. 117–121 (2011).
3. Elrod, T., Winer, R. S.: An empirical evaluation of aggregation approaches for developing market segments, The Journal of Marketing, 65–74 (1982).
4. Pearl, J.: Probabilistic reasoning in intelligent systems: networks of plausible inference, Morgan Kaufmann (1988).
5. Haralick, R. M.: The table look-up rule. Communications in Statistics-Theory and Methods, 512, pp. 1163–1191 (1976).
6. Kanti, T.: Market Segmentation and Customer Focus Strategies and Their Contribution towards Effective Value Chain Management. International Journal of Marketing Studies, 43 (2012).
7. Kulkarni, A. V. Kanal, L. N.: An optimization approach to hierarchical classifier design, Proceedings of 3rd International Joint Conference on Pattern Recognition (1976).
8. Myers, J. H., Tauber, E.: Market structure analysis, Marketing Classics Press (2011).
9. Wedel, M.: Market segmentation: Conceptual and methodological foundations, Springer (2000).
10. Pham, D. T., Dimov, S. S., Nguyen, C. D.: Selection of K in K-means clustering. Proceedings of the Institution of Mechanical Engineers, Part C: Journal of Mechanical Engineering Science, 2191, pp. 103–119 (2005).
11. Pelleg, D., Moore, A. W.: X-means: Extending K-means with Efficient Estimation of the Number of Clusters. In ICML, pp. 727–734 (2000).
12. Koonsanit, K., Jaruskulchai, C., & Eiumnoh, A.: Parameter-free K-means clustering algorithm for satellite imagery application. In Proceedings of Information Science and Applications (ICISA-2012 International Conference on, pp. 1–6, IEEE, (2012).
13. Corral, G., Garcia-Piquer, A., Orriols-Puig, A., Fornells, A., & Golobardes, E.. Analysis of vulnerability assessment results based on CAOS. Applied Soft Computing, Vol. 11 no. 7, pp. 4321–4331, Elsevier, (2011).
14. Mardia, K. V., Kent, J. T., Bibby, J. M.: Multivariate analysis. Academic press, (1979).
15. Hu, X.: A data mining approach for retailing bank customer attrition analysis, Applied Intelligence, 221, pp. 47–60 (2005).
16. Li, W., Wu, X., Sun, Y., Zhang, Q.: Credit card customer segmentation and target marketing based on data mining. In Computational Intelligence and Security CIS, 2010 International Conference on, pp. 73–76. IEEE (2010).
17. Ling, C. X., Li, C.: Data Mining for Direct Marketing: Problems and Solutions. In KDD Vol. 98, pp. 3–79 (1998).

Offline Writer Identification and Verification—A State-of-the-Art

Chayan Halder, Sk.Md. Obaidullah and Kaushik Roy

Abstract In forensic science different unique bio-metric information of humans are being used to analyses forensic evidence like finger print, signature, retina scan etc. The same can be used applied on handwriting analysis. The Automatic Writer Identification and Verification (AWIV) is a study which combines forensic analysis field and computer vision and pattern recognition field. This paper presents a survey of literature on the offline handwritten writer identification/verification with the type of data, features and classification approaches attempted till date in different languages and scripts. The analysis of the approaches has been described for further enhancement and adaptation of these techniques in different languages and scripts.

Keywords Writer identification · Writer verification · Handwriting forensic · Handwriting features · Classifiers

1 Introduction

Handwritten documents have been playing a long role in day-to-day life and also in forensic. There are systems to automatically identify/verify writers in non-Indic scripts mostly in Roman, but same is not available for Indic scripts. Mostly, the analysis has been done manually by experts but authenticity of the analysis is not

---

C. Halder (✉) · K. Roy
Department of Computer Science, West Bengal State University,
Kolkata 126, WB, India
e-mail: chayan.halderz@gmail.com

K. Roy
e-mail: kaushik.mrg@gmail.com

Sk.Md. Obaidullah
Department of Computer Science & Engineering, Aliah University,
Kolkata, WB, India
e-mail: sk.obaidullah@gmail.com

© Springer India 2016
S.C. Satapathy et al. (eds.), *Information Systems Design and Intelligent Applications*, Advances in Intelligent Systems and Computing 435,
DOI 10.1007/978-81-322-2757-1_17

conclusive. So, an automated system to identify/verify writer with high confidence is very important for Indic scripts.

Currently due to the rapid development of technology and its applications in wide areas the identification of writer through handwriting is highly essential. AWIVS has such potential that it can be utilized in copy rights management, financial sector, forensic science, access control, graphology, historical document analysis [1] etc. Due to these opportunities, the number of researchers involved in this area are increasing in current time [2–12]. AWIVS has its challenges of handwriting variations and nature of scripts. The AWIVS can be both online and offline depending on the input method of writing. The online Writer Identification and Verification needs special equipment like tablet, stylus or digital pen for writing. The offline methods involve handwritten characters and text using conventional pen and paper. The collected sample documents have been digitized using scanners or cameras with different resolution according to the need of the work. Both methods can be text-dependent or text-independent regarding the input database.

In this paper the emphasis is on the survey of different offline writer identification and verification techniques. The work on offline writer identification/verification have been carried out mainly in non-Indic scripts like Roman, Chinese and Arabic [2–8, 13–20]. So it is certain that most databases are on those scripts. Although there are some bits and pieces of works on Indic scripts like Devanagari, Oriya, Telugu, Kannada and Bangla etc. [9–12, 21–26] but there is no standard database for these scripts. So creation of standard databases on different Indic scripts consisting of writer information is exigent. As the interest is more towards writer identification/verification on Indic scripts, so in the following survey more emphasis has been given towards the works of Indic scripts and only most significant works of non-Indic scripts have been presented. Here firstly, details of the database, then applied features and finally classification and results for all approaches are described.

2 Offline Writer Identification/Verification on non-Indic Scripts

Among writer identification/verification on non-Indic scripts, Roman, Chinese and Arabic got major attention. In this section only the most recent and significant works of Schomaker et al. [2, 19], Djeddi et al. [4, 5], Jain and Doermann [6], Srihari et al. [18] etc. are described.

2.1 *Writer Identification*

Dependent methods Al-Maadeed et al. [13] have presented writer identification method in Arabic script using 32,000 text from 100 people. The features used include some edge-based directional features such as height, area, length and three edge-direction distributions with different sizes. They have used 75 % data for training and 25 % for testing on WED classifier to achieve 90 % accuracy for top-10 choice.

Independent methods Schomaker et al. in [2] have used a total of four data sets, two of which are medieval handwriting datasets namely Dutch charter dataset and the English diverse dataset; the other two databases of contemporary handwritings are IAM and Firemaker databases. Siddiqi and Vincent [14] have used IAM and RIMES databases. Fiel and Sablatnig [3] have worked on Roman script using IAM database and TrigraphSlant dataset for both writer retrieval and writer identification. Djeddi et al. have worked on IFN/ENIT database and GRDS database in [4] and on ICFHR 2012 Latin/Greek database in [5]. Jain and Doermann in [6] worked with the IAM, ICFHR 2012 Writer Id dataset, DARPA MADCAT database. In [7] Ding et al. proposed multi script method of writer identification on Chinese, Greek and English language on database of 45 documents from 15 writers and ICDAR 2011 writer identification database. He et al. [15] have proposed their scheme on a database of Chinese script containing 1000 documents written by 500 writers. Each sample has 64 Chinese characters. Yuan et al. [16] have presented writer identification on Chinese script on a database of 200 Chinese handwriting text collected from 100 writers. Ram and Moghaddam [17] have worked on PD100 dataset. More work on in this respect can be found in [8].

In [2] Schomaker et al. have used directional ink trace based Quill feature. They have also tried a complex variant named Quill-Hinge feature. Siddiqi and Vincent in [14] have calculated seven simple features namely: chain code histogram (f1), 1st order differential chain code histogram (f2), 2nd order differential chain code histogram (f3), curvature index histogram (f4), local stroke direction histogram (f5), f2 computed locally and f3 computed locally. Combinations of different features among these seven features are also applied by them. The Scale Invariant Feature Transform has been used by Fiel and Sablatnig [3]. Djeddi et al. [4, 5] have computed probability distributions of gray level run-length, edge-hinge features for both [4, 5] and also edge-direction, combination of codebook, visual features, chain code, polygonized representation of contours, AR coefficient feature for [5]. Jain and Doermann in [6] have extracted Connected Component, Vertical cut, Seam cut and Contour gradient based features. The Local Contour Distribution Features using sliding windows has been applied by Ding et al. [7]. He et al. [15], have used Gabor filter based features. A spectral feature extraction method based on Fast Fourier Transformation has been applied in [16] by Yuan et al. In another approach, Ram and Moghaddam [17] have used grapheme based features. The nearest-neighbour (instance-based) classifier has been used by Schomaker et al. in [2] and for Quill-Hinge feature they got 97 % for top-1 and 98 % for top-10 choices. Siddiqi et al. have computed distance measures like: Euclidean Distance, Bhattacharyya and Hamming distance etc. in [14] and achieved accuracies of 77 and 93 % for top-1 and top-10 choices respectively in case of IAM database and 75 and 95 % for top-1 and top-10 choices respectively in case of RIMES database. By combining features f3, f4, f5, f6 together of the seven features they have achieved highest accuracies of 86 and 97 % for top-1 and top-10 choice respectively for IAM datasets and 79 % (top-1) and 93 % (top-10) in case of RIMES dataset. In [3], Fiel and Sablatnig have used K-means clustering for writer retrieval and K-NN classifier for writer identification and achieved highest accuracy of 98.9 % for both writer retrieval and writer identification on IAM database, TrigraphSlant dataset. The Similarity measure using Manhattan distance has been used by Djeddi et al. in [4] and in [5] the K-NN and SVM classifiers have been used in their work. In [4], highest accuracy of 91.5 % has been achieved on IFN/ENIT database for combination of features and on GRDS database highest accuracy of 99.5 % has been obtained for edge hinge feature and in case of combination of features and two databases 92.4 % accuracy has been achieved. The highest accuracy of 92.06 % for only Greek, 83.33 % for only English and 73.41 % for Greek+English has been found in [5]. Jain and Doermann in [6] have used K-NN classifier to get highest accuracy of 96.5, 98 % for English on IAM and ICFHR datasets respectively and 97.5 and 87.5 % for Greek and Arabic respectively. Ding et al. [7] have applied the Similarity measure using weighted Manhattan distance to get 80 and 94.2 % accuracy for their and ICDAR database respectively. A Hidden Markov Tree (HMT) in wavelet domain has been used by He et al. [15]. The top-1, top-15 and top-30 results that have been achieved are 40, 82.4 and 100 % respectively. Yuan et al. [16] got 98 % accuracy for top-10 and 64 % for top-1 using the Euclidean and WED classifiers. Though it has higher computation cost this scheme has the advantage of stable feature and reduced the randomness in Chinese character. In [17], Ram and Moghaddam have used fuzzy clustering method and got about 90 % accuracy in average on 50 people that have been selected randomly from the database.

2.2 *Writer Verification*

Dependent methods In [18], Srihari et al. proposed a statistical model for writer verification using the CEDAR database. Srihari et al. used the Macro and Micro features and focused on extraction of characteristics from the documents along with likelihoods for two classes assuming statistical independence of the distances, etc. They have achieved an accuracy of 94.6 % for same writers and 97.6 % for different writers.

2.3 *Combined (Identification+Verification)*

Independent methods Schomaker et al. [19] have used three different databases for their work namely the IAM; Firemaker and ImUnipen database. They have also combined the IAM and Firemaker databases to form a database namely Large with 900 writers. In another work [20] they have worked on IFN/ENIT dataset of Arabic script.

Various features based on Probability Distribution Functions, grapheme emission, run-length etc. used by Schomaker et al. [19]. In [20] they have also combined some textural and allographic features for Arabic script. After extracting textural features probability distribution function has been generated. For the allographic features, a codebook of 400 allographs has been generated from the handwritings of 61 writers and the similarities of these allographs have been used as another feature.

The X^2, Euclidean and Hamming distance measures are used for writer identification and verification by Bulacu and Schomaker [19]. The performance of Grapheme Emission PDF feature has been best among all features producing 94 % accuracy for identification/verification. The combination of features contour-hinge PDF, Grapheme Emission PDF, run-length on background PDFs produced writer identification accuracies of 92 % for top-1 and 98 % for top-10 choices. In [20] the nearest neighbourhood classifier has been used by Schomaker et al. The best accuracies that have been seen in experiments are 88 % in top-1 and 99 % in top-10. In the following Table 1, a summarized view of all the non-Indic script writer identification/verification techniques are given.

3 Offline Writer Identification and Verification Methods on Indic Scripts

Indic scripts like Devanagari, Bangla, Oriya, Telugu and Kannada etc. are very popular though there exist very few works on these. Some works of Halder et al. [21] on Devanagari script; Pal et al. on Oriya script [9]; Chanda et al. on Telugu script [22]; Hangarge et al. on Kannada script [23]; Garain and Paquet [24], Pal et al. [25], Biswas and Das [10] and Halder and Roy [11, 12] on Bangla script are among the few works available in literature mainly on identification. Till now according to our knowledge there is no method in literature on offline writer verification considering Indic scripts. Only Gupta and Namboodiri [26] proposed a online writer verification system using boosting method on Devanagari script.

3.1 *Writer Identification*

Dependent methods Halder et al. [21] have used 250 documents from 50 writers for writer identification on isolated Devanagari characters. Halder and Roy in [11] used only isolated Bangla numerals for writer identification on 450 documents from 90 writers. In [12] they worked on all isolated Bangla characters (alphabets+numerals +vowel modifiers) from same set of writers like [11].

In [21], Halder et al. have used 64 dimensional directional chain code feature. In [11, 12] Halder and Roy have used 400 and 64 dimensional features.

Halder et al. [21] have used LIBLINEAR and LIBSVM classifiers and obtained highest individuality of 61.98 % for characters JA ("ज") and identification accuracy of 99.12 %. Halder and Roy in [11, 12] have used LIBLINEAR and MLP

Table 1 A summarised table for non-Indic writer identification/verification methods

Citation	Database	Language	Features	Classification techniques	Results	Comments
Al-Ma'adeed et al. [25]	100 writers	Arabic	Height area, length and edge direction distribution	WED classifier	90 %	Small dataset and the drawback is text-dependency
Schomaker et al. [9, 19, 20]	250 writers (500 documents), IAM, firemaker, unipen, 1FN/ENIT	Dutch, English, Arabic	Edge based directional PDFs (textural and allograph prototype approach), Quill, Quill-Hinge	K-NN, feed forward neural network, distance measures (linear, Hamming etc.), SVM	Identification: 87 % [9], 97 % [19], 88 % [20]; Verification: 97.4 % [9], 94.4 % [20]	Different datasets used. Also tried a unique quill based feature. In Arabic achieved promising results
Siddiqi et al. [24]	IAM, RIMES	English, French	Writer specific local features	Bayesian classifier, Euclidean, Bhattacharyya and Hamming distance	86 %	Writer specific features achieved reliable accuracies
Fiel and Sablatnig [3]	IAM, TrigraphSlant	English, Dutch	Scale invariant feature transform (SIFT)	K-means clustering, K-NN classifier	Retrieval: IAM: 93.1 %, TrigraphSlant: 98.9 %; Identification: IAM: 90.8 %, TrigraphSlant: 98.9 %	Used different databases, the results are promising
Djeddi et al. [4, 5]	IFN/ENIT and GRDS [4]; ICFHR 2012 Latin/Greek database [5]	Arabic [4], German [4], English [4, 5], French [4], and Greek [5]	Probability distributions and edge-hinge features [4, 5]; edge-direction, combination of codebook and visual features, autoregressive (AR) coefficients [5]	Manhattan distance [4]; K-NN, SVM [5]	IFN/ENIT; Run lengths: 83.5 %, Edge hinge: 89.2 %, Combination: 91.5 %; GRDS: Run lengths: 97.6 %, Edge hinge; 99.5 %, Combination: 98.6 %; IFN/ENIT+GRDS: Run lengths: 85.3 %, Edge hinge: 90.5 %, Combination; 92,4 % [4]; Greek: SVM: 92.06 %, KNN: 92.06 %; English: SVM: 83.33 %,KNN: 80.95 %; Greek+English: SVM: 73.02 %, KNN: 73.41 % [5]	Multi script writer identification and the run-length feature gives better result from other in case of [5]

(continued)

Table 1 (continued)

Citation	Database	Language	Features	Classification techniques	Results	Comments
Jain and Doermann [6]	IAM, ICFHR 2012, Arabic-MADCAT	English, Greek, Arabic	Connected component, vertical cut, seam cut, contour gradient features	K-NN classifier	Identification: IAM: 96.5 %; ICFHR 2012 English: 98 %, Greek: 97.5 %; MADCAT: 87.5 %	Out of six different features vertical cut and Seam cut are most promising
Ding et al. [7]	15 Writers(45 documents), ICDAR 2011	Chinese, Greek, English	Local contour distribution features (LCDF)	Weighted Manhattan distance	80.0 %; ICDAR 2011: 94.2 %	Very small dataset used for experiment on their database
He et al. [15]	500 writers (1000 documents)	Chinese	Gabor filter, General Gaussian model (GGD)	Hidden Markov tree (HMT), wavelet TRANSFORM	40 %	Used a large database but results arc low
Ram et al. [17]	50 writers selected from PD100	Persian	Grapheme based, gradient based	Fuzzy clustering, MLP	90 %	Small database, got promising results
Srihari et al. [20]	1568 writers (4704 documents)	English	Different macro and micro features	LLR and CDFs	94.6 % (same writers), 97.6 % (different writers)	Used a very large database, lot freely available and got promising results

classifiers. For [11] numeral 5 proved to be most individual with accuracy of 43.41 % and highest writer identification accuracy of 97.07 % has been achieved. In case of [12] highest individuality of 55.85 % for the character GA ("গ") and writer identification accuracy of 99.75 % have been achieved.

Independent methods Pal et al. in [9] have proposed an offline text independent Writer Identification procedure on Oriya script. Their dataset consists of two sets of handwriting from each of 100 writers. One set has been used for training and other set for testing. Both set contains different text with varied number of words, from each writer. Hangarge et al. [23] in their offline text independent writer identification on Kannada language have used a database consists of 400 images collected from 20 writers. Garain and Paquet in [24] have used the RIMES database and ISI database for multi script writer identification namely Roman and Bangla. Pal et al. [25] have worked on the database of 208 documents from 104 writers for text independent writer identification using the Bangla characters. In the work of Das et al. [10] on text independent writer identification of Bangla handwriting, they have used their own BESUS Database. Pal et al. [9] have used the directional chain-code and curvature feature for their work. The curvature feature has been calculated using bi-quadratic interpolation method. Four different kinds of features extraction techniques like Discrete Cosine Transform (DCT), Gabor Filtering, Gabor-Energy and Gray Scale Co-Occurrence Matrices (GSCM) have been used in the work of Hangarge et al. [23]. The two dimensional autoregressive (AR) coefficients has been used by Garain and Paquet [24]. Pal et al. [25] have used 64 dimensional directional feature and 400 dimensional gradient features. Das et al. [10] have fragmented the writing into two different techniques and then used the Radon transform projection profile to extract the feature.

SVM classifier has been used for the work of Pal et al. [9] and they have achieved an accuracy of 94 % on writer identification. Hangarge et al. [23] have utilized the K-NN classifier for their work and they have achieved an accuracy of 77.0 % for DCT, 88.5 % for Gabor Energy and 79.5 % for GLCM. They have presented that the Gabor Filtering has more potential in this respect for Kannada script than DCT and GLCM. Garain and Paquet in [24] have used Euclidean distance measure to get highest 97 % and 100 % for top-10 choice in RIMES and ISI database respectively. For both the two database combined they got 95 % for top-10 choice. With the use of SVM classifier Pal et al. [25] have achieved an accuracy of 95.19 % for top-1 choice in case of 400 dimensional gradient feature. The accuracy of 99.03 % has been achieved by their work when top-3 choices and 400 dimensional gradient feature has been considered. The Euclidean distance has been calculated on the features by Das et al. [10]. When the two sets have been combined together they have achieved 83.63 % accuracy for top-1 choice in case of test script containing 90–110 words. For the same factors with top-3 choice the result has been improved to 92.72 %. But when the words are 20–30 in case of combined fragmented test sets the results became 61.8 and 80 % for the top-1 and top-3 choices respectively. In the following Table 2, a summarized view of all the Indic script writer identification/verification techniques is given.

Table 2 A summarised table for indic writer identification methods

Citation	Database	Language	Features	Classification techniques	Results	Comments
Halder et al. [11, 12, 21]	50 Writers (250 documents) [21], 90 writers (450 documents) [11, 12]	Hindi, Bangla	64 dimensional [11, 12, 21], 400 dimensional [11, 12]	LIBLINEAR [11, 12, 21], LIBSVM [21], MLP [11, 12]	Individuality: 61.98 % [21], 43.41 % [11], 55.85 % [12], Identification: 99.12 % [21], 97.07 % [11], 99.75 % [12]	Got promising results with small database, used isolated characters
Pal et al. [9, 25]	100 writers (200 documents) [9], 104 writers (208 documents) [25]	Oriya, Bangia	64,400 dimensional features	SVM classifier	94 % [9], 95.19 % [25]	Got promising results with small database
Hangarge et al. [23]	20 writers (400 documents)	Kannada	DCT, Gabor Filter, Gabor-Energy and GLCM	K-NN classifier	DCT: 77.0 %, Gabor energy: 88.5 %, GLCM: 79.5 %	Gabor filtering has more potential in this respect for Kannada script than DCT and GLCM
Das et al. [10]	BESUS database	Bangia	Radon transform	Euclidean distance	83.63 %	Promising result, small database, different type of feature

4 Conclusion and Future Scopes

The paper presents a survey on the writer identification and verification procedures available in literature. From this study it is evident that script-dependent features perform better than the script-independent features. It can also be considered that recently the trend of research in this area is moving towards multi scripts and non-Roman scripts. The lack of standard database on Indic scripts like Bangla is one of the prime concerns in this respect. In current situation standard databases with sufficient number of documents and writers having proper preservation of inter writer and intra writer variations on Indic scripts can be a solution for database related problem. For other difficulties like text, word and character segmentation methods, researchers need to develop methods with conscious and collective approaches which will bring the success. It can be concluded that there is a pressing need to work with Indic scripts and languages on the above area.

Acknowledgments One of the author would like to thank Department of Science and Technology (DST) for support in the form of INSPIRE fellowship.

References

1. Fornes, A. et al.: Writer identification in old handwritten music scores. In: Proceedings of 8th IAPR workshop on DAS, pp. 347–353 (2008).
2. Brink, A. A., et al.: Writer identification using directional ink-trace width measurements. IEEE Trans. Pattern Recogn. **45**(1), 162–171 (2012).
3. Fiel, S., Sablatnig, R.: Writer retrieval and writer identification using local features. In: Proceedings of 10th IAPR workshop on DAS, pp. 145–149 (2012).
4. Djeddi, C., et al.: Multi-script writer identification optimized with retrieval mechanism. In: Proceedings. of ICFHR, pp. 509–514 (2012).
5. Djeddi, C., et al.: Text-independent writer recognition using multi-script handwritten texts. Pattern Recogn. Lett. **34**, 1196–1202 (2013).
6. Jain, R., Doermann, D.: Writer identification using an alphabet of contour gradient descriptors. In: Proceedings of 12th ICDAR, pp. 550–554 (2013).
7. Ding, H., et al.: Local contour features for writer identification. Advanced Sc. Tech. Lett. **28**, 66–71 (2013).
8. Ahmed, A. A., Sulong, G.: Arabic writer identification: a review of literature. J. Theo. Appl. Info. Tech., **69**(3), 474–484 (2014).
9. Chanda, S., et al.: Text independent writer identification for Oriya script. In: Proceedings of 10th IAPR workshop on DAS, pp. 369–373 (2012).
10. Biswas, S., Das, A. K.: Writer identification of Bangla handwritings by radon transform projection profile. In: Proceedings of 10th IAPR workshop on DAS, pp. 215–219 (2012).
11. Halder, C., Roy, K.: Individuality of isolated Bangla numerals. J. Netw. Innov. Comput. **1**, 33–42 (2013).
12. Halder, C., Roy, K.: Individuality of isolated Bangla characters. In: Proceedings of ICDCCom, pp. 1–6 (2014).
13. Al-Maadeed, et al.: Writer identification using edge-based directional probability distribution features for Arabic words. In: Proceedings of AICCSA, pp. 582–590 (2008).

14. Siddiqi, I., Vincent, N.: A set of chain code based features for writer recognition. In: Proceedings of 10th ICDAR, pp. 981–985 (2009).
15. He, Z., et al.: Writer identification of Chinese handwriting documents using hidden Markov tree model. IEEE Trans. Pattern Recogn. **41**(4), 1295–1307 (2008).
16. Yan, Y., et al.: Chinese handwriting identification based on stable spectral feature of texture images. Int. J. Intel. Engg. Sys. **2**(1), 17–22 (2009).
17. Ram, S. S., Moghaddam, M. E.: Text-independent Persian writer identification using fuzzy clustering approach. In: Proceedings of ICIME, pp. 728–731 (2009).
18. Srihari, S. N., et al.: A statistical model for writer verification. In: Proceedings of 8th ICDAR, vol. 2, pp. 1105–1109 (2005).
19. Bulacu, M., Schomaker, L.: Text-independent writer identification and verification using textural and allographic features. IEEE Trans. PAMI. **29**(4), 701–717 (2007).
20. Bulacu, M., et al.: Text-independent writer identification and verification on offline Arabic handwriting. In: Proceedings of 9th ICDAR, pp. 769–773 (2007).
21. Halder, C., et al.: Writer identification from handwritten Devanagari script. In: Proceedings of INDIA-2015, pp. 497–505 (2015).
22. Purkait, P., et al.: Writer identification for handwritten Telugu documents using directional morphological features. In: Proceedings of 12th ICFHR, pp. 658–663 (2010).
23. Dhandra, B. V., et al.: Writer identification by texture analysis based on Kannada handwriting. Int. J. Comm. Netw. Secur. **1**(4), 80–85 (2012).
24. Garain, U., Paquet, T.: Off-line multi-script writer identification using AR coefficients. In: Proceedings of 10th ICDAR, pp. 991–995 (2009).
25. Chanda, S., et al.: Text independent writer identification for Bengali script. In: Proceedings of ICPR, pp. 2005–2008 (2010).
26. Gupta, S., Namboodiri, A.: Text dependent writer verification using boosting. In Proceedings of 11th ICFHR, (2008).

Handwritten Oriya Digit Recognition Using Maximum Common Subgraph Based Similarity Measures

Swarnendu Ghosh, Nibaran Das, Mahantapas Kundu and Mita Nasipuri

Abstract Optical Character Recognition have attracted the attention of lots of researchers lately. In the current work we propose a graph based approach to perform a recognition task for handwritten Oriya digits. Our proposal includes a procedure to convert handwritten digits into graphs followed by computation of the maximum common subgraph. Finally similarity measures between graphs were used to design a feature vector. Classification was performed using the K-nearest neighbor algorithm. After training the system on 5000 images an accuracy of 97.64 % was achieved on a test set of 2200 images. The result obtained shows the robustness of our approach.

Keywords Handwritten digit recognition · Subgraph isomorphism · Maximum common subgraph · Graph based similarity measures

1 Introduction

Writing on paper can be counted as the most commonly used method of formal communication. With the progress of digital method of communication in the recent decades, a bridge is required which will connect these two domains of communication process. Handwritten character recognition is one such field of research that is trying to achieve that goal. In this research we will introduce a novel approach for handwritten Oriya digit recognition. The procedure starts with conversion of images into graphs. Then the feature vector for the image is created using maximum

S. Ghosh (✉) · N. Das · M. Kundu · M. Nasipuri
Department of Computer Science, Jadavpur University, Kolkata, India
e-mail: swarbir@gmail.com

N. Das
e-mail: nibaran@ieee.org

M. Kundu
e-mail: mahantapas@gmail.com

M. Nasipuri
e-mail: mitanasipuri@yahoo.com

© Springer India 2016
S.C. Satapathy et al. (eds.), *Information Systems Design and Intelligent Applications*, Advances in Intelligent Systems and Computing 435,
DOI 10.1007/978-81-322-2757-1_18

common subgraph based similarity measures with respect to the training images. Finally K-nearest neighbor algorithm was used for classification.

The next section introduces some of the related works to the readers. Section 3 mainly focuses on the related theoretical aspects whose understanding is essential for the understanding of the methodology. Section 4 describes the workflow of the current work in full details. In Sect. 5 the results obtained are shown and critically analyzed. Finally the work is concluded in Sect. 6 with a bit of light shed on possible ways for future research.

2 Related Work

Handwritten Character Recognition has attracted the focus of researcher for a long period of time. Over the last couple of decades we have seen numerous works on handwritten character recognition for many different scripts like English, German, Chinese, Japanese, Korean, Arabic, Bangla, Devanagari and so on. But there are many local languages in India like Oriya which has not received the same amount of focus. In the year 2001, an OCR system was proposed [1] for printed Oriya character recognition. In 2007, a system [2] was developed to recognize handwritten Oriya character using a curvature based feature along with a quadratic classifiers. Initially 1176 features were used which was finally reduced to 392 features using principal component analysis. A success rate of 94.60 % was achieved on 18,190 test samples. Later in 2012, a Hopfield Neural Network based methodology [3] was used for handwritten Oriya numerals. Graph based matching techniques are becoming very popular in various fields. Such methods which uses maximum common subgraph based similarity analysis can be seen in the works of Das et al. [4].

3 Theoretical Aspects

Our current approach deals with the conversion of the digits into graphical structures and performs feature extraction using maximal common subgraph based similarity measures. Finally classification is performed using K-Nearest Neighbor classifier. For understanding of the methodologies described in the paper it is essential to understand the concept of maximal common subgraph and the K-Nearest Neighbor classifier.

3.1 Maximum Common Subgraph

The maximal common subgraph of two graphs G and H is the largest graph in terms of edges that is isomorphic to a subgraph of G and H. The Maximal Common Subgraph has been formally defined in a work of Welling [5] as follows:

Let $G = (V, E, \alpha, L)$ and $G' = (V', E', \alpha', L')$ be graphs. A graph isomorphism between G and G' is a bijective mapping $f : V \rightarrow V'$ such that

- $\alpha(v) = \alpha'(v) \forall v \in V'$
- For any edge $e = (u, v, l) \in E, \exists$ an edge $e' = (f(u), f(v), l') \in E'$ such that $l = l'$

Let $G_1 = (V_1, E_1, \alpha_1, L_1)$ and $G_2 = (V_2, E_2, \alpha_2, L_2)$ be graphs. A common subgraph of G_1 and G_2 is a graph $G = (V \cdot E, \alpha, L)$ such that there exist subgraph isomorphism from G to G_1 and from G to G_2. We call G a maximal common subgraph of G_1 and G_2 if there exists no other common subgraph of G_1 and G_2 that has more edges than G.

From the definition it is apparent that with greater similarity in between the images their maximal common subgraph would be larger.

3.2 *K-NN Classifier*

K-Nearest Neighbor is an algorithm for classification or regression. It's a non-parametric method which involves finding the k training samples which are closest to the test sample in the feature space, where k is a user-defined constant. The training examples are n-dimensional vectors in a feature space. Each example is marked with a class label. For a test example in the feature space, we look for the nearest k training samples, where the distance is usually computed as Euclidian distance, and assign the class according to a majority voting technique.

4 Methodologies

Our current work starts with the collection of the dataset containing 7200 handwritten Oriya digits. After preprocessing the digits they are converted to graphs, followed by feature extraction using graph based similarity measures and classification using K-nearest neighbor classifier. Figure 1 shows the basic workflow.

4.1 *Data Collection*

The numeral samples used for this study are developed at the Laboratory of Center for Microprocessor Applications for Training Education and Research, Computer Science and Engineering Department, Jadavpur University. The numeral samples

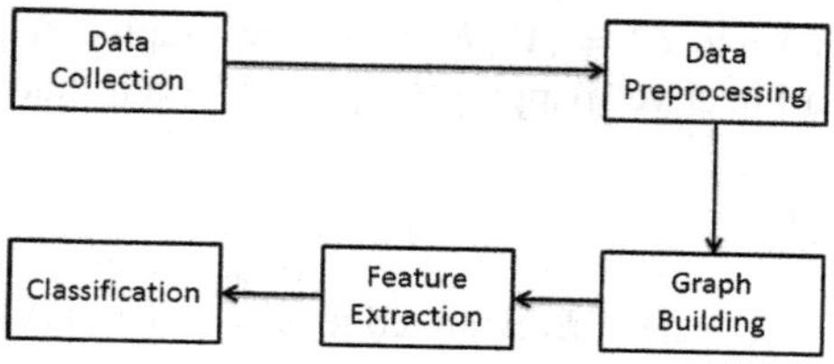

Fig. 1 Basic workflow

were collected from the students of National Institute of Science and Technology Orissa. Each participant was asked to write naturally in a sequence of digits in a specially designed form.

4.2 Data Preprocessing

After collecting the handwritten samples, the sheets were scanned and cropped manually. This phase of cropping was not a perfect cropping. It was done just to separate each digits from the sheet. After obtaining the cropped digits they were binarized using a threshold of 80 %. The next phase is the region of interest (ROI) selection. This phase involves finding the perfect boundaries of the image. The top boundary is selected by searching for the first row from the top which has the occurrence of a black pixel. The three other boundaries are similarly found using the occurrence of the first black pixel while moving row or column wise from the respective side. After selecting the ROI the image was cropped according to the selected boundaries. After cropping the image is resized into the size of 64 × 64 pixels. The resized image is then skeletonized. Skeletonization is the process that removes pixels on the boundaries of objects but does not allow objects to break apart. The pixels remaining make up the image skeleton. The final image shows the character in only single pixel width. Figure 2 shows the flowchart of the data preprocessing phase and a sample of the processed digits are shown in Fig. 3.

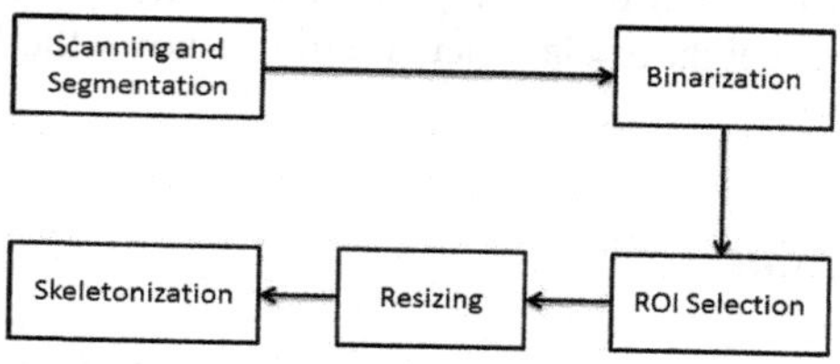

Fig. 2 The flowchart showing the data preprocessing phase

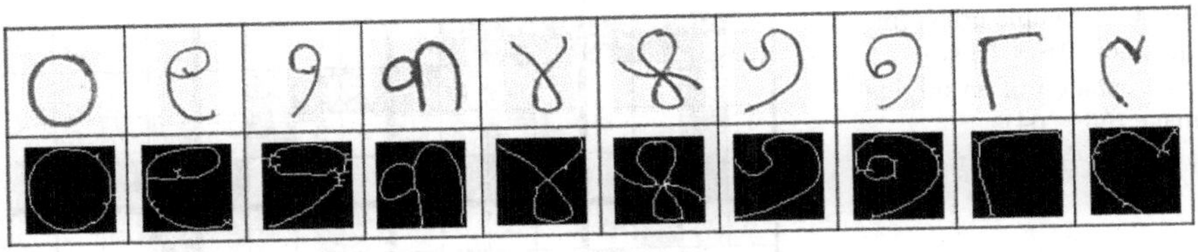

Fig. 3 *Top row* showing 10 handwritten unprocessed Oriya digits starting from zero to nine and *bottom row* showing the processed skeletonized image

4.3 Graph Building

Here a new of graph building from the image matrix is introduced. Here for our experiment the image is of size 64 × 64 pixel units. Now the image is segmented into 4 × 4 segments. Each segment is of size 16 × 16 pixel units. Each segment is considered as a node for the graph. So there are 16 nodes for each graph. Now let us consider the background as black or '0' and the foreground as white or '1'. As we here we traverse in the column major direction, a segment can be connected to the immediate right and or below and or south-east, south-west segment depending on pixels that occur in the border region. Each (16) W × (16) H segment is consider as the node of the graph. Connection between the nodes are depends on the last row or and column of the segment. The conditions are as follows:

- A segment is connected to its immediate right segment if in the last column of pixels one value is '1' in the former and in the immediate right and or immediate N-E and or immediate S-E pixel in the later is '1'.
- A segment is connected to its immediate below segment if in the last row of pixels one value is '1' in the former and in the immediate below and or immediate S-W and or immediate S-E pixel in the later is '1'.
- A segment is connected to its immediate south-east segment if the extreme S-E pixel value is '1' in the former and the immediate S-E pixel in the later is '1'.
- A segment is connected to its immediate south-west segment if the extreme S-W pixel value is '1' in the former and the immediate S-W pixel in the later is '1'.

Figure 4 shows the concept behind the graph formation. The image is not an exact representation of the current concept. In the image instead of a 64 × 64 image we show a 16 × 16 version of it for the sake of simplicity.

Based on these connections the edges may be defined as follows:

- For each node i (i = 1 to 16) if it connects to it immediate right node then it I is connected to i + 1 node.
- For each node i (i = 1 to 16) if it connects to it immediate below node then it I is connected to i + 4 node.
- For each node i (i = 1 to 16) if it connects to it immediate S-E node then it I is connected to i + 5 node.
- For each node i (i = 1 to 16) if it connects to it immediate S-W node then it I is connected to i + 3 node.

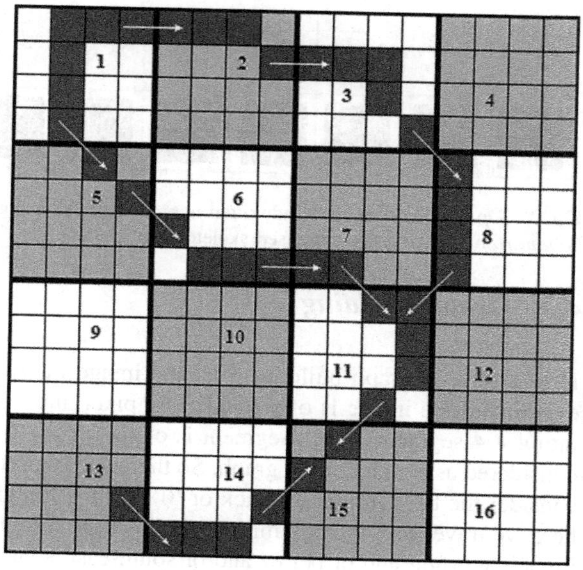

Fig. 4 A sample graph for an equivalent 16 × 16 size representation of an original image (64 × 64)

The graph for the image shown in Fig. 4 would be as follows:

1 → 2 5 −1 −1	9 → −1 −1 −1 −1
2 → 3 −1 −1 −1	10 → −1 −1 −1 −1
3 → −1 −1 −1 8	11 → −1 15 −1 −1
4 → −1 −1 −1 −1	12 → −1 −1 −1 −1
5 → 6 −1 −1 −1	13 → 14 −1 −1 −1
6 → 7 −1 −1 −1	14 → 15 −1 −1 −1
7 → −1 11 −1 −1	15 → −1 −1 −1 −1
8 → −1 −1 11 −1	16 → −1 −1 −1 −1

The graph format shown above depicts the connectivity of each of the 16 nodes with the nodes to each of their East, South, South-west and South-east neighboring nodes respectively. '−1' depicts the absence of any edges, whereas any other digits shows the node number of the neighbor in the corresponding direction.

4.4 *Feature Extraction*

The features extracted for each test sample is dependent on similarity of the graph corresponding to the test image and the graphs corresponding to the training images. There are various kinds of graph similarity measure involving maximal common subgraph. Usage of some of them was nicely demonstrated in the works of

Das et al. [6]. The similarity measure used for the current task can be described as follows.

$$sim(G_1, G_2) = \frac{v_{mcs}}{\max(v_{G_1}, v_{G_2})} \times \frac{e_{mcs}}{\max(e_{G_1}, e_{G_2})} \quad (1)$$

where, $sim(G_1, G_2)$ is the similarity score between the graph G_1 and G_2. v_{mcs} is the number of vertices in the maximum common subgraph. e_{mcs} is the number of edges for maximum common subgraph. v_{G_1} and v_{G_2} are the number of vertices in graphs G_1 and G_2 and e_{G_1} and e_{G_2} are the number of edges in graphs G_1 and G_2.

The feature vector for each sample can be described as a 10 dimensional vector V where the ith attribute of the vector corresponding to the graph G_k of the kth sample is given by,

$$V_i^k = \frac{1}{D}\sum_{j=1}^{D} sim(G_k, G_j) \quad (2)$$

where, D is the number of images corresponding to the ith digit in the training set. In simple terms we can say the 10 dimensional vector V is basically the average similarity of the sample with each of the 10 digits from the training set. The feature matrix for K samples appear as follows:

$$M = \begin{bmatrix} V_0^1 & V_1^1 & \cdots & V_9^1 \\ V_0^2 & V_1^2 & \cdots & V_9^2 \\ \vdots & \vdots & \ddots & \vdots \\ V_0^K & V_i^K & \cdots & V_9^K \end{bmatrix}_{K\times 10}$$

4.5 *Classification*

After obtaining the feature matrix the K-Nearest Neighbor Classifier was used for classification of the test samples. The data was split into training and testing set. The Classifier performance was evaluated using different values of K.

5 Results and Analysis

The experiments were conducted on a handwritten Oriya digit database consisting of 7200 images (720 images for each of the 10 digits). Our experiments involved the classification of 2200 test instances of images of digits (220 images for each of the 10 digits) using the K-Nearest Neighbor classifier trained over 5000 training instances (500 images for each of the 10 digits). The experiments were performed

Table 1 Class based accuracy measures

Class	TP rate	FP rate	Precision	Recall	F-measure
0	1	0	1	1	1
1	0.982	0.002	0.982	0.982	0.982
2	0.959	0.01	0.913	0.959	0.936
3	1	0	1	1	1
4	0.995	0.003	0.978	0.995	0.986
5	0.977	0.001	0.995	0.977	0.986
6	0.964	0.005	0.959	0.964	0.961
7	0.891	0.006	0.942	0.891	0.916
8	1	0	1	1	1
9	0.995	0.001	0.995	0.995	0.995
Avg.	0.976	0.003	0.977	0.976	0.976

using Weka Machine Learning Toolkit. The K-NN Classifier gave an accuracy of 97.72 % over the training set and 97.64 % over the test set. The experiment has been conducted with 3 values of K, i.e. 1, 3 and 5. But the accuracy remains the same for all the cases. Table 1 shows the precision recall and F-Measure for each of the classes. Maximum F-Measure was shown by the digits 0, 3 and 8. While the digit 7 exhibits the least value of F-Measure. The confusion among the digits are more clearly demonstrated in Table 2. Some examples of incorrect classification are shown in Fig. 5.

Table 2 Class based confusion matrix

Predicted		0	1	2	3	4	5	6	7	8	9
Actual	0	220	0	0	0	0	0	0	0	0	0
	1	0	216	2	0	0	0	0	1	0	1
	2	0	3	211	0	0	0	0	6	0	0
	3	0	0	0	220	0	0	0	0	0	0
	4	0	0	0	0	219	1	0	0	0	0
	5	0	0	0	0	5	215	0	0	0	0
	6	0	0	3	0	0	0	212	5	0	0
	7	0	0	15	0	0	0	9	196	0	0
	8	0	0	0	0	0	0	0	0	220	0
	9	0	1	0	0	0	0	0	0	0	219

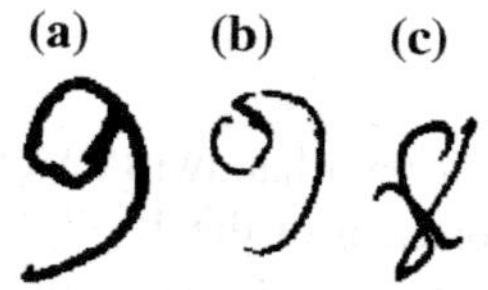

Fig. 5 Some misclassifications shown, **a** 7 misclassified as 2, **b** 6 misclassified as 7, **c** 5 misclassified as 4

6 Conclusion

In the current work we have shown the use of maximum common subgraph based similarities for the purpose of handwritten Oriya digit recognition. The system attained an accuracy of 97.64 % which establishes the soundness of the system. Further research may be performed by using the method on more datasets that includes different scripts. It is also worthy to note that the algorithm is dependent on a third party Skeletonization algorithm implemented through MATLAB. Further customization of the thinning algorithm may boost results. Another viable avenue of research may be to find a way to apply graph kernels for similarity measures. Overall, the procedure of graph based similarity analysis, having proven its worth in the current field of digit recognition, can be seen as viable tool for further research.

References

1. Chaudhuri, B.B., Pal, U., Mitra, M.: Automatic recognition of printed Oriya script. Proc. Sixth Int. Conf. Doc. Anal. Recognit. (2001).
2. Pal, U., Wakabayashi, T., Kimura, F.: A system for off-line Oriya handwritten character recognition using curvature feature. Proceedings—10th International Conference on Information Technology, ICIT 2007. pp. 227–229 (2007).
3. Sarangi, P.K., K Sahoo, A., Ahmed, P.: Recognition of Isolated Handwritten Oriya Numerals using Hopfield Neural Network, (2012).
4. Das, N., Ghosh, S., Gonçalves, T., Quaresma, P.: Comparison of Different Graph Distance Metrics for Semantic Text Based Classification. Polibits. 51–57 (2014).
5. Welling, R.: A Performance Analysis on Maximal Common Subgraph Algorithms. 15th Twente Student Conference on IT, Enschede, (2011).
6. Das, N., Ghosh, S., Quaresma, P., Gonçalves, T.: Using graphs and semantic information to improve text classifiers. PolTAL 2014—9th International Conference on Natural Language Processing, Warsaw, Poland (2014).

Design of Non-volatile SRAM Cell Using Memristor

Soumitra Pal and N.S. Ranjan

Abstract Emerging chip technologies employ power-off mode to diminish the power dissipation of chips. Non-volatile SRAM (NvSRAM) enables a chip to store the data during power–off modes. This non-volatility can be achieved through memristor memory technology which is a promising emerging technology with unique properties like low-power, high density and good-scalability. This paper provides a detailed study of memristor and proposes a memristor based 7T2M NvSRAM cell. This cell incorporates two memristors which store the bit information present in the 6T SRAM Latch, and a 1T switch which helps to restore the previously written bit in situations of power supply failures, thereby making the SRAM non-volatile.

Keywords Memristor · NvSRAM · Read access time · Write access time

1 Introduction

In 1971, Chua theoretically predicted the presence of Memristors as fourth important circuit component using symmetry arguments [1]. Memristor establishes the relationship between magnetic flux and charge as per the equation: $d\phi = Mdq$ Stanley William and his team at HP Labs physically realized memristor for first time in 2008 [2]. Memristor has attracted lots of research interests due to its scalability, non-volatility, lower power consumption, and 3-D stacking capability [3].

S. Pal (✉)
Applied Electronics and Instrumentation Engineering, C. V. Raman College of Engineering, Bidya Nagar, Mahura, Janla, Bhubaneswar 752054, India
e-mail: soumitra10028.13@bitmesra.ac.in

N.S. Ranjan
Electronics and Communication Engineering, Birla Institute of Technology, Mesra, Ranchi, India
e-mail: be1062313@bitmesra.ac.in

© Springer India 2016
S.C. Satapathy et al. (eds.), *Information Systems Design and Intelligent Applications*, Advances in Intelligent Systems and Computing 435,
DOI 10.1007/978-81-322-2757-1_19

These attributes have resulted in application of memristors in fields of non-volatile memory design [4], neuromorphic applications [5], programmable logic [6], analyzing non-linear circuits [7], digital and analog circuits [8].

The non-volatile characteristics of memristors make it a promising contestant for future memory equipment. Memristor based memory also called as Resistive RAM (RRAM) offers several advantages over conventional memories. Memristor can store multiple bits of information thereby increasing the amount of information that can be stored in single memory cell [9]. As RRAM is non-volatile, it is possible to turn off during hold mode to reduce power consumption. As proved, it is possible to scale down the memristor devices to 10 nm or even below. Furthermore, memristor memories can accomplish an integration density of 100 Gbits/cm which is few times higher than that of today's advanced flash memory.

In light of above discussions this paper provides the following benefits:

(1) It carries out a detailed study of memristor and memristor based memory.
(2) It proposes a novel 7-Transistor 2-Memristor (7T2M) NvSRAM cell.
(3) Furthermore, variability analysis and comparison of design metrics such as read/write access time are also carried out.

To authenticate the proposed NvSRAM cell, wide simulations on HSPICE are carried out using 16 nm PTM [10].

The sections are organized as follows. Section 2 gives a brief idea about memristor. Section 3 presents the proposed 7T2M NvSRAM cell. Simulation results along with the comparison are presented in Sect. 4. Finally, the conclusion of this work appears in Sect. 5.

2 Memristor

Voltage and current are related through resistance ($dV = RdI$), charge and voltage are expressed through capacitance and inductance relates flux and voltage. Using symmetry arguments Chua postulated that there is a missing link in between flux (ϕ) and charge (q) ($d\phi = Mdq$) which he called as memristance (M) [1].

A physical model of memristor is proposed by Strukov et al. in [2] which is comprise of an electrically switchable thin semiconductor film inserted between two metal contacts (see Fig. 1). This semiconductor film has length D and comprises of two layers of titanium dioxide films. Out of those, one layer is of pure TiO_2 (un-doped layer) having high resistivity and other one is highly conductive doped layer filled with oxygen vacancies. In Fig. 1, 'w' is doped region's (TiO_{2-x} layer) width. When an external voltage is supplied across the memristor, the electric field repels oxygen vacancies (positively charged) in the doped layer resulting in the length change [2], thereby changing the total resistance of the memristor. When the doped area is extended to the complete length (i.e. w/D = 1), then its resistance is denoted by R_{ON}. Similarly if the un-doped area is extended to the complete length (i.e. w/D = 0), then its resistance is denoted by R_{OFF}. Oxygen vacancies, which are positively charged, do not move around by themselves. These Oxygen vacancies become absolutely immobile until voltage are supplied again, thereby making memristor non-volatile.

A linear charge controlled memristor can be modeled as two resistors (un-doped and doped region resistance) in series [2] as per the following equation:

$$R(w) = R_{ON} \times \frac{w}{D} + R_{OFF}\left(1 - \frac{w}{D}\right) \quad (1)$$

$$M(q) = R_{OFF}\left(1 - \frac{\mu_v R_{ON}}{D^2} q(t)\right) \quad (2)$$

where, M(q) is the overall memristance of the memristor, ROFF and RON are the OFF state and ON state resistances, μv is the mobility of oxygen ions and 'D' is the device length.

The I–V relationship of a memristor can be expressed as [2]:

$$V(t) = \left(R_{ON}\frac{w(t)}{D} + R_{OFF}\left(1 - \frac{w(t)}{D}\right)\right) i(t) \quad (3)$$

where the value of w(t)/D is between 0 and 1 and w(t) is given as:

$$w(t) = \frac{\mu_v R_{ON}}{D} q(t) \quad (4)$$

Figure 1 shows the device structure and symbol of memristor used in electrical circuits. Figure 2a shows the applied square wave input and change in memristance of the memristor with the appliesd input voltage is displayed in Fig. 2b. Memristor's I–V response to 0.6 V square wave is presented in Fig. 3. This I–V curve resembles a pinched hysteresis curve which is the memristor fingerprint [1].

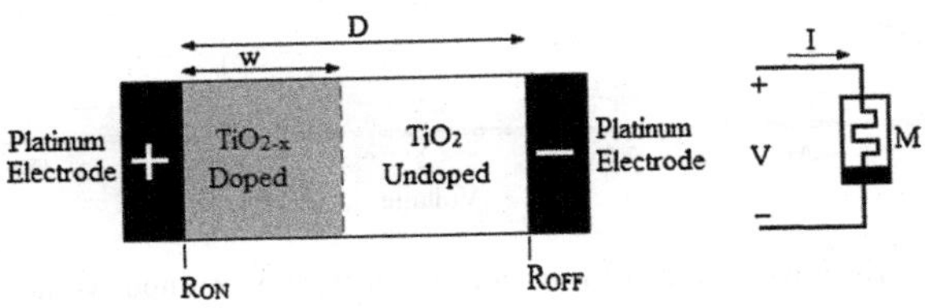

Fig. 1 Device structure of memristor and its circuit symbol

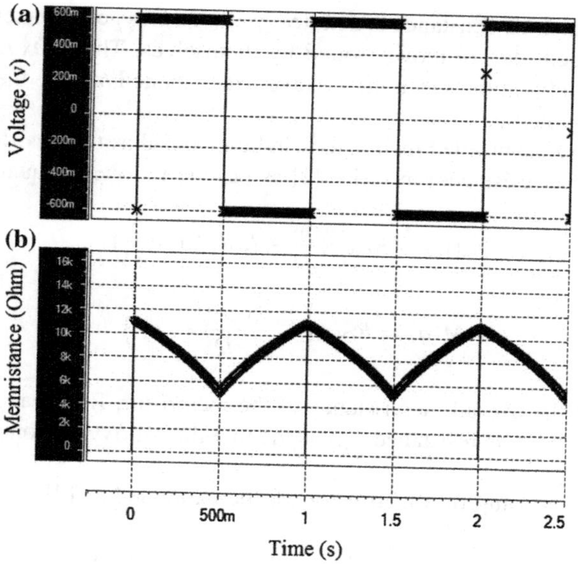

Fig. 2 **a** Applied input voltage, **b** change of memristance with input voltage

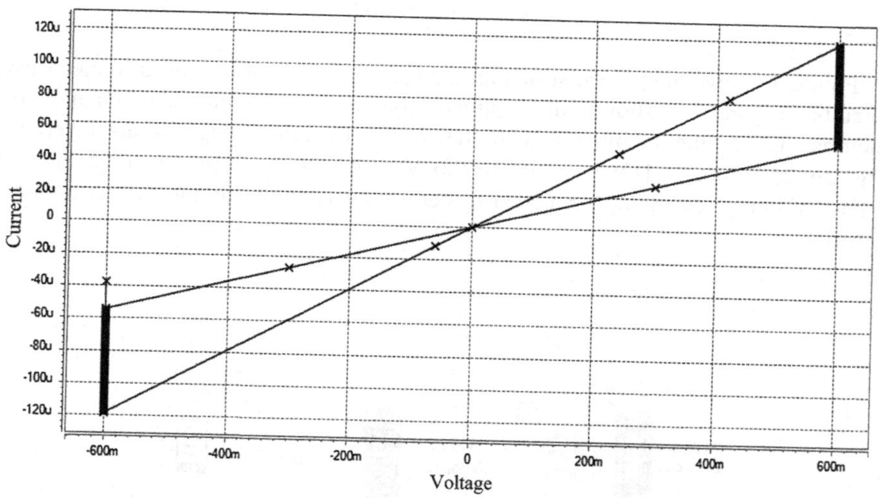

Fig. 3 I–V characteristic of memristor in response to square wave input voltage

3 Proposed NvSRAM Cell Design

3.1 Structure

The proposed circuit of 7T2M NvSRAM is shown in Fig. 4. It comprises of basic 6T SRAM cell, two memristors (M1, M2) and a NMOSFET (N5) for restore operation. Two memristors are connected to node Q (M1) and QB (M2) with polarity as shown in the Fig. 4. The NMOSFET, N5, whose gate is connected to ML (Memory Line), is used in restore operation in case of power supply failure (Figs. 5 and 6).

3.2 Read/Write Operation

The 7T2M NvSRAM cell performs read and write operations similar to standard 6T SRAM except during both read and write operations ML is kept low to turn off the NMOSFET N5.

During write operation data is latched on nodes Q and QB depending on the BL and BLB voltage levels. The memristnce of the two memrisots changes to either High resistance State (HRS) or Low Resistance State (LRS) corresponding to

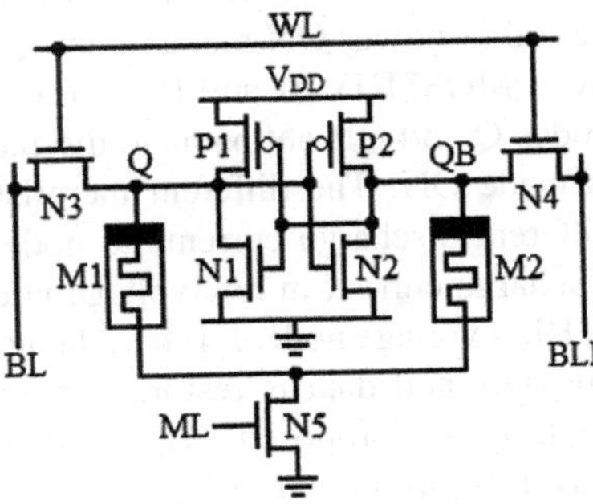

Fig. 4 Proposed 7T2M NvSRAM cell

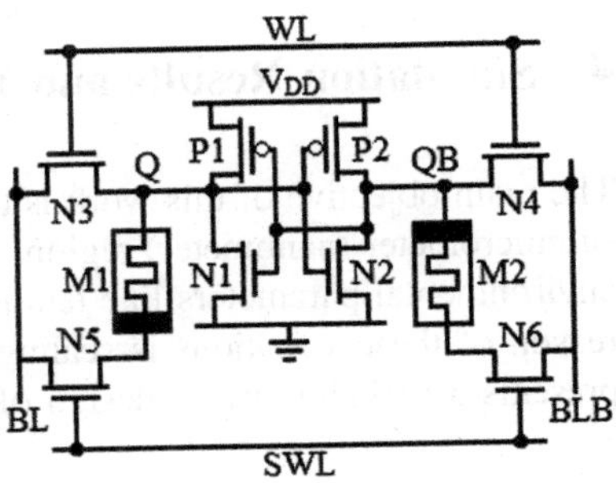

Fig. 5 8T2M NvSRAM cell [11]

Fig. 6 6T SRAM cell

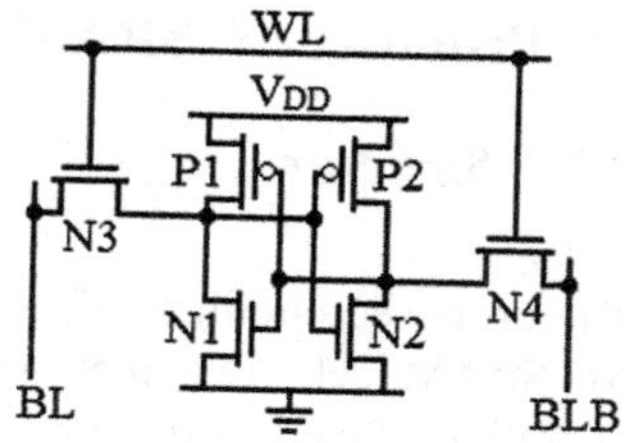

voltage levels at nodes Q and QB. For example, after successful write operation if node Q = ‘0’ (QB = ‘1’) a significant amount of current flows from node QB to Q. This current is into the negative polarity of M2 and positive polarity of M1. Hence, the memristance of M2 is changed to HRS and that of M1 is changed to LRS. Similarly if node Q is set to ‘1’ i.e. Q = ‘1’ (QB = ‘0’), current flows from node Q to QB and memristances of M1 and M2 become opposite to that of previous case.

3.3 Restore Operation

In case of the power failure i.e. V_{DD} = ‘0’, Q and QB nodes voltages are pulled to ground. During power-ON, data are recollected from two memristors (M1 and M2) into the SRAM cell data storage nodes. In restore mode ML is pulled up to turn on the NMOSFET N5. BL and BLB are turned low to ensure that during the initial period of power-ON mode nodes Q and QB are at same state. As V_{DD} increases the two PMOSFETs P1 and P2 charge the nodes Q and QB equally. At the same time nodes Q and QB get paths to discharge through M1 and M2 respectively since, N5 is made ON. The different memristance of two memristors M1 and M2 provide different discharge currents to node Q and QB. For example, LRS provides large discharge current in low voltage node and HRS provides smaller discharge current in High voltage node. Further, the node voltages are amplified by two cross-coupled inverters and data is restored in the corresponding nodes successfully. The non-valitility of proposed NvSRAM is experimentally proved through simulation waveform as shown in Fig. 7.

4 Simulation Results and Discussion

The main objective of this work is to design a non-volatile SRAM cell in the deep submicrometer (nanometer) region, where variations occur because of process and environmental parameters like temperature, operating voltage and noise. The main reason of these variations is scaling down of technology node. Hence, this paper presents a variation aware design of NvSRAM cell.

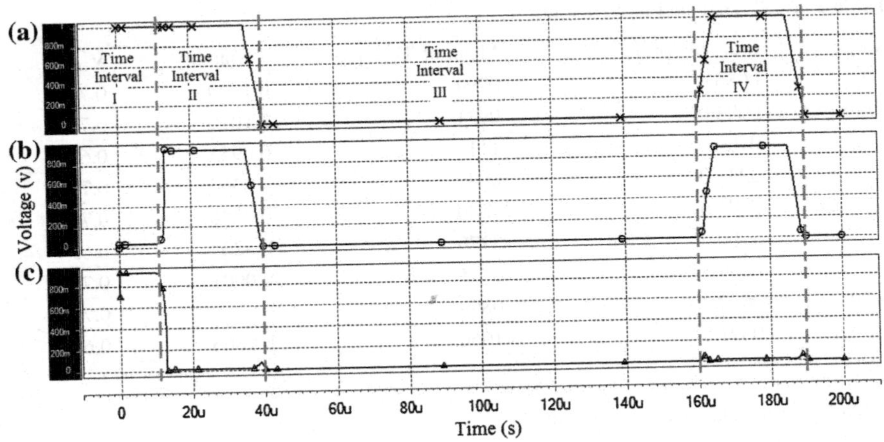

Fig. 7 Restore operation: **a** power supply (V_{DD}) **b** node Q voltage level, **c** node QB voltage level. *Time interval I* indicates the initial state of NvSRAM storage nodes Q and QB. During *time interval II* write operation is performed opposite to previous state. *Time interval III* shows the power-off period. *Time interval IV* shows the successful restoration of previously written data at NvSRAM storage nodes Q and QB

4.1 Simulation Setup

Various design metrics (read and write delay) of an SRAM cell are estimated using 16 nm PTM at different supply voltages with 5000 Monte Carlo simulation. It also carries out the variability (variability is calculated as the ratio of standard deviation (σ) of a design metric to mean (μ) of a design metric) analysis of read and write access time at different supply voltages. The design metrics of proposed cell is compared with 8T2M NvSRAM cell (Fig. 5) and conventional 6T SRAM cell (Fig. 6).

4.2 Read Delay and Its Variability

Read delay or read access time (T_{RA}) is estimated in the same way as done in [12, 13]. From the simulation result it is noticed that theroposed 7T2M NVSRAM cell exhibits 16 % (20 %) longer T_{RA} at nominal V_{DD} (0.7 V) compared to 6T SRAM (8T2M NVSRAM) (see Table 1). Our proposed cell exhibits tighter spread in T_{RA} at nominal V_{DD} (see Table 1). At nominal V_{DD} it shows 16 % (20 %) narrower spread in T_{RA} when compared with 6T (8T2M) (see Table 1).

Table 1 Read delay and its variability

Cell	Std. dev. of T_{RA} (ps)	Mean of T_{RA} (ps)	Variability	V_{DD}
6T	0086.6	0794	0.1090	0.77
	0094.0	0931	0.1010	0.70
	0109.0	1110	0.0955	0.63
	0233.0	0989	0.2360	0.77
7T2M	0081.6	1080	0.0756	0.70
	3100.0	3770	0.8228	0.63
	0069.8	0775	0.0901	0.77
8T2M	0072.9	0898	0.0812	0.70
	0091.8	1060	0.0866	0.63

Table 2 Write delay and its variability

Cell	Std. dev. of T_{WA} (ps)	Mean of T_{WA} (ps)	Variability	V_{DD}
6T	73.1	596	0.123	0.77
	59.3	621	0.095	0.70
	59.7	656	0.091	0.63
	59.3	554	0.107	0.77
7T2M	64.8	589	0.110	0.70
	61.4	613	0.100	0.63
	66.2	498	0.133	0.77
8T2M	85.3	499	0.171	0.70
	181	401	0.451	0.63

4.3 *Write Delay and Its Variability*

Write delay or write access time (T_{WA}) is estimated in the same way as done in [14–16]. Proposed cell exhibits shorter write delay (5 %) compared to 6T SRAM (@ 0.7 V) (see Table 2). 8T2M cell exhibits shortest write delay compared to all as it has two extra access NMOSFETs which provide additional write current, thereby reducing write delay. however, the proposed cell provides narrower spread in write delay distribution (see Table 2) at all considered voltages (except at 0.7 V and 0.63 V compared with 6T).

5 Conclusion

In this paper a non-volatile 7T2M memory cell has been presented. The simulation results depicts that the proposed memory cell has better read and write access time variability than the conventional 6T SRAM and 8T2M cell across 16, 22, 32 and

45 nm technology nodes. Hence, the proposed 7T2M celll would make a better variation aware memristor based non-volatile memory cell.

References

1. L. O. Chua, "Memristor—the missing circuit element," *IEEE Trans.Circuit Theory*, vol. CT-18, no. 5, pp. 507–511, Sep. 1971.
2. D. B. Strukov, G. S. Snider, D. R. Stewart, and R. S. Williams. "The missing memristor found". *Nature*, 453(7191):80–83, 2008.
3. J. Hutchby and M. Garner, "Assessment of the potential & maturity of selected emerging research memory technologies," in *Proc. Workshop & ERD/ERM Working Group Meeting*, Barza, Italy, Apr. 2010, pp. 1–6.
4. Y. Ho, G. Huang, and P. Li. Dynamical properties and design analysis for nonvolatile memristor memories. *Circuits and Systems I: Regular* Papers, IEEE Transactions on, 58 (4):724–736, 2011.
5. S. H. Jo, T. Chang, I. Ebong, B. B. Bhadviya, P. Mazumder and W. Lu, "Nanoscale memristor device as synapse in neuromorphic systems," American Chemical Society, Nano Lett., vol. 10, no. 4, pp. 1297–1301, March 1, 2010.
6. J. Borghetti, G. S. Snider, P. J. Kuekes, J. J. Yang, D. R. Stewart, and R. S. Williams, "'Memristive' switches enable 'stateful' logic operations via material implication," *Nature*, vol. 464, pp. 873–876, Apr. 2010.
7. K. Murali and M. Lakshmanan, "Effect of sinusoidal excitation on the Chua's circuit," IEEE Trans. Circuits Syst. I, Fundam. Theory Appl., vol. 39, no. 4, pp. 264–270, Apr. 1992.
8. F. Corinto, A. Ascoli, and M. Gilli. Nonlinear dynamics of memristor oscillators. Circuits and Systems I: Regular Papers, IEEE Transactions *on*, 58(6):1323–1336, 2011.
9. Leon O. Chua and Sung Mo Kang, "Memristive Devices and Systems," Proceedings of the IEEE, vol. 64, no. 2, pp. 209–223, Feb, 1976.
10. Nanoscale Integration and Modeling (NIMO) Group, Arizon State University (ASU). [Online]. Available: http://ptm.asu.edu/.
11. R. Vaddi, S. dasgupta, and R. P. Agarwal, "Device and circuit co-design robustness studies in the subthreshold logic for ultralow-power applications for 32 nm CMOS," *IEEE Trans. Electron Devices*, vol. 57, no. 3, pp. 654–664, Mar. 2010.
12. J. P. Kulkarni, K. Kim, and K. Roy, "A 160 mV robust Schmitt trigger based subthreshold SRAM," *IEEE J. Solid-State Circuits*, vol. 42, no. 10, pp. 2303–2313, Oct. 2007.
13. H. Noguchi, S. Okumura, Y. Iguchi, H. Fujiwara, Y. Morita, K. Nii, H. Kawaguchi, and M. Yoshimoto, "Which is the best dual-port SRAM in 45-nm process technology?—8T, 10T single end, and 10T differential," in *Proc. IEEE ICICDT*, Jun. 2008, pp. 55–58.
14. Pal, S.; Arif, S., "A single ended write double ended read decoupled 8-T SRAM cell with improved read stability and writability," *IEEE Int. Conf. on Computer Communication and Informatics (ICCCI),* vol., no., pp. 1–4, 8–10 Jan. 2015.
15. Arif, S.; Pal, S., "Variation-resilient CNFET-based 8T SRAM cell for ultra-low-power application," *IEEE Int. Conf. on Signal Processing And Communication Engineering Systems (SPACES),* vol., no., pp. 147–151, 2-3 Jan. 2015.
16. Pal, S.; Arif, S., "A Fully Differential Read-Decoupled 7-T SRAM Cell to Reduce Dynamic Power Consumption", *ARPN Journal of Engineering and Applied Sciences,* VOL. 10, NO. 5, Mar. 2015.

Evolutionary Algorithm Based LFC of Single Area Thermal Power System with Different Steam Configurations and Non-linearity

K. Jagatheesan, B. Anand and Nilanjan Dey

Abstract Load Frequency Control (LFC) of single area thermal power system is presented in this work. Commonly used industrial Proportional-Integral-Derivative (PID) controller is considered as a supplementary controller and parameters are optimized by using evolutionary algorithm (Ant Colony Optimization (ACO)). Three cost functions are considered to optimize controller gain values. Such as, Integral Absolute Error (IAE), Integral Time Absolute Error (ITAE) and Integral Square Error (ISE) and also three different stem configurations (Non Reheat turbine, Single Stage Reheat turbine and Double stage reheat turbine) are considered in this work. Further the performance of proposed algorithm is proved by adding non-linearity (Generation Rate Constrain, Governor Dead Band and Boiler Dynamics) into the same power system and value of Step Load Perturbation (SLP) in all three steam configurations. Time domain analysis is used to study the performance of power system with different scenarios.

Keywords Cost functions · Evolutionary algorithm · Load frequency control (LFC) · Non-linearity · Proportional-Integral-Derivative · Steam configuration

K. Jagatheesan (✉)
Department of EEE, Mahendra Institute of Engineering and Technology, Namakkal, Tamil Nadu, India
e-mail: jaga.ksr@gmail.com

B. Anand
Department of EEE, Hindusthan College of Engineering and Technology, Coimbatore, Tamil Nadu, India

N. Dey
Department of CSE, BCET, Durgapur, India

© Springer India 2016
S.C. Satapathy et al. (eds.), *Information Systems Design and Intelligent Applications*, Advances in Intelligent Systems and Computing 435,
DOI 10.1007/978-81-322-2757-1_20

1 Introduction

The From the literature survey is found that many research works has been reported related to single/multi area thermal and hydro power system for delivering good quality of power to the consumers [1–33]. In power system LFC plays major role for effective operation and quality of power supply [1–3]. The objective of LFC in power system is to keep frequency deviations and power flow between control areas with in the limit by adjusting output of generators to match the power generation with fluctuating load demand. The well designed and operated power plant having the ability to handle power generation with load demand by maintaining system frequency and voltage across the generator terminals [1–3]. LFC issue in a power system is solved by implementing suitable controller. Generally, in power system input of controller is Area Control Error (ACE) and it generate proper control output signal, which is given to the power system as a control signal. From the past few decades variety of controller has been reported such as classical controllers [4], Proportional-Integral controller (PI) [30], Proportional-Integral-Derivative controller (PID) [5, 6], Integral-Double derivative controller (IDD) [7], Fractional Order Proportional-Integral-Derivative controller (FOPID) [8, 9], and 2DOF [10] controllers are successfully developed and implemented.

The selection of controller is a prime effort; apart from this proper selection of controller gain and objective function is more crucial for the design of better controlled power system. From the literature recently different bio-inspired algorithms are created to optimize different controller gain values in LFC/AGC issue in power system. Such as, Artificial Bee colony (ABC) [20], Stochastic Particle Swarm Optimization (SPSO) [5], Imperialist Competitive Algorithm (ICA) [21], Firefly Algorithm (FA) [22], Quasi-Oppositional Harmony Search algorithm (QOHS) [23], Cuckoo search [24, 31], Particle Swarm Optimization based optimal Sugeno Fuzzy Logic Control (PSO-SFLC) [25], Direct Synthesis (DS) [26], Differential Evolution (DE) algorithm [27].

The proper selection of cost/objective function is more important during the tuning of controller parameter optimization issue in LFC/AGC of multi-area power system [30–33]. The commonly used cost functions are Integral Absolute Error (IAE), Integral Time Absolute Error (ITAE) and Integral Square Error (ISE). The rest of paper is organized as follows: the introduction about LFC and different optimization technique with cost function are discussed in section "Introduction". The proposed single area power system with different steam configurations and components with proposed algorithm are given the section "Single area thermal power system", results of simulink modelled investigation power system shown and discussed in the section "Simulation results and observations" and finally performance of different cost and steam configurations are discussed in "conclusion".

2 Single Area Thermal Power System

2.1 *Thermal Power System*

The investigated single area reheat thermal power system shown in Fig. 1 is equipped with appropriate governor unit, turbine with reheater unit, generator, non-linearity, boiler dynamics and speed regulator unit.

The transfer function model of investigated power system is shown in Fig. 1 and nominal parameters are taken from [28–30]. where R represents the self speed regulation parameter for the governor in p.u. Hz; Tg represents speed governor time constant in sec; Tr is the reheat time constant in sec; Kr is the reheat gain; Tt is the steam chest time constant in sec; Tp, Kp is the load frequency constant (Tp = 2H/f * D, Kp = 1/D); delXE represent incremental governor valve position change; delPg incremental generation change; delF incremental frequency deviation; ACE stands for area control error.

In thermal power system water is converted into steam and with help of turbine, high pressure and high temperature steam is converted into useful mechanical energy. Based on the steam stages the turbines units are classified into three different types. Such as non reheat turbine, single stage reheat turbine and double stage reheat turbine [28–30]. The generation rate constraint and Governor dead band physical constraints are included in this system. For the thermal power system 0.0017 puMW/s value is considered as a GRC constraint and the transfer function of Governor dead band non-linearity is given in the expression as [28–30]:

$$\text{Governor with the dead band } (\mathrm{G_g}) = \frac{-\frac{0.2}{\pi}S + 0.8}{T_g S + 1} \tag{1}$$

In this work drum type boiler is considered for investigation and it is shown in produce steam under pressure. In power system, when load demand occurs it affects system stability as well as system parameters (frequency and power flow between connected areas) from its specified value. In order to conquer aforementioned shortcomings are solved by proper selection supplementary controller for power system is more crucial issue. In this work Proportional-Integral-Derivative controller is considered as a secondary controller.

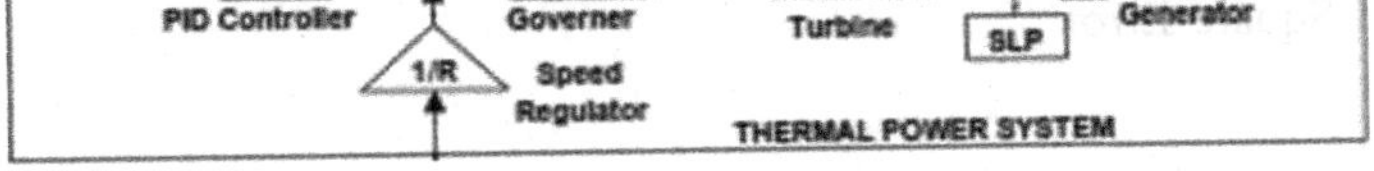

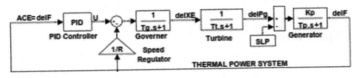

Fig. 1 Transfer function model of non-reheat thermal power system

2.2 Proportional-Integral-Derivative (PID) Controller

The commonly used industrial PID controller is considered because of simplicity and easy implementation. Transfer function of PID controller is given by [5, 6]:

$$G_{PID}(s) = K_P + K_I S + \frac{K_D}{S} \tag{2}$$

The PID controller parameters are: Kp, Ki, and Kd-Proportional, Integral and derivative gain values of controller respectively. The input of PID controller is Area Control Error (ACE—It is defined as linear combination of frequency deviation and tie line power flow deviations). The area control error expression is given in Eq. (6).

$$ACE = B \cdot \Delta F + \Delta P_{tie} \tag{3}$$

In single area power system there no tie line power flow. So the value of frequency deviation is equals to area control error. The area control error of single area thermal power system and control output signal generated by PID controller is given in the expression:

$$ACE = B \cdot \Delta F \tag{4}$$

$$u(t) = K_P \cdot ACE + K_I \int_0^t ACE\,dt + K_D \frac{d\,ACE}{dt} \tag{5}$$

The design of PID controller is obtained by using three different cost functions and controller gain values are optimized by Ant Colony Optimization technique (ACO). The design procedure of PID controller using ACO and gain values of PID controller and different steam configurations, cost function are given in the Sect. 2.3. In this work PID controller parameters are optimized using three different cost functions. Such as Integral Time Absolute error (ITAE), Integral Absolute Error (IAE) and Integral Square Error (ISE) cost functions are used [28–30]. The cost functions are

Integral Absolute Error

$$J_1 = \int_0^\infty \left|\{\Delta f_i + \varDelta P_{tiei-j}\}\right| dt \tag{6}$$

Integral Square Error

$$J_2 = \int_0^\infty (\{\Delta f_i + \varDelta P_{tiei-j}\})^2 dt \tag{7}$$

Integral Absolute Time Error

$$J_3 = \int_0^\infty t\left|\{\Delta f_i + \Delta P_{tiei-j}\}\right|dt \tag{8}$$

2.3 *Ant Colony Optimization Technique (ACO)*

Recently evolutionary computational technique plays very crucial role in combinatorial optimization problem in different issues. In this work Ant Colony Optimization (ACO) technique is consider for LFC issue of power system. Initially Ant Colony Optimization (ACO) initially technique proposed by Marco Dorigo in 1992 in his Ph.D thesis [6, 32, 33]. The natures of real ants are: during the food searching process initially all the ants are spread around the nest in a random manner, then it walks different paths. During the ant movement it stores the pheromone chemical into the ground. The strength of the chemical round is based on the quality and quantity of the food source searched by each ant. Finally, shortest path and good quality and quantity path having large pheromone concentration over poor path and chemical stored in the poor path evaporate very fast compare to shortest path. These behaviors of real ants are inspired by many researchers to develop ACO technique [6, 32, 33].

The transition probability from town i and j for the kth ant as follows

$$p_{ij}(t) = \frac{\tau_{ij}(t)^{\alpha}(\eta_{ij})^{\beta}}{\sum_{j \in nodes} \tau_{ij}(t)^{\alpha}(\eta_{ij})^{\beta}} \tag{9}$$

The value of pheromone versus heuristic information η_{ij} is given by:

$$\eta_{ij} = \frac{1}{d_{ij}} \tag{10}$$

The global updating rule is implemented in ant system as follows, where all ants starts their tours, pheromone is deposited and updated on all edges based on

$$\tau_{ij}(t+1) = (1-\rho)\tau_{ij}(t) + \sum_{\substack{k \in colony\,that \\ used\,edge(i,j)}} \frac{Q}{L_k} \tag{11}$$

where Pij—Probability between the town i and j; τ_{ij}—Pheromone associated with the edge joining cities i and j; dij—distance between cities i and j; Q—Constant related to quality of pheromone; Lk—length of the tour performed by Kth ant;

α, β—constant that find the relative time between pheromone and heuristic values on the decision of the ant; ρ—Evaporation rate. The Ant Colony Optimization (ACO) optimization technique optimized PID controller gain values (Proportional Gain (Kp), Integral gain (Ki and Derivative gain (Kd)) of different steam turbine with cost functions are: Non-Reheat Turbine (ISE-Kp = 0.88, Ki = 0.86, Kd = 0.95; IAE-Kp = 0.96, Ki = 1, Kd = 0.2; ITAE-Kp = 0.51, Ki = 0.97, Kd = 0.18), Single Reheat Turbine (ISE-Kp = 0.95, Ki = 0.56, Kd = 0.89; IAE-Kp = 1, Ki = 0.97, Kd = 0.1; ITAE-Kp = 0.98, Ki = 0.93, Kd = 0.01), Double Reheat Turbine (ISE-Kp = 0.61, Ki = 0.34, Kd = 0.75; IAE-Kp = 1, Ki = 1, Kd = 0.09; ITAE-Kp = 0.44, Ki = 0.98, Kd = 0.13).

3 Simulation Results and Observations

The performance of proposed ant colony optimization technique is applied in single area power system to optimize PID controller parameters with different scenarios with different objective functions and steam configurations.

Scenario 1: Thermal Power System without non-linearity and different cost functions by considering 1 % SLP.

In this, scenario thermal power is equipped with non-reheat, single reheat and double reheat turbine with PID controller by considering one percent Step Load Perturbation (1 % SLP) load demand. The gain values of controller parameters are optimized by using ACO with ISE, IAE and ITAE cost functions. Frequency deviations in non-reheat, single reheat and double reheat turbine equipped power system response given in Fig. 2.

It is evident from the response ITAE cost function based PID controller give quickly settled response compared to IAE and ISE optimized PID controller response in all turbine units. But peak under shoots in system response is effectively reduced by ISE cost function optimized PID controller compared to IAE and ITAE cost function optimized PID controller performance. The numerical values of settling time with different cost and seam configuration are given Table 1.

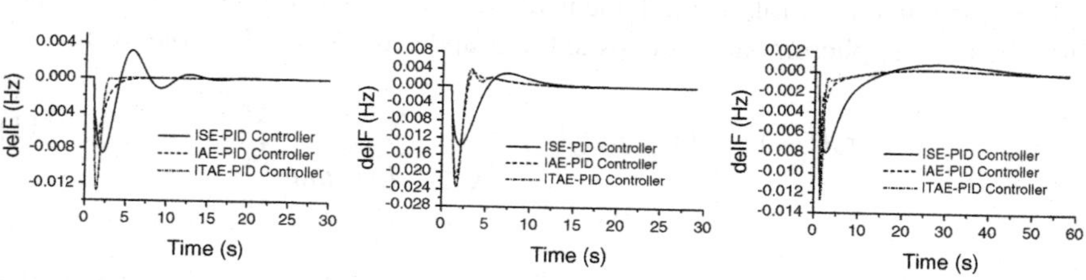

Fig. 2 Frequency deviations of non-reheat turbine, single stage reheat turbine and double stage reheat turbine thermal power system with different cost functions

Table 1 Performance comparison of settling time for scenario-1

	Steam configuration	Objective function	Settling time (s)	Peak overshoot (Hz)	Peak undershoot (Hz)
Thermal power system without non-linearity by considering 1 % SLP	Non-reheat turbine	ISE	22.15	0.003	−0.0084
		IAE	9.79	0	−0.0076
		ITAE	5.75	0	−0.0127
	Single reheat turbine	ISE	21.88	0.0029	−0.013
		IAE	16.2	0.0027	−0.0212
		ITAE	15.83	0.0035	−0.023
	Double reheat turbine	ISE	84	0.00087	−0.0079
		IAE	47.12	0.00029	−0.011
		ITAE	47.07	0.0003	−0.0125

Scenario 2: Thermal Power System non-linearity, boiler dynamics and different cost functions by considering 1 % SLP.

In this scenario generation rate constraint, governor dead band non-linearity and boiler dynamics is considered into the power system. When non linearity added into the power system, it yield more damping oscillations with overshoots, undershoots and takes more time to settle. Figure 3 shows the frequency deviations with different steam turbine equipped power system and different objective functions optimized PID controller response.

The settling time of thermal power system frequency deviation is effectively reduced by implementing ITAE cost function optimized PID controller. The response of ITAE cost function based is more superior compared to ISE and IAE based PID controller response in terms of settling time. From the response shown in Fig. 3, Clearly evident that ISE cost function optimized PID controller takes more time to settle compare to IAE and ITAE cost function based PID controller, but it yield very minimal peak under shoot compare to other for all different steam configurations.

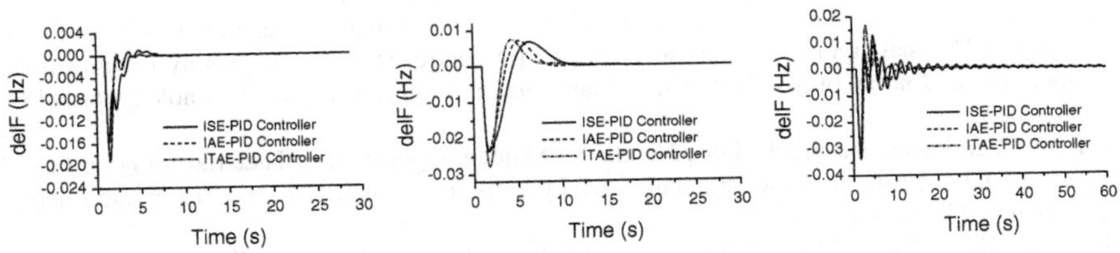

Fig. 3 Frequency deviations of non-reheat turbine, single stage reheat turbine and double stage reheat turbine thermal power system with different cost functions, non-linearity and boiler dynamics

4 Conclusion

In this work, load frequency control of single area with Proportional-Integral-Derivative (PID) controller equipped thermal power system is investigated. The controller parameters are optimized by using evolutionary computation based Ant Colony Optimization (ACO) technique. The PID controller is designed by using three different cost functions. Such as, Integral Absolute Error (IAE), Integral Time Absolute Error (ITAE) and Integral Square Error (ISE). These different cost function optimized PID controller is implemented in different stem configuration based turbine equipped power system. Steam configurations are: non-reheat turbine, Single stage reheat turbine and double stage reheat turbine.

Observation from the simulation result shows that, ITAE cost function based PID controller give fast settled response in all three different scenarios compared to IAE and ISE cost function based PID controller response (Scenario 1-non-reheat turbine: $5.75 < 9.79 < 22.15$; Single reheat turbine: $15.83 < 16.2 < 21.88$; double reheat turbine: $47.07 < 47.12 < 84$). Further peak undershoots in the system response for all three different scenarios effectively reduced by implementing ISE cost function based PID controller compared to IAE and ITAE cost function optimized PID controller response.

References

1. Nagrath, I.J., Kothari, D.P.: Power System Engineering. Tata Mc-Graw Hill Publishing Company limited, New Delhi (1994).
2. Kundur, P.: Power system stability and control. Tata Mc-Graw Hill Publishing Company limited, New Delhi (1994).
3. Elgerd, O.I.: Electric Energy System Theory: An Introduction. Tata Mc-Graw Hill Publishing Company limited. New York (1970).
4. Nandha, J., Mishra, S.: A novel classical controller for Automatic generation control in thermal and hydro thermal systems. PEDES, 1–6, 2010.
5. Jagatheesan, K. Anand, B. and Ebrahim, M.A.: Stochastic Particle Swarm Optimization for tuning of PID Controller in Load Frequency Control of Single Area Reheat Thermal Power System. International Journal of Electrical and Power Engineering, 8(2), 33–40, 2014.
6. Jagatheesan, K., Anand, B., and Dey, N.: Automatic generation control of Thermal-Thermal-Hydro power systems with PID controller using ant colony optimization. International Journal of Service Science, Management, Engineering, and Technology, 6(2),18–34, 2015.
7. Lalit Chandra saikia, Nidul Sinha, Nanda, J.: Maiden application of bacterial foraging based fuzzy IDD controller in AGC of a multi-area hydrothermal system. Electric Power and Energy Systems, 45, 98–106, 2013.
8. Seyed Abbas Taher, Masoud Hajiakbari Fini, Saber Falahati Aliabadi.: Fractional order PID controller design for LFC in electric power systems using imperialist competitive algorithm. Ain Shams Engineering journal, 5, 121–135, 2014.
9. Indranil Pan, Saptarshi Das,: Fractional-order load frequency control of interconnected power systems using chaotic multi-objective optimization. Applied Soft Computing, 29, 328–344, 2015.

10. Puja Dash, Lalit Chandra Saikia, Nidul Sinha: Comparison of performance of several FACTS devices using Cuckoo search algorithm optimized 2DOF controllers in multi-area AGC. Electric power and Energy systems, 65, 316–324, 2015.
11. Ashmole PH, Battebury DR, Bowdler RK.: Power-system model for large frequency disturbances. Proceedings of the IEEE, 121(7), 601–608, 1974.
12. Pan CT, Liaw CM.: An adaptive controller for power system load–frequency control," IEEE Transactions on Power Systems. 4(1), 122–128, 1989.
13. Wang Y, Zhou R, Wen C. Robust, "load–frequency controller design for power systems," IEE Proceeding-C 1993, Vol. 140, No. 1, 1993.
14. Wang Y, Zhou R, Wen C.: New robust adaptive load-frequency control with system parametric uncertainties. IEE Proceedings—Generation Transmission and Distribution, 141 (3), 1994.
15. Jiang,En L, Yao W, Wu QH, Wen JY, Cheng SJ.: Delay-dependent stability for load frequency control with constant and time-varying delays. IEEE Transactions on Power Systems, 27(2), 932–94, 2012.
16. Singh Parmar KP, Majhi S, Kothari DP.: Load frequency control of a realistic power system with multi-source power generation. Electrical Power and Energy Systems, 42, 426–433, 2012.
17. Foord TR.: Step response of a governed hydro-generator. Proceedings of the IEEE, 125(11), 1978.
18. Kusic GL, Sutterfield, J.A.; Caprez, A.R.; Haneline, J.L.; Bergman, B.R.: Automatic generation control for hydro systems. IEEE Transactions on Energy Conversion, 3(1), 33–39, 1988.
19. Doolla S, Bhatti TS.: Load frequency control of an isolated small-hydro power plant with reduced dumped load. IEEE Transactions on Power Systems, 21(4), 1912–1919, 2006.
20. Haluk gozde, M.cengiz Taplamacioglu, Ilhan kocaarslan.: Comparative performance analysis of artificial bee colony algorithm in automatic generation control for interconnected reheat thermal power system. Electric power and energy systems, 42,167–178, 2012.
21. Seyed Abbas Taher, Masoud Hajiakbari Fini, Saber Falahati Aliabadi.: Fractional order PID controller design for LFC in electric power systems using imperialist competitive algorithm. Ain Shams Engineering Journal, 5, 121–135, 2014.
22. K.Naidu, H.Mokhlis, A.H.A.Bakar.: Application of Firefly Algorithm (FA) based optimization in load frequency control for interconnected reheat thermal power system. IEEE Jordan conference on Applied Electrical Engineering and Computing Technologies, 2013.
23. Chandan Kumar Shiva, G.Shankar, V.Mukherjee.: Automatic generation control of power system using a novel quasi-oppositional harmony search algorithm. Electric power and energy systems, 73, 767–804, 2015.
24. Puja Dash, Lalit Chandra Saikia, Nidul Sinha: Comparison of performance of several FACTS devices using Cuckoo search algorithm optimized 2DOF controllers in multi-area AGC. Electric Power and Energy Systems, 65, 316–324, 2015.
25. Nattapol S-ngwong, Issarachai Ngamroo.: Intelligent photovoltaic farms for robust frequency stabilization in multi-area interconnected power system based on PSO-based optimal sugeno fuzzy logic control. Renewable Energy, Vol. 72, pp: 555–567, 2015.
26. Md Nishat Anwar, Somnath Pan.: A new PID load frequency controller design method in frequency domain through direct synthesis approach. Electrical power and Energy Systems, 67, 560–569, 2015.
27. Tulasichandra Sekar Gorripotu, Rabindra Kumar Sahu, Sidhartha Pand.: AGC of a multi-area power system deregulated environment using redox flow batteries and interline power flow controller. Engineering Science and Technology, an International Journal, 1–24, 2015 (In press).
28. Jagatheesan, K. and Anand, B.: Dynamic Performance of Multi-Area Hydro Thermal Power Systems with integral Controller Considering Various Performance Indices Methods. IEEE International Conference on Emerging Trends in Science, Engineering and Technology, Tiruchirappalli, December 13–14, 2012.

29. Jagatheesan, K. and Anand, B.: Performance Analysis of Three Area Thermal Power Systems with Different Steam System Configurations considering Non Linearity and Boiler Dynamics using Conventional Controller. Proceedings of 2015 International Conference on Computer Communication and Informatics (ICCCI-2015), 130–137, Jan. 08–10, 2015, Coimbatore, India.
30. Jagatheesan, K. and Anand, B.: Load frequency control of an interconnected three area reheat thermal power systems considering non linearity and boiler dynamics with conventional controller. Advances in Natural and Applied Science, 8(20), 16–24, 2014. ISSN: 1998-1090.
31. Dey N, Samanta S, Yang XS, Chaudhri SS, Das A.: Optimization of Scaling Factors in Electrocardiogram Signal Watermarking using Cuckoo Search," International Journal of Bio-Inspired Computation (IJBIC), 5(5), 315–326, 2014.
32. M. Omar, M. Solimn, A.M. Abdel ghany, F. Bendary: Optimal tuning of PID controllers for hydrothermal load frequency control using ant colony optimization. International Journal on Electrical Engineering and Informatics, 5(3), 348–356, 2013.
33. Jagatheesan, K. and Anand, B.: Automatic Generation Control of Three Area Hydro-Thermal Power Systems considering Electric and Mechanical Governor with conventional and Ant Colony Optimization technique. Advances in Natural and Applied Science, 8 (20), 25–33, 2014. ISSN: 1998-1090.

Game Theory and Its Applications in Machine Learning

J. Ujwala Rekha, K. Shahu Chatrapati and A. Vinaya Babu

Abstract Machine learning is a discipline that deals with the study of algorithms that can learn from the data. Typically, these algorithms run by generating a model built from the observed data, and then employ the generated model to predict and make decisions. Most of the problems in machine learning could be translated to multi-objective optimization problems where multiple objectives have to be optimized at the same time in the presence of two or more conflicting objectives. Mapping multi-optimization problems to game theory can give stable solutions. This paper presents an introduction of game theory and collects the survey on how game theory is applied to some of the machine learning problems.

Keywords Game theory · Machine learning · Multioptimization

1 Introduction

Game theory is a sub-field of mathematics that deals with the study of modeling conflict and cooperation. It is used to mathematically model the situations in which a decision-maker is faced with a challenge of taking a decision in the presence of conflicts.

J. Ujwala Rekha (✉) · A. Vinaya Babu
Department of Computer Science and Engineering, JNTUH College of Engineering, Hyderabad, India
e-mail: ujwala_rekha@yahoo.com

A. Vinaya Babu
e-mail: dravinayababu@gmail.com

K. Shahu Chatrapati
Department of Computer Science and Engineering, JNTUH College of Engineering, Jagitial, India
e-mail: shahujntu@gmail.com

© Springer India 2016
S.C. Satapathy et al. (eds.), *Information Systems Design and Intelligent Applications*, Advances in Intelligent Systems and Computing 435,
DOI 10.1007/978-81-322-2757-1_21

According to the chronology of game theory given in [1], game theory ideas can be traced back to thousands of years [2]. An earliest example is the letter authored by James Waldegrave in 1713, in which a minimax mixed strategy solution for a two-person version of the gambling game Le Her is presented [3]. Even though, the games were studied by many scholars over the years, the credit goes to Émile Borel for conceptualizing the games into a mathematical theory, who provided the proof for minimax theorem for zero-sum two-person matrix games in 1921. Yet, game theory is not considered a discipline in its own right until the publication of "Theory of Games and Economic Behavior" by John von Neumann and Oskar Morgenstern [4]. During the 1950s and 1960s, the study of games advanced greatly by the works of many scholars, and discovered the applications to the problems of war and politics. Subsequently, in the 1970s, it brought forth a great revolution in economics. Besides, it also found applications in sociology and biology. Particularly, it received great attention when ten game theorists were honored with the Nobel Prize in economics.

Around the past few years, enormous research is done at the intersection of algorithm design and game theory, bringing forth the algorithmic game theory [5]. During the same period, machine learning also has made constant progress in the research of developing new techniques for learning from data. While these disciplines seem to be different, there are a numerous connections between them: the issues in machine learning have strong link to the issues in algorithmic game theory and vice versa, and the techniques used in one can be used to solve the problems of the other [6].

In this paper, Sect. 2 presents the formal definitions of games, strategies and payoffs. It then presents examples of games studied widely in game theory in Sect. 3 and gives a classification of games in Sect. 4. Besides, Sect. 5 explains some of the solution concepts applied predominantly in game theory. Finally, this paper collects the survey on how game theory is applied to some of the machine learning problems in Sect. 6 and concludes with Sect. 7. The survey includes the details of how game theory is applied by the authors in automatic speech recognition in Sect. 6.2.

2 Games, Strategies, and Payoffs

A game contains three elements: a collection of players $N = \{1,\ldots,n\}$; a collection of strategies S_i for each player $i \in N$; and the payoff each player $i \in N$ gets towards the end of the game, which is a real-valued function expressed as $u_i : S_1 \times \cdots \times S_n \rightarrow R$.

If $s_i \in S_i$ is the strategy taken by the player i, then $s = s_1,\ldots,s_n$, a vector of strategies taken by all players is known as the strategy profile of the game and the set $S = S_1 \times \cdots \times S_n$ represents the set of all possible manners in which the players can select the strategies. It can be noticed that the payoff of each player is a function of his own strategy and the strategies of remaining players. Therefore, if s_{-i} represents $(n-1)$—dimensional vector of the strategies played by all other players excluding player i, then the payoff function $u_i : S_1 \times \cdots \times S_n \rightarrow R$ of a player i can also be written as $u_i(s_i, s_{-i}) :\rightarrow R$.

3 Examples of Games

We will discuss what are perhaps the most well-known and well-studied games, viz. Prisoner's Dilemma, Battle of the Sexes and Matching Pennies in this section.

3.1 Prisoner's Dilemma

Prisoner's Dilemma, one of the well-known games was introduced by Professor Tucker in 1950 [7, 8]. This game consists of two players who are on a trial for a crime. During interrogation they are not allowed to communicate with each other and the payoffs for each prisoner of the available choices are stated as follows:

- If the two accused chooses to remain silent, they will be sentenced to imprisonment for 2 years each.
- If one of them admits and the other chooses to remain silent, then the term of confessor will be reduced by 1 year and the silent accomplice will be sentenced to imprisonment for 5 years.
- Finally, if the two accused admit, they will be sentenced to imprisonment for a term of 4 years each.

The choices available to the players and the payoffs of the corresponding choices can be summarized concisely by means of the 2×2 matrix as follows (Fig. 1).

The game of Prisoner's Dilemma is an illustration of the simultaneous game, and the only stable solution for this game is that the two accused admit the crime, because in the remaining three possibilities, one of the accused can alter his choice from "silence" to "confession" and increase his own payoff. The solution of this game is known as the dominant strategy solution, one of the solution concepts of game theory.

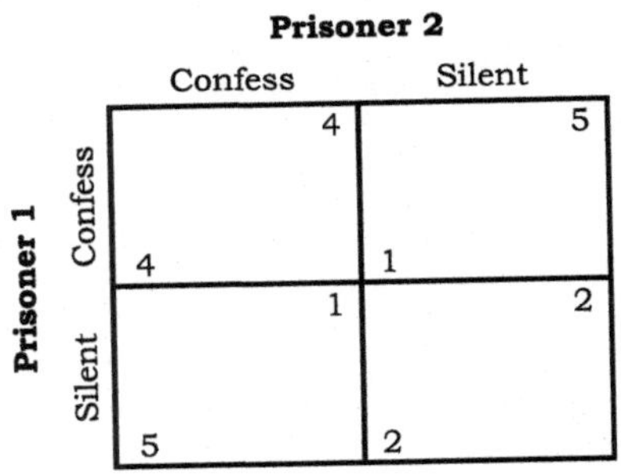

Fig. 1 Payoff matrix of Prisoner's Dilemma game

3.2 *Battle of the Sexes*

Another popular game widely studied in game theory is the "Battle of the Sexes" presented by Luce in 1959 [9]. This game consists of a boy and a girl, who have two alternatives for spending evening: going to a football game or a musical show and the payoff matrix of the available choices is given in Fig. 2.

Definitely, the solutions in which two of them choose to spend the evening individually are not stable because in both of the cases either of them can improve his/her own payoff by altering his/her choice. Conversely, the solutions in which both of them wish to spend the evening jointly are stable. This is an illustration of the game that does not have a dominant strategy solution for either of the players. On the contrary, there is more than one solution to the game; the boy prefers football game, but the girl prefers music show depending on their payoffs.

3.3 *Matching Pennies*

The Matching Pennies is yet another popular and widely studied games in the game theory. The game consists of two players who must flip a coin secretly to either the head or the tail and disclose the choices at the same time. If both pennies match, then Player 1 wins a penny from the other and holds two pennies. On the other hand, if the two pennies do not match, then Player 2 wins and receives a penny from the other and thus holds two pennies. In this game one player's gain is absolutely equal to the other player's loss and is therefore called a zero-sum game. The payoff matrix for the game is given below.

Certainly, this game has no stable solution. As an alternative, the best way to play the game is by randomizing their choices (Fig 3).

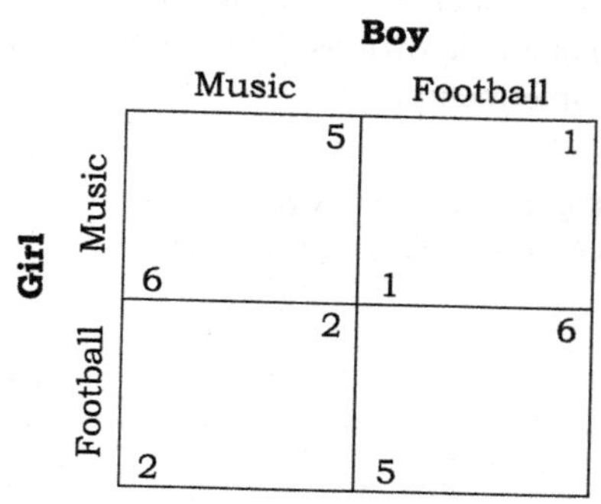

Fig. 2 Payoff matrix of Battle of the Sexes game

Fig. 3 Payoff matrix of Matching Pennies game

		Player 2	
		Head	Tail
Player 1	Head	-1 1	1 -1
	Tail	1 -1	-1 1

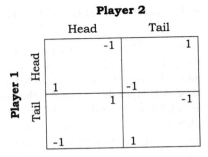

4 Types of Games

In game theory, games are categorized depending on several criteria. We describe below the taxonomy of games in game theory.

4.1 Cooperative and Non-cooperative Games

In a non-cooperative game, a player is an individual and each player strives to improve his own payoff and is not concerned about others' payoff. Alternatively, in a cooperative game, the player is not an individual, but groups of players form coalitions and strive to achieve a common goal by increasing his own payoff and the payoff of the coalition. Therefore, a cooperative game can be thought of as a competition among the groups of players instead of individual players.

Another feature that distinguishes cooperative and non-cooperative games is: the players in cooperative games can interact with each other and make binding agreements, whereas the players in non-cooperative games may interact but cannot make binding agreements.

4.2 Simultaneous and Sequential Games

Simultaneous games are also known as strategic or normal form games. In these games all the players make their choices simultaneously and if they do not play simultaneously, the latter players are uninformed of the former players' strategies. Conversely, in sequential games also known as dynamic or extensive games, the players make their choices successively and the latter players are informed of the strategies of the former players. Moreover, simultaneous games are represented using normal form representation, whereas sequential games are represented using extensive form. While Chess is an example of a sequential game, the Prisoner's dilemma and the Battle of the Sexes are the examples of simultaneous games.

4.3 Zero-Sum and Non-zero-sum Games

If the sum of each player's payoff is equal to zero toward the end of the game, then the game is called a zero-sum game; otherwise the game is called a non-zero-sum game. In the zero-sum game a team or individual's gain is absolutely the loss of another team or individual. On the other hand, in non-zero-sum game, the total gain or total loss toward the end of the game is distributed among all the players. Therefore, while the players in zero-sum games do not have any common interest, the players in non-zero-sum games can have some conflicting and some common interests.

4.4 Games with Perfect and Imperfect Information

A game in which all the players are informed about other players' moves during his chance is called a game with perfect information; otherwise the game is said to have imperfect information. While all sequential games like backgammon and chess have perfect information, simultaneous games, namely the Prisoner's Dilemma and the Battle of the Sexes have imperfect information.

4.5 Games with Complete and Incomplete Information

Frequently, perfect information is incorrectly interpreted as complete information, because they appear to be analogous. However, in games with complete information it is essential that every player has the information on the strategies and the payoffs available to other players, but not essentially the information on the other players' moves.

5 Solution Concepts in Game Theory

A solution concept is a formal rule that stipulates how the game should be played. Frequently, a solution concept can lead to multiple solutions, in which case, a game theorist must narrow down the solutions by restructuring the rule. The list of solution concepts in game theory is very huge; hence, we present some of the most commonly employed solution concepts, viz., the dominant strategy solution, the pure strategy Nash equilibrium, the mixed strategy Nash equilibrium, the core and the Shapley value.

5.1 *Dominant Strategy Solution*

A strategy profile $s \in S$ is the dominant strategy solution, if each player's strategy yields him a greater payoff than any other strategy, regardless of other players' strategy. That is,

$$u_i(s_i, s'_{-i}) \geq u_i(s'_i, s'_{-i}) \tag{1}$$

for all players i and each alternate strategy $s' \in S$.

It can be observed that the dominant strategy solution may not yield an optimal payoff to any of the players. For example, in the Prisoner's Dilemma, confession yields a greater payoff for both the players, irrespective of other player's choice and therefore the dominant strategy solution is mutually confessing. However, if both the players confess, the payoff is not optimal for either of them.

5.2 *Pure Strategy Nash Equilibrium*

Games rarely have dominant strategy solutions; hence a less stringent and widely acceptable solution is required. A more promising solution is a Nash equilibrium, which became a predominant solution concept in game theory with varied applications. Mathematically, a strategy profile $s \in S$ is the Nash equilibrium, if for each player i and each alternative strategy $s'_i \in S_i$,

$$u_i(s_i, s_{-i}) \geq u_i(s'_i, s_{-i}) \tag{2}$$

Alternatively, it can be stated that a player cannot increase his payoff by altering his chosen strategy from s_i to s'_i when all other players' choices at equilibrium are unaltered.

5.3 *Mixed Strategy Nash Equilibrium*

The Matching Pennies game is an illustration of a game that does not have any pure strategy Nash equilibrium solution. Yet, a stable solution can be realized in the Matching Pennies game if the players randomize their choices. Specifically, if each player chooses one of the two strategies with a probability of ½, then the expected payoff of each player is 0 and neither of them can increase his payoff by choosing an alternate probability distribution.

Usually, a strategy in which a player chooses his strategies by assigning a probability distribution over his set of possible choices is called a mixed strategy; and a strategy is a mixed strategy Nash equilibrium, if no player can improve his expected payoff by altering his choice when all other players' choices at the equilibrium are unchanged. It is to be noted that, expected payoffs are taken into account to determine the most favorable choice.

5.4 Core

It is stated earlier that, in a cooperative game, a group of players form coalitions, and a collective payoff is given to each coalition that specifies the worth of the coalition. Mathematically, a cooperative game is denoted by a tuple (N, v), where $N = \{1, ..., n\}$ is a set of players, and $v(S)$ is the collective payoff of the coalition $S \subseteq N$, which is a real valued function defined as $v : 2^N \to R$.

One of the problems in a coalition game is to equitably distribute $v(N)$ among the players of the coalition in such a way that the individual players prefer to be in the grand coalition N than to abandon it. Alternatively, a set of payoffs to the individual players of the grand coalition N are to be determined in such a way that no group of players can improve their payoff by abandoning the grand coalition.

Core is one of the solution concepts in cooperative game theory that resolves this problem. It is presented by Edgeworth in 1881 and is expressed as a set of feasible payoffs that are 'coalition optimal' [10]. In other words, the core is a set of feasible payoffs that cannot be improved by the collective action of any subset of the players, acting by themselves.

Formally, the Core is a set of feasible payoffs $x \in R^N$ that satisfies the following properties:

1. Budget Balance Property: $\sum_{i=1}^{N} x_i = v(N)$ and
2. Core Property: $\forall S \subseteq N, \sum_{i \in S} x_i \geq v(S)$

According to the budget balance property, all players jointly cannot obtain better payoff than the net payoff assigned to the grand coalition. Besides, the core property guarantees that the players do not threaten to abandon the grand coalition because their payoffs cannot be improved by playing in any other coalition.

5.5 Shapley Value

Although, Core solutions are Pareto optimal and stable, they can either be empty or not unique. Besides, they do not regard all the players in a fair way. Shapley value is an alternative solution concept in cooperative games that is single-valued and fair to all players. It is based on calculating the marginal contribution of each player i, and is expressed as the average extra value player i contributes to each of the possible coalition.

Mathematically, the Shapley value of a player i, in a coalition game (N, v) with N players is given as follows

$$\Phi_i(v) = \sum_{S \subseteq P \setminus \{i\}} \frac{|S|\,!(N - |S| - 1)!}{N\,!}(v(S \cup \{i\}) - v(S)) \quad (3)$$

A coalition S can be formed in $|S|!$ ways and can generate a payoff of $v(S)$. Adding a player i to the coalition S generates a payoff of $v(S \cup \{i\})$. Hence, the marginal contribution of a player i in the coalition S can be computed as $v(S \cup \{i\}) - v(S)$. The grand coalition N can be obtained by adding other $N - |S| - 1$ players to S which can be formed in $(N - |S| - 1)!$ ways. Consequently, the extra value the player i adds in each situation is $v(S \cup \{i\}) - v(S)$ multiplied by $|S|!(N - |S| - 1)!$ for each of the possible ways this particular situation arises. Summing over all possible subsets S and dividing by $N!$ gives the average excess value player i contributes to a coalition.

6 Applications of Game Theory in Machine Learning

Game theory being a branch of applied mathematics found its applications in economics, business, politics, biology, computer science and many other fields. Specifically, in this section, we illustrate a few of the important applications of game theory in machine learning that is a subfield of computer science and statistics.

6.1 Spam Detection

A spam detector or filter is a software used to discover unwanted email, and then disallow it from getting into a user's inbox. Spam detectors can either be aggressive by blocking the occasional legitimate email in order to keep more spam out of a user's inbox, or permit few extra spam messages so as to make sure that all legitimate emails pass through. In [11], interaction between spammer and spam detector is demonstrated as a sequential two-person non cooperative Stackelberg game. A spam that passes through the inbox is a cost to the spam detector, whereas a benefit to the spammer. A Stackelberg equilibrium is obtained if the spammer and spam detector play their best stratagem at the same time.

6.2 Natural Language Processing

Game theoretic modeling for deriving the meaning of an utterance is given in Language and Equilibrium [12] where the author demonstrates a distinct viewpoint on the natural language semantics and pragmatics. He begins with the idea that form and meaning correspondences are established as a consequence of balancing conflicting syntactic, informational, conventional, and flow constraints, and shows how to extract the meaning of an utterance by constructing a model as a system of mutually dependent games, that is a generalization of the Fregean Principle of Composition [13].

6.3 Gene Expression Analysis

Gene expression is the process by which genetic information is used in protein synthesis. Gene expression analysis is observing the picture of gene expressions (activity) in the sample of cells under investigation. In the thesis [14], the author employs game theory to quantitatively estimate the significance of each gene in controlling or provoking the condition of interest, for example, a tumor, taking into consideration how a gene associates in all subgroups of genes.

6.4 Robot Navigation

It is an important discipline of research to construct and execute a multi-robot system to do tasks such as searching, exploring, rescuing, and map-building. One of the key problems in multi-robot systems is determining how robots coordinate or cooperate among themselves in such a way that the global performance is optimized. The problem of coordination is further complicated, if the environment is dynamic. In [15], a dynamic-programming formulation is proposed for determining the payoff function for each robot, where travel costs as well as interaction between robots are taken into account. Depending on this payoff function, the robots play a cooperative, non zero-sum game, and both pure Nash Equilibrium and Mixed-Strategy Nash Equilibrium solutions are presented to help the robots navigate in such a way so as to achieve an optimal global performance.

6.5 Image Segmentation

Image segmentation decomposes an image into segments or regions containing "identical" pixels. It is used to identify objects and trace the boundaries within images. The problem of image segmentation is formulated as an evolutionary cluster game played by two players in [16]. At the same time, each player chooses a pixel that must be clustered, and obtains a payoff according to the similarity that the chosen pixel has with respect to the pixel chosen by the opponent. The evolutionary stable algorithm finds a cluster of pixels in the given image. Next, the similarity matrix is recomputed for the unlabeled pixel set and the evolutionary stable algorithm is run to find another segment. The process is iterated until all pixels have been labeled.

6.6 *Disaster Management*

In emergency management after a major disaster, it is necessary that organizations and individuals who are familiar and unfamiliar with disaster environment be able to manage and use the resources effectively to respond in time. Determining when, where, how, and with whom these resources should be distributed is a difficult problem in emergency or disaster management due to cross-cultural differences and interoperability issues. In [17], a game theoretic formulation is proposed to maximize the efficiency of actors involved in emergency management.

6.7 *Information Retrieval Systems*

In the information retrieval system, when the user inputs a query, it searches for the objects in the collection that match the query. In the process, more than one object may match the query, perhaps with varying degrees of relevance. Therefore, information retrieval systems compute rankings of the objects depending on how well an object in the collection matches the query. The top ranking objects are then presented to the user. In [18], the information retrieval process is modeled as a game between two abstract players. The "intellectual crowd" that uses the search engines is one player and the community of information retrieval systems is another player. The authors apply game theory by treating the search log as Nash equilibrium strategies and solve the inverse problem of calculating the payoff functions using the Alpha model, where the search log contains the statistics of users' queries and search engines' replies.

6.8 *Recommendation Systems*

Recommendation systems are the software tools that provide recommendations to the user where the recommendations are related to decision making processes such as what items to be bought, what music to be heard, what online news to read and so on. Recommendation systems seek ratings of the items from the users that purchased or viewed them, and then aggregate the ratings to generate personalized recommendations for each user. Consequently, the quality of the recommendations for each user depends on ratings provided by all users. However, users prefer to maintain privacy by not disclosing much information about their personal preferences. On the other hand, they would like to get high-quality recommendations. In [7], the tradeoff between privacy preservation and high-quality recommendations is addressed by employing game theory to model the interaction of users and derives a Nash equilibrium point. At Nash equilibrium the rating' strategy of each user is such that no user can benefit in terms of improving his privacy by unilaterally deviating from that point.

6.9 Automatic Speech Recognition

Conventional automatic speech recognition systems take into account all features of an input feature vector for training and recognition. But, most of the features are redundant and irrelevant. Consequently, training an automatic speech recognition system on such a feature vector can be ineffective due to over-fitting. Furthermore, each sound is characterized by a different set of features. To deal with these problems, feature subset that is relevant for identifying each sound is investigated in [19–21]. In [19–21] a cooperative game theory based framework is proposed, where the features are the players and a group of players form a coalition that maximizes the accuracy of the system.

7 Conclusions

Game theory deals with the systems in which there are multiple agents having possibly conflicting objectives. Most problems in machine learning can be modeled as a game consisting of two or more agents, trying to maximize his own payoff. Consequently, various solution concepts of game theory can be applied in machine learning, and this paper presented some of the applications of game theory in machine learning.

References

1. Walker, Paul. "A chronology of game theory." (1995).
2. Aumann, Robert J., and Michael Maschler. "Game theoretic analysis of a bankruptcy problem from the Talmud." Journal of Economic Theory 36.2 (1985): 195–213.
3. Bellhouse, David. "The problem of Waldegrave." Electronic Journal for the History of Probability and Statistics 3.2 (2007): 1–12.

4. Von Neumann, John, and Oskar Morgenstern. Theory of Games and Economic Behavior (60th Anniversary Commemorative Edition). Princeton university press, 2007.
5. Nisan, Noam, Tim Roughgarden, Eva Tardos, and Vijay V. Vazirani, Eds. Algorithmic game theory. Cambridge University Press, 2007.
6. A. Blum (PI), et al. "Machine Learning, Game Theory, and Mechanism Design for a Networked World". Carnegie Mellon University School of Computer Science, 22 July 2014. https://www.cs.cmu.edu/~mblum/search/AGTML35.pdf/.
7. Kuhn, Steven, "Prisoner's Dilemma", The Stanford Encyclopedia of Philosophy (Fall 2014 Edition), Edward N. Zalta (ed.), http://plato.stanford.edu/archives/fall2014/entries/prisoner-dilemma/.
8. Rapoport, Anatol. Prisoner's dilemma: A study in conflict and cooperation. Vol. 165. University of Michigan Press, 1965.
9. Luce, R. Duncan, and Howard Raiffa. Games and decisions: Introduction and critical survey. Courier Dover Publications, 2012.
10. Edgeworth, Francis Ysidro. Mathematical psychics: An essay on the application of mathematics to the moral sciences. C. Keagann Paul, 1881.
11. Brückner, Michael, and Tobias Scheffer. "Stackelberg games for adversarial prediction problems." Proceedings of the 17th ACM SIGKDD international conference on Knowledge discovery and data mining. ACM, 2011.
12. Prashant Parikh. 2010. Language and Equilibrium. The MIT Press.
13. Szabó, Zoltán Gendler, "Compositionality", The Stanford Encyclopedia of Philosophy (Fall 2013 Edition), Edward N. Zalta (ed.), http://plato.stanford.edu/archives/fall2013/entries/compositionality/.
14. Moretti, Stefano. "Game Theory applied to gene expression analysis." 4OR 7.2 (2009): 195–198.
15. Meng, Yan. "Multi-robot searching using game-theory based approach." International Journal of Advanced Robotic Systems 5.4 (2008): 341–350.
16. Shen, Dan, et al. "A clustering game based framework for image segmentation." Information Science, Signal Processing and their Applications (ISSPA), 2012 11th International Conference on. IEEE, 2012.
17. Coles, John, and Jun Zhuang. "Decisions in disaster recovery operations: a game theoretic perspective on organization cooperation." Journal of Homeland Security and Emergency Management 8.1 (2011).
18. Parfionov, George, and Roman Zapatrin. "Memento Ludi: Information Retrieval from a Game-Theoretic Perspective", Contributions to game theory and management. Vol. IV. Collected papers: 339.
19. Rekha, J. Ujwala, K. Shahu Chatrapati, and A. Vinaya Babu. "Game theoretic approach for automatic speech segmentation and recognition." Electrical & Electronics Engineers in Israel (IEEEI), 2014 IEEE 28th Convention of. IEEE, 2014.
20. Rekha, J. Ujwala, Shahu, K. Chatrapati, Babu, A. Vinaya. "Feature selection using game theory for phoneme based speech recognition." Contemporary Computing and Informatics (IC3I), 2014 International Conference on. IEEE, 2014.
21. Rekha, J. Ujwala, Shahu, K. Chatrapati, Babu, A. Vinaya. "Feature selection for phoneme recognition using a cooperative game theory based framework". Multimedia, Communication and Computing Application: Proceedings of the 2014 International Conference on Multimedia, Communication and Computing Application (MCCA 2014), 191–195, CRC Press, 2015.

A Study on Speech Processing

J. Ujwala Rekha, K. Shahu Chatrapati and A. Vinaya Babu

Abstract Speech is the most natural means of communication in human-to-human interactions. Automatic Speech Recognition (ASR) is the application of technology in developing machines that can autonomously transcribe a speech into a text in the real-time. This paper presents a short review of ASR systems. Fundamentally, the design of speech recognition system involves three major processes such as feature extraction, acoustic modeling and classification. Consequently, emphasis is laid on describing essential principles of the various techniques employed in each of these processes. On the other hand, it also presents the milestones in the speech processing research to date.

Keywords Speech processing · ASR · Modelling

1 Introduction

The research on speech processing can be traced back to the 18th century and the initial emphasis is laid on producing speech rather than recognizing speech. For instance, in 1769, a German physicist and an engineer named Christian Gottlieb Kratzenstein, constructed resonant cavities that produced the vowel sounds when excited by a vibrating reed [1]. At the same time in 1769, Wolfgang von Kempelen an Austro-Hungarian initiated the construction of a physically operated speaking

---

J. Ujwala Rekha (✉) · A. Vinaya Babu
Department of Computer Science and Engineering,
JNTUH College of Engineering, Hyderabad, India
e-mail: ujwala_rekha@yahoo.com

A. Vinaya Babu
e-mail: dravinayababu@gmail.com

K. Shahu Chatrapati
Department of Computer Science and Engineering,
JNTUH College of Engineering, Jagtial, India
e-mail: shahujntu@gmail.com

© Springer India 2016
S.C. Satapathy et al. (eds.), *Information Systems Design and Intelligent Applications*, Advances in Intelligent Systems and Computing 435,
DOI 10.1007/978-81-322-2757-1_22

engine that modeled the human vocal tract and produced connected words. After two failed attempts, and nearly 20 years of experiments, he unveiled the final version of the engine in the year 1791 [2]. The machine is very ingenious and could generate 19 consonant sounds; however, it requires substantial skill and proficiency in operating. Subsequently, in late 1800s, Sir Charles Wheatstone, an English scientist enhanced the von Kempelen's speaking machine [3]. Furthermore, machines of analogous type, like Euphonia by Joseph Faber (1830–40s) [4] and talking device by Riesz [5] were developed in the 19th century, but there was no real progress in the field later.

In the beginning of the 20th century, Harvey Fletcher and fellow scientists at Bell Laboratories showed that the characteristics of a speech signal extend over a broad range of frequencies, and stated the correlation between the speech characteristics and the spectrum [6]. Influenced by Fletcher's work in 1939, Homer Dudley built an electrical speech synthesizer called VODER (Voice Operating Demonstrator) [7] by spectral synthesis method, which is analogous to Wheatstone's mechanical device. Thus, the investigations of Fletcher and Dudley ascertained that the sound produced is characterized by the spectrum and have laid the basis for speech recognition. As a result, the algorithms of contemporary speech recognition systems are dependent predominantly on extraction of features from either the spectrum or its variant Cepstrum.

The first speech recognition system was pioneered jointly by Davis, Biddulph, and Balashek at Bell laboratories in 1952, which could recognize isolated digits and is called Audrey (Automatic Digit Recognizer) [8]. Subsequently, after 10 years, IBM demonstrated the "SHOEBOX" device in the World Fair in 1962, which can recognize ten digits and six mathematical operations [9].

During the 1970s, U.S. Department of Defense sponsored DARPA Speech Understanding (SUR) program that resulted in the development of Carnegie Mellon's "Harpy" [10], a speech recognizer that can recognize 1,011 words. The Harpy's development is remarkable as it pioneered a cost-effective search algorithm, namely beam search. In addition, the 1970s witnessed significant progress in speech recognition research with the introduction of Hidden Markov Models (HMMs) that were applied in speech processing applications, for instance DRAGON developed by Jim Baker at CMU 1952 [11]. Concurrently, Jelinek at IBM built a continuous speech recognition system by employing statistical methods [12].

In the following decade, the lexicon of speech recognizers grew from about a few hundred words to several thousand words. For example, while, Kurzweil's speech-to-text program [13] could understand 1,000 words, the IBMs Tangora [14] could understand 20,000 words. In the later years, besides, increasing the size of the vocabularies, the researchers focused on the issues like speaker independence, noisy environment and conversational speech.

In the 1990s, with the increasing potential of information processing systems, speech recognition software turned out to be viable for common masses. For example, Dragon Dictate and Dragon Naturally speaking were available in the commercial market.

In 2008, speech recognition industry made foray into the mobile market with Google Voice Search app developed for the iPhone. The voice requests are sent to the Google's cloud, where they are processed and matched against the huge pool of human speech samples. In 2011, Siri entered the market and captured the world by storm. Similar to Google Voice Search app, Siri employs cloud-based processing and provides contextual reply by using AI techniques. While Google released another app in 2012, namely Google Now to compete with Apple's Siri, Microsoft released Cortana in 2014 for Windows phone analogous in function to Siri and Google Now. Even though, these intelligent assistants' dependence on the cloud is an added benefit for the source of its power, it is an inconvenience because the speech is transmitted to remote servers for processing which is time-consuming.

On the other hand, with more advances in speech processing algorithms and information processing systems, the inadequacies of the present intelligent personal assistants will be resolved leading to ubiquitous speech. For instance, we may witness speech recognition technology embedded in household items as well.

2 Basic Model of a Speech Recognizer

The basic model of an automatic speech recognition step is shown in the Fig. 1. Fundamentally, three processes are required for designing a speech recognition system, namely the feature extraction, acoustic modeling and classification. Initially, the feature vectors $X = x_1, \ldots x_N$ are obtained from the input speech

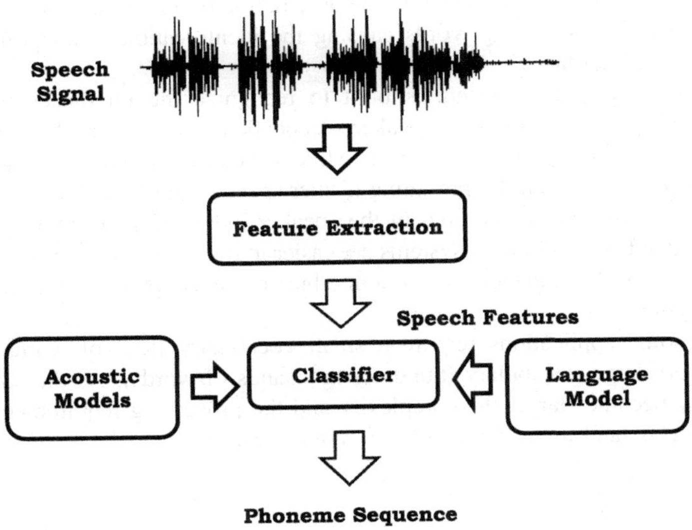

Fig. 1 Basic model of a speech recognizer

waveform, then the classification algorithm selects a label l_i with the maximum likelihood as follows

$$l_i = \arg\max_{\mathrm{i}} P(l_i|X) = \arg\max_{\mathrm{i}} \frac{P(X|l_i)P(l_i)}{P(X)} \tag{1}$$

where $P(X|l_i)$ is an acoustic model that defines label l_i and $P(l_i)$ is the language model. Prior to classification, an acoustic model for each linguistic label must be generated, by running the training algorithm on the training set that contains feature vectors and their associated labels, where an acoustic model represents the statistical relationship between the feature vectors and the associated label.

3 Classification of Speech Recognition Systems

Speech recognition systems can be classified depending on various parameters, some of which are listed below:

- Isolated-word/Continuous speech recognition systems
- Speaker-dependent/Speaker-independent recognition systems
- Small/Medium/Large/Very-large vocabulary recognition systems

Isolated-word speech recognition system requires the utterances to be separated by pauses, and is the simplest to implement since the endpoints can be detected effortlessly and the articulation of an utterance is not affected by other utterances. On the contrary, in a continuous speech recognition system the utterances are not isolated by the pauses. Consequently, the articulation of an utterance is affected by the preceding and following words causing the identification of endpoints of an utterance a difficult task.

A speaker dependent system is built to recognize the utterances of certain speakers only, on the contrary a speaker independent system is built to recognize the utterances of any speaker. Therefore, while a speaker is required to register in a speaker dependent system by providing his/her speech samples before utilizing the system, no registration is required in the speaker independent system. Generally, speaker dependent recognition systems are easier to develop, economical, and more precise than speaker independent systems, but not as adaptable as speaker independent systems.

While some applications require a small vocabulary (tens of words), others demand very large vocabulary (hundreds/thousands of words). As the size of the vocabulary becomes large, the complexity and the processing requirements of the system ascends and the accuracy of the system descends.

4 Feature Extraction

Feature extraction is a process that maps a continuous speech waveform into a sequence of feature vectors. Initially, the continuous speech waveform is split into frames of equal length, for instance 25 ms, as the speech signal is considered to be stationary within this period. Moreover, the frames are sampled so that the frames are overlapping (for instance, every 10 ms for frames of length 25 ms). The sampled time varying speech frames are then transformed into the spectral domain by estimating the energy of the signal across a range of filters or frequency bands, that is to say, a Fourier transform of the continuous speech wave form is obtained. The pursuit for better speech parameterization persuaded the discovery of several features, which are advantageous in particular settings and applications. Often, the speech features have multiple implementations that differed either in the number of filters employed, the shape of the filters, the spacing between the filters, the bandwidth of the filters, or the manner in which the spectrum is warped.

Speech recognition systems are either developed from all features in a feature set or a subset of features as in [15, 16]. In addition, features from multiple feature streams can be combined to form an augmented feature vector. While certain recognition systems operate on the augmented feature set [17], others subject the augmented feature set to dimensionality reduction before modeling [18, 19].

We present below some of the widely used features in developing a speech recognition system.

4.1 Linear Predictive Coding

The source-filter model was proposed by Gunner Fant in 1960 [20], and according to the proposed model, the production of speech constitutes a two-step process: generation of periodic or aperiodic sound from the sources such as vocal chords and/or glottis, and modulation of the generated sound by the resonances of the vocal tract. Consequently, the shape of the spectrum and the sound produced can be differentiated by the properties of the sources and filters.

In the implementation of the source-filter model, the voiced sound is modeled as a periodic impulse train and an unvoiced sound is modeled as a white noise. Moreover, the filter is approximated by an all-pole model, in which the coefficients of the model are calculated by linear prediction analysis. In linear prediction, a speech sample at a specific point is approximated as a linear combination of the previous speech samples [21].

4.2 Cepstral Coefficients

According to the source-filter model, the produced speech is the response of the vocal tract filter to the excitation source. Consequently, to analyze and model the source and the filter properties of the speech individually, the source and the filter components have to be separated from the speech. The cepstrum introduced by Bogert et al. [22] separates the source and the filter components from the speech without any a priori knowledge about either of them by taking the inverse discrete Fourier transform of the log spectrum.

4.3 Mel-Frequency Cepstral Coefficients (MFCCs)

Generally, filters spread on the Mel scale approximate the response of the human auditory system more accurately than the linearly-spread filters employed in the normal cepstrum. Therefore, Davis and Mermelstein introduced Mel-frequency cepstrum [23, 24] that warps the filters on the Mel-scale, and the coefficients of the Mel-frequency ceptstrum can be computed as follows:

i. Compute the Fourier transform of the short-term signal.
ii. Transform the obtained spectrum by mapping onto the Mel scale and calculate the energy in each filter bank.
iii. Obtain the logarithm of the calculated filter bank energies.
iv. Apply the discrete cosine transform to the log filter bank energies.

The coefficients of the discrete cosine transform are the Mel-frequency Cepstral coefficients. There are other implementations for computing MFCCs, that vary in the shape or spacing of the filters employed in the mapping of the scale [25].

4.4 Perceptual Linear Predictive Coefficients

The Perceptual Linear Prediction (PLP) analysis is based on the psychophysics of hearing and was presented by Hermansky [26]. PLP is a linear prediction method, and prior to prediction, it converts the short-term signal by various psychophysical transformations and the method for computing PLP coefficients is given as follows:

i. Obtain the Fourier transform of the short-term signal.
ii. Warp the spectrum obtained along its axis of frequency into the Bark frequency.
iii. Convolve the warped spectrum with the simulated critical band filter to model the critical-band integration of the human auditory system.
iv. Downsize the samples of the convolved spectrum at intervals of approximately 1 Bark.

v. Pre-emphasize the sampled output by the simulated equal-loudness curve to model the sensitivity of hearing
vi. According to the power law of Stevens [27], raise the equalized values to the power of 0.33.
vii. Compute the linear predictive coefficients of the resulting auditory warped spectrum.
viii. Finally, obtain the cepstral coefficients from the linear predictive coefficients that are computed.

4.5 RASTA Filtering

Relative Spectral Processing (RASTA) is used to remove the steady-state background noise from the speech. Principally, RASTA filtering assumes that the steady-state background noise varies gradually compared to the speech. Therefore, it operates by replacing the critical-band short-term spectrum of the PLP processing, with a spectrum realized through filtering each frequency band by a band-pass filter, containing a sharp spectral zero at the zero frequency. Thus, the band-pass filters in each channel, filter the steady-state or slowly varying background noise in each channel. The detailed mechanism of RASTA filtering is described in [28, 29] and is described below. For each short-term frame do the following:

i. Obtain the critical-band power spectrum as in PLP processing.
ii. Compress the spectral amplitudes by non-linear static transformation
iii. Get the time trajectory of each compressed spectral component, and do band-pass filtering.
iv. Decompress the filtered spectrum by reverse non-linear static transformation.
v. Multiply the resultant spectrum by equal-loudness curve and raise to the power of 0.33 analogous to PLP analysis.
vi. Finally, obtain LPC coefficients for the resultant spectrum.

4.6 Spectral Entropy Features

In signal processing, the "peakiness" of a distribution is determined by a measure called "spectral entropy". Particularly, in speech recognition applications, the spectral entropy features are employed to obtain the formants of the speech [17, 30]. At first, the spectrum is transformed to a Probability Mass Function (PMF) by normalization in each sub-band, then the spectral entropy features are derived by calculating the entropy in each sub-band.

5 Acoustic Modeling and Classification

Acoustic modeling is the mechanism of constructing a model that is able to infer the labels of a speech signal given the set of features. The mechanism involves, training corpus consisting of the pairs of the set of features and the associated label to be supplied to the training algorithm, which examines the training corpus and approximates the parameters of the model by iteratively adapting the parameters, such that the objective criterion is met.

In this section, some of the widely employed acoustic models and the classification algorithms are presented.

5.1 *Dynamic Time Warping Algorithm*

In the template-based speech recognition, a template for each word in the vocabulary is stored, and the incoming feature vectors are compared against the feature vectors of each template by summing the pairwise distances to find the best match. However, different renditions of the same word are rarely spoken at the same speed resulting in variable-length spectral sequences. Therefore, differences in the speaking rate should not influence the dissimilarity score between different renditions of the same word.

Dynamic Time Warping (DTW) algorithm presented by Sakoe and Chiba [31], determines an optimal match between two sequences of feature vectors with variable-length as follows. Let the two sequences of feature vectors, be positioned on the edges of a lattice as displayed in the Fig. 2, with the test pattern on the bottom edge and the template or reference pattern on the left edge of the lattice. Each cell of the lattice contains the distance between corresponding vectors of the two sequences. The path through the lattice that minimizes the aggregate distance between the two sequences is the best match between the sequences.

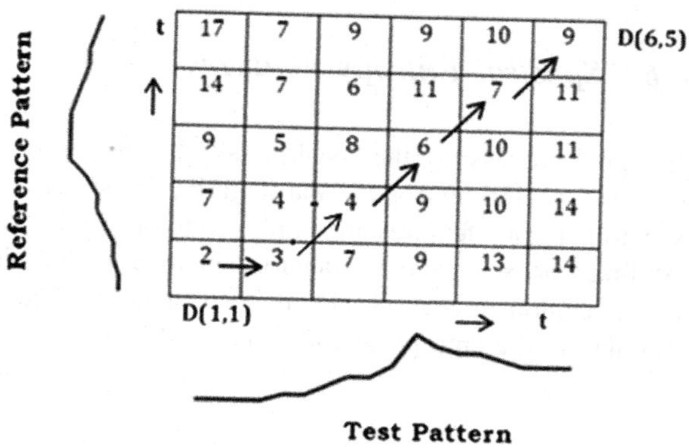

Fig. 2 An illustration of dynamic time warping

For example, the pathway along the cursors is the best path in the Fig. 2. Together the first and the second vectors of the test pattern pair with the first vector of the template, while the third vector of the test pattern pairs with the second vector of the template, the fourth pairs with the third, the fifth to the fourth, and the sixth vector of the test pattern pairs with the fifth vector of the template.

It can be observed that, as the length of the feature vector sequences increases, the number of possible paths through the lattice also increases. Yet, many of the associations can be avoided, for instance the first element of the test pattern is implausible to match with the last element of the template. The DTW algorithm exploits some inferences about the likely solution, making the comparison cost-effective.

5.2 *Hidden Markov Models*

An HMM is a finite automaton, in which the sequence of observation vectors produced is visible while the sequence of the states the model made transitions in order to produce the observed sequence of vectors is hidden. The HMMs were pioneered by Baum and his team [32] around 1960s and 1970s, and was employed in speech processing applications by Baker at CMU [11] and Jelinek at IBM [12] around 1970s.

There are quite a number of variations of and the extensions to HMMs, for instance, discrete-density HMMs [33], continuous-density HMMs [34–36], context-dependent HMMs [37], context-independent HMMs [38], tied-state HMMs [39], variable-duration HMMs [40], segmental HMMs [41], conditional Gaussian HMMs [42], and long temporal context HMMs [43] employed in speech recognition tasks. In this section, only the basic model of HMM, and the issues to be resolved while using HMMs in speech recognition task are described.

In acoustic modeling based on HMMs, each linguistic label is modeled by an HMM of the form shown in Fig. 3. Formally, an HMM moves from the state s_i to

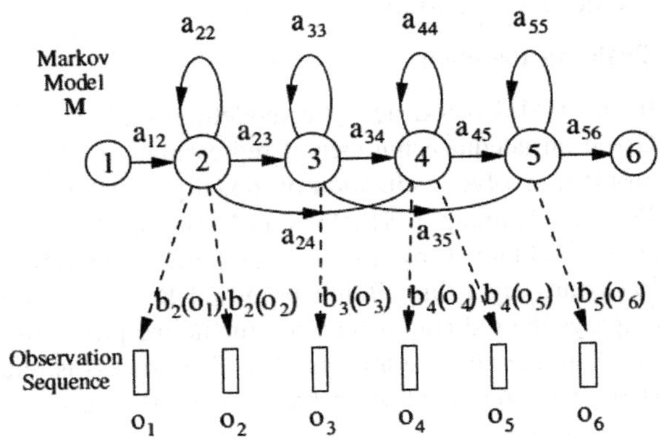

Fig. 3 An illustration of HMM [44]

the state s_j with a probability of a_{ij} every unit of time, and generates an observation vector o_t with a probability distribution $b_j(o_t)$ corresponding to the state s_j.

When applying HMMs to speech recognition tasks, the following three basic issues have to be resolved:

- Given a corpus of training data that contains a sufficient number of representative examples of each linguistic label, how to calculate the parameters of the HMM that describe the linguistic label.
- How to calculate the probability that a trained HMM generates a sequence of observations.
- Given an observation sequence and a trained HMM, how to decode the most likely sequence of states that produced the observation sequence.

Recognition and Decoding Problem

The second and third issues listed above can be paraphrased as the recognition and decoding problems respectively. Let $O = o_1, o_2, \ldots, o_t$ be an observation sequence, and $X = x(1), x(2), \ldots, x(T)$ be the sequence of states, then the probability that an observation sequence O is generated by HMM M going through state transitions defined by X can be computed as follows:

$$P(O,X|M) = a_{x(0)x(1)}b_{x(1)}(o_1)a_{x(1)x(2)}b_{x(2)}(o_2)a_{x(2)x(3)}b_{x(3)}(o_3)\ldots \quad (2)$$

As the sequence of state transitions is hidden in an HMM, the probability can be computed by determining the most likely state sequence as follows:

$$P(O|M) = \max_x \left\{ a_{x(0)x(1)} \prod_{t=1}^{T} b_{x(t)}(o_t)a_{x(t)x(t+1)} \right\} \quad (3)$$

The estimation of Eq. 3 and the most probable state sequence that produced the given observation sequence, though apparently simple are not computationally straightforward. Fortunately, dynamic programming methods, namely, Forward-Backward and Viterbi algorithms exist [45] that solve the two issues of recognition and decoding respectively.

Training Problem

In the HMM-based acoustic modeling, each HMM represents a linguistic label. Given an adequate number of representative instances of each label, the training problem involves estimation of model parameters $\{a_{ij}\}$ and $b_j(.)$ that maximize the $P(O|M)$. Training an HMM is an NP-complete problem, and hence the parameters can be obtained by locally maximizing $P(O|M)$. Centered on the Forward-Backward algorithm, Baum proposed Baum-Welch Re-estimation algorithm that employs the old parameters to estimate the probability $P(O|M)$ which is later used to obtain the new parameters [44]. The process is reiterated until a local maximum is attained and is called an Expectation-Maximization method.

5.3 *Stochastic Segment Models*

In contrast to the HMMs that model the observation sequences at the frame-level, in segment modeling, the observation sequences are modeled at the segment-level, where a segment is a variable-length sequence of observations constituting one linguistic label [46, 47]. The mechanism for stochastic segment modeling is given in [47], where a variable-length observation sequence is warped on a time-scale to generate a fixed-length representation. Subsequently, assuming that the samples in a segment are independent, it employs a multivariate Gaussian density function to model the samples of a time-warped segment. Let $Y = [y_1, \ldots, y_m]$ be the re-sampled and time-warped segment, then assuming that the samples in a segment are independent the classification algorithm selects the label l_i with the maximum a posteriori probability as follows

$$P(Y|l_i) = \prod_{j=1}^{m} p_j(y_j|l_i) \tag{4}$$

5.4 *Conditional Random Fields*

A Conditional Random Field (CRF) is a discriminative model, and contrary to the HMMs that estimate the joint probability of observing a sequence of observations and a label, a CRF estimates the a posteriori probability of the label sequence given the sequence of observations. To put it another way, a CRF uses the features of the observed sequence to determine the likelihood of the label sequence. In general, the process of speech recognition task can be represented as a linear-chain CRF [48] as depicted in Fig. 4, where $X = x_1, \ldots, x_n$ is a sequence of observations and $Y = y_1, \ldots, y_n$ is the sequence of labels given to X. As can be seen, the label of a sample is dependent on the label of directly preceding sample and the observations of the sample. Accordingly, the dependencies in a linear-chain can be represented via two types of feature functions: "state feature function" $s_i(y_t, x_t)$ defined over the pairs of label and observations at time t; and "transition feature function" $f_i(y_{t-1}, y_t, x_t)$ defined over pairs of labels at time t and $t - 1$, and observations at t. Furthermore, the likelihood of the label sequence Y given an observation sequence X can be calculated as follows:

$$P(Y|X) = \frac{\exp \sum_t \left(\sum_i \lambda_i s_i(y_t, x_t) + \mu_i f_i(y_t, y_{t-1}, x_t) \right)}{Z(X)} \tag{5}$$

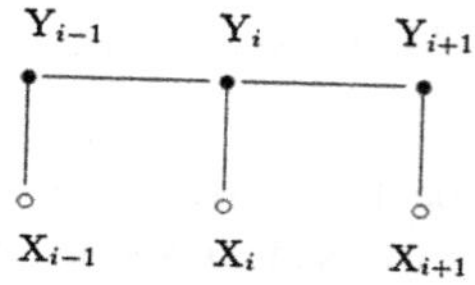

Fig. 4 An illustration of linear CRF

where t extends over all samples of a given observation sequence and $Z(X)$ is a normalization constant got by performing summation over all possible label sequences of Y calculated as follows:

$$Z(X) = \sum_{y} \left(\exp \sum_{t} \left(\sum_{i} \lambda_i s_i(y_t, x_t) + \mu_i f_i(y_t, y_{t-1}, x_t) \right) \right) \quad (6)$$

where the weights λ_i and μ_i are the corresponding weights of the state feature function and the transition feature function for the ith feature in the feature vector. Thus, a CRF is modeled by the feature functions $s_i(.)$, $f_i(.)$, and the corresponding weights λ_i and μ_i respectively.

Training Conditional Random Fields

Training a conditional random field involves determining feature functions $s_i(.)$, $f_i(.)$, and the associated weights λ_i and μ_i respectively that maximize the conditional probability $P(Y|X)$ over a set of training data. Methods like quasi-Newton gradient descent and stochastic gradient descent algorithms are applied in speech recognition systems [49] for obtaining the feature functions and the associated weights of a CRF.

Classification Using Conditional Random Fields

Finally, classification using CRFs consists of determining the label sequence $\hat{Y}$ that maximizes the Eq. 5 given an observation sequence X calculated as follows:

$$\hat{y} = \arg\max_{Y} P(Y|X) \quad (7)$$

which can be obtained by the application of the Viterbi algorithm.

5.5 *Artificial Neural Networks*

The Artificial neural network (ANN) is a computational model based on the biological neural network applied in machine learning algorithms. Generally, it comprises of a simple, interconnected non-linear computational elements of the form shown in Fig. 5, where $x_1, x_2, x_3, .., x_n$ are the inputs and $w_1, w_2, w_3, \ldots, w_n$ are the corresponding weights. As can be seen, the sum of weighted inputs $x_1, x_2, x_3, .., x_n$ with corresponding weights $w_1, w_2, w_3, \ldots, w_n$ is thresholded

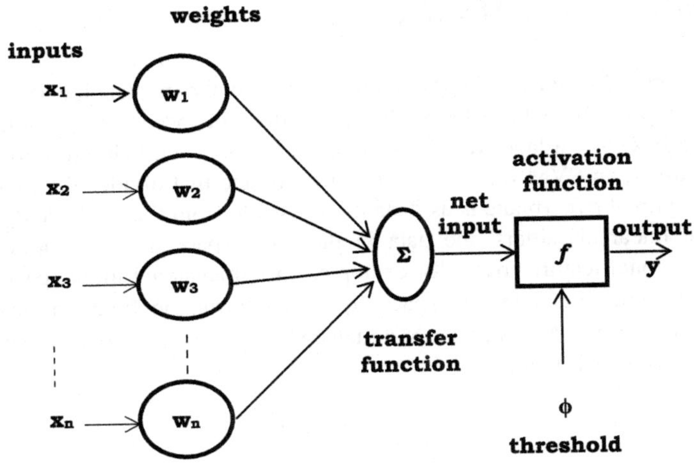

Fig. 5 A typical computational element of an ANN

with an off-set ϕ and non-linearly compressed to generate the output y as shown below:

$$y = f\left(\sum_{i=1}^{n} w_i x_i - \phi\right) \quad (8)$$

where f is a non-linear function.

In automatic speech recognition, the input neurons are supplied with acoustic features, that are then weighted, added, and transformed by a nonlinear function, which in turn are supplied to successive neurons. This activity is repeated until an output neuron is triggered, which specifies the sound produced.

An ANN is generally defined by three types of parameters:

i. The number of layers in the network and the topology of the interconnection network.
ii. The learning algorithm employed for adapting the weights of each computation element.
iii. The activation function that maps the weighted input of a computational element to its output activation.

Once the number of layers, the topology of the interconnection network, the learning algorithm and the activation function are well-defined, training the neural network constitutes executing the learning algorithm on a corpus of labeled training data, to determine the weights and the threshold of each computational element.

A wide variety of neural networks are used in speech recognition, for instance, recurrent networks [50], multi-layer perceptrons [51], Hopfield networks [52], time-delay neural networks [53], self-organizing maps [54], and so on.

5.6 Support Vector Machines

Support Vector Machines (SVMs) applied extensively in machine learning and datamining applications have found the applications in speech recognition tasks as well [55–58]. Given a labeled corpus of training data, each belonging to either of the two classes, an SVM finds an optimal hyperplane that divides the two classes, where an optimal hyperplane is as distant as possible from the example data. Since SVMs are linear classifiers, the data in the input space are transformed into a high-dimensional feature space, for example, a n-dimensional input space is mapped to a N-dimensional feature space by a transformation function such as $\phi : R^n \rightarrow R^N$, where $n > N$. Later, an N-dimensional optimal hyperplane w and a bias b are calculated such that

$$\arg\min_{w,b} \frac{1}{2}||w||^2 \tag{9}$$

$$\text{subject to } y_i(w \cdot \phi(x_i) - b) \geq 1 \tag{10}$$

where x_i is a known data in the training set and $y_i \in \{-1, 1\}$ is the label of the x_i and w is a linear combination of a set of vectors in the N-dimensional feature space expressed as follows:

$$w = \sum \alpha_i y_i \phi(x_i) \tag{11}$$

where the summation is done over all vectors in the training set. Consequently, finding an optimal hyperplane that is as distant as possible from the examples depends on $\alpha' s$.

Once, an optimal hyperplane is obtained, the class of an unknown data x can be determined as one of the two classes, depending on whether $w \cdot \phi(x) - b \geq 1$ or $w \cdot \phi(x) - b \leq 1$.

5.7 Vector Quantization

Vector Quantization (VQ) is a technique that maps an N-dimensional feature vector x consisting of real-valued and continuous-random variables onto another real and discrete-valued N-dimensional feature vector y_i. Formally, we can express $y_i = q(x)$ where the N-dimensional feature vector y_i is derived from a codebook consisting of a finite set of feature vectors $Y = \{y_i,\ 1 \leq i \leq L\}$ such that the quantization error between x and y_i is minimized. Generally, the quantization error in speech recognition applications is the dissimilarity or the distortion score between the vectors. Therefore, $q(x)$ can be expressed as follows

$$q(x) = \arg\min_{y_i} d(x, y_i) \tag{12}$$

The construction of the codebook is called training or populating the codebook and k-means clustering algorithm for codebook design as described in [59] is presented below:

i. Arbitrarily select L vectors from the corpus of training data as the initial collection of codewords in the codebook, where L is the size of the codebook, which is equal to the number of distinct labels.
ii. For every vector in the training set, identify the closest codeword in the codebook, and assign the vector to the corresponding cluster.
iii. Compute the centroid of each cluster and update the codeword with the centroid.
iv. Repeat steps ii and iii until the overall distortion is below a certain threshold.

Since the computational complexity of VQ is low when compared to HMMs and DTW, it is used extensively in speech recognition applications [60]. But, one of the main shortcomings is that, it does not deal with the temporal aspects of a speech. To mitigate this problem, often a hybrid approach combining DTW and VQ is used to enhance the efficiency and accuracy of the system [61].

6 Conclusions

An interdisciplinary speech recognition research employs techniques from the works of varied fields such as biology, physics, linguistics, psychology, computer science, information theory, signal processing, probability theory, and so on. This paper is an attempt to present a comprehensive survey of the widely used techniques in constructing a speech recognition system.

References

1. Christian Gottlieb Kratzenstein, Sur la naissance de la formation des voyelles, J. Phys., Vol 21, pp. 358–380, 1782.
2. Dudley, Homer, and Thomas H. Tarnoczy. "The speaking machine of Wolfgang von Kempelen." The Journal of the Acoustical Society of America 22.2 (1950): 151–166.
3. Bowers, Brian, ed. Sir Charles Wheatstone FRS: 1802–1875. No. 29. IET, 2001.
4. Lindsay, David. "Talking Head: In the mid-1800s, Joseph Faber spent seventeen years working on his speech synthesizer." American Heritage of Invention and Technology 13 (1997): 56–63.
5. Cater, John C. "Electronically Speaking: Computer Speech Generation." Sams, 1983.
6. Fletcher, Harvey. "The Nature of Speech and Its Interpretation1." Bell System Technical Journal 1.1 (1922): 129–144.

7. Dudley, Homer, R. R. Riesz, and S. S. A. Watkins. "A synthetic speaker." Journal of the Franklin Institute 227.6 (1939): 739–764.
8. Davis, K. H., R. Biddulph, and S. Balashek. "Automatic recognition of spoken digits." The Journal of the Acoustical Society of America 24.6 (1952): 637–642.
9. Dersch, W.C. SHOEBOX- a voice responsive machine, DATAMATION, 8:47–50, June 1962.
10. Lowerre, Bruce T. "The HARPY speech recognition system." (1976).
11. Baker, James. "The DRAGON system–An overview." Acoustics, Speech and Signal Processing, IEEE Transactions on 23.1 (1975): 24–29.
12. Jelinek, Frederick. "Continuous speech recognition by statistical methods." Proceedings of the IEEE 64. 1976.
13. Kurzweil, Raymond. "The Kurzweil reading machine: A technical overview." Science, Technology and the Handicapped (1976): 3–11.
14. Averbuch, Ar, et al. "Experiments with the TANGORA 20,000 word speech recognizer." Acoustics, Speech, and Signal Processing, IEEE International Conference on ICASSP'87. Vol. 12. IEEE, 1987.
15. Rekha, J. Ujwala, Shahu, K. Chatrapati, Babu, A. Vinaya. "Feature selection for phoneme recognition using a cooperative game theory based framework". Multimedia, Communication and Computing Application: Proceedings of the 2014 International Conference on Multimedia, Communication and Computing Application (MCCA 2014), 191–195, CRC Press, 2015.
16. Rekha, J. Ujwala, Shahu, K. Chatrapati, Babu, A. Vinaya. "Feature selection using game theory for phoneme based speech recognition." Contemporary Computing and Informatics (IC3I), 2014 International Conference on. IEEE, 2014.
17. Toh, Aik Ming, Roberto Togneri, and Sven Nordholm. "Spectral entropy as speech features for speech recognition." Proceedings of PEECS 1 (2005).
18. Gelbart, David, Nelson Morgan, and Alexey Tsymbal. "Hill-climbing feature selection for multi-stream ASR." In: INTERSPEECH, pp. 2967–2970 (2009).
19. Paliwal, K. K. Dimensionality reduction of the enhanced feature set for the HMM-based speech recognizer, Digital Signal Processing 2(3) (1992): 157–173.
20. Fant, Gunnar. Acoustic Theory of Speech Production. No. 2. Walter de Gruyter, 1970.
21. Makhoul, John. "Linear prediction: A tutorial review." Proceedings of the IEEE 63.4 (1975): 561–580.
22. Bogert, B. P., and G. E. Peterson. "The acoustics of speech." Handbook of speech pathology (1957): 109–173.
23. Davis, Steven, and Paul Mermelstein. "Comparison of parametric representations for monosyllabic word recognition in continuously spoken sentences." Acoustics, Speech and Signal Processing, IEEE Transactions on 28.4 (1980): 357–366.
24. Mermelstein, Paul. "Distance measures for speech recognition, psychological and instrumental." Pattern recognition and artificial intelligence 116 (1976): 374–388.
25. Zheng, Fang, Guoliang Zhang, and Zhanjiang Song. "Comparison of different implementations of MFCC." Journal of Computer Science and Technology 16.6 (2001): 582–589.
26. Hermansky, Hynek. "Perceptual linear predictive (PLP) analysis of speech." the Journal of the Acoustical Society of America 87.4 (1990): 1738–1752.
27. Stevens, Stanley S. "On the psychophysical law." Psychological review 64.3 (1957): 153.
28. Hermansky, Hynek, and Nelson Morgan. "RASTA processing of speech." Speech and Audio Processing, IEEE Transactions on 2.4 (1994): 578–589.
29. Hermansky, Hynek, et al. "RASTA-PLP speech analysis technique." Acoustics, Speech, and Signal Processing, IEEE International Conference on. Vol. 1. IEEE, 1992.
30. Misra, H., Ikbal, S., Bourlard, H., & Hermansky, H. "Spectral entropy based feature for robust ASR." Acoustics, Speech, and Signal Processing, 2004. Proceedings. (ICASSP'04). IEEE International Conference on. Vol. 1. IEEE, 2004.
31. Sakoe, Hiroaki, et al. "Dynamic programming algorithm optimization for spoken word recognition." Readings in speech recognition 159 (1990).

32. Baum, Leonard E., and Ted Petrie. "Statistical inference for probabilistic functions of finite state Markov chains." The annals of mathematical statistics (1966): 1554–1563.
33. Debyeche, Mohamed, Jean Paul Haton, and Amrane Houacine. "Improved Vector Quantization Approach for Discrete HMM Speech Recognition System." Int. Arab J. Inf. Technol. 4.4 (2007): 338–344.
34. Cheng, Chih-Chieh, Fei Sha, and Lawrence K. Saul. "Matrix updates for perceptron training of continuous density hidden markov models." Proceedings of the 26th Annual International Conference on Machine Learning. ACM, 2009.
35. Gauvain, Jean-Luc, and Chin-Hui Lee. "Bayesian learning for hidden Markov model with Gaussian mixture state observation densities." Speech Communication 11.2 (1992): 205–213.
36. Rabiner, L. R., et al. "Recognition of isolated digits using hidden Markov models with continuous mixture densities." AT&T Technical Journal 64.6 (1985): 1211–1234.
37. Razavi, Marzieh, and Ramya Rasipuram. On Modeling Context-dependent Clustered States: Comparing HMM/GMM, Hybrid HMM/ANN and KL-HMM Approaches. No. EPFL-REPORT-192598. Idiap, 2013.
38. Dupont, Stéphane, et al. "Context Independent and Context Dependent Hybrid HMM/ANN Systems for Training Independent Tasks." Proceedings of the EUROSPEECH'97. 1997.
39. Woodland, Philip C., and Steve J. Young. "The HTK tied-state continuous speech recogniser." Eurospeech. 1993.
40. Levinson, Stephen E. "Continuously variable duration hidden Markov models for automatic speech recognition." Computer Speech & Language 1.1 (1986): 29–45.
41. Russell, Martin. "A segmental HMM for speech pattern modelling." Acoustics, Speech, and Signal Processing, 1993. ICASSP-93., 1993 IEEE International Conference on. Vol. 2. IEEE, 1993.
42. Kenny, Patrick, Matthew Lennig, and Paul Mermelstein. "A linear predictive HMM for vector-valued observations with applications to speech recognition." Acoustics, Speech and Signal Processing, IEEE Transactions on 38.2 (1990): 220–225.
43. Petr Schwarz, and Jan Cernocky. (2008) "Phoneme Recognition Based on Long Temporal Context.", Ph.D. Thesis, Brno University of Technology, Czech Republic.
44. Evermann, Gunnar, et al. The HTK book. Vol. 2. Cambridge: Entropic Cambridge Research Laboratory, 1997.
45. Rabiner, Lawrence. "A tutorial on hidden Markov models and selected applications in speech recognition." Proceedings of the IEEE 77.2 (1989): 257–286.
46. Ostendorf, M., and V. Digalakis. "The stochastic segment model for continuous speech recognition." Signals, Systems and Computers, 1991. 1991 Conference Record of the Twenty-Fifth Asilomar Conference on. IEEE, 1991.
47. Ostendorf, Mari, and Salim Roukos. "A stochastic segment model for phoneme-based continuous speech recognition." Acoustics, Speech and Signal Processing, IEEE Transactions on 37.12 (1989): 1857–1869.
48. Morris, Jeremy J. "A study on the use of conditional random fields for automatic speech recognition." PhD diss., The Ohio State University, 2010.
49. Gunawardana, Asela, et al. "Hidden conditional random fields for phone classification." INTERSPEECH. 2005.
50. Graves, Alex, Abdel-rahman Mohamed, and Geoffrey Hinton. "Speech recognition with deep recurrent neural networks." Acoustics, Speech and Signal Processing (ICASSP), 2013 IEEE International Conference on. IEEE, 2013.
51. Pinto, Joel Praveen. Multilayer perceptron based hierarchical acoustic modeling for automatic speech recognition. Diss. Ecole polytechnique fédérale de Lausanne, 2010.
52. Fukuda, Yohji, and Haruya Matsumoto. "Speech Recognition Using Modular Organizations Based On Multiple Hopfield Neural Networks", Speech Science and Technology (SST-92), 1992: 226–231.
53. Minghu, Jiang, et al. "Fast learning algorithms for time-delay neural networks phoneme recognition." Signal Processing Proceedings, 1998. ICSP'98. 1998 Fourth International Conference on. IEEE, 1998.

54. Venkateswarlu, R. L. K., and R. Vasantha Kumari. "Novel approach for speech recognition by using self—Organized maps." Emerging Trends in Networks and Computer Communications (ETNCC), 2011 International Conference on. IEEE, 2011.
55. Ganapathiraju, Aravind, Jonathan E. Hamaker, and Joseph Picone. "Applications of support vector machines to speech recognition." Signal Processing, IEEE Transactions on 52.8 (2004): 2348–2355.
56. N.D. Smith and M. Niranjan. Data-dependent Kernels in SVM Classification of Speech Patterns. In Proceedings of the International Conference on Spoken Language Processing (ICSLP), volume 1, pages 297–300, Beijing, China, 2000.
57. A. Ganapathiraju, J. Hamaker, and J. Picone. Hybrid SVM/HMM Architectures for Speech Recognition. In Proceedings of the 2000 Speech Transcription Workshop, volume 4, pages 504–507, Maryland (USA), May 2000.
58. J. Padrell-Sendra, D. Martın-Iglesias, and F. Dıaz-de-Marıa. Support vector machines for continuous speech recognition. In Proceedings of the 14th European Signal Processing Conference, Florence, Italy, 2006.
59. Makhoul, John, Salim Roucos, and Herbert Gish. "Vector quantization in speech coding." Proceedings of the IEEE 73.11 (1985): 1551–1588.
60. Furui, Sadaoki. "Vector-quantization-based speech recognition and speaker recognition techniques." Signals, Systems and Computers, 1991. 1991 Conference Record of the Twenty-Fifth Asilomar Conference on. IEEE, 1991.
61. Zaharia, Tiberius, et al. "Quantized dynamic time warping (DTW) algorithm." Communications (COMM), 2010 8th International Conference on. IEEE, 2010.

Forest Type Classification: A Hybrid NN-GA Model Based Approach

Sankhadeep Chatterjee, Subhodeep Ghosh, Subham Dawn, Sirshendu Hore and Nilanjan Dey

Abstract Recent researches have used geographically weighted variables calculated for two tree species, *Cryptomeria japonica* (Sugi, or Japanese Cedar) and *Chamaecyparis obtusa* (Hinoki, or Japanese Cypress) to classify the two species and one mixed forest class. In machine learning context it has been found to be difficult to predict that a pixel belongs to a specific class in a heterogeneous landscape image, especially in forest images, as ground features of nearly located pixel of different classes have very similar spectral characteristics. In the present work the authors have proposed a GA trained Neural Network classifier to tackle the task. The local search based traditional weight optimization algorithms may get trapped in local optima and may be poor in training the network. NN trained with GA (NN-GA) overcomes the problem by gradually optimizing the input weight vector of the NN. The performance of NN-GA has been compared with NN, SVM and Random Forest classifiers in terms of performance measures like accuracy,

S. Chatterjee (✉)
Department of Computer Science & Engineering,
University of Calcutta, Kolkata, India
e-mail: sankha3531@gmail.com

S. Ghosh · S. Dawn
Department of Computer Science & Engineering,
Academy of Technology, Aedconagar, Hooghly 712121, India
e-mail: subhodeepghosh7@gmail.com

S. Dawn
e-mail: knightdawnxp@gmail.com

S. Hore
Department of Computer Science & Engineering,
Hooghly Engineering & Technology College, Hooghly, India
e-mail: shirshendu.hore@hetc.ac.in

N. Dey
Department of Computer Science & Engineering,
BCET, Durgapur, India
e-mail: neelanjandey@gmail.com

© Springer India 2016
S.C. Satapathy et al. (eds.), *Information Systems Design and Intelligent Applications*, Advances in Intelligent Systems and Computing 435,
DOI 10.1007/978-81-322-2757-1_23

precision, recall, F-Measure and Kappa Statistic. The results have been found to be satisfactory and a reasonable improvement has been made over the existing performances in the literature by using NN-GA.

Keywords Image classification · Neural network · Genetic algorithm · Random forest · SVM

1 Introduction

Image classification has drawn attention of the remote-sensing researchers due to the complexities and challenges in the context of machine learning and soft computing. Most of the remote sensing research is mainly focused on reducing the classification error due to several natural, environmental, and other effects. Several research works have been found in this regard [1–8]. The research works have proposed image classification using several methods and have tried to reduce the classification error to a greater extent. In remote sensing context the image classification can formally be defined as to classify pixels of a given image into different regions, each of which is basically a landcover type. In satellite image, each pixel represents a specific landcover area though it is highly likely that it may belong to more than one land cover type. With increasing number of instances the problem becomes more severe as the number of such pixel increases. Thus, a large amount of uncertainty may get involved during the classification task. Several Unsupervised learning methods have been proposed to handle this problem efficiently [9–12]. Unsupervised techniques like Fuzzy c-means [13], split and merge [14] and ANN [15–20] based methods [21] have been applied for the task. The problem of uncertainty during classification becomes more challenging in the Forest images as the level of variation is higher in a very small geographic area. Thus, using traditional machine learning techniques it becomes quite challenging to achieve reasonable amount of classification accuracy. The problem has been overcome by applying a SVM [22] based model with assistance of geographically weighted variables [23]. The authors have shown slight improvement over the existing methods (without using geographically weighted variables).

In the present work the authors have proposed a GA trained Neural Network model to tackle the task. The input weight vector of the NN has been gradually optimized using GA to increase the performance of NN. The application of GA in training NN has already been found to be quite satisfactory in several real life applications [24]. The proposed model has been compared with two well-known classifiers SVM and Random Forest [25, 26]. The performances of the proposed models and the other models have been measured using accuracy, precision, recall, F-Measure and Kappa Statistic.

2 Proposed NN-GA Model

GA was proposed by Holland (1975) to implement the theory of natural selection to solve optimization problems. The GA starts solving a problem by using a set of initial solutions. And it continuously applies crossover and mutations on the solutions to produce better offspring. The survival of any offspring depends on the fitness which is decided by the problem definition of the problem being solved. GA has lesser chance of getting trapped into local optima. Thus, it can be better choice than the traditional methods. The method of applying GA can be summarized as follows;

1. **Generation of initial population** 'N' numbers of chromosomes are randomly generated. Each chromosome is actually an array of random real weight values, biologically genes; they vary in between '0' to '1'.
2. **Calculating fitness values** A fitness function has to be defined, using it the fitness of each individual solution (chromosome) has to be evaluated. RMSE of NN training is used as the fitness function.
3. **Selection** The smaller the RMSE, higher is the chance of getting selected for the next generation. $RMSE_i$ denotes the fitness function value of *i*th solution. The selection procedure works as follows:

   3.1. $RMSE_i$ is calculated for each solution in population.
   3.2. All $RMSE_i$ are aggregated or averaged together to find $RMSE_{ave}$
   3.3. A random value ($RMSE_r$) is selected from predefined closed interval $[0, RMSE_{ave}]$
   3.4. For all solutions $RMSE_r - RMSE_i$ is calculated and if the result of the subtraction is less than or equal to '0' the *i*th individual is selected.
   3.5. The process goes on until the number of solutions selected for next generation (mating pool) is equal to the number of solutions in the population initially.

4. **Cross-over** The selected chromosomes take part in cross-over where the after selecting cross-over points on the chromosome the genes at the right of that point for both the chromosomes taking part get exchanged. And it creates two new individual.
5. **Mutation** Genes of same chromosome take part in this phase. Genes from randomly selected position are swapped to create new individual solution.
6. **Termination condition** Finally the termination condition is checked. In the present work number of generation has been selected as terminating condition. When the user given number of generation is reached the best possible individual is selected as the optimized weight vector, otherwise it starts from step 2 again.

Figure 2 depicts the flowchart of NN trained with GA. The GA block in the flowchart is separately depicted in Fig. 1.

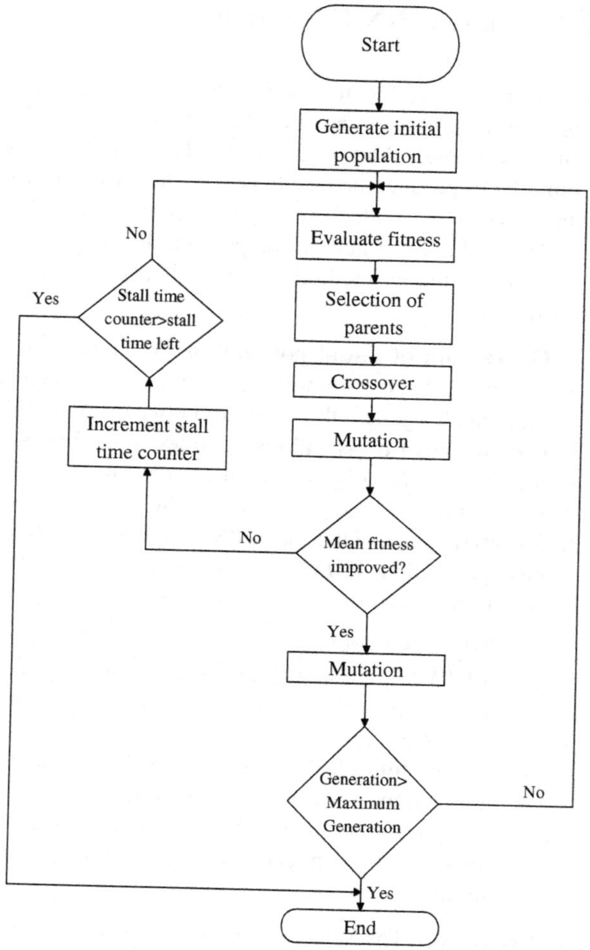

Fig. 1 The genetic algorithm which has been followed to optimize the input weight vector

3 Dataset Description

The dataset [23] used in the current work includes information of forested area in Ibaraki Prefecture, Japan (36° 57_N, 140° 38_E), approximately 13 km × 12 km. The landscape consists mainly of *Chamaecyparis obtusa* (Hinoki, or Japanese Cypress) planted forest ('h' class), *Cryptomeria japonica* (Sugi, or Japanese Cedar) planted forest ('s' class) and mixed natural forest, along with other land cover types (agriculture, roads, buildings, etc.) [23] which have been mapped based on the spectral characteristics at visible-to-near infrared wavelengths of the satellite images taken by ASTER satellite imagery. There are all total 27 attributes which are; spectral information in the green, red, and near infrared wavelengths for three dates (Sept. 26, 2010; March 19, 2011; May 08, 2011 (Total 9 attributes). Predicted spectral values (based on spatial interpolation) minus actual spectral values for the 's' class (Total 9 attributes) and Predicted spectral values (based on spatial interpolation) minus actual spectral values for the 'h' class (Total 9 attributes).

4 Experimental Methodology

The experiment is conducted on the dataset [23] obtained from UCI Machine Learning Repository. The experiments are performed by using Support Vector Machine (LibSVM) [27], Random Forest and real coded NN, NN-GA classifiers. For NN scaled conjugate gradient algorithm [28] has been used as the learning algorithm. The algorithm is well known and benchmarked against traditional back-propagation and other algorithms. The basic flow of experiment opted in the present work is as follows

1. **Preprocessing** The following preprocessing is done on the dataset before the classification
   (a) **Data Cleaning**—The data might contain missing values or noise. It is important to remove noise and fill up empty entries by suitable data by means of statistical analysis.
   (b) **Data Normalization**—the needs to be normalized before classification task is carried on to reduce distance between attribute values. It is generally achieved by keeping the value range in between −1 to +1.
2. After preprocessing the datasets are divided into two parts. One of which is used as training dataset and the other as testing dataset. In the present work two third (70 %) of the data is used as training data and rest (30 %) as testing data.
3. In the training phase the training dataset is supplied to different algorithms respectively to build the required classification model.
4. In the testing phase the classification models obtained from the training phase is employed to test the accuracy of the model.

To measure the performance and to compare the performances we use several statistical performance measures like accuracy, precision, recall, Kappa statistic [29], True positive rate (TP rate), and F-measure. The performance measuring parameters are calculated from the confusion matrix [30] which is a tabular representation that provides visualization of the performance of a classification algorithm. The objective function of Genetic algorithm (RMSE) is defined as follows; RMSE [31] of a classifier prediction with respect to the computed variable v_{c_k} is determined as thesquare root of the mean-squared error and is given by:

$$RMSE = \sqrt{\frac{\sum_{k=1}^{n} \left(v_{d_k} - v_{c_k}\right)^2}{n}}$$

Fig. 2 Flowchart of NN training using genetic algorithm depicted in Fig. 3

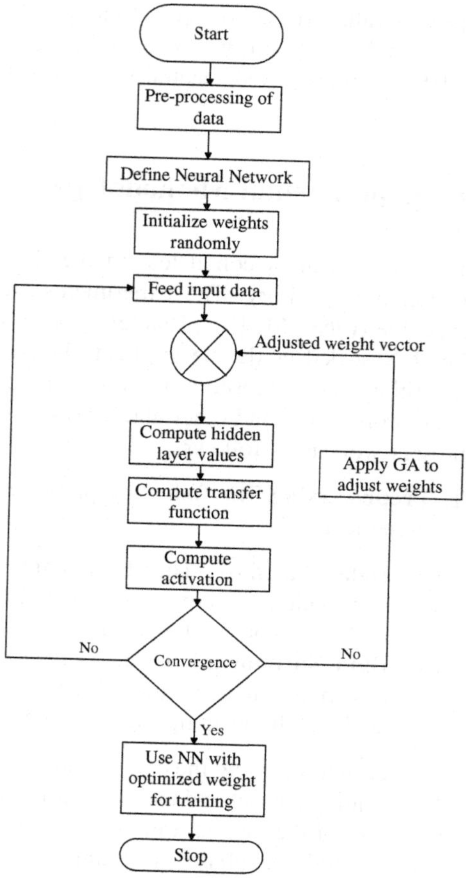

where, v_{d_k} denotes the originally observed value of kth object and v_{c_k} denotes the predicted value by the classifier. The genetic algorithm based optimization of input weight vector has been implemented by following the Fig. 2. The different parameters used as inputs are as follows (Table 1).

5 Results and Discussion

The experiments have been carried out on the dataset described in Sect. 3. The experimental methodology has been described in Sect. 4. Table 2 reports the experimental results for Neural Network, Support Vector Machine (LibSVM), Random Forest and NN-GA classifiers. The experimental results suggest that the performance of Neural Network (trained with scaled conjugate gradient descent

Table 1 Genetic algorithm setup for input weight vector optimization

Maximum number of generation	1000
Population size	500
Crossover probability	0.2
Mutation	Gaussian
Crossover	Single point crossover
Selection	Roulette
Stall Time Limit	75 s

Table 2 Performance measures of different algorithms

	NN	SVM	Random forest	NN-GA
Accuracy	85.35	85.99	82.17	95.54
Precision	86.31	84.67	86.37	94.66
Recall	82.32	82.78	78.83	95.76
F-Measure	84.27	83.71	82.59	95.21
Kappa statistic	0.793	0.804	0.746	0.938

algorithm) is moderate with an accuracy of 85.35 %, precision 86.31 %, recall 82.32 %, F-Measure 84.27 % and Kappa Statistic 0.793 while the SVM (LibSVM) has performed almost same as the NN but with negligible. The experimental results suggest that Random Forest (which is considered to be one of the best tree based classifiers) may not be suitable for the classification of image classification of Forest or other areas which involves sufficient amount of uncertainty during classification. Though, the precision (86.37 %) of the classifier is better than NN and SVM. In the column four of Table 2 the experimental result of the proposed model has been shown. The objective of the GA was to reduce the RMSE (as described in Sect. 4). The model has performed significantly well than all the other classifiers in this study with an accuracy of 95.54 %, precision 94.66 %, recall 95.67 %, F-Measure 95.21 % and Kappa Statistic 0.938. Figure 3 depicts the different performance measures for the classifiers under consideration. The comparative analysis has revealed that the NN-GA is superior not only in terms of accuracy but also in terms

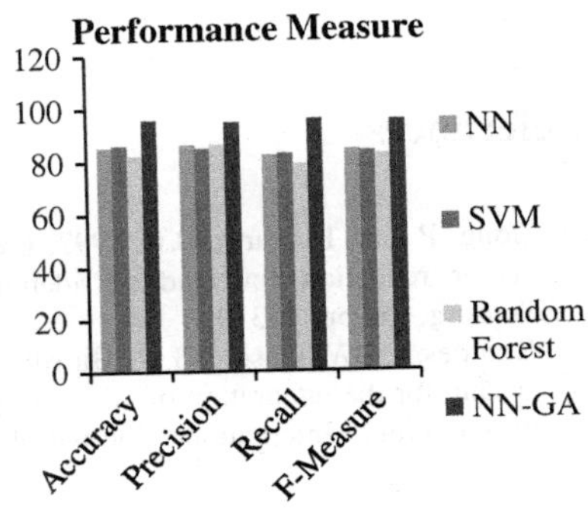

Fig. 3 Comparison of performance measure of NN, SVM, Random Forest and NN-GA classifiers

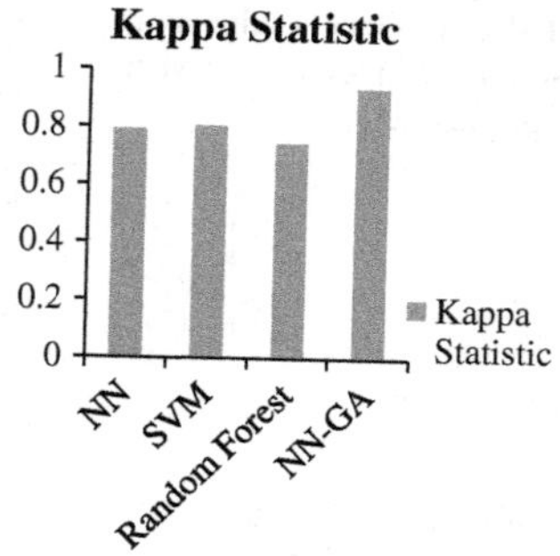

Fig. 4 Comparison of kappa statistic of NN, SVM, random forest and NN-GA classifiers

of precision, recall and F-Measure. Figure 4 depicts the same for Kappa Statistic. The plot establishes the superiority of NN-GA once again.

6 Conclusion

The present work has proposed a Genetic algorithm trained Neural Network model to classify two forest type along with one mixed class of forest of Japanese Cedar Japanese Cypress. The experimental results have suggested the superiority of NN-GA over the NN (trained with scaled conjugate gradient algorithm) for classification of pixels in Forest images. The study has also established that the NN-GA is a better classifier than support vector machines for this task. A satisfactory amount of improvement has been found over the existing work in literature. Besides the previous works have compared the performances of algorithms mainly based on accuracy which is not a good metric for performance measure as, accuracy of an algorithm varies greatly if number of instances varies in different classes. The present work have analyzed the performances of algorithms in terms of several performance measuring parameters like accuracy, precision, recall, F-Measure and Kappa Statistic to provide a vivid picture of the exact performance of the algorithms and to have a fair comparison of algorithms.

References

1. Gong, P. and Howarth, P.J., 1992, Frequency-based contextual classification and gray-level vector reduction for land-use identification. Photogrammetric Engineering and Remote Sensing, 58, pp. 423–437.
2. Kontoes, C., Wilkinson, G.G., Burrill, A., Goffredo, S. and Megier, J., 1993, An experimental system for the integration of GIS data in knowledge-based image analysis for remote sensing of agriculture. International Journal of Geographical Information Systems, 7, pp. 247–262.

3. Foody, G.M., 1996, Approaches for the production and evaluation of fuzzy land cover classification from remotely-sensed data. International Journal of Remote Sensing, 17, pp. 1317–1340.
4. San Miguel-Ayanz, J. and Biging, G.S., 1997, Comparison of single-stage and multi-stage classification approaches for cover type mapping with TM and SPOT data. Remote Sensing of Environment, 59, pp. 92–104.
5. Stuckens, J., Coppin, P.R. and Bauer, M.E., 2000, Integrating contextual information with per-pixel classification for improved land cover classification. Remote Sensing of Environment, 71, pp. 282–296.
6. Franklin, S.E., Peddle, D.R., Dechka, J.A. and Stenhouse, G.B., 2002, Evidential reasoning with Landsat TM, DEM and GIS data for land cover classification in support of grizzly bear habitat mapping. International Journal of Remote Sensing, 23, pp. 4633–4652.
7. Pal, M. and Mather, P.M., 2003, An assessment of the effectiveness of decision tree methods for land cover classification. Remote Sensing of Environment, 86, pp. 554–565.
8. Gallego, F.J., 2004, Remote sensing and land cover area estimation. International Journal of Remote Sensing, 25, pp. 3019–3047.
9. Bandyopadhyay, S., Maulik, U.: Genetic Clustering for Automatic Evolution of Clusters and Application to Image Classification, Pattern Recognition, 35(2), 2002, 1197–1208.
10. Bandyopadhyay, S., Maulik, U., Mukhopadhyay, A.: Multiobjective Genetic Clustering for Pixel Classification in Remote Sensing Imagery, IEEE Transactions on Geoscience and Remote Sensing, 45(5), 2007, 1506–1511.
11. Bandyopadhyay, S., Pal, S. K.: Pixel Classification Using Variable String Genetic Algorithms with Chromosome Differentiation, IEEE Transactions on Geoscience and Remote Sensing, 39 (2), 2001, 303– 308.
12. Maulik, U., Bandyopadhyay, S.: Fuzzy partitioning using a real-coded variable-length genetic algorithm for pixel classification, IEEE Transactions on Geoscience and Remote Sensing, 41 (5), 2003, 1075– 1081.
13. Cannon, R. L., Dave, R., Bezdek, J. C., Trivedi, M.: Segmentation of a Thematic Mapper Image using Fuzzy c-means Clustering Algorithm, IEEE Transactions on Geoscience and Remote Sensing, 24, 1986, 400– 408.
14. Laprade, R. H.: Split-and-merge Segmentation of Aerial Photographs, Computer Vision Graphics and Image Processing, 48, 1988, 77–86.
15. Hecht-Nielsen, R., Neurocomputing. Addison-Wesley, Reading, MA, 1990.
16. Schalkoff, R.J., Artificial Neural Networks. McGraw-Hill, New York, 1997.
17. Jain, A.K., Mao, J., Mohiuddin, K.M., Artificial neural networks: a tutorial. Comput. IEEE March, pp. 31–44, 1996.
18. K.J. Hunt, R. Haas, R. Murray-Smith, “Extending the functional equivalence of radial basis function networks and fuzzy inference systems”, IEEE Trans Neural Networks Vol.7, No.3, pp.776–781, 1996.
19. D.S. Broomhead, L. David, “Radial basis functions, multivariable functional interpolation and adaptive networks”, In Technical report Royal Signals and Radar Establishment, No. 4148, p. 1–34,1998.
20. J. Han, M. Kamber, “Data mining: concepts and Techniques”, 2nd ed. Morgan and Kaufmann, pp. 285–378, 2005.
21. Baraldi, A., Parmiggiani, F.: A Neural Network for Unsupervised Categorization of Multivalued Input Pattern: An Application to Satellite Image Clustering, IEEE Transactions on Geoscience and Remote Sensing, 33, 1995, 305–316.
22. Cortes, Corinna, and Vladimir Vapnik. “Support-vector networks.” Machine learning 20, no. 3 (1995): 273–297.
23. Johnson, B., Tateishi, R., Xie, Z. (2012). Using geographically weighted variables for image classification. Remote Sensing Letters, 3(6), 491–499.

24. S. Chatterjee, R. Chakraborty, S. Hore: "A Quality Prediction method of Weight Lifting Activity". Proceedings of the Michael Faraday IET International Summit - 2015 (MFIIS-2015), An IET International Conference On September 12 – 13, 2015 in Kolkata, India, (in press).
25. V. Y., Kulkarni, and P. K. Sinha, "Random forest classifiers: a survey and future research directions", Int. J. Adv. Comput, Vol.36, No. 1, pp. 1144–1153, 2013.
26. L. Brieman, "Random Forests", Machine Learning, Vol. 45, pp. 5–32, 2001.
27. Chang, C. C., & Lin, C. J. (2011). LIBSVM: A library for support vector machines. ACM Transactions on Intelligent Systems and Technology (TIST), 2(3), 27.
28. Møller, Martin Fodslette. "A scaled conjugate gradient algorithm for fast supervised learning." Neural networks 6, no. 4 (1993): 525–533.
29. Carletta J. Assessing agreement on classification tasks: the kappa statistic. Comput Linguist, Cambridge, MA, USA: MIT Press, 22(2):249–54, 1996.
30. Stehman SV. Selecting and interpreting measures of thematic classification accuracy. Remote SensEnviron, 62(1):77–89, 1997.
31. Armstrong JS, Collopy F. Error measures for generalizing about forecasting methods: empirical comparisons. Int. J Forecast, 8:69–80, 1992.

Optimizing Technique to Improve the Speed of Data Throughput Through Pipeline

Nandigam Suresh Kumar and D.V. Rama Koti Reddy

Abstract High speed data processing is very important factor in super computers. Parallel computing is one of the important elements mostly used in fast computers. There are different methods actively involved to satisfy the concept parallel computing. In the present paper pipeline method is discussed with its flaws and different clock schemes. In the present paper data propagation delay is discussed at different existing techniques and presented a new method to optimize the data propagation delay. The new method is compared with different existing methods and designs of new technique with simulation results are presented. The simulation results are obtained from Virtual Digital Integrator.

Keywords Pipeline · Parallel computing · Parallel processor · Fast fetching · Propagation delay · Clock · Latches

1 Introduction

Pipeline is the important element used in parallel computing. Pipeline will help the processor in processing the data quickly. The fetching, decoding and processing the data can be performed simultaneously with the help of pipeline. This can be achieved with multi stages of pipeline. The multi stages in the pipeline will hold the data while processing the data by processor. Similarly the data can be processed simultaneously in arithmetic pipeline at different stages.

In many fast computers the pipeline itself helps the processor in computing the data quickly and to do the operation simultaneously. In conventional pipeline clock

N. Suresh Kumar (✉)
GITAM Institute of Technology, GITAM University, Visakhapatnam, Andhrapradesh, India
e-mail: nskpatnaik@gmail.com

D.V. Rama Koti Reddy
College of Engineering, Andhra University, Visakhapatnam, Andhrapradesh, India
e-mail: rkreddy_67@yahoo.co.in

© Springer India 2016
S.C. Satapathy et al. (eds.), *Information Systems Design and Intelligent Applications*, Advances in Intelligent Systems and Computing 435,
DOI 10.1007/978-81-322-2757-1_24

skew is a severe problem encountered when the stages are increased. The clock skew is a severe problem exists in many digital elements. The clock skew causes poor data organization and further it leads to data loss at end element. If the clock skew is manageable the data propagation can be finished in definite time. The high speed devices are facing two main challenges, the clock skew and jitter [1]. The clock skew is due to component mismatch. The jitter is of two types, one is random and deterministic jitter. The jitter occurs due to power fluctuation or other environmental conditions. Jitter, because of its random nature, does not create spurs at any specific frequency but degrades the signal-to-noise ratio (SNR) by raising the noise floor [2]. In high speed and slow devices the clock skew creates hold issues between neighbouring flip-flops. Especially this problem is high in Multiplexer D-Flip-Flop (Mux-DFF), because a common clock pulse is given to master and slave flip-flops. This problem was mitigated with the clock scan method proposed by McCluskey [3]. But clocked scan method also faces hold time problems, since common scan clock is used for both master and slave latch. The front end MUX is eliminated in Level Sensitive Scan Design (LSSD) [4]. The hold time problems can be minimized by controlling clock skew.

Improper clock synchronization causes data latency and poor data propagation in pipeline. This is because; many flip flops are integrated in the pipeline logic path. This leads delay in data propagation when pipeline stages increases. There is chance of data latency as the multiple data is passing through pipeline stages. These all can be organized in good fashion only when proper clock is applied between pipeline stages.

In traditional pipeline two types of clock schemes are generally used to synchronize the multiple stages. Synchronous clock system is used when same clock is used in all stage domains as shown in Fig. 1. It is very easy to handle synchronous clock scheme and it is very old method used in almost all circuits. But mostly used in slow devices and where throughput is not a concern. In synchronous systems, this problem is particularly serious. Since all sequential elements are clocked, huge current peaks are observed in correspondence of the clock edges. These peaks are also produced by sequential elements, data waves [5].

In Asynchronous clock system an individual clock is distributed throughout the pipeline stages. But managing and supplying individual clock scheme is difficult. Here the clock pulse at next stage always depends upon handshaking signals. It saves the power consumption, but hardware and signalling overhead involved in local communication [6–9].

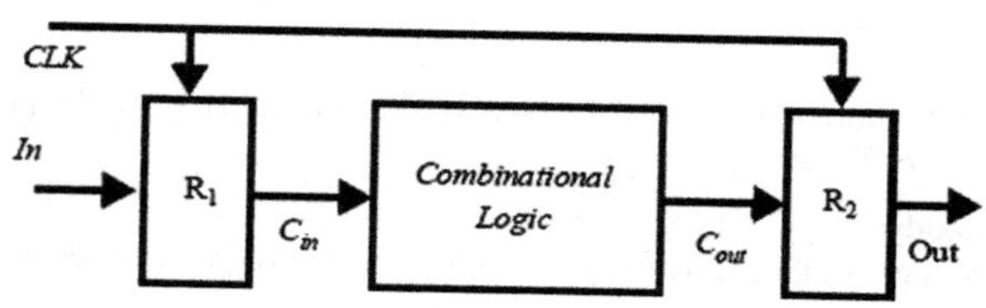

Fig. 1 Conventional pipeline

In 1969, Cotton first proposed these problems in the name of ‘maximum rate propagation’ [10]. There are some solutions proposed to solve these problems by introducing buffers, delay elements, stage splitting and etc. Since the data input is generating in nanoseconds, the time synchronization is required to produce output with high accuracy, minimum propagation delay by controlling the clock skew. However the scheduled clock skew minimizes total clock period [11, 12].

2 Existing Method

In the pipeline stages when the logic contains different length of logic paths it is very difficult to achieve the right data at the end element. These logic paths create problems at the output in producing correct output. In conventional pipeline a clock pulse is chosen greater than the Maximum logic propagation path (D_{max}). But this creates delay differences in logic path and creates data latency. This problem is balanced by introducing a delay balancing circuit in pipeline clock path as shown in Fig. 2. The delay balancing element in the clock path compensated the difference between maximum data propagation (D_{max}) and minimum data propagation (D_{min}) [13, 14]. This is called wave pipeline. In the wavepipelined system, the clock period is chosen to be ($D_{max} - D_{min}$) + set up time (t_s), hold time (t_h), and skew element [15]. This is further improved by hybrid pipeline. In hybrid pipeline the stages are splitted into small stages, so that the ($D_{max} - D_{min}$) can be minimized [1, 8]. On the other hand, a new wave pipeline method named as ‘Wave Component Sampling’ is used to sample the excess values produced in the delay differences [16]. In this case, minimum number of registers is used in synchronization and so consumes less power.

The clock signal is derived in the wave pipelining is

$$T_{clk.w} \geq (D_{\max} - D_{\min}) + T_h + T_s + 2\Delta_{clk} \quad (1)$$

In Mesochronous pipeline delay elements are introduced which is equal to the delay created by logic paths between pipeline stages as shown in Fig. 3 and the waves are separated based on physical properties of internal nodes in the logic stage [10, 17–19].

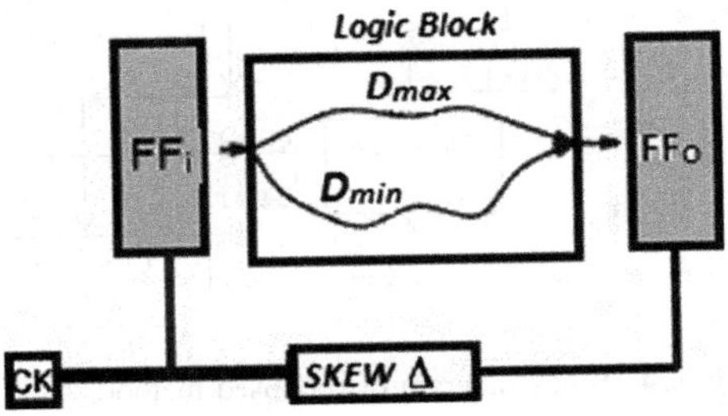

Fig. 2 Wave pipeline

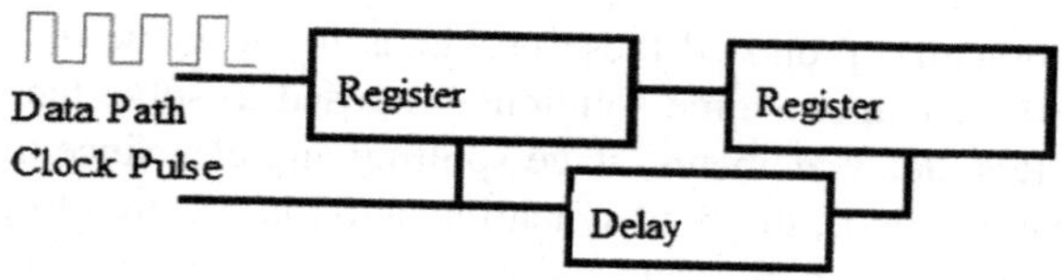

Fig. 3 Mesochronous pipeline

An Adjustable delay buffers (ADB) are inserted to control the clock skew. Some ADBs are inserted or deleted to control the clock skew. To control the clock skew appropriate number of delay buffers need to be inserted or removed and location of ADB is also represented [20].

3 Enhanced Method

In the proposed work a pulse detector is used to trigger the next stage. Until the first stage output is detected the pulse detector does not send any clock pulse to the next stage as shown in Fig. 4. This saves power consumption. In the proposed work the clock input of the first stage is connected with external clock source. The clock pulse of the next stage is depending upon the previous stage output. As the output of the first stage generates the Check binary Information (CBI) unit detects it and activates the Clock gate. Then the clock gate will enable the second stage. In the same way the next stage clock is always depend upon the previous stage output. As the CBI detects the data wave from previous stage and it send this information to Clock gate. Then the clock gate triggers the next stage. In this method low propagation delay is observed when compared with other methods (Figs. 5, 6, 7, 8, 9, 10, 11 and 12).

The Check Binary Information minimizes the data latency and controls the clock skew when compared with other existing methods discussed in the present paper. When compared with wave pipeline and Mesochronous pipeline the complications in delay buffers in clock path is recognisably reduced. The accuracy of data after propagation is significantly increased in the proposed method, when compared with

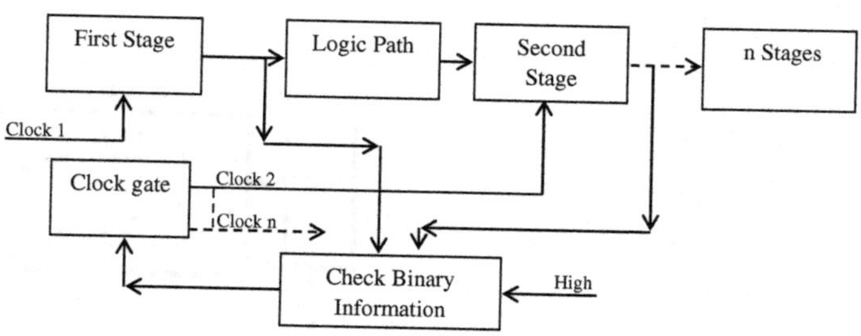

Fig. 4 Block diagram of proposed method

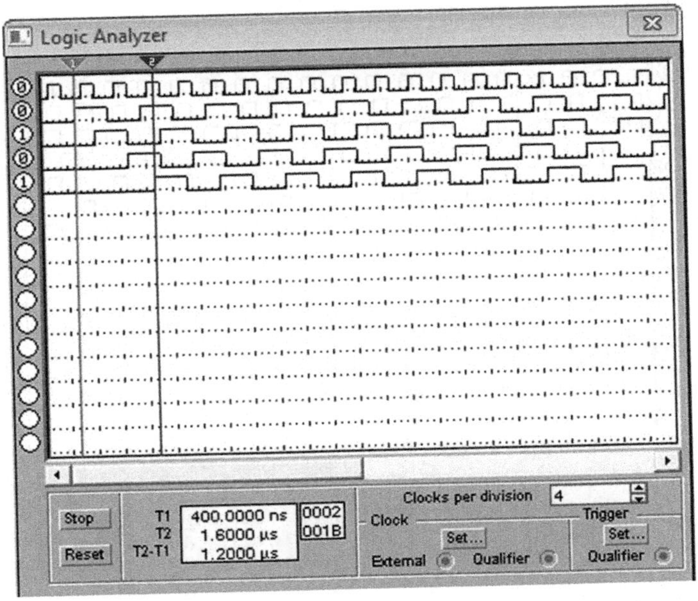

Fig. 5 Wave propagation through 3 stage conventional pipeline

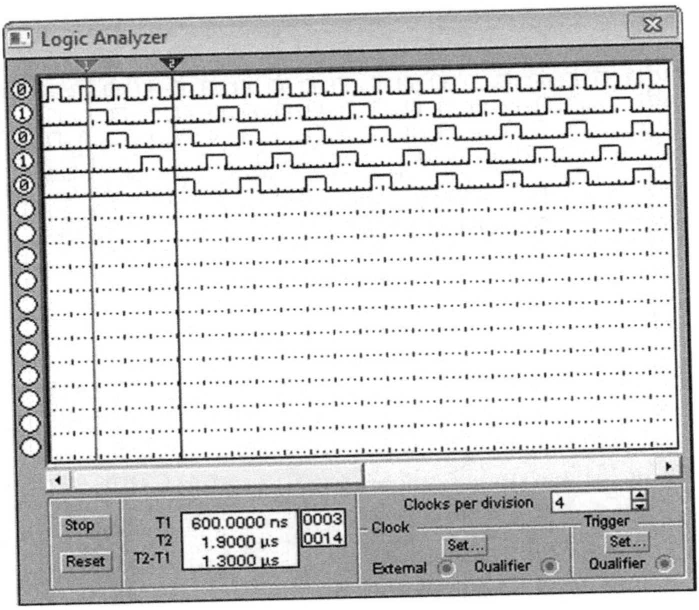

Fig. 6 Wave propagation through 3 stage wave pipeline

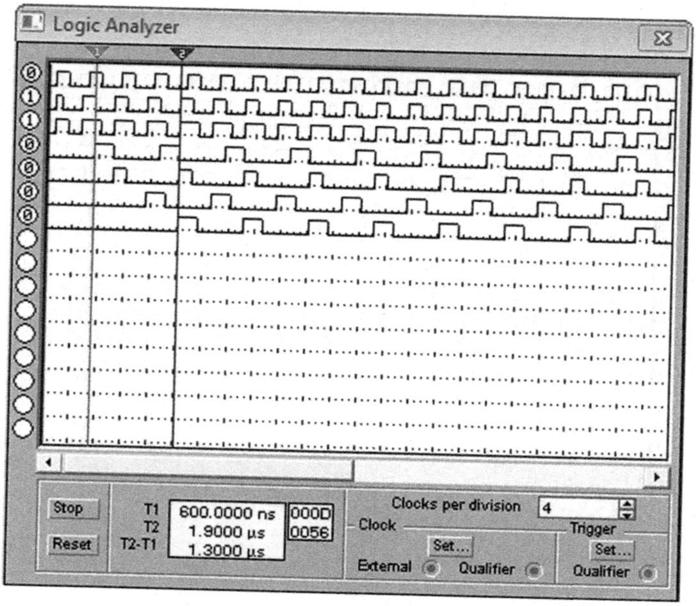

Fig. 7 Wave propagation through 3 stage mesochronous pipeline

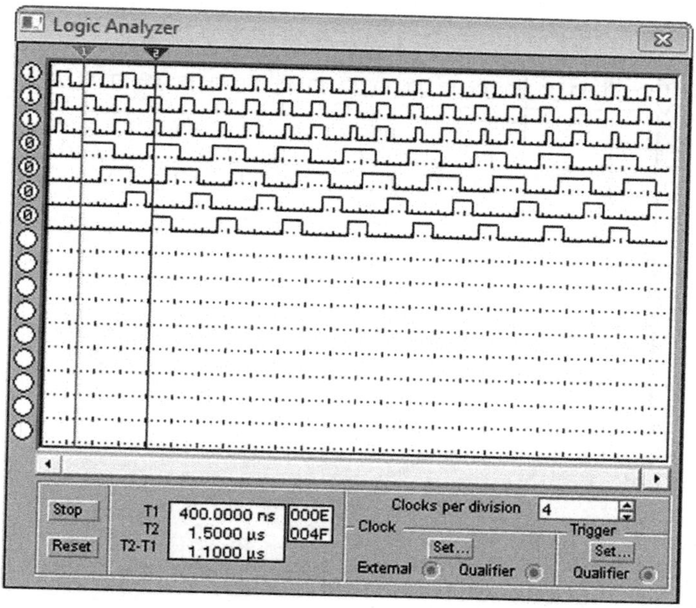

Fig. 8 A three stage new pipeline method where data propagation delay of 1.1 μs

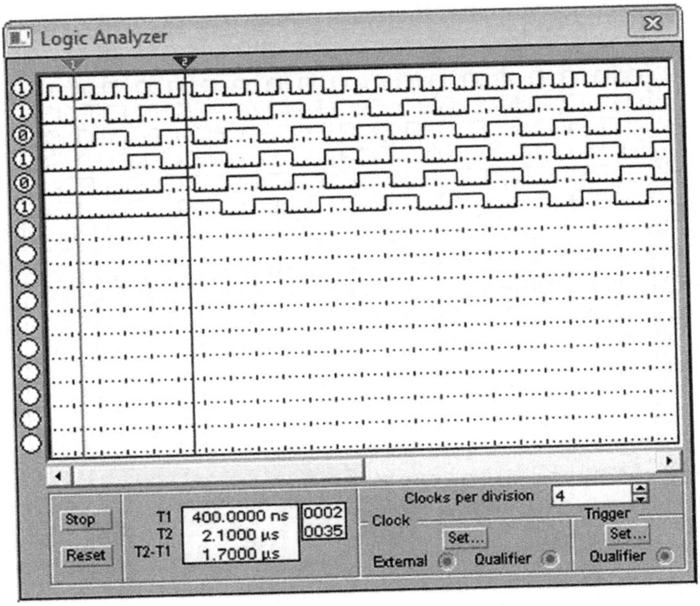

Fig. 9 Wave propagation through 4 stage conventional pipeline

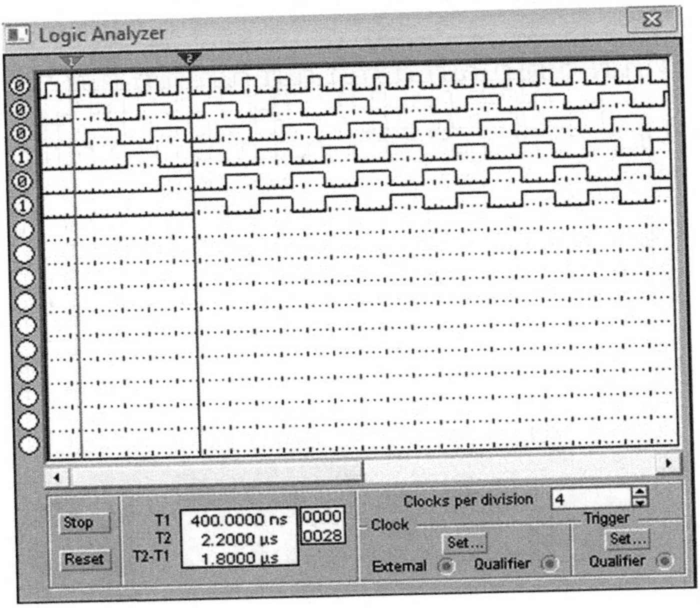

Fig. 10 Wave propagation through 4 stage wave pipeline

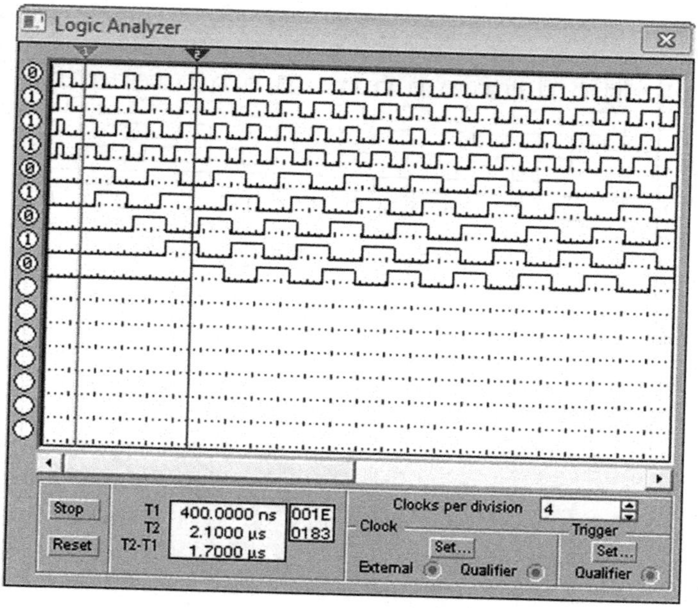

Fig. 11 Data propagation through four stage mesochronous pipeline

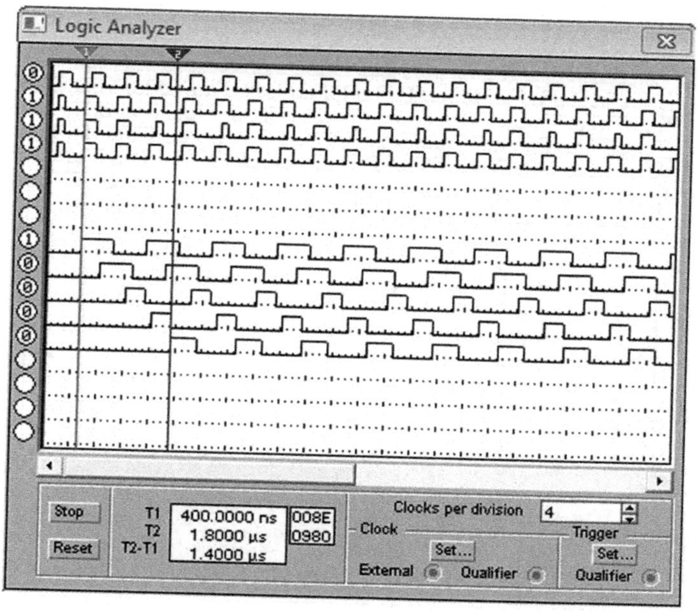

Fig. 12 Data propagation through four stage new pipeline method

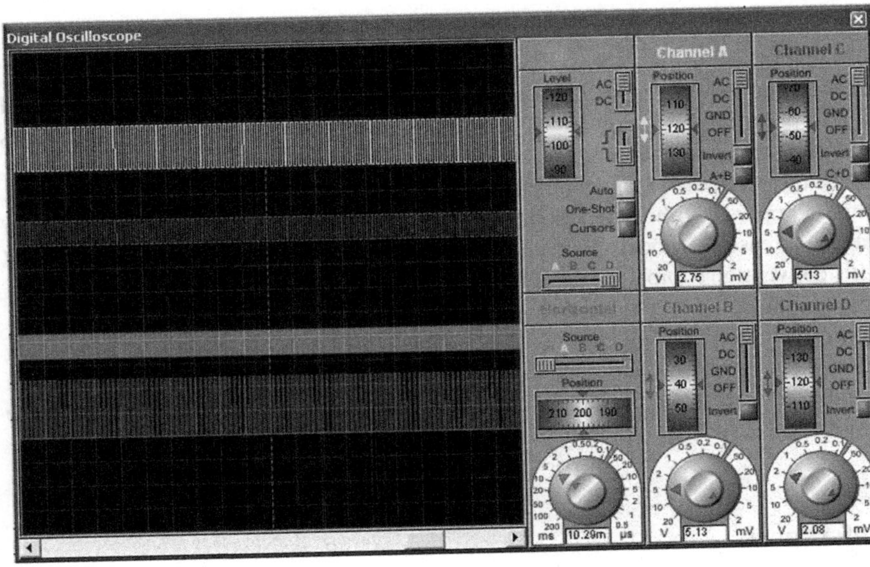

Fig. 13 A four stage new pipeline method which is designed in proteus

conventional, wave and mesochronous pipeline methods. The data propagation results are also observed in proteus simulator. In proteus simulator, it is found a difficulty to measure the distance between input and output at higher data rates as the pulse width is not observable. For readers benefit, a sample wave form is shown in Fig. 13.

4 Results

In this section the enhance method is compared with existing methods and presents their simulation results produced in Virtual Digital Integrator. In the following figures the first wave form is clock pulse and in the last four wave forms the first one is data input wave at the first stage and the last three are data out waves at first three stages of the pipeline. There are two vertical lines crossing across the timing diagram, if the first one is T1 and the second one is T2, then the difference (T1 − T2) will be the propagation delay through the pipeline. A three stage pipeline is constructed and tested in Virtual Digital Integrator and obtained results of clock pulses applied at each stage and data input and data outputs at each stage. In Fig. 1 the data is arrived at 400 ns after clock pule is applied and data output is observed in the last stage after 1.6 µs. Hence there is 1.2 µs propagation delay is observed in traditional pipeline. The propagation delay is the time taken to travel the data from stage one input to stage 3 output terminal in this case. In the same way the propagation delay is observe in wave pipeline, Mesochronous, and new pipeline.

5 Conclusions

After through comparisons it is observed that 1.1 μs propagation delay is observed in the proposed method. In three stage pipeline the conventional pipeline is recording best propagation delay when compared with other two existing methods. But as the number of stages increases the wave pipeline has best propagation delay when compared with conventional and mesochronous pipeline. as the number of stages increases Mesochronous pipeline has less propagation delay when compared with conventional pipeline. In all the cases the proposed method producing less propagation delays at any number of stages, when compared with all other methods. Hence the proposed pipeline will help the high speed computer in evaluating arithmetic operations and real time operations quickly. With this pipeline high speed computers can be interfaced to synchronize the speed with the real time appliances.

References

1. Hyein lee et al., (2010), "Pulse width allocation and clock skew Scheduling: Optimizing Sequential Circuits Based on Pulsed Latches", IEEE Transaction on Computer aided design of integrated circuits and systems, vol 29, No 3, March.
2. Arto Rantala (2007) et al., "A DLL clock generator for a high speed A/D-converter with 1 ps jitter and skew calibrator with 1 ps precision in 0.35 μm CMOS", Springer Science, Analog Integr Circ Sig Process, pp 69–79.
3. McCluskey EJ (1986) Logic design principles: with emphasis on testable semicustom circuits. Prentice Hall, Englewood Cliffs.
4. Li HH (2009) Cypress Semiconductor, USA, personal communication on LSSD.
5. Benini (1997) et al., "Clock Skew Optimization for Peak Current Reduction", Springer, Journal of VLSI Signal Processing,, Kluwer Academic Publishers, pp 117–130.
6. Jacobs. G.M et al., (1990), "A fully asynchronous digital signal processor using self-timed circuits" IEEE J. Solid-State Circuits, vol. SC-25, dec.
7. Lu. S.L. (1988), "Implementation of iterative network with CMOS differential logic", IEEE J. Solid-State Circuits, vol. SC-23, pp. 1013–1017, Aug.
8. Afghahi. M and C. Svensson, (1992), "Performance d synchronous and asynchronous schemes for VLSI" IEEE Trans. Computer, vol. (2–41, pp. 858–872, July.
9. Hong-Yi Huang et al., "Low-voltage Low-Power CMOS True Single-Phase Clocking scheme with Locally Asynchronous Logic Circuits", IEEE xplore.
10. Stephen h. unger," Clocking Schemes for High-Speed Digital Systems", IEEE Transactions on computers, vol. c-35, no. 10, october 1986, pp 880–895.
11. Masoume Jabbarifar (2014) et al., "Online Incremental Clock Synchronization",Springer, J Netw Syst Manage, doi:10.1007/s10922-014-9331-7.
12. Xinjie wei (2006) etal., "Legitimate Skew Clock Routing with Buffer Insertion", Springer, Journal of VLSI Signal Processing, pp 107–116.
13. Wayne P. Burleson, (1998), "Wave-Pipelining: A Tutorial and Research Survey", IEEE Transactions on very large scale integration (vlsi) systems, vol. 6, no. 3, September.
14. Hauck. O et al., (1998), "Asynchronous wave pipelines for high throughput datapaths", ISBN:0-7803-5008-1/98/ $10.000 IEEE.

15. G. Seetharaman, B. Venkataramani and G. Lakshminarayanan, (2008) "Design and FPGA implementation of self-tuned wave-pipelined filters with Distributed Arithmetic Algorithm", Springer, Research journal on circuits, systems and signal processing, 27: 261–276.
16. Refik sever et al., (2013), "A new wave-pipelining methodology: wave component sampling method", International journal of Electronics, Published by Taylor&Francis, doi:10.1080/007217. 2013794479.
17. Suryanarayana B. Tatapudi, Student Member, IEEE and José G. Delgado-Frias, Senior Member, IEEE, (2006) "A Mesochronous high performance digital systems", VOL. 53, NO. 5, May.
18. Suryanarayana B. Tatapudi and José G. Delgado-Frias, "Designing Pipelined Systems with a Clock Period Approaching Pipeline Register Delay", School of Electrical Engineering and Computer Science Washington State University, Pullman, WA 99164–2752.
19. Chuan-Hua Chang et al., "Delay Balancing using Latches", Research gate publications, University of Michigan, http://www.researchgate.net/publication/229039585_Delay_Balancing_using_Latches.
20. Chi-Chou Kao (2015) et AL., "Clock Skew Minimization in Multiple Dynamic Supply Voltage with Adjustable Delay Buffers Restriction", Springer, J Sign Process Syst, pp: 99–104.

An Improved Data Hiding Scheme in Motion Vectors of Video Streams

K. Sridhar, Syed Abdul Sattar and M. Chandra Mohan

Abstract An inter frame data hiding scheme in motion vectors of the compressed video streams is proposed in this paper. The main objective of this paper is to achieve a relatively high embedding capacity while preserving the encoding and the decoding schemes where the data hiding is based on the changing of motion vectors. However this approach tries to preserve the perceptual quality of the compressed video streams there by reflecting the usage of this algorithm to real time applications. The proposed approach is also been compared against the conventional LSB based approach both subjective and objective quality analysis were recorded for different experimental conditions.

Keywords Video coding · Motion vectors · Data hiding

1 Introduction

Digital data hiding of multimedia documents is growing activity, many application and approaches were proposed so far in the literature to explain this activity and of great interest in the field of research. Various approaches aimed for authentication, copy right protection and secret data communication are expected to appear with new challenges and constraints day by day [1].

Data hiding in video streams can be divided into two categories; one of them embeds the secret message in the pixels of the video frames, while the other hides the message into motion vectors of the compressed or encoded video streams [2–4].

K. Sridhar (✉)
Department of ECE, VREC, Nizambad, India
e-mail: rahulmani_147@yahoo.com

S.A. Sattar
Department of ECE, RITS, Chevella, India

M. Chandra Mohan
Department of CSE, SDC, JNTUH, Kukatpally, India

© Springer India 2016
S.C. Satapathy et al. (eds.), *Information Systems Design and Intelligent Applications*, Advances in Intelligent Systems and Computing 435,
DOI 10.1007/978-81-322-2757-1_25

For information concealing in spatial domain or in picture element directly; so as to ensure the in recovery of hidden message once potential distortion attributable to lossy compression or channel noise, it always encodes the message with error correcting code and embeds them into multiple locations, that resists the video compression to some extent however limits the embedding capability however preserves a high sensory activity quality once reconstructed.

In general, most of the work done so far on data hiding in motion vectors focus on modifying the motion vectors based on their properties such as their magnitude, phase angle, etc. In [3, 5], the message streams is hidden in some of the motion vectors whose magnitude is above a predefined threshold, and are called candidate motion vectors (CMVs). The larger component of each CMV is considered for data embedding and a single bit is embedded into hide. In this paper also there is slight change of these CMVs is performed with improved. This paper is organized as, Sect. 1 explains about the need and necessity of data hiding in the motion vectors, while the Sect. 2 presents about few earlier methods, their approaches benefits and demerits. Sections 3 and 4 presents the proposed methodology, its mathematical analysis in a step by step manner. Section 5 presents the experimental results that are obtained using the proposed approach using different video streams.

2 Background

Most of the approaches implemented for data hiding in motion vectors are tend to change the values of motion vectors, some of the approaches are as follows:

In [6], a simple way of selecting a region is performed in which the data is encoded and embedded into motion vectors. In [5] authors used every 2 bits from the message bit stream to select one of the four sizes (16×16, 8×8, 16×8, 8×16) for the motion estimation process.

In [7] and also in [8] authors tried to hide the data into a video stream using the phase angle between two successive CMV. These CMV are taken into account based on the magnitude of the motion vectors as in [3]. So the message bit which is ought to hide is encoded as phase angle difference in regions between CMV. The block matching scheme is selected to look among the chosen sector for which the magnitude to be larger than the predefined threshold. In [9] Kutter proposed an data hiding method which embeds secret message into the least significant bit (LSB) of the horizontal and vertical components of CMVs. Moreover, Xu in [3] embedded the secret message into MVs according to the CMV magnitude, and either component of the horizontal component or the vertical component is modified in terms of the phase angle.

In [10] Wang et al. proposed a mechanism to hide secret information in motion vectors. The selected region of motion vector is confined to hide watermark info, so as to get fascinating video quality when watermark is embedded. An intra-mode is employed to cipher those macro blocks badly suffering from proscribing the region of the chosen motion vectors in inter-mode.

This paper mainly focuses on the method proposed by Xu in [3] and tries to improve its performance.

3 Modeling of Data Hiding in Motion Vectors

Motion estimation (ME) and compensation is an integral procedure to reduce temporal redundancy between adjacent P-frames and spatial redundancy within I-frame. Each inter-frame MB in the P-frame is represented by forward MVs and quantized DCT (Discrete Cosine Transform) coefficients of the prediction residual after motion compensation. MV describes the coarse motion of current MB from the reference MB of adjacent frames. Similarly, each inter-frame MB in the B-frame is represented by forward and backward MVs and quantized DCT coefficients of the prediction residual. The number of MVs for a MB is determined by the prediction mode and the number of reference frame. For simplicity, in the theoretical analysis of the paper, each MB sized 16 × 16 consists of only one block and has only one reference frame.

The embedding process can be modeled as adding an additive independent noise to the horizontal and vertical components of MVs.

$$\begin{cases} SV^h_{k,l} = V^h_{k,l} + \eta^h_{k,l} \\ SV^v_{k,l} = V^v_{k,l} + \eta^v_{k,l} \end{cases} \tag{1}$$

where $V^h_{k,l}$ and $V^v_{k,l}$ are the horizontal and vertical components of the motion vectors in the cover video. In inter frame coding the video encoder compresses the temporal redundancies between the frames by motion compensation with motion vector prediction. In motion compensation prediction the best motion prediction means that we can get the minimum residual error between the current macro block and the predicted macro block. Therefore one can replace the original motion vector with another local optimal motion vector to embed the data in the motion vectors of inter frame.

4 Proposed Approach

Assume that $S = \{MV_0, MV_1, \ldots MV_{n-1}\}$, Compute its phase angle as $\theta_i = \arctan\left(\frac{MV_v}{MV_h}\right)$ where MV_v and MV_h are the motion vectors along vertical and horizontal components. Alternative for embedding the data in the magnitude of motion vectors it more suitable to embed in the phase of two motion vectors.

The algorithm for embedding is as followed:

- Read a H.264 coded video stream
- Initialize the starting frame for embedding (STF) and the number of frames (NF) for embedding
- Read the message or bit stream to be embedded
- Consider the general video format IBBPBBI of frames and select' P' and 'B' frames for embedding
- Apply YUV transformation to all the frames required for processing and consider only 'Y' component.
- Calculate MV_v and MV_h from the processed video stream
- Calculate the magnitude and phase of these motion vectors
- Evaluate the positions or index of the motion vectors whose magnitude is greater than a already defined threshold
- Divide the messages that are to be embedded in horizontal and vertical components based on the index.
- Apply the embedding rule for embedding based on the phase like
  If the phase angle ≤90° and if there is available message for embedding then
  If mod (MV_{h},2) = message bit (Horizontal message stream) the $MV_h = MV_h$ else $MV_h = MV_h + C$
  Where 'C' is a constant for embedding ranges in [0 1]
- Similarly for Vertical motion vector components.
- At the extraction the process is reverse
  if the Phase angle ≤90° then extract from horizontal components or else extract from vertical components

5 Experimental Results

The experiments were conducted on Matlab Software tool with different video streams like foreman, Suzie, flower, playground, scenery taken from [11]. One of the result is depicted in the below figures.

The experiments are conducted with evaluation of PSNR for different frames, at different data rates. The proposed approach is being compared against the conventional LSB based substitution approach. The experimental results are found to be satisfactory and shown a considerable amount of improvement in terms of hiding capacity and PSNR. In Fig. 1 a sample frame is displayed for the better understanding of the embedding process which is followed by the Fig. 2 where the motion vectors are shown with a vector flow marks. As stated this approach is being compared against the conventional LSB substitution approach and the analysis is shown in the performance analysis Fig. 3, from which it can be stated that the present approach on an average leading 20–22 dB of improvement in PNSR which is a greater achievement (Fig. 4).

Fig. 1 'Y' components of IBP frames

Fig. 2 Predicted motion vectors

Fig. 3 Performance analysis of proposed approach for different frames

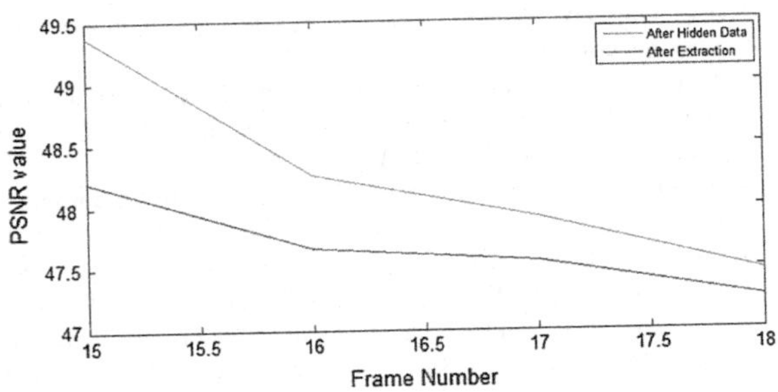

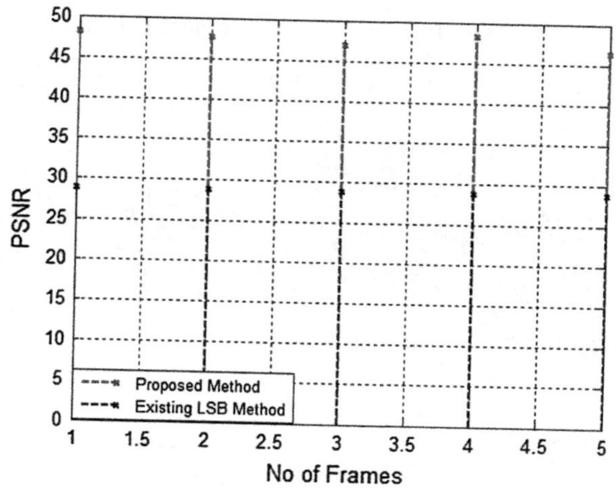

Fig. 4 Comparison of the proposed approach with LSB substitution method

6 Conclusion

An improved data hiding scheme depend on the modification of motion vectors is proposed in this paper, the algorithm is tested on several video streams and found that this method is leading a considerable improvement both perceptually and capacity wise. The algorithm suffers from a limitation that this approach is consuming more processing time than earlier. The deployment tool or language can be changed in order to improvise this. This algorithm can also be extended in future with changing of transformation domain and also decreasing the processing time.

References

1. J.L. Dugelay and G. Doerr, "A guide tour of video watermarking," Signal Process Image Comm, vol. 18, pp. 263–82, 2003.
2. F. Jordan, M. Kutter, and T. Ebrahimi, "Proposal of a watermarking technique for hiding data in compressed and decompressed video," ISO/IEC Document, JTC1/SC29/WG11, Stockholm, Sweden, Tech. Rep. M2281, Jul. 1997.
3. C. Xu, X. Ping, and T. Zhang, "Steganography in compressed video stream," in Proc. 1st Int. Conf. Innov. Comput., Inf. Control, Sep. 2006, pp. 269–272.
4. H. Aly, "Data hiding in motion vectors of compressed video based on their associated prediction error," IEEE Trans. Inf. Forensics Security, vol. 6, no. 1, pp. 14–18, Mar. 2011.
5. J. Zhang, J. Li, and L. Zhang, "Video watermark technique in motion vector," in Proc. XIV Symp. Computer Graphics and Image Processing, Oct. 2001, pp. 179–182.
6. P. Wang, Z. Zheng, and J. Ying, "A novel video watermark technique in motion vectors," in Int. Conf. Audio, Language and Image Processing (ICALIP), Jul. 2008, pp. 1555–1559.
7. D.-Y. Fang and L.-W. Chang, "Data hiding for digital video with phase of motion vector," in Proc. Int. Symp. Circuits and Systems (ISCAS), 2006, pp. 1422–1425.
8. X. He and Z. Luo, "A novel steganographic algorithm based on the motion vector phase," in Proc. Int. Conf. Comp. Sc. and Software Eng., 2008, pp. 822–825.

9. F. Jordan, M. Kutter, and T. Ebrahimi, "Proposal of a watermarking technique for hiding data in compressed and decompressed video," ISO/IEC Document, JTC1/SC29/WG11, Stockholm, Sweden, Tech. Rep. M2281, Jul. 1997.
10. Li L, Wang P, Zhang ZD, "A video watermarking scheme based on motion vectors and mode selection". In: Proceedings of IEEE International Conference Computer Science and Software Engineering, vol. 5. 2008.
11. http://trace.eas.asu.edu/.

Hiding Sensitive Items Using Pearson's Correlation Coefficient Weighing Mechanism

K. Srinivasa Rao, Ch. Suresh Babu and A. Damodaram

Abstract Data mining algorithms extract high level information from massive volumes of data. Along with advantage of extracting useful pattern, it also poses threats of revealing sensitive information. We can hide sensitive information by using privacy preserving data mining. As association rules are a key tool for finding the pattern, so certain rules can be categorized as sensitive if its disclosure risk is above a specific threshold. In literature, there are different techniques exist for hiding sensitive information. Some of these techniques are based on support and confidence framework, which suffers with limitations including choosing a suitable value of these measures which cause losing useful information, generation of large number of association rules and loss of database accuracy. We propose correlation based approach which uses measures other than support and confidence such as correlation among items in sensitive item sets to hide the sensitive items in the database.

Keywords Sensitive information · Data mining · Frequent item set · Association rule mining · Pearson's correlation coefficient · Support and confidence

K. Srinivasa Rao (✉)
Department of Computer Science and Engineering,
Malineni Lakshmaiah Women's Engineering College,
Guntur, Andhra Pradesh, India
e-mail: ksrao517@gmail.com

Ch. Suresh Babu
Department of Computer Science and Engineering,
Sree Vaanmayi Institute of Science & Technology, Bibinagar,
Nalgonda, India
e-mail: sureshbabuchangalasetty@gmail.com

A. Damodaram
Department of Computer Science and Engineering, JNTUH,
Kukatpally, Hyderabad, India
e-mail: damodarama@rediffmail.com

© Springer India 2016
S.C. Satapathy et al. (eds.), *Information Systems Design and Intelligent Applications*, Advances in Intelligent Systems and Computing 435,
DOI 10.1007/978-81-322-2757-1_26

1 Introduction

Computers have guaranteed us a wellspring of wisdom but delivered a downpour of information. This immense measure of data makes it crucial to create devices to find hidden knowledge. These instruments are known as data mining instruments. Data mining is the information revelation procedure of investigating valuable data and patterns out of large databases. Of late, data mining has procured much importance as it clears approach to oversee and acquire hidden information and utilize this data in decision-making. In that capacity, data mining ensures to find hidden knowledge, but what happens if that hidden knowledge is sensitive and owners are not willing to expose such knowledge to the public or to adversaries? This question motivates researchers to develop techniques, algorithms and protocols to assure privacy of data for data owners while at the same time fulfilling their need to share their data for a typical reason. Whilst dealing with sensitive information, it turns out to be exceptionally crucial to ensure data against unapproved access. A noteworthy issue that emerged is the need to have a balance between the confidentiality of the disclosed data and the needs of the data clients. In course of doing this, it gets to be quite essential to adjust the data values and relationships.

In our research, we will investigate measures other then support and confidence. Then Pearson's Correlation Coefficient based weighing mechanism will be applied to hide the sensitive items.

2 Motivation

A decent motivational illustration is described in Agrawal et al. [1]. Consider the accompanying setting. Assume we have a server and numerous customers. Every customer has an arrangement of things (books or music or pages or TV programs). The customers need the server to accumulate statistical data about associations among different items, perhaps in order to provide recommendations to the customers/clients. However, the clients do not want the server to know which customer has got which items. When a client sends its set of items to the server, it modifies the set according to some specific randomization policy. The server then gathers statistical information from the modified transactions and recovers from it the actual associations.

Accomplishing a genuine harmony between the exposure and concealing is a tricky issue. Association rule poses a great threat to the data if the data is in control of a malicious user. Therefore, there is a need of an effective strategy to release the database by hiding sensitive rules. Motivation of proposed approach is based on the fact that mostly the techniques that are reviewed in the literature are based on support and confidence. Other measures are also discussed in literature like lift, leverage, conviction and all—confidence. (Naeem et al. [2]) We will use these measures for privacy preservation.

In our research, a solution is proposed where we will examine measures other than support and confidence. At that point Pearson's Correlation Coefficient based weighing mechanism is applied to hide the sensitive items. It is described in coming sections.

3 Related Work

Privacy Preserving Data Mining (PPDM) has become a well known research topic in data mining for the past few years. Clifton et al. [3] analyzed that data mining brings threat against databases and he addressed some possible solutions for protecting privacy in data mining. Rizvi et al. [4] discussed the issue of preserving privacy in association rule mining. Normally the techniques discussed in literature and have a problem of side effects and lack of tradeoff between privacy and accuracy. Based on this idea, Jieh-Shan et al. in 2010 [5] proposed PPUM, Privacy Preserving Utility Mining and modeled two novel algorithms. Performance of these two algorithms is analyzed with the help of three parameters Hiding Failure, Missing Cost and Data Difference Ratio (DDR) between original and modified database.

4 Problem Statement and Proposed Work

After literature survey we conclude that mostly techniques are based on support and confidence framework. Related work uses utility value among sensitive item set and count of conflicts among sensitive item sets. In this paper, we will hide the sensitive items using framework other than conventional support and confidence based on Pearson's Correlation Coefficient. Related work somehow gives better performance regarding privacy preserving data mining. Our work considers the correlation among items in transactions. In addition, we tried to minimize the impact of hiding sensitive item sets on the sanitized database i.e. reducing the side effects of privacy preserving and maintain the tradeoff between privacy and Accuracy. To prove this our framework is also executed on the same IBM synthetic dataset considered by Jieh-Shan et al. [5].

4.1 Proposed Approach

In this work, we will use Column-wise correlation among data items in the original database to hide sensitive items. Firstly frequent item sets are mined from original database using minimum support threshold. A correlation matrix is created by placing frequent item sets on x-axis and individual items present in frequent item

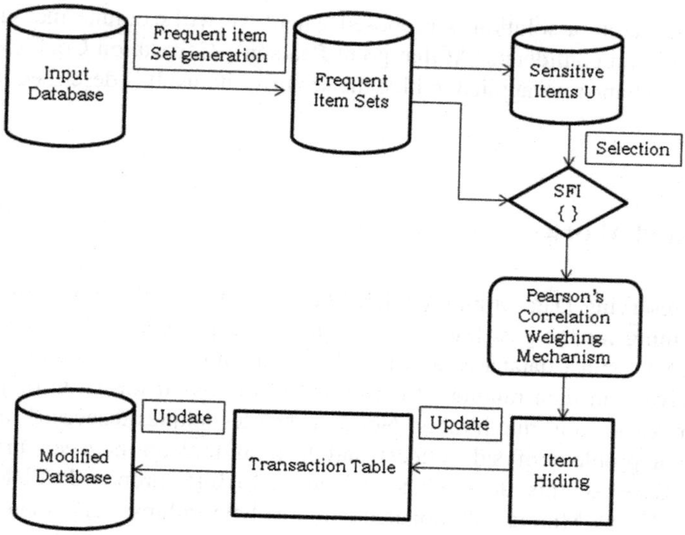

Fig. 1 Proposed solution for hiding sensitive items

sets on y-axis. Here frequent item sets represent rows and individual items represent columns. Columns (items) having minimum correlation threshold are selected for hiding process. So instead of hiding sensitive items directly as in the case of conventional PPDM techniques, this approach selects items having minimum correlation threshold for hiding process. Hence this framework uses support threshold as well as correlation threshold leading to a hybrid approach. Proposed framework is given below (Fig. 1).

5 Effectiveness Measurement

Effectiveness of our proposed algorithm is measured by adopting the evaluation measures used by Oliveira and Zaiane [6]. We describe the evaluation measures as follows:

(a) **Hiding Failure** It is the measure of sensitive items that are even mined from sanitized database.
(b) **Miss Cost** It is the measure of non-sensitive items that are hidden by accident during hiding process.
(c) **DDR** It is called Data Difference Ratio between modified database and original database. It is the measure of database accuracy.

Table 1 Characteristics of the dataset considered for the work

Dataset name	No. of transactions	Distinct items	Avg. length of transactions
T10I4D100K	1000	795	7

6 Experimental Results Discussion

The experiment was conducted on IBM synthetic dataset (http://fimi.ua.ac.be/data/T10I4D100K.dat) [7] which was used by Jieh-Shan et al. [5] and Saravanabhavan et al. [8]. Performance of the proposed approach is compared with that of Jieh-Shan et al. [5]. Three parameters were considered i.e. Hiding Failure, Missing Cost and Data Difference Ratio (DDR) for analyzing the effectiveness of the approach (Table 1).

6.1 Experimental Results of the Proposed Work

Experiments were conducted on U1 = {274,346,753,742,809} *U2 = {70,132,192, 204,381}* |U| = 5 NS = {415 676 825 834 177 424 490 204}m where U1 is sensitive item set, NS is non sensitive item set.

From the above graphs (Figs. 2 and 3) it is evident that for Sensitive Itemsets U1, Hiding Failure is almost zero in the negative values of Correlation threshold, which implies that the items from our sensitive itemsets U1 are negatively correlated and if

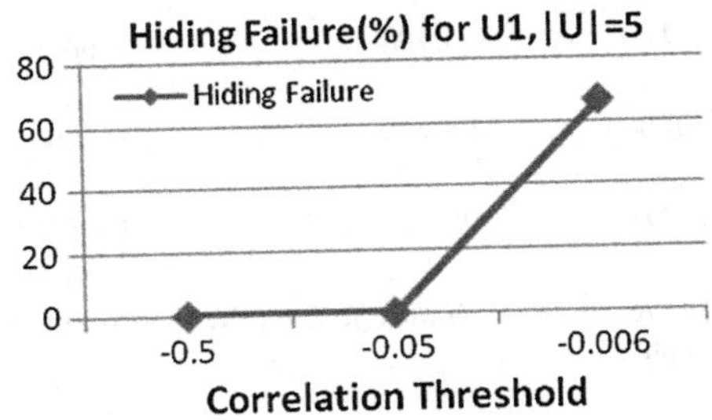

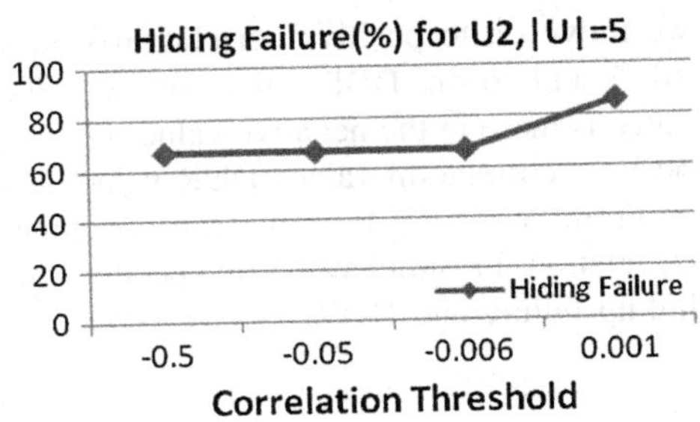

Fig. 2 Hiding failure curve for various correlation thresholds on U1 and U2

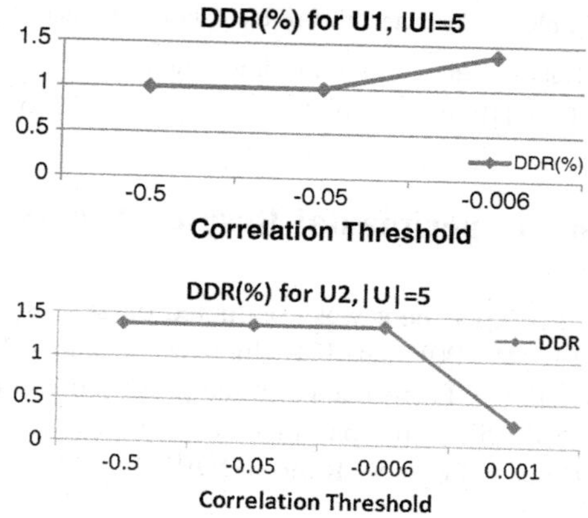

Fig. 3 Data difference ratio curve for various correlation thresholds on U1 and U2

Table 2 Hiding failure, miss cost and data difference ratio for various correlation thresholds

Corr. thres	(Initial FI's, new FI's)	Hiding failure	Miss cost	DDR
Results for T10I4D100K data set with Num trans = 1000 mining thres = 0.01				
U1 = {274,346,753,742,809} |U| = 5 NS = {415 676 825 834 177 424 490 204}				
−0.5	*505,483*	0 (0/22)	0 (0/10)	0.01 (101/10098)
−0.05	*505,483*	0 (0/22)	0 (0/10)	0.01 (101/10098)
−0.006	*505,482*	0.6667 (14/21)	0.1 (1/10)	0.01386 (140/10098)
U2 = {70,132,192,204,381} |U| = 5 NS = {415 676 825 834 177 424 490 204}				
−0.5	*505,498*	0.6666 (14/21)	0.1 (1/10)	0.01377 (139/10098)
−0.05	*505,498*	0.6666 (14/21)	0.1 (1/10)	0.01377 (139/10098)
−0.006	*505,498*	0.6666 (14/21)	0.1 (1/10)	0.01377 (139/10098)
0.001	*506,503*	0.85714 (18/21)	0 (0/10)	0.00247 (25/10098)

Above values of *Hiding failure*, *Miss cost* and *DDR* are represented as percentage in following graphs

we move on to positive correlation values Hiding failure increases to 30 % and 70 % and so on. DDR value follows reverse direction of Hiding Failure i.e. it's value is more in the negative values of Correlation threshold, and if we move on to positive correlation values DDR value decreases it is observed in Fig. 3.

From above Table 2 one can conclude that there is a flexibility of tuning the correlation threshold value on specific sensitive item set to obtain better results of hiding failure and DDR.

Table 3 Performance Comparison with existing work

Algorithm	Sensitive item set U and it's size	Threshold	HF (%)	MC (%)	DDR (%)
HHUIF	5	5,000 and 5,000 (utility)	0	0	2.95
MSICF	5	5,000 and 5,000 (utility)	0	0	2.94
Proposed algorithm using pearson's correlation coefficient	U1 = {274,346,753,742,809} |U| = 5	−0.5	0	0	1
	U2 = {*70,132,192,204,381*} |U| = 5	0.001	85	0	2.40

6.2 *Experimental Results Comparison*

Following table shows the comparison between experimental of the proposed work with existing work.

From the Table 3, we know that we are ensuring the lower DDR value and the other two measures reach the minimum value.

7 Conclusion

Here privacy preserving data mining is done by considering correlation among items in transactions in database. Pearson's Correlation Coefficient based weighing mechanism is applied on IBM synthetic data set to hide the sensitive items. Performance of the approach is compared with Jieh-Shan et al. [5]. Our approach also results in similar values as Jieh-Shan et al. [5], but our approach causes less DDR value which is the measure of database accuracy and has a flexibility of tuning between getting higher as well as lower values of hiding failure and DDR value i.e. maintains the tradeoff between privacy and accuracy of the original database. DDR value of our approach is compared with that of Saravanabhavan [8] also, which shows that ours got better results.

References

1. Agrawal R., Imielinski T. and Swami A., "Mining associations between sets of items in large databases", SIGMOD93, Washington D.C, USA, 207–216 (1993).
2. Naeem, M., Asghar, S, "A Novel Architecture for Hiding Sensitive Association Rules", Proceedings of the 2010 International Conference on Data Mining (DMIN 2010). Las Vegas, Nevada, USA, CSREA Press 2010. ISBN: 1-60132-138-4, Robert Stahlbock and Sven Crone (Eds.), (2010).

3. Saygin Y., Verykios V. S. and Clifton C., "Using unknowns to prevent discovery of association rules". ACM SIGMOD Record, 30(4): pp. 45–54 (2001).
4. S. Rizvi, and J. Haritsa, "Maintaining data privacy in association rule mining," in Proceedings of 28th Intl. Conf. on Very Large Databases (VLDB), Morgan Kaufmann, San Francisco, pp. 682– 693 (2002).
5. Jieh-Shan Yeh, Po-Chiang Hsu, "HHUIF and MSICF: Novel algorithms for privacy preserving utility mining", Expert Systems with Applications, Vol: 37, pp: 4779–4786 (2010).
6. Oliveira, S. R. M., & Zaine, O. R, "A framework for enforcing privacy in mining frequent patterns", Technical Report, TR02-13, Computer Science Department, University of Alberta, Canada, (2000).
7. Frequent Itemset Mining Dataset Repository, http://fimi.ua.ac.be/data/T10I4D100K.dat.
8. C.Saravanabhavan, R.M.S.Parvathi, "Privacy Preserving Sensitive Utility Pattern Mining", Journal of Theoretical and Applied Information Technology, Vol. 49, No.2, pp.496–506 (2013).

OpenCV Based Implementation of Zhang-Suen Thinning Algorithm Using Java for Arabic Text Recognition

Abdul Khader Jilani Saudagar and Habeeb Vulla Mohammed

Abstract The aim of this research work is to implement Zhang-Suen thinning algorithm on openCV based java platform. The novelty lies in the comparative study of the obtained results using the proposed implementation with the existing implementations of Zhang-Suen thinning algorithm viz. using Matlab, C++ and compare the performance factor viz computation time with others. The experimental results achieved by openCV based java platform are faster when compared to Matlab and C++.

Keywords Arabic text recognition · Thinning algorithm · Arabic text extraction

1 Introduction

Image thinning is a signal transformation that converts a thick digital image into a thin digital image. The skeleton conveys the structural connectivity of the main component of an object and is of one pixel in size. Text thinning has been a part of morphological image processing for a wide variety of applications. Thinning phase has a major role in text processing of optical character recognition (OCR) systems mainly in the Arabic Character Recognition (ACR) system. It abridges the text shape and minimizes the amount of data that needs to be handled for recognition.

A.K.J. Saudagar (✉) · H.V. Mohammed
College of Computer and Information Sciences, Al Imam Mohammad Ibn Saud Islamic University (IMSIU), Airport Road, Riyadh 13318, Saudi Arabia
e-mail: saudagar_jilani@ccis.imamu.edu.sa

H.V. Mohammed
e-mail: habeebvulla@ccis.imamu.edu.sa

© Springer India 2016
S.C. Satapathy et al. (eds.), *Information Systems Design and Intelligent Applications*, Advances in Intelligent Systems and Computing 435,
DOI 10.1007/978-81-322-2757-1_27

Many thinning algorithms have been proposed in the past and one such algorithm is the fast parallel thinning algorithm [1]. Another modified algorithm is proposed by [2] where obtained results were compared against results of [1], the obtained results show that the proposed modified approach is much faster. A methodology and thinning algorithm employed for preprocessing the archaeological images is presented by [3]. An iterative approach for skeletonization which combines parallel and sequential methods is proposed by [4]. An image thinning operation [5] has been implemented on a binary image of 128 × 128 pixels using Zhang Suen's thinning algorithm using MATLAB 7.12 and also synthesized on Virtex 5 in Xilinx ISE for understanding the hardware complexity. Parallel thinning algorithms presented by [6] generates one-pixel-wide skeletons with a difficulty in maintaining the connectivity between components.

Balinese papyrus [7] is one of the media used to identify the literature that was written in papyrus. The application of Zhang-Suen thinning method is very effective in improving the quality of the image making it much easier to read text on the papyrus. An innovative approach for recognition of Arabic text based on feature extraction and on dynamic cursor sizing is presented by [8]. The issues that must be considered when choosing or designing a thinning algorithm for the Arabic Character Recognition (ACR) is discussed by [9]. The work done by [10] addresses an efficient iterative thinning algorithm based on boundary pixels deletion using color coding for different pixel types in Arabic OCR systems.

An OCR system for Arabic [11] was developed which includes: preprocessing step followed by segmentation, thinning phase, feature extraction and finally classification. A general concept on the recognition processes involved in the entire system of the off-line Arabic character recognition is explained by [12]. A thinning algorithm for Arabic handwriting using color coding for both thinning processing and gap recovery in the final output skeleton is proposed by [13]. Few thinning algorithms [14] are implemented, tested and analyzed based on their performance patterns.

In spite of rigorous efforts in developing an Arabic OCR with similar interpretation competence as humans, a limitation still exists. Due to the availability of numerous thinning algorithms, the selection of appropriate algorithm for specific platform is also an important question.

2 Materials and Methods

Zhang-Suen thinning method for hauling out the skeleton of an image consists of eliminating all the outline points of the picture except those points that actually belong to the skeleton. For maintaining the connectivity in the skeleton, each iteration has to undergo two sub-iterations.

Table 1 Designations of the nine pixels in a 3 × 3 window

S9 (m − 1, n − 1)	S2 (m − 1, n)	S3 (m − 1, n + 1)
S8 (m, n − 1)	S1 (m, n)	S4 (m, n + 1)
S7 (m + 1, n − 1)	S6 (m + 1, n)	S5 (m + 1, n + 1)

In the initial iterative step, the outline point S1 is erased from the digital pattern if it obeys the following conditions:

$$2 <= K(S1) <= 6. \quad (1)$$

$$J(S1) = 1. \quad (2)$$

$$S2 * S4 * S6 = 0. \quad (3)$$

$$S4 * S6 * S8 = 0. \quad (4)$$

Here J(S1) is the count of 01 patterns in the prearranged set S2, S3, S4, …, S8, S9 which are the eight adjacent coordinates of S1 as shown in Table 1.

K(S1) is the count of nonzero adjacent coordinates of S1, that is,

$$K(S1) = S2 + S3 + S4 + \ldots + S8 + S9. \quad (5)$$

If any one of the condition is false, e.g., the values of S2, S3, S4, …, S9 as depicted in Fig. 1, then J(S1) = 2.

Hence, S1 is not erased from the input image.

In the subsequent sub-iteration, the conditions (3) and (4) are altered as shown in Fig. 2.

$$S2 * S4 * S8 = 0. \quad (6)$$

$$S2 * S6 * S8 = 0. \quad (7)$$

And the others are same.

From conditions (3) and (4) of initial sub-iteration, it can be seen that the initial sub-iteration eliminates only the south-east peripheral coordinates and the north-west corner coordinates which cannot fit into ideal skeleton.

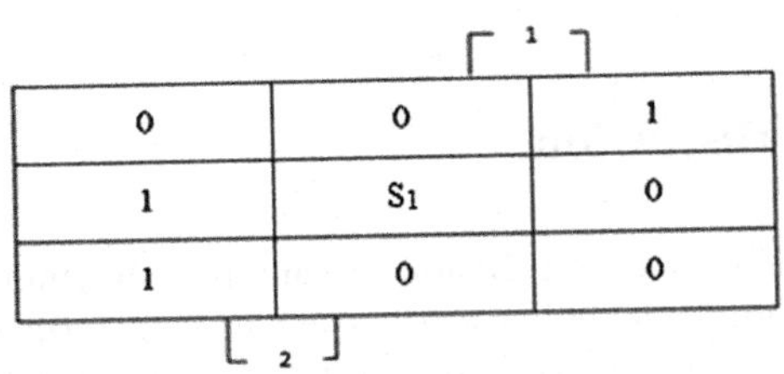

Fig. 1 Counting the 0 and 1 patterns in the prearranged set S1, S2, S3, S4, …, S8, S9

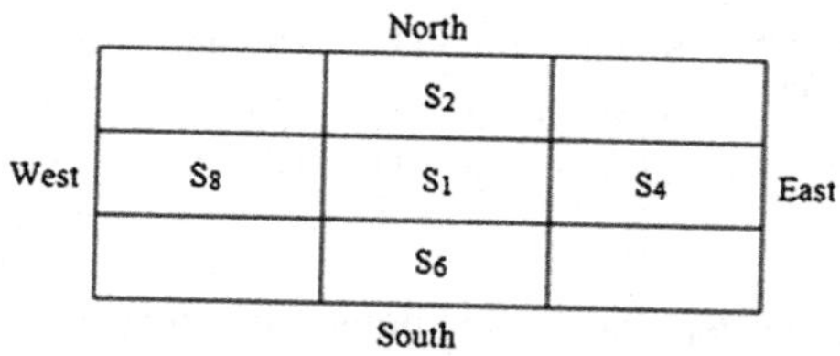

Fig. 2 Points under construction and their locations

The testimony for the initial sub-iteration is given, that is, the coordinates to be deleted which obey the conditions

$$S2 * S4 * S6 = 0. \tag{8}$$

$$S4 * S6 * S8 = 0. \tag{9}$$

The explanation to Eqs. (8) and (9) are S4 = 0 or S6 = 0 or (S2 = 0 and S8 = 0). So the coordinate S1, which has been eliminated, might be an east or south boundary coordinate or a north-west angle coordinate. In a similar manner, it can be proved that the point S1 eliminated in the subsequent sub-iteration might be a north-west boundary coordinate or a south-east angle coordinate.

From condition (1), the boundary points of a skeleton line are maintained. Also, condition (2) avert the elimination of those coordinates that loll between the boundary points of a skeleton line surviving the iterations until no further coordinates are removed.

2.1 *Performance Measurement*

The implementation of Zhang-Suen thinning algorithm on various platforms is performed and performance measurement [15] is calculated using computation time. As such there is no research undertaken which shows the comparative study of Zhang-Suen thinning algorithm on different platforms, tools such as Java, Matlab, C++ and tabulate the values based on performance criteria in particular for thinning Arabic characters.

3 Results and Discussion

The implementation is tested on 20 image samples containing Arabic text. The images used in the implementation process are in jpeg format. The system configuration used for coding and testing is on Intel i5 × 64 processor with 4 GB RAM

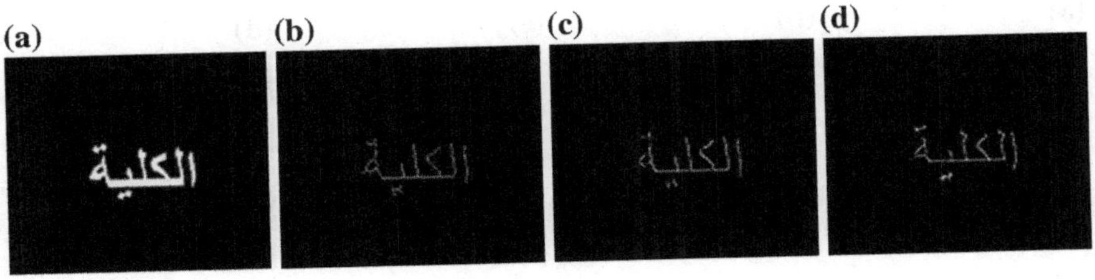

Fig. 3 **a** Original input image 1. **b** Open CV—Java. **c** Matlab. **d** C++

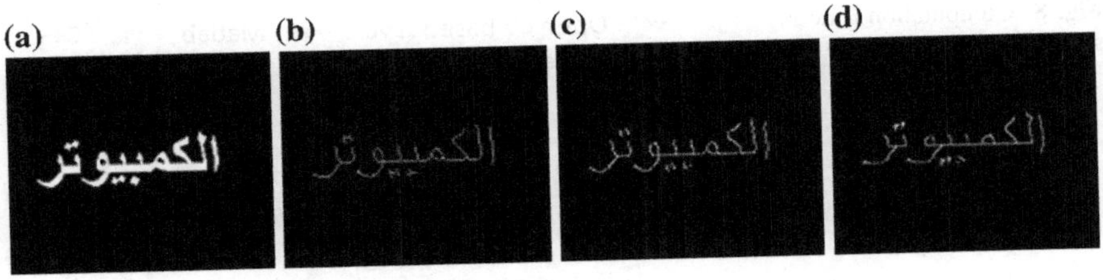

Fig. 4 **a** Original input image 2. **b** Open CV—Java. **c** Matlab. **d** C++

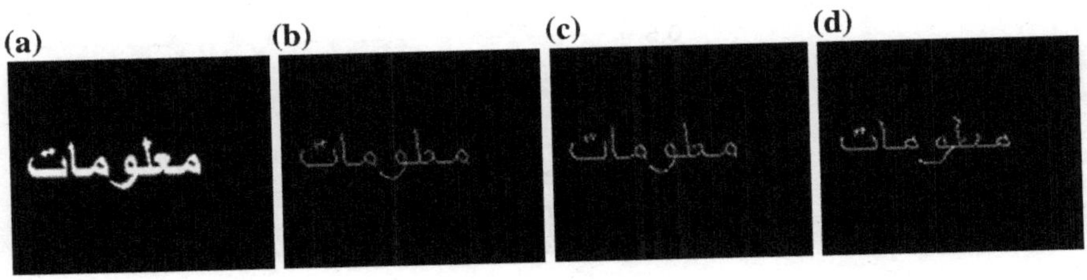

Fig. 5 **a** Original input image 3. **b** Open CV—Java. **c** Matlab. **d** C++

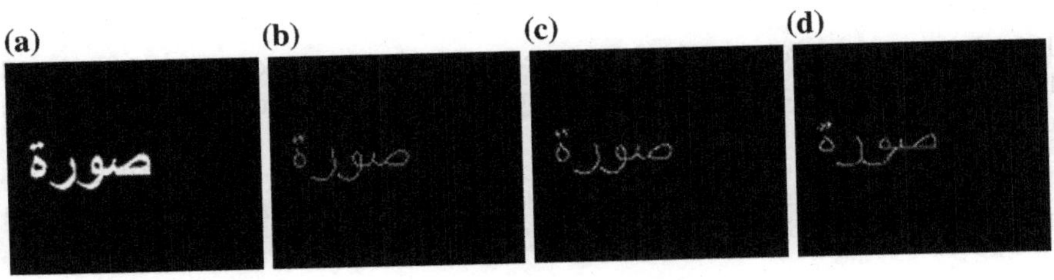

Fig. 6 **a** Original input image 4. **b** Open CV—Java. **c** Matlab. **d** C++

and 512 GB hard disk. The output thinned images were obtained using openCV based Java implementation, Matlab and C++. Five image samples are as shown in Figs. 3, 4, 5, 6, and 7 and their corresponding graphs for computation time in seconds is plotted as shown in Fig. 8.

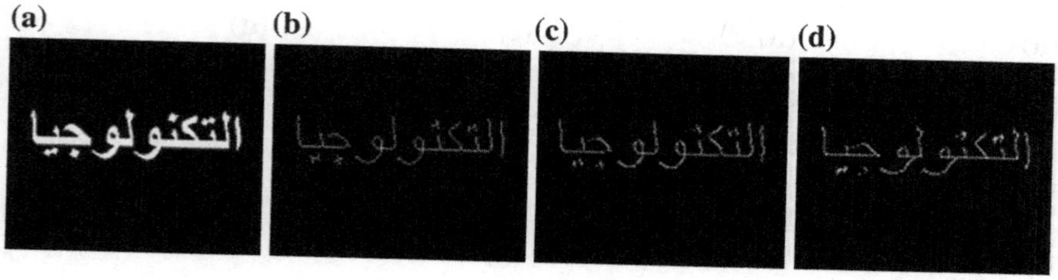

Fig. 7 **a** Original input image 5. **b** Open CV—Java. **c** Matlab. **d** C++

Fig. 8 Computation time in seconds

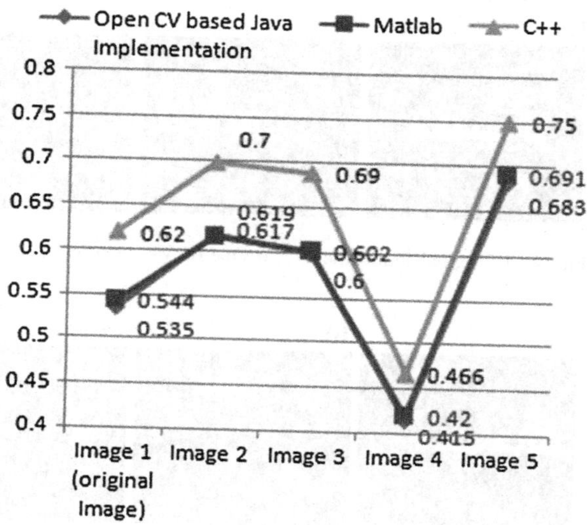

4 Conclusion

It is observed from results that the thinned output images processed and produced by openCV based Java implementation are faster when compared to results obtained from other implementations.

Acknowledgments This research project is supported by King Abdulaziz City for Science and Technology (KACST), Saudi Arabia, vide grant no. AT-32-87.

References

1. Zhang, T. Y., Suen, C. Y.: A Fast Parallel Algorithm for Thinning Digital Patterns. Communications of ACM. 27(3), 236–239 (1984)
2. Zhang, Y. Y., Wang, P. S. P.: A Modified Parallel Thinning Algorithm. In: Proc. of Int. Conf. on Pattern Recognit, Rome, Italy. 2. pp. 1023–1025 (1988)

3. Devi, H. K. A.: Thinning: A Preprocessing Technique for an OCR System for Brahmi Script. Anc. Asia. 1, 167–172 (2006)
4. Abu-Ain, T., Abdullah, S. N. H. S., Bataineh, B., Omar, K.: A Fast and Efficient Thinning Algorithm for Binary Images. J. ICT Res. Appl. 7(3), 205–216 (2013)
5. Karnea, A. S., Navalgunda, S. S.: Implementation of an Image Thinning Algorithm using Verilog and Matlab. In: Proc. of Natl. Conf. on Women in Science Engineering, Dharwad, India, pp. 333–337 (2013)
6. Kumar, G., Oberoi, A.: Improved Parallel Thinning Algorithm for Numeral Patterns. International J. of Res. in Commerce, IT & M. 3(8), 43–47 (2013)
7. Sudarma, M., Sutramiani, N. P.: The Thinning Zhang-Suen Application Method in the Image of Balinese Scripts on the Papyrus. Int. J. of Computer Applications. 91(1), 9–13 (2014)
8. Al-A'ali, M., Ahmad, J.: Optical Character Recognition System for Arabic Text using Cursive Multi Directional Approach. J. of Computer Science. 3(7), 549–555 (2007)
9. Al-Shatnawi, A. M., Omar, K., Zeki, A. M.: Challenges in Thinning of Arabic Text. In: Proc. of Int. Conf. on Artificial Intell. and Machine Learning, Dubai, UAE, pp. 127–133 (2011)
10. Ali, M. A.: An Efficient Thinning Algorithm for Arabic OCR Systems. Signal & Image Processing an Int. J. 3(3), 31–38 (2012)
11. Supriana, I., Nasution, A.: Arabic Character Recognition System Development. In: Proc. of 4th Int. Conf. on Electrical Engineering and Informatics, Selangor, Malaysia. pp. 334–341 (2013)
12. Aljuaid, H. A., Muhamad, D.: Offline Arabic Character Recognition using Genetic Approach: A Survey. http://comp.utm.my/publications/files/2013/04/Offline-Arabic-Character-Recognition-using-Genetic-Approach-A-Survey.pdf
13. Ali, M. A., Jumari, K. B.: Skeletonization Algorithm for an Arabic Handwriting. WSEAS Trans. on Computers. 2 (3), 662–667 (2003)
14. Snehkunj, R. K.: A Comparative Research of Thinning Algorithms. Int. J. of Computer Applications.12–15 (2011)
15. Taraek, P.: Performance Measurements of Thinning Algorithms. J. of Inf. Control and Management Syst. 6(2), 125–132 (2008)

Numerical Modeling of Twin Band MIMO Antenna

Vilas V. Mapare and G.G. Sarate

Abstract Over the last decade multiple input multiple output (MIMO) systems have received considerable attention. There is some limitation while obtaining the most from MIMO, such as mutual coupling between antenna elements. Mutual coupling and therefore inter-elements spacing have important effects on the channel capacity of a MIMO communication system, its error rate and ambiguity of MIMO radar system. The effect of mutual coupling on MIMO system has been studied and then different array configurations are considered. Different configuration show different mutual coupling behaviour. After modelling and simulation, the array was designed, implemented and finally verified the result using Vector Network Analyzer. In this paper, a compact Twin bands MIMO antenna with low mutual coupling, operating over the range of 2.1–2.845 GHz is proposed.

Keywords Multi band · Rectangular microstrip antenna · Gap coupled rectangular microstrip antenna · Multiple resonators · Parasitic resonators · Multi band MIMO antenna

1 Introduction

Nowadays, Antenna plays an important role in today technology because of development in MIMO system. Good antenna array design with low mutual coupling that satisfies the space constraints and provides good MIMO performance has become the motivation. Several antenna designs technique has been proposed to reduce mutual coupling. Different ground structures are used and presented in [1],

V.V. Mapare (✉)
Sant Gadge Baba Amravati University, Amravati, India
e-mail: mapare.vilas@gmail.com

G.G. Sarate
Government Polytechnic, Amravati, India
e-mail: ggsanshu@gmail.com

© Springer India 2016
S.C. Satapathy et al. (eds.), *Information Systems Design and Intelligent Applications*, Advances in Intelligent Systems and Computing 435,
DOI 10.1007/978-81-322-2757-1_28

lump circuits or neutralization lines are used in [2], parasitic element are used to create reverse coupling to reduce mutual coupling and discussed in [3]. In this paper, we proposed simple but effective antenna diversity techniques to reduce mutual coupling between array elements without changing the antenna's structure. In [4–7], the mutual coupling was effectively reduced by using defected ground structures (DGS). Antenna array with low mutual coupling, operating at 2.45 GHz is reported in [4], and the improvement is achieved by etching two λ/2 slots on the ground plane. In this method antenna structure remains unchanged and without any extra cost antenna can be fabricated. The idea is to change array configuration to find a configuration with the lowest amount of mutual coupling between elements.

1.1 Effect of Distributed Parameter on Six Patch Configuration

The single RMSA is splitted into six elements along the width and changing the lengths of individual Twin elements as well varying the spacing between the individual Twin elements and properly adjusting the feed point results in Twin band frequency. As there are six radiating elements in the system, which is asymmetrical about the feed point.

Of the six equal width splited elements of different individual Twin lengths, L1 and L6 STRIP radiate at frequencies closer to each other. The strip of length L2 and L5 is decreased to suppress the frequency band and increase the frequency difference between the two resonant frequencies in turn increase the frequency ratio. The length L4 is increased to bring it's resonance frequency closer to that of the fed element. The two resonant frequencies corresponding to minimum return loss are fr1 = 2.085 and fr2 = 2.53 with bandwidth of 43 MHz at each band. Table 1 gives the details of various parameters, for this configuration. The frequency ratio is increased from 1.19 to 1.21 due to this configuration as shown in Fig. 1. Thus, it can be noted that by properly adjusting the lengths of the elements, Twin frequency operation with frequency ratio higher than that of five patch configuration is obtained with six patch configuration.

In the Smith chart, a resonant circuit shown up as a circle. The larger the circle, the stronger the coupling. In six patches the radius of resonance circle is smaller. Hence the centre of the chart lies outside the circle. At critical coupling i.e. 2.085 and 2.53 GHz the resonance circle touches the centre of the Smith chart as shown in Fig. 2 (Fig. 3).

Table 1 Twin frequency response of configuration with six strips

Lengths L (mm)	Gaps, S (mm)	x (mm)	fr1 (GHz)	RL1 (dB)	BW1 (MHz)	fr2 (GHz)	RL2 (dB)	BW2 (MHz)	fr2/fr1
L1 = 33	S_1 = 2.3	7	2.085	−24	43	2.53	−25	43	1.21
L2 = 31	S_2 = 1								
L3(fed) = 33	S_3 = 0.1								
L4 = 36	S_4 = 0.35								
L5 = 31	S_5 = 2.1								
L6 = 33									

ϵr = 4.3, h = 1.59 mm, tanδ = 0.02, W = 7.68 mm

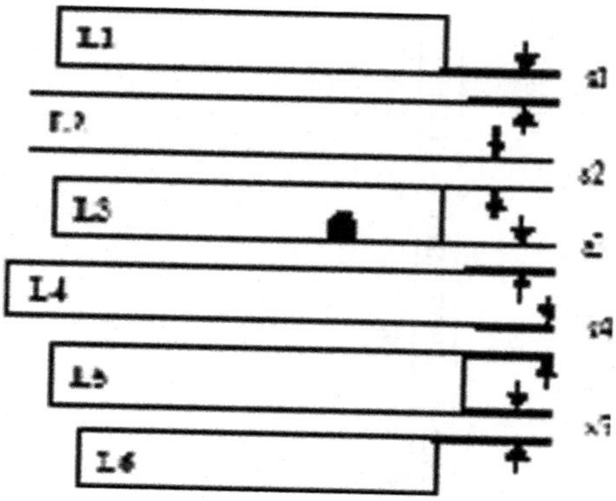

Fig. 1 Six strip configuration [8]

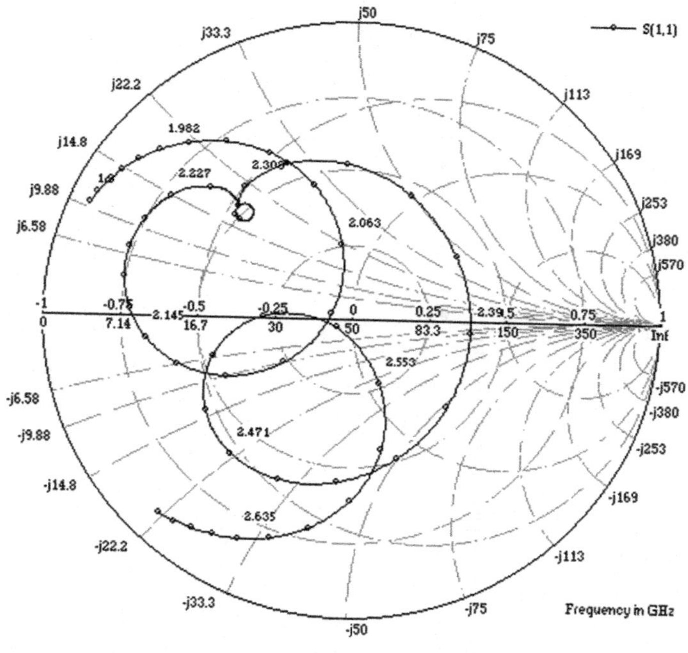

Fig. 2 Smith chart of six patch antenna

1.2 *Effect of Distributed Parameter on Seven Patch Configuration*

For Twin frequency operation, the lengths of strip L2 and L6 are increased to decrease their resonance frequencies, so that they resonant at frequencies closer to

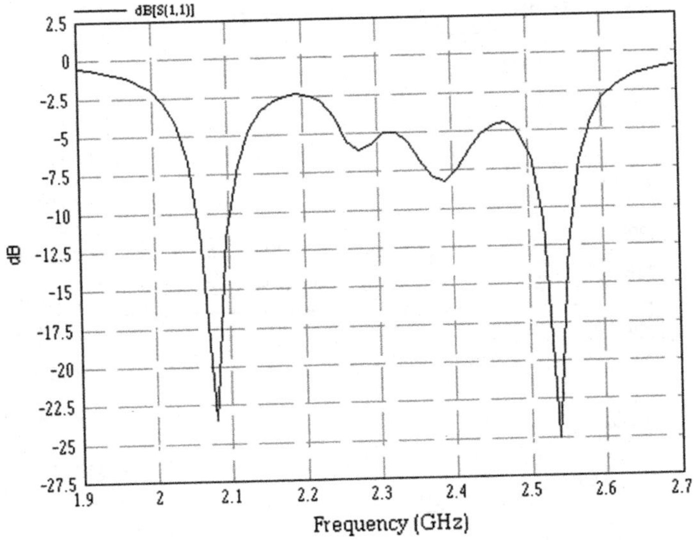

Fig. 3 S11 of six patch antenna

that of L1 and L7 strip, while the lengths L1 and L7 are decreased to increase their resonance frequencies. The lengths L3 and L5 is decreased to increase their resonant frequencies so that they resonant at frequencies closer to that of the driven element (L4) also to have sufficient separation between the two frequencies for Twin frequency operation. When the difference between the lengths of the different elements is large, the frequency ratio also increases. With increase in the number of elements and proper spacing between the elements of different length the bandwidth of individual Twin band increases as well. Table 2 yields the lumped parameter values of seven patch antenna.

In the Smith chart, a resonant circuit shown up as a circle. The larger the circle, the stronger the coupling. In three patches the radius of resonance circle is smaller. Hence the centre of the chart lies outside the circle. At critical coupling i.e. 1.99 and 2.47 GHz the resonance circle touches the centre of the Smith chart as shown in Fig. 4 (Fig. 5).

Table 2 Lumped model of seven patch antenna

Freq (GHz)	Q	L (nH)	R (Ω)
1.94e+00	6.00e+00	2.67e+00	5.46e+00
1.98e+00	4.32e+00	3.28e+00	9.43e+00
1.99e+00	1.45e−02	−1.68e−01	1.33e+01
2.31e+00	3.94e+00	3.76e−01	1.38e+00
2.41e+00	−3.14e+00	−3.53e+00	1.69e+00
2.47e+00	−1.82e+00	−1.32e+00	1.06e+01
2.57e+00	−4.42e+00	−2.71e+00	9.95e+00

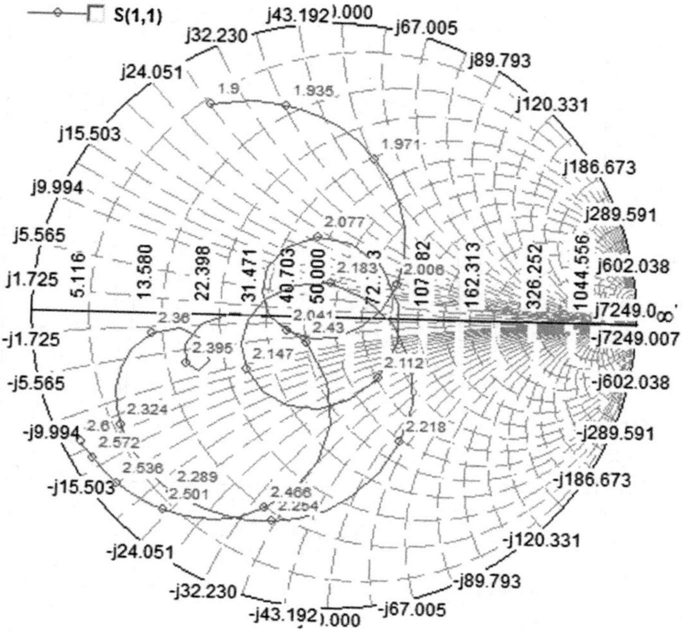

Fig. 4 Smith chart of seven patch antenna

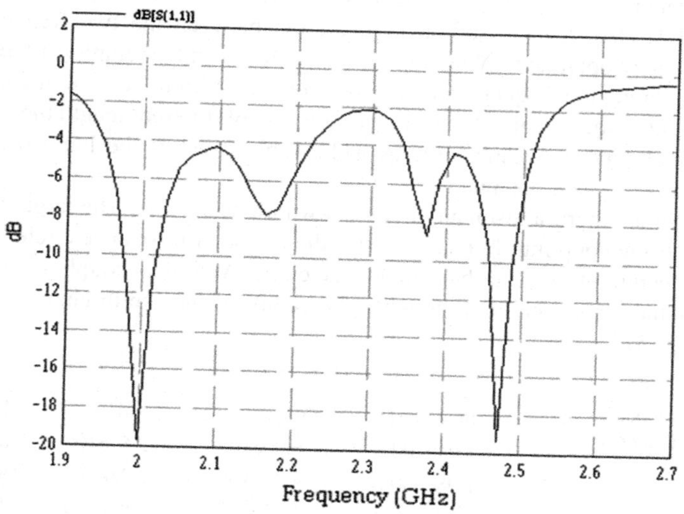

Fig. 5 S11 of seven patch antenna

Table 3 Lumped model of eight patch antenna

Freq (GHz)	Q	L (nH)	R (Ω)
2.00e+00	3.76e+00	2.41e+00	8.05e+00
2.05e+00	1.2e+00	3.48e+00	3.57e+00
2.11e+00	1.93e−02	−1.64e−01	1.15e+01
2.33e+00	1.42e+00	6.77e−01	6.95e+00
2.41e+00	−2.00e+00	−5.85e+00	4.44e+00
2.53e+00	−1.85e+00	−1.32e+00	1.13e+01
2.66e+00	−4.31e+00	−2.11e+00	8.19e+00

1.3 Eight Patch Configurations

For Twin frequency operation with eight elements the lengths L2 and L7 are decreased to increase their resonance frequencies, so that they resonant at frequencies closer to that of L1 and L8 element, while the lengths of strip L1 and L8 are decreased to increase their resonance frequencies. The lengths of strip L3 and L6 are increased to decrease their resonant frequencies as well suppress the other frequency bands. The length L5 is decreased so that it resonant frequency matches with that of the fed element (L4). As the difference between the different elements is small the bandwidth as well as the frequency ratio has also increased to 1.27. Table 3 yields the lumped parameter values of eight patch antenna.

In the Smith chart, a resonant circuit shown up as a circle. The larger the circle, the stronger the coupling. In three patches the radius of resonance circle is smaller. Hence the centre of the chart lies outside the circle. At critical coupling i.e. 2.1 and 2.65 GHz the resonance circle touches the centre of the Smith chart as shown in Fig. 6 (Fig. 7).

1.4 Nine and Ten Patch Configuration

As the number of elements increase, the flexibility for Twin frequency operation also increases, as there is more number of parameters that can be varied to obtain the desired performance. With the increase in the number of elements, properly varying the lengths of different elements also the separation between the elements leads to the increase in frequency ratio. The Twin frequency response for more than eight elements is tabulated.

The single RMSA is further divided along its width into nine strips, results in strip of smaller widths. All nine strips radiate equally if lengths unchanged which gives multi frequency operation, so for Twin band the lengths are varied. As the number of strips increases so does the resonance frequencies. The two frequencies are to be achieved with maximum separation between them and suppressing the other frequency bands. The lengths L1 and L2 are same so that they resonate at frequencies closer to each other. Similarly lengths L7 and L9 are same so that they resonate at frequency closer to that of L1 and L2 element. The lengths L3 and L8

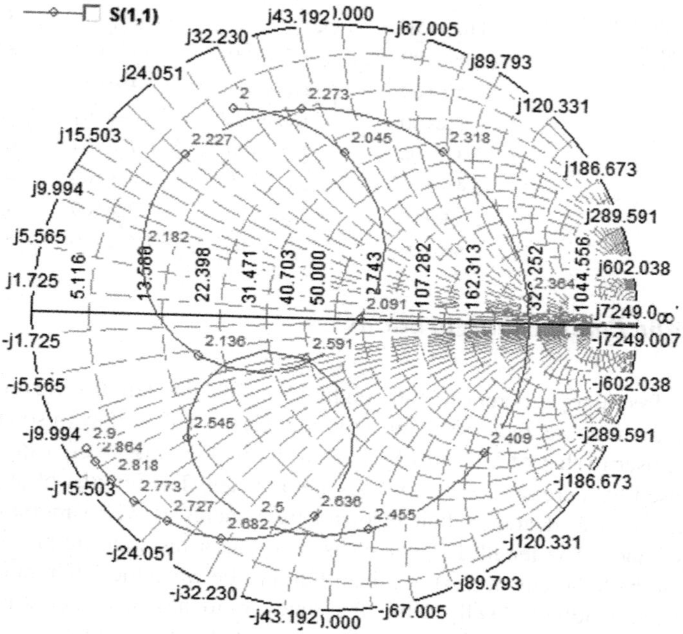

Fig. 6 Smith chart of eight patch antenna

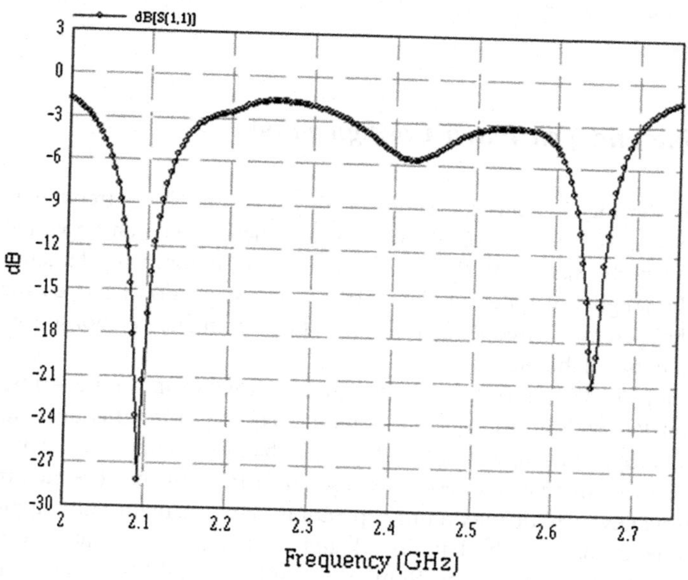

Fig. 7 S11 of eight patch antenna

Table 4 Lumped model of nine patch antenna

Freq (GHz)	Q	L (nH)	R (Ω)
2.00e+00	3.61e+00	2.44e+00	8.46e+00
2.06e+00	7.77e+00	2.88e+00	4.80e+00
2.07+00	−1.06e−02	−1.61e−01	1.93e+01
2.31e+00	2.28e+00	4.65e−01	2.96e+00
2.68e+00	6.95e+00	−1.60e+00	1.60e+00
2.85e+00	8.22e+00	−1.07e+00	2.32e+01
2.90e+00	8.42e+00	−8.97e+00	1.95e+00

Table 5 Lumped model of ten patch antenna

Freq (GHz)	Q	L (nH)	R (Ω)
2.00e+00	2.61e+00	2.38e+00	1.14e+00
2.03e+00	1.24e+00	2.88e+00	2.93e+00
2.05e+00	−1.31e−02	−1.39e−01	1.39e+01
2.38e+00	1.23e+00	6.83e−01	8.28e+00
2.55e+00	4.49e+00	−2.53e+00	9.05e+00
2.72e+00	1.32e+00	−1.97e+00	1.82e+01
2.865e+00	−0.54e+00	−1.06e+00	2.53e+00

are increased to decrease their resonant frequencies. The length of strip L4 is decreased and L6 is increased as compared to the fed element (L5) so as to increase the frequencies ratio and resonate to the frequency closer to that of the fed element. For the case when the difference between the lengths of the different elements is large, frequency ratio of 1.31 is obtained, which is significantly more than the previous cases, but the bandwidths at the two frequency bands are reduced. Tables 4 and 5 yields the lumped parameter values of nine and ten patch antenna.

In the Smith chart, a resonant circuit shown up as a circle. The larger the circle, the stronger is coupling. Hence the centre of the chart lies outside the circle. At critical coupling i.e. 2.07 and 2.722 GHz the resonance circle touches the centre of the Smith chart as shown in Fig. 8 (Fig. 9).

For Twin frequency operation, the length of strip L1, L3 and L8, L10 are same so that they resonate at frequencies closer to each other. While the lengths of L2 and L9 are increased to decrease the resonant frequencies of these elements so that it is add ups to the resonant frequencies of the above mentioned four elements. The lengths L4 and L7 are increased to decrease their resonant frequencies so as suppress the other bands. The length of strip L6 element is decreased to increase its resonant frequency closer to that of the fed element (L5). The frequencies ratio has increased from 1.31 to 1.33. As the difference between the lengths of the different elements is large, the bandwidth of the two frequency bands is reduced to 47 and 30 MHz. In the Smith chart, a resonant circuit shown up as a circle. The larger the circle, the stronger the coupling. In three patch the radius of resonance circle is smaller. Hence the centre of the chart lies outside the circle. At critical coupling i.e. 2.06 and 2.72 GHz the resonance circle touches the centre of the Smith chart as shown in Fig. 10 (Fig. 11).

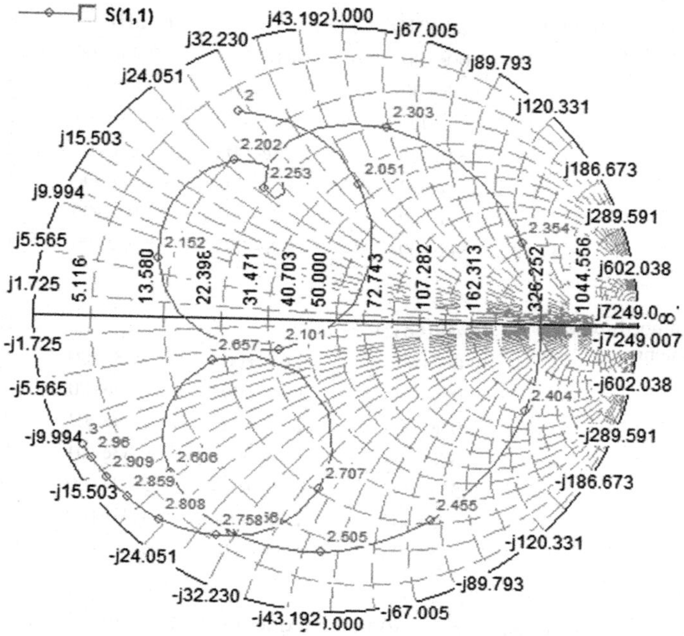

Fig. 8 Smith chart of nine patch antenna

1.5 Eleven Patch Configurations

In eleven patches configuration by optimizing the length and the gap frequencies ratio has increased from 1.33 to 1.36. As the difference between the lengths of the different elements is small, the bandwidth of the two frequency bands is increased to 87 and 36 MHz. The results are summarized in the Table 6.

In the Smith chart, a resonant circuit shown up as a circle. The larger the circle, the stronger the coupling. In three patches the radius of resonance circle is smaller. Hence the centre of the chart lies outside the circle. At critical coupling i.e. 2.052 and 2.78 GHz the resonance circle touches the centre of the Smith chart as shown in Fig. 12.

1.6 Twelve Patch Configurations

In twelve patches configuration by optimizing the length and the gap frequencies ratio has increased from 1.36 to 1.4.

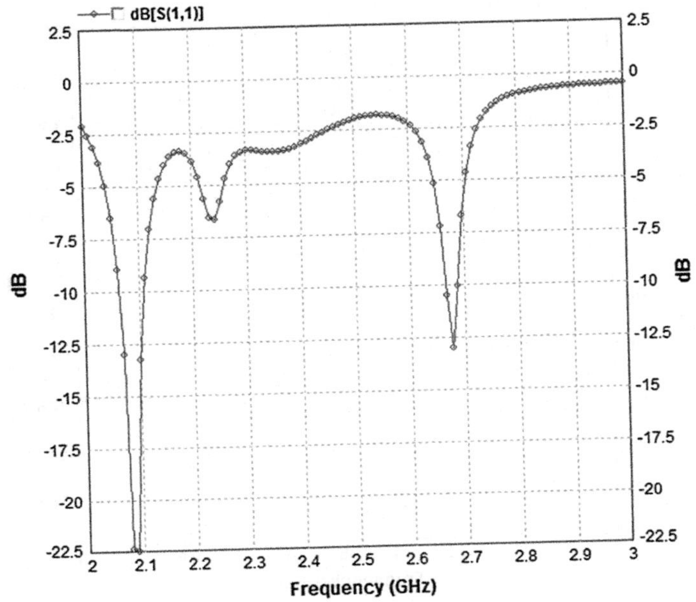

Fig. 9 S11 of nine patch antenna

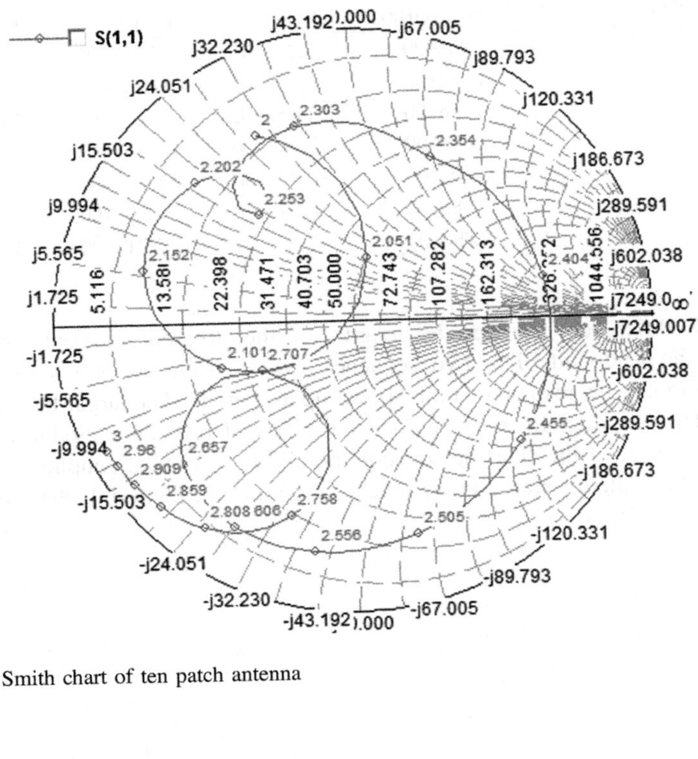

Fig. 10 Smith chart of ten patch antenna

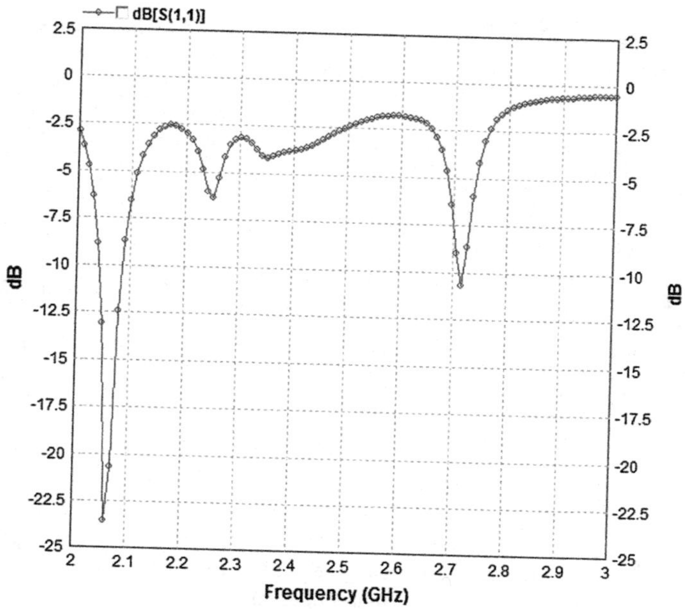

Fig. 11 S11 chart of ten patch antenna

Table 6 Lumped model of eleven patch antenna

Freq (GHz)	Q	L (nH)	R (Ω)
1.90e+00	7.29e+00	2.33e+00	3.28e+00
1.98e+00	3.67e+00	3.82e+00	1.29e+00
2.05e+00	−1.45e−02	−1.00e−01	1.65e+01
2.33e+00	1.98e+00	4.16e-01	3.07e+00
2.50e+00	−0.42e+00	−6.70e+00	3.07e+00
2.78e+00	−0.09e+00	−1.32e+00	1.06e+01
2.89e+00	−0.42e+00	−2.71e+00	3.75e+00

As the difference between the lengths of the different elements is further decrease, the bandwidth of the two frequency bands is decreased to 32 and 35 MHz. In the Smith chart, a resonant circuit shown up as a circle. The larger the circle, the stronger the coupling. In three patches the radius of resonance circle is smaller. Hence the centre of the chart lies outside the circle. At critical coupling i.e. 2.033 and 2.845 GHz the resonance circle touches the centre of the Smith chart as shown in Fig. 13 (Table 7).

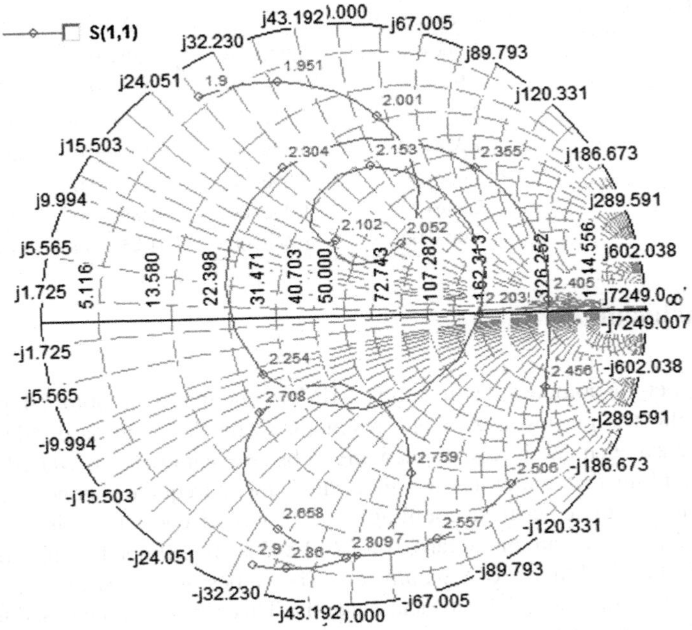

Fig. 12 Smith chart of eleven patch antenna

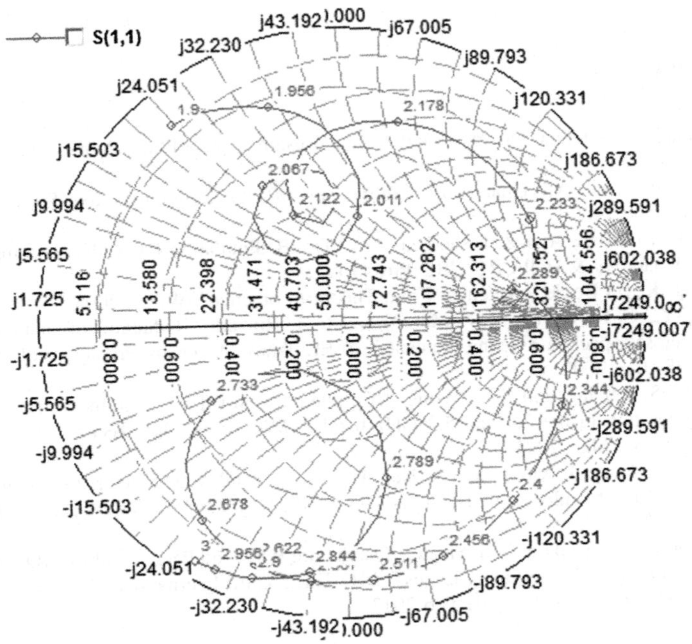

Fig. 13 Smith chart of twelve patch antenna

Table 7 Lumped model o f twelve patch antenna

Freq (GHz)	Q	L (nH)	R (Ω)
1.90e+00	5.93e+00	1.98e+00	4.00e+00
1.95e+00	3.62e+00	2.91e+00	9.86e+00
2.03e+00	−1.45e−02	−1.03e−01	1.33e+01
2.32e+00	−5.24e+00	−9.39e−01	2.61e+00
2.48e+00	−7.06e+00	−3.87e+00	8.58e+00
2.84e+00	−1.06e+00	−1.89e+00	1.00e+01
2.99e+00	−1.48e+00	−1.45e+00	1.84e+00

2 Conclusion

In this paper, a compact Twin bands MIMO antenna with optimizing mutual coupling effect is proposed. A single resonator along the width is divided so that width of each element is equal, then the substrate parameter is varied. Due to mutual coupling that combines the individual Twin frequency bands result into overcritical and under critical coupling. The proposed configurations have Twin band with VSWR ≤ 2 are in the range of 2.1–2.845 GHz. Lumped model equivalent circuit is just a fitted model of S parameters. The values of the circuit elements may or may not have or match any physical meaning. This concept can be used to design multiband antenna, which can have lowest mutual coupling and used in multiple Input Multiple Output (MIMO) technology.

References

1. C.A. Balanis, "Antenna Theory: Analysis and Design", 3rd edition, John Wiley and Sons,. 2005.
2. Tolga M. Duman, Ali Gharayeb, "Coding for MIMO Communication System", Wiley, December, 10, 2007.
3. Bassem R. Mahafza, Atef Z. Elsherbeni, "MATLAB Simulation for Radar Systems Design," 2004 by Chapman & Hall/CRC CRC Press LLC.
4. Yi Huang, Kevin Boyle, "Antenna: From Theory to practice", Wiley, 2008.
5. Chiu, C.-Y., C.-H. Cheng, R.D. Murch, and C.R. Rowell, "Reduction of mutual coupling between closely-packed antenna elements", IEEE Trans. Antennas Propag., Vol. 55, No. 6, 1732–1738, Jun. 2007.
6. Wu, T.-Y., S.-T. Fang, and K.-L. Wong, "A printed diversity Twin band monopole antenna for WLAN operation in the 2.4 and 5.2 GHz bands," Microwave Opt. Technol. Lett., Vol. 36, No. 6, 436–439, Mar. 2003.
7. Zhang, S., Z. Ying, J. Xiong, and S. He, "Ultra-wide band MIMO/diversity antennas with a tree-like structure to enhance wideband isolation," IEEE Antennas Wireless Propag. Lett., Vol. 8, 1279–1283, 2009.
8. S.V. Mapare, V.V. Mapare, G.G. Sarate, "Quadrature Band and Penta-band MIMO Antenna By Mutually Coupling Non-radiating Edges". 2013 IEEE Third International Advance Computing Conference (IACC), pp. 198–201 Feb 2013.

Performance Evaluation of Video-Based Face Recognition Approaches for Online Video Contextual Advertisement User-Oriented System

Le Nguyen Bao, Dac-Nhuong Le, Le Van Chung and Gia Nhu Nguyen

Abstract In this research, we propose the online video contextual advertisement user-oriented system. Our system is a combination of video-based face recognition using machine learning models from the camera with multimedia communications and networking streaming architecture using Meta-data structure to video data storage. The real images captured by the camera will be analyzed based on predefined set of conditions to determine the appropriate object classes. Based on the defined object class, the system will access the multimedia advertising contents database and automatically select and play the appropriate contents. We analyse existing face recognition in videos and age estimation from face images approaches. Our experiment was analyzed and evaluated in performance when we integrate analyze age from the face identification in order to select the optimal approach for our system.

Keywords Contextual advertising · Face recognition · Age estimation · Real-time tracking · Real-time detection

1 Introduction

The type of advertising have strongly grown in both width and depth in recently years. About its width, we see advertising everywhere, from the developed capitalist countries to developing socialist country. Advertising has become an important

L.N. Bao (✉) · L.V. Chung · G.N. Nguyen
Duy Tan University, Da Nang, Vietnam
e-mail: baole@duytan.edu.vn

L.V. Chung
e-mail: levanchung@duytan.edu.vn

G.N. Nguyen
e-mail: nguyengianhu@duytan.edu.vn

D.-N. Le
Hai Phong University, Hai Phong, Vietnam
e-mail: nhuongld@hus.edu.vn

© Springer India 2016
S.C. Satapathy et al. (eds.), *Information Systems Design and Intelligent Applications*, Advances in Intelligent Systems and Computing 435,
DOI 10.1007/978-81-322-2757-1_29

economic activity in the market economy. This is the inevitable trend, as one of the effective means of showing the commercial competitiveness, is a driving force stimulating economic development. Regarding depth, advertising not only disfiguring the way of living of the users but also to change their thinking, deeply influenced the culture of many classes of people in the society. In this research, we want to recognize the third dimension; it is the technical development of the advertising model. Initially, the advertising techniques go from the primitive means such as rumors, words of mouth between friends, an interpretation of the sales person to these new forms such as gifts, awards,... has been taking the advantage of the power of mass media such as newspapers, magazines, radio, television, cinema. Today, advertising has been a convergence and new development step when moving to multimedia communicating model with the high interoperability via the Internet. This is also a challenge for the dynamic new forms of advertising, in contrast with the unilateral form of direct advertising as people have become passive with the advertising systems. The advertising customization, with the ability to update in real time to change advertising text according to the full context of a search or the website someone is seeing, like: *Google AdWords, Google AdSense*. This is contextual display advertising or targeting advertising technology and is very effective.

In this paper, we propose the online video contextual advertisement user-oriented system. The system is a combination of face recognition in videos using machine learning models from the camera with multimedia communications and networking streaming architecture used Meta-data structure to storage video data. We analysis existing video-based face recognition and the age difference approaches. The rest of our paper is organized as follows: Our online video contextual advertisement user-oriented system introduces in Sect. 2. The existing face recognition in videos and age estimation from face image approaches are analyzed in Sect. 3. The experiments and results are presented in Sect. 4, and finally, the conclusion and our future works are mentioned in Sect. 5.

2 Our Online Video Contextual Advertisement User Oriented System

In this section, we will focus on analyzing our model of automatically customized advertising system to customer objects in real-time context applied in customization of advertising video contents with target directed to make the content of ads relevant and truly useful for customers in each specific contexts through object analysis identifier acquired from the camera. The system is a combination of image processing and identification with multimedia communications and networking streaming architecture used Meta-data structure to store video data plus machine learning models to deliver maximum efficiency to the system. Our system represented in Fig. 1. The system consists of three major phases as follows:

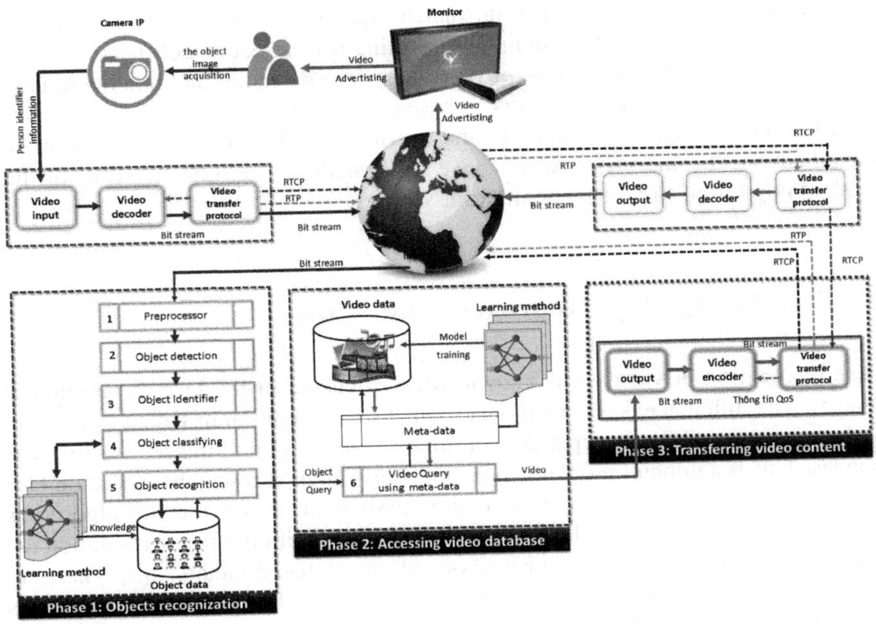

Fig. 1 Online video contextual advertisement user-oriented system

- *Phase* 1: Identifying and classifying objects based on images captured from the camera
- *Phase* 2: Accessing video database under classified objects
- *Phase* 3: Transferring video contents

In the first phase, the object image acquisition directly from the camera, then analyzed for detection and separation of objects from the background to select the specific characters of the object selected. The characteristics of the object will be layered and then identified object. During the implementation process of image acquisition from video often deformed by the receiver with low quality leading to calibration very complex because it depends too much on the surrounding environment. Video is due to a string of successive motional images for creation. To be able to transmit video, information and data encrypted. Therefore, to be able to process the video signal in addition to handling the problem objectively as lighting conditions, camera placement, etc…We also need to consider issues of video compression standards. Therefore, the workings listed as decontamination, gray-scale calibration usually determined through limens (*Thresholds*) in program and decided by the user (*refining*) and no ability to automatically balance. Pretreatment stage whose function is to standardize images to be found making search more efficient. The work in the pretreatment step may include: Standardize sizes between images in the database to be found, adjust brightness, darkness of the image, noise filtering, normalize the position and posture of face photo, etc.

After the video data eliminated unwanted ingredients from pretreatment will be transitioned to object identification. Our targeting is to detect objects based on face. Face recognition section consists of two steps: face detection and automatic identification of objects. The main work is based on the characteristic extraction techniques from images of objects and performed benchmark for automatic identification. In the next section, we will analyze the effectiveness of the methodology used to face recognition in videos.

3 Face Recognition Approaches for Videos

Face recognition in videos was known as an extension of the face recognition in images. The difference is the input data of the face recognition in videos system are real-time videos while the input data of the face recognition in images are still face images. This is a difficult problem, there are many different researches focus on face recognition in videos. These approaches are classified in [1]. The tracking algorithms tracked faces in videos. The high quality face images with the better resolution are more advantageous for matching based on still image based methods [2]. There are many efficient and effective researches in this area have been done for still images only. So we could not been extended to apply its to video sequences directly. The human faces in the video scenes can not be limited orientations and positions. Therefore, this is a challenge for researchers [1–4]. The first approach is improved face recognition in the videos by using temporal information, and the intrinsic property available in videos. The second approach is represented video like a complex face manifold to extract a variety of features such as exemplars and image sets are called the local models or subspaces. There are two key issues in the face detection in video process are: what features are used to extract and which learning algorithm is applied? [1, 2, 5]. Face detection in real time and multi-tier is very important in the face detection in videos. We could be classified the real time face detection methods into two groups: cascade Ada-Boost approach and validate faces used color information [3, 6, 7]. General, the tracking algorithms tracked faces can be classified into three categories: head, facial feature and combination tracking. The real time is its foremost feature for video processing are considered in [7–9].

There are many approaches for face recognition in videos. Summary of the existing face recognition in videos approaches are listed in Table 1.

The linear methods are effectively for simple Euclidean structure in learning data. PCA algorithm aims to a projection to maximizes its variance [10]. MDS algorithm conserves the pairwise distances between data points in the new projection space [11]. LDA algorithm used a linear projection to maximize ratio of the between-class to the within-class scatter with the class information additional [12]. LLE is a nonlinear dimensionality reduction algorithm [13], while Isomap [14] opened a new paradigm of manifold learning model. These methods are allow discovery on the non-linear structure high-dimensional of the manifold in the lower dimensional space. The major disadvantage of these algorithms are can not be done

Table 1 Summary of face recognition in videos approaches

Approaches	Algorithms	References
Linear dimensionality reduction	PCA (*principal component analysis*)	[10]
	LDA (*linear discriminant analysis*)	[12]
	MDS (*multidimensional scaling*)	[11]
	LNDM (*learning neighborhood discriminative manifolds*)	[1]
Nonlinear dimensionality reduction	LLE (*locally linear embedding*)	[13]
	Isomap	[14]
Learning	LPP (*locality preserving projections*)	[15]
	OLPP (*orthogonal locality preserving projections*)	[4]
	MFA (*marginal fisher analysis*)	[4]
	NPE (*neighborhood preserving embedding*)	[16]
	KPA (*kernel principal angles*)	[17]
	DCC (*discriminative canonical correlations*)	[18]
	NDMP (*neighborhood discriminative manifold projection*)	[19]

for the problem with out-of-sample. Because, new data points in the out-of-sample failed projection onto the embedded space. The learning methods resolved this limitation based on the optimal linear approximations algorithm. The neighborhood information are embedded in the neighborhood adjacency graphs [4, 15–19]. Nowadays, a hot research topic is face recognition in the 3D environment [20, 21]. After detecting the faces, our system should recognize age of the face to choice advertisements suitable. The approaches for age recognition will be presented in Sect. 4.

4 Age Estimation from Face Image Approaches

The first approach for age simulation algorithm and FaceVACS tool used shape and texture. Then create facial images for each different age groups. The authors in [22] were measured and analyzed the facial cues be affected by age variations. The wavelet transform is based on the age simulation to the prototype of facial images proposed in [23]. The face simulation based on statistical models are used to training image to find out the correlation between coded face representation and the actual age of faces [24, 25]. This correlation is the basic to estimate the age of an person and reconstruct the face of all ages. The recently studies can be listed in Table 2.

These approaches have overcome the limitation of age invariant face recognition. The existing age invariant face recognition methods can be divide into two categories: generative and discriminative methods [26–30].

Table 2 Summary of age estimation from face image approaches

Authors	Approaches	References
FaceVACS	Used shape and texture, and create facial images for different age groups	[22]
Park (2010)	Used SVM and features to classify with the gradient orientation pyramid	[26]
Li (2011)	Used SIFT/LBP features and variation of RS-LDA	[27]
Klare (2011)	Proposed on learning feature subspaces for face recognition across time lapse	[28]
Du (2012)	Used NMF algorithm with sparseness constraints	[29]
Gong (2013)	Presented HFA method for age invariant in face recognition	[30]

5 Experiment and Results

In the experiment, we used two standard test datasets:

- The Honda/UCSD [31] datasets consists of 20 individuals moving heads combinations of 2D/3D rotation, expression and speed. The resolution of video sequence is 640 × 480 and recorded at 15 fps by at least two video sequences. The person rotates and turns their head preferred order and speed, and typically in about 15 s, the individual is able to provide a wide range of different poses in each video.
- The CMU Motion of Body (MoBo) [32] datasets consists of 25 individuals walking on fours patterns: slow, fast, incline and with a ball a treadmill in the CMU 3D room. The subjects are captured by six high resolution color cameras distributed evenly around the treadmill. The video databases for face recognition have common drawbacks that they are always fixed on the same scene. The faces have been identified will be stored in the standardized image size 200 × 150 pixel.

In the first experiment, we analyze and evaluate the effectiveness of different approaches for video face recognition with two datasets. Figure 2 shows the recognition rate (%) results from some typical approaches for face recognition in video on the two datasets. The comparison showed that the DDC algorithm has the best performance in both datasets.

Figure 3 shows a sample face images extracted from the Honda/UCSD Video Databases. The face patterns are used for training and identification of age difference.

In the second experiment, we analyze and evaluate the effectiveness of age difference in face recognition approaches with two datasets. Figure 4 shows the experimental results of identification rates on the face recognition from Honda/UCSD and CMU-MoBo databases. The experiment results show that the Hidden Factor Analysis approach has the best performance.

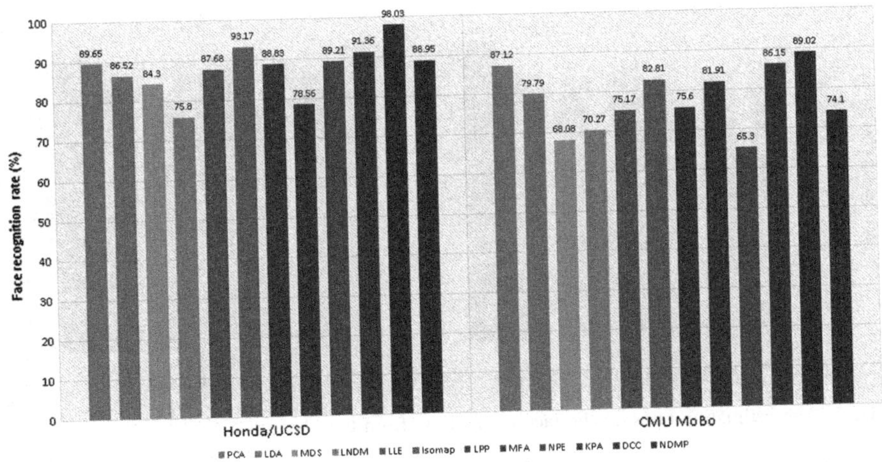

Fig. 2 Evaluation performance of video face recognition approaches on two datasets

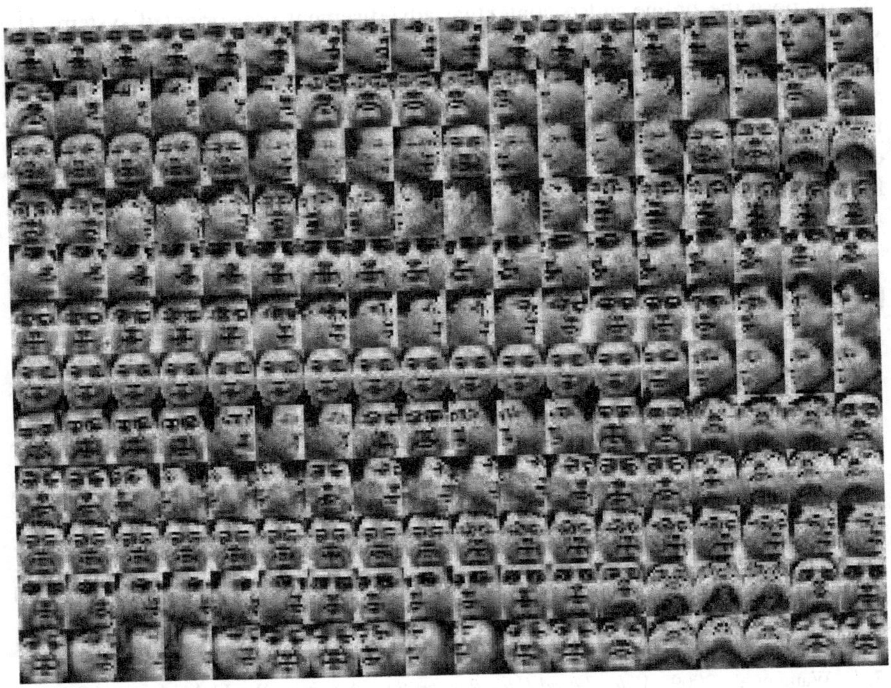

Fig. 3 Sample face images extracted from the Honda/UCSD dataset

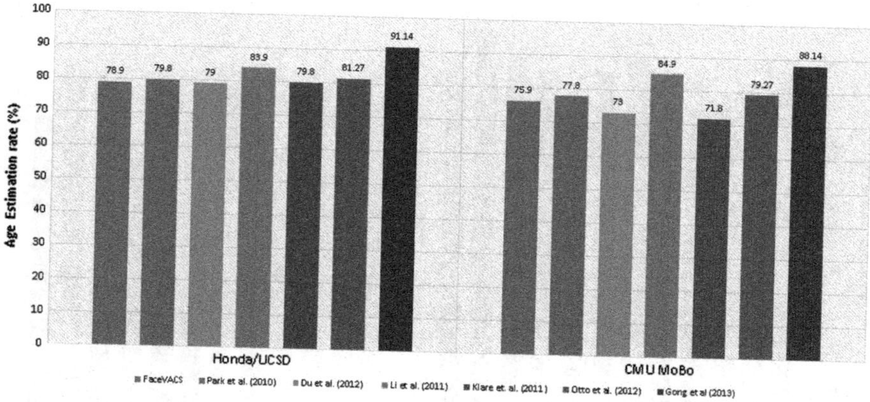

Fig. 4 Age estimation rates on the face recognition from two datasets

6 Conclusions

We propose the online video contextual advertisement user-oriented system. The system is a combination of face recognition in video using machine learning models from the camera with multimedia communications and networking streaming architecture used Meta-data structure to storage video data. We analysis existing face recognition in videos and age estimation from face images approaches. Our experiment was analyzed and evaluated performance when we integrate analyze age from the face identification in order to select the optimal approach for our system.

References

1. Z. Zhang et al. (2011), *Video-Based Face Recognition: State of the Art*, LNCS Vol. 7098, pp. 1–9, Springer.
2. W.Y. Zhao et al. (2003), *Face Recognition: A Literature Survey*, ACM Computing Surveys, Vol. 35.
3. P. Viola and M.J. Jones (2001), *Robust real-time object detection*, IEEE ICCV Workshop on Statistical and Computational Theories of Vision. Vancouver, Canada.
4. S. Yan et al. (2007), *Graph embedding: A general framework for dimensionality reduction*. IEEE Transactions on PAMI 29(1), pp. 40–51.
5. C. Zhang and Z. Zhang (2010), *A Survey of Recent Advances in Face Detection*, Technical Report MSR-TR-2010-66.
6. Z. Zhang et al. (2002), *Realtime multiview face detection*, Pro. of 5*th* IEEE International Conference on Automatic Face and Gesture Recognition, pp. 142–147.
7. H. Wang et al. (2009), *Video-based Face Recognition: A Survey*, World Academy of Science, Engineering and Technology Vol. 3, pp. 273–283.
8. Y. Jin, F. Mokhtarian (2005), *Data Fusion for Robust Head Tracking by Particles*, Pro. 2*nd* Joint IEEE International Workshop on VS-PETS.

9. W. Zheng, S.M. Bhandarkar (2009), *Face detection and tracking using a Boosted Adaptive Particle Filter*, Journal Of Visual Communication And Image Representation, Vol. 20(1), pp. 9–27.
10. M. Turk, A. Pentland (1991), *Eigenfaces for recognition*. J. Cogn. Neurosc. 3(1),71–86.
11. T.F. Cox et al. (2001), *Multidimensional Scaling*. Chapman and Hall, Boca Raton.
12. P.N. Belhumeur et al. (1997), *Eigenfaces vs Fisherfaces: Recognition using class specific linear projection*. IEEE Trans. PAMI 19, pp. 711–720.
13. J. See, M.F.A. Fauzi (2011), *Learning Neighborhood Discriminative Manifolds for Video-Based Face Recognition*. LNCS vol. 6978, pp. 247–256. Springer.
14. W. Fan et al. (2005), *Video-based face recognition using bayesian inference model*. LNCS, vol. 3546, pp. 122–130. Springer.
15. X.F. He, P. Niyogi (2003), *Locality preserving projections*. Proc. NIPS 16, pp. 153–160.
16. X. He et al. (2005), *Neighborhood reserving embedding*. Pro. of IEEE ICCV, pp. 1208–1213.
17. L. Wolf, A. Shashua (2003): *Learning over sets using kernel principal angles*. Journal of Machine Learning Research Vol. 4, pp. 913–931.
18. T. Kim et al. (2007), *Discriminative learning and recognition of image set classes using canonical correlations*. IEEE Trans. PAMI Vol. 29(6), pp. 1005–1018.
19. J. See and M.F.A. Fauzi (2011), *Learning Neighborhood Discriminative Manifolds for Video-Based Face Recognition*, LNCS 6978, pp. 247–256, Springer.
20. A.F. Abate et al. (2007), *2D and 3D face recognition: A survey*; Pattern Recognition Letters, Vol. 28(15), pp. 1885–1906.
21. J.Y. Choi et al. (2008), *Feature Subspace Determination in Video-based Mismatched Face Recognition*, *8th* IEEE International Conference on Automatic Face and Gesture Recognition.
22. FaceVACS software developer kit, cognitec systems GbmH.
23. B. Tiddeman et al. (2001): Prototyping and transforming facial textures for perception research, IEEE Computer Graphics and Applications, Vol. 21(5), pp. 42–50.
24. A. Lanitis et al. (2002): Toward automatic simulation of aging effects on face images, IEEE Trans on Pattern Analysis and Machine Intelligence, Vol. 24(4), pp. 442–455.
25. A. Lanitis et al. (2004), Comparing different classifiers for automatic age estimation, IEEE Transactions on Systems, Man, and Cybernactics, Vol. 34(1), pp. 621–628.
26. U. Park et al. (2010). *Age-invariant face recognition*. IEEE Trans. Pattern Anal. Mach. Intell., Vol. 32(5), pp. 947–954.
27. Z. Li et al. (2011). *A discriminative model for age invariant face recognition*. IEEE Transactions on Information Forensics and Security, 6(3–2), pp. 1028–1037.
28. B. Klare and A.K. Jain. *Face recognition across time lapse: On learning feature subspaces*. In IJCB, pp. 1–8, 2011.
29. J.X. Du, C.M. Zhai, and Y.Q. Ye (2012). *Face aging simulation based on NMF algorithm with sparseness constraints*. Neurocomputing.
30. D. Gong, Z. Li, D. Lin (2013), *Hidden Factor Analysis for Age Invariant Face Recognition*, IEEE International Conference on Computer Vision (ICCV), pp. 2872–2879.
31. K.C. Lee et al. (2005): *Visual tracking and recognition using probabilistic appearance manifolds*. Computer Vision and Image Understanding 99, pp. 303–331.
32. R. Gross, J. Shi (2001), The CMU motion of body (MOBO) database.

A Survey on Power Gating Techniques in Low Power VLSI Design

G. Srikanth, Bhanu M. Bhaskara and M. Asha Rani

Abstract The most effective technique to reduce dynamic power is the supply voltage reduction by technology scaling which reduces threshold voltage. Under deep submicron technology, reduction in threshold voltage increases leakage currents, gate tunneling currents and leakage power in standby mode. Most of the handheld devices have long standby mode cause leakage current contributing to leakage power dissipation. In this paper, various leakage power reductions, charge recycling techniques, data retention of memories. Various Power gating techniques are discussed in detail.

Keywords Dynamic voltage and frequency scaling (DVFS) · Power gating (PG) · Multi-threshold CMOS (MTCMOS) · Dual-threshold CMOS (DTCMOS)

1 Introduction

The dynamic power (DP) in VLSI circuits depends on the clock frequency, capacitance and square of the supply voltage. The most effective technique to reduce DP is reducing the supply voltage by scaling the device. Supply voltage scaling increases the circuit delay and reduces threshold voltage, minimize

G. Srikanth (✉)
Department of Electronics and Communication Engineering,
CMR Technical Campus, Kandlakoya, Medchal, Hyderabad, Telangana, India
e-mail: gimmadisrikanth79@gmail.com

B.M. Bhaskara
Department Electrical Engineering, Majmaah University,
Majmaah City, Riyad, Saudi Arabia

M. Asha Rani
Department of Electronics and Communication Engineering,
Jawaharlal Nehru University College of Engineering, Hyderabad, Telangana, India
e-mail: ashajntu1@jntu.ac.in

© Springer India 2016
S.C. Satapathy et al. (eds.), *Information Systems Design and Intelligent Applications*, Advances in Intelligent Systems and Computing 435,
DOI 10.1007/978-81-322-2757-1_30

switching power in active mode. Scaling down in deep submicron technology results increase in gate tunneling currents, leakage current (I_{leak}) and leakage energy dissipation.

Handheld devices are operated by a low power battery. They are idle for most of time, and in standby mode the sub-threshold leakage current is the dominant component in total power dissipation. Sub-threshold leakages cause a small current flow through the devices in OFF state.

Total power dissipation majorly due to leakage power (P_{leak}) and P_{leak} is reduced by employing various techniques like stack forcing, power gating, multi-supply voltage optimization, self-biasing techniques, etc. Muti-voltage or dynamic voltage and frequency scaling (DVFS) techniques can be employed to various blocks of chip, depending on the requirement, to reduce both the dynamic and leakage power [1]. Chen et al. [1] proposed floor planning methodology with DVFS for power gating.

This paper mainly focused on the various leakage power reductions, the charge recycling techniques, data-retention techniques in memories, and scope for work are discussed. Various Power gating techniques are discussed in detail.

This paper organizes as follows: In Sect. 2, various low power dissipation techniques stack forcing, power gating, self-bias transistor and Memories are discussed. Conclusions are drawn in Sect. 3.

2 Low Power Dissipation Techniques

Several low power dissipation techniques are described below:

2.1 *Stack Forcing*

In PMOS and NMOS devices, I_{leak} increases exponentially with the voltages at the transistor terminals [2]. For NMOS device, more voltage at the source (V_S) terminal exponentially decreases sub-threshold current (I_{sub}) due to:

- V_{GS} is negative, then I_{sub} decreases exponentially.
- V_{BS} is negative, cause rise in body effect which results rise in V_{th} and decrease in I_{sub}.
- Decrease in V_{DS} reduces the barrier and decreases I_{sub}.

V_S increases when several transistors in the stack are turned off and are called as transistor self- reverse biasing. Various blocks with stack forcing techniques and input-vector control method are discussed in [2]. Leakage power in active mode cannot be reduced by using this technique.

2.2 Power Gating

A power Gating (PG) or Multi-threshold CMOS (MTCMOS) technique supply voltage to unutilized blocks is deactivated using on or off- chip voltage supply techniques and reduces standby power and support high speed in active mode. At sub-micron technology, logic circuits use low threshold MOSFETs. MTCMOS use a high threshold MOSFET to change the supply to the logic core as shown in Fig. 1. In NMOS MTCMOS, sleep transistor (ST) is connected in between ground and virtual ground (VG) of the circuit by charging to VDD and effectively suppress the leakage current [3]. During its wake up time, current spikes produced during discharging results ground bouncing at the near module. The limitations of power gating due to wake up latency and power penalty can be overcome by multiple sleep modes by reducing static power dissipation during inactive time [3]. Different MTCMOS implementations are shown in Fig. 1 and their operation is shown in Table 1. MTCMOS with multiple sleep modes is used to control different modes of data path partition based on the flow, to minimize power consumption [3] as shown in Fig. 2.

MTCMOS technique can be used to develop various circuits and low power carry look-ahead adder design is described in [4]. Mutoh et al. developed, a low power digital signal processor chip for mobile phone application using the MTCMOS technique with power management scheme [5]. Various static and dynamic approaches for PG are available in literature to generate the sleep signal. Youssef et al. made a comparative study on the static and dynamic sleep generation approaches for PG and dynamic approach is more effective to regulate PG blocks and more accurate compared to the static approach [6].

The other popular PG techniques apart from Generic PGs are Dual-Threshold CMOS (DTCMOS) and stack forcing PGs as shown in Fig. 3. In DTCMOS, Gate and body terminals are joined together to minimize delay degradation factor [7]. In stack forcing transistors are stacked to reduce leakage current and have smaller delay degradation factor. Optimization for ultra-low V_{dd} with PG transistor sizing is discussed in [7].

MTCMOS techniques according to granularity can be classified into two types: (a) fine-grain and (b) coarse-grain PGs. MTCMOS with the fine-grain approach using optimum sleep transistor sizing is discussed in [8]. In this approach, one sleep transistor is used at each gate. In coarse-grain power gating, a few sleep transistors are used to supply power to PG devices. Hariyama et al. developed asynchronous fine-grain PG pipeline architecture on FPGAs [9]. Fine-grain PG for synchronous FPGA architecture desire more overhead and unfeasible to realize. Rahman et al. studied fine-grain and coarse-grain PG approaches using Field Programmable Gate Arrays (FPGAs) architecture. The fine-grain PG approach decreases leakage current more effectively compared to coarse-grain PG but coarse-grain PG requires optimum area [10]. Nair et al. compared coarse-grain and fine-grain PG techniques for SRAM cells and SRAM arrays and proved that coarse-grain provide more leakage power saving compared to fine-grain PG since part of the circuit connected to the supply rails while the remaining circuit is OFF [11].

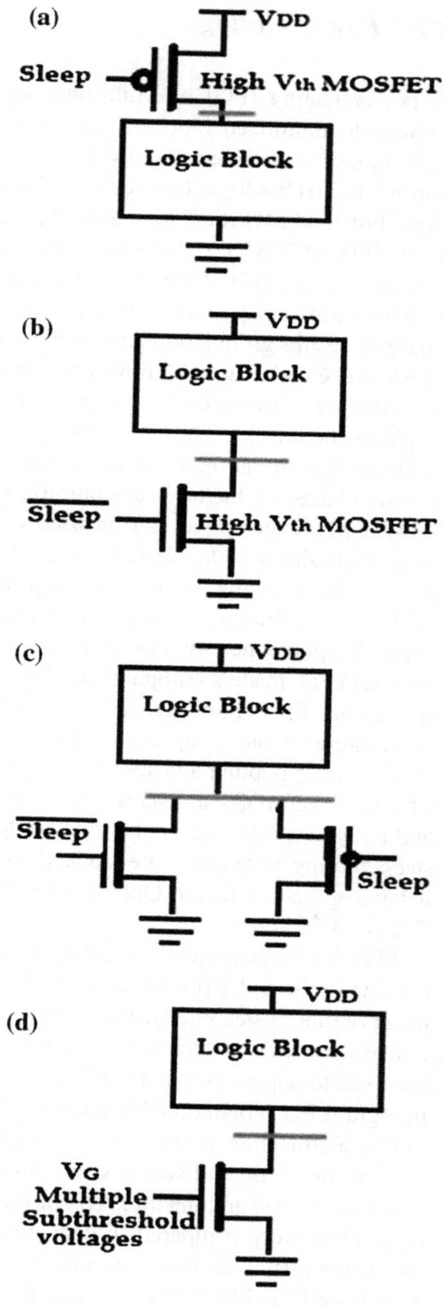

Fig. 1 Various MTCMOS implementation techniques

Table 1 Operation of MTCMOS techniques

MTCMOS with NMOS	MTCMOS with intermediate mode	MTCMOS with multiple sleep modes
Virtual ground at $\simeq V_{DD}$ standby mode $\simeq$ GND active mode	Virtual ground at $\simeq V_{DD}$ standby mode $\simeq V_{th\_p}$ intermediate mode $\simeq$ GND active mode	V_G at $\simeq 0$ snore $\simeq V_1$ dream $\simeq V_2$ sleep $\simeq$ V_{DD} active mode

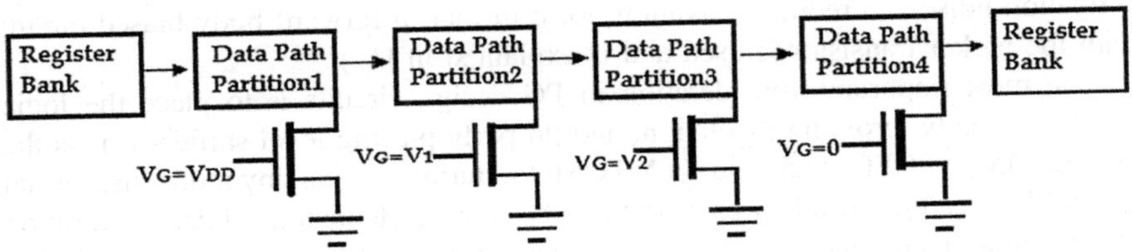

Fig. 2 Data path partition with MTCMOS multiple sleep modes

Fig. 3 DTCMOS and stack forcing PGs architecture

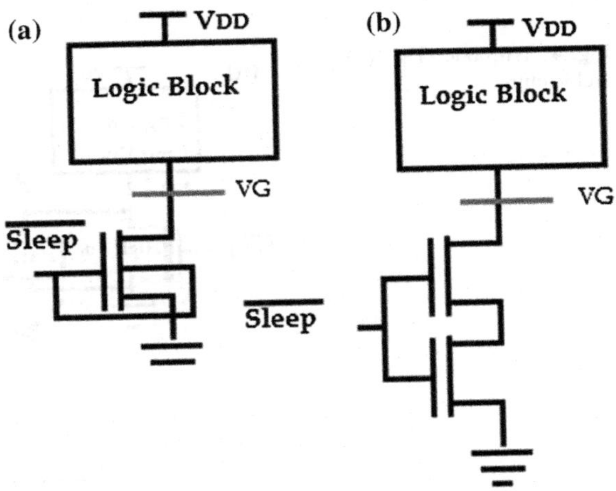

Yeh et al. discussed chip level floor-plan design for coarse-grain PG to save leakage power [12]. Pakbaznia et al. proposed delay budgeting using heuristic sizing algorithm for sleep transistor sizing in coarse-grain PG [13]. Compared to other methods, this method reduces width of sleep transistor to a minimum of 40 %. Mu et al. proposed testing method for coarse-grain PGs and SRAMs for detecting stuck-open power switches [14]. Saito et al. used fine grained PG in coarse-grain dynamically reconfigurable processor arrays [15]. PG technique is employed for larger modules like Central Processing Unit, intellectual property core to reduce dynamic power consumption and fine grain PG are applicable to smaller logic blocks like Arithmetic Logic Unit to reduce leakage power.

In PG circuits, during the wake-up process, a large instantaneous current through ST and generates activation noise which perturb nearby active logic blocks. To reduce activation noise Jiao et al. proposed a tri-mode PG architecture using the threshold voltage tuning method with small activation delay using small STs [16]. In tri-mode (TM) PGs, a parker transistor with zero-body-bias is connected in parallel with ST as shown in Fig. 4. A TM low threshold voltage PMOS parker (TML) transistor is used in place of high threshold voltage PMOS parker transistor to reduce the voltage swing of VG during PARK to the active mode which reduces activation noise. To reduce activation noise further, a forward body-biased circuit with the parker transistor is used and is explained in [16].

The most important consideration in PG design circuits is to place the logic blocks correctly, providing isolation, and properly placing level shifters across the voltage ON and OFF blocks etc. In MTCMOS, during the standby mode, the virtual ground is at $\simeq V_{DD}$, results loss of data in the storage element need data restoration or recycling techniques. Hence, power gating technique is not appropriate to sequential circuits and memory cells. In some applications, state retention power gating, can be used to preserve the states of FFs or memories when the supply

Fig. 4 Tri-mode (TM) PG architectures

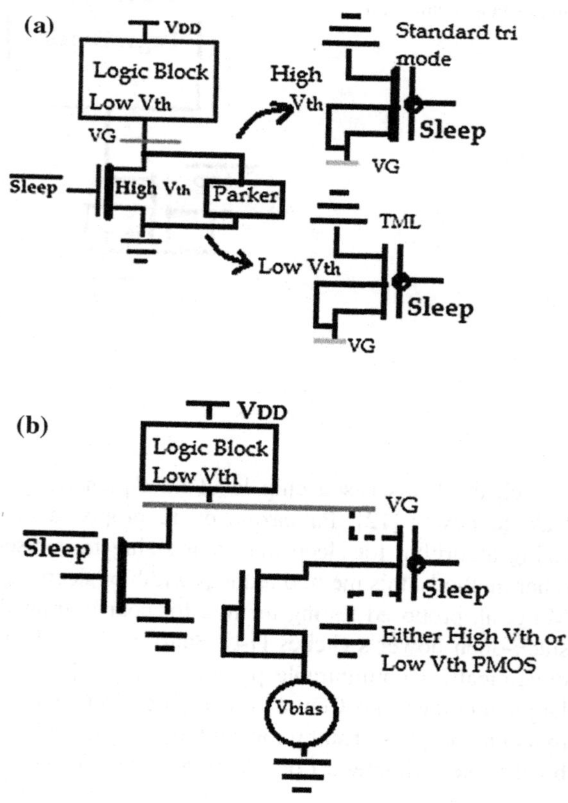

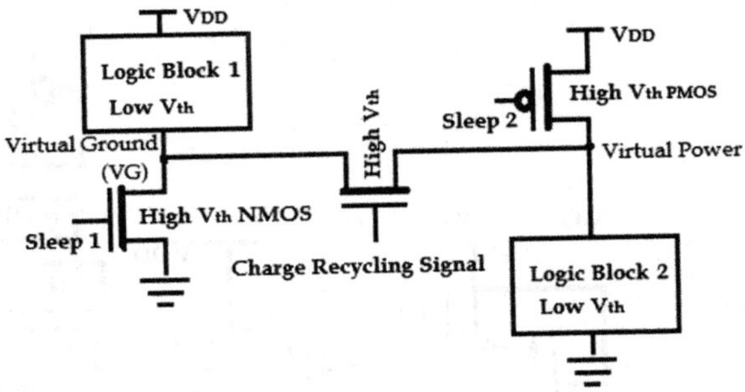

Fig. 5 Charge recycling MTCMOS

voltage of the block is deactivated. Retention flip-flops (FFs), can be used in place of the conventional FFs, and require more area overhead. Liu et al. proposed a charge recycling technique using both PMOS and NMOS high threshold sleep transistors [17] as shown in Fig. 5. The energy consumed during the active-to-standby-active mode transition is reduced by recycling charge stored in virtual power and VG rails and is explained in [17]. This technique reduces energy overhead during the transition and also the leakage power significantly. Pakbaznia et al. proposed charge recycling methods for both MTCMOS and Super cutoff CMOS (SCCMOS) and the proposed technique employs smaller charge recycling transistors; and reduces peak voltage, settling time and leakage power [18]. Kim et al. proposed virtual power/ground rail clamp (VRC) technique which retains and eliminates restoration methods used with power gating [19]. Pakbhaznia et al. developed tri-modal MTCMOS switch operates in active, drowsy and sleep modes and eliminates the use of retention FFs for data retention [20].

2.3 *Self-Bias Transistor (SBT)*

Gopalakrishnan et al. [21] proposed SBT technique to reduce P_{leak} by reducing I_{leak}. NMOS SBT is connected to the pull-down network and GND, PMOS SBT is connected in between the pull-up network and the V_{DD} as shown in Fig. 6. SBT controls current in OFF state transistors and the operation is explained in [21].

2.4 *Memories*

2.4.1 Gated-V_{DD} for SRAM

SRAM cells are connected to NMOS gated V_{DD} transistor as shown in Fig. 7 and PMOS gated V_{DD} transistor is placed between V_{DD} and PMOS transistors of the SRAM cell. The unused SRAM cells in the cache enters into the standby mode

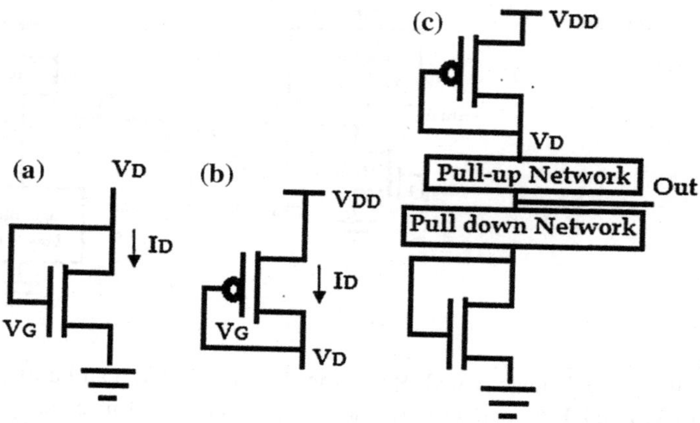

Fig. 6 SBT techniques

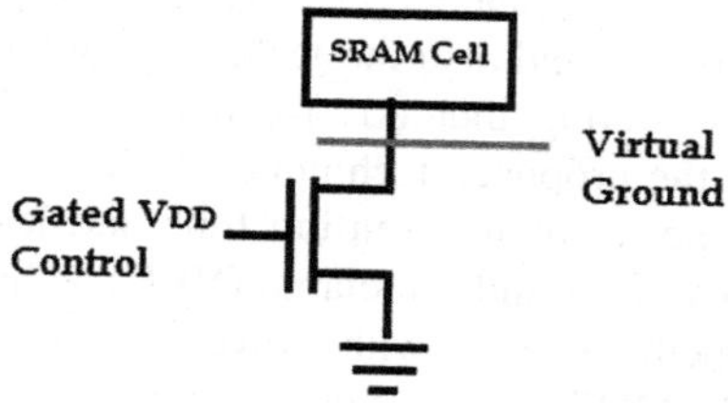

Fig. 7 SRAM cell using gated-V_{DD}

when the NMOS transistor is OFF and used SRAM cells in the cache enters into the active mode when the NMOS transistor is ON [22]. NMOS dual- V_t gated V_{DD} with charge pump reduces the energy delay and leakage power and explained in [22].

Nourivand et al. proposed standby supply voltage tuning data retention technique for SRAMs as shown in Fig. 8 to reduce leakage currents and avoid data retention failures [23].

Dynamic random access memory (DRAM) requires frequent refreshing and total energy dissipation rely on the frequency of the refreshing signal. Optimum data – retention time of DRAM cell depends on the period of the refreshing signal. Most handheld devices have a long sleep mode, hence frequent refreshing cause more power dissipation. Kim et al. proposed DRAM cells refreshing, on the block based instead of conventional row refreshment, and a multiple refreshment periods are used depending on the requirement of block refreshment. This technique reduces the power by a factor of 4 when compared to the conventional refreshing DRAM technique [24].

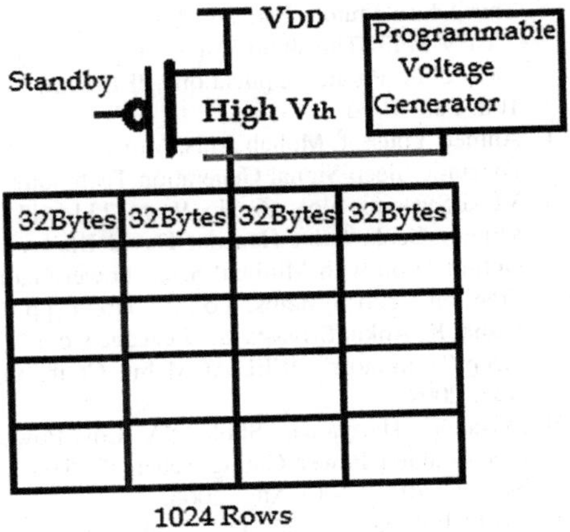

Fig. 8 Standby supply voltage tuning data retention technique for SRAMs

3 Conclusions

Dynamic power consumption can be reduced by scaling, which decrease supply voltage. Reduced threshold voltage under deep submicron technology results leakage current and it is the dominant component in total power dissipation. In this paper, a brief review of the leakage power reduction techniques, and limitations are disused in brief. This paper majorly focused on the power gating techniques which is a prominent technique to reduce leakage power consumption. Various MTCMOS implementation techniques, application in data path partition, DTCMOS and stack forcing power gating, fine-grain and coarse-grain power gating techniques, tri-mode power gates, charge-recycling techniques and data-retention techniques for SRAMs are discussed in brief. This paper gives a scope to work in these directions.

References

1. Shi-Hao Chen, Jiing-Yuan Lin, "Implementation and verification practices of DVFS and Power Gating", Int. Symposium on VLSI Design, Automation and Test, pp. 19–22, 28–30 April 2009.
2. Anup Jalan and Mamta Khosla, "Analysis of Leakage power reduction techniques in digital circuits", 2011 Annual IEEE India Conference, pp. 1–4, 16–18 Dec. 2011.
3. Harmander Singh, Kanak Agarwal, Dennis Sylvester, Kevin J. Nowka, "Enhanced Leakage Reduction Techniques Using Intermediate Strength Power Gating", IEEE Tran. on VLSI Systems, Vol. 15, No. 11, pp. 1215–1224, Nov. 2007.
4. Jeong Beom Kim, Dong Whee Kim, "Low-Power Carry Look-Ahead Adder with Multi-Threshold Voltage Voltage CMOS Technology", Int. Conference on Semiconductor, Vol. 2, pp. 537–540, Oct. 15 2007-Sept. 17 2007.

5. Shin'lchiro Mutoh, Satoshi Shigematsu, Yasuyuki Matsuya, Hideki Fukuda, Junzo Yamada, "A 1 V Multi-Threshold Voltage CMOS DSP with an Efficient Power Management Technique for Mobile Phone Application", IEEE Int. Conference on Solid-State Circuits, pp. 168–169, 10th Feb. 1996.
6. Ahmed Youssef, Mohab Anis, Mohamed Elmasry, "A Comparative Study between Static and Dynamic Sleep Signal Generation Techniques for Leakage Tolerant Designs", IEEE Tran. on VLSI Systems, Vol. 16, No. 9, pp. 1114–1126, Sept. 2009.
7. Mingco Scok, Scott Hanson, David Blaauw, Dennis Sylvester, "Sleep Mode Analysis and Optimization with Minimal-Sized Power Gating Switch for Ultra-Low V_{dd} Operation", IEEE Tran. on VLSI Systems, Vol. 20, No. 4, pp. 605–615, April 2012.
8. Vishal K, Ankur Srivastava, "Leakage Control Through Fine-Grained Placement and Sizing of Sleep Transistors", IEEE/ACM Int. Conf. on Computer Aided Design, pp. 533–536, 7–11 Nov. 2004.
9. Masanori Hariyama, Shota "A Low-Power Field Programmable VLSI Based on a Fine-Grained Power Gating Scheme", 51st Midwest Symposium on Circuits and Systems, pp. 702–705, 10–13 Aug. 2008.
10. Arifur Rahman, Satyaki Das, Tim Tuan, Steve Trimberger, "Determination of Power Gating Granularity for FPGA Fabric", IEEE Conference Custom Integrated Circuits pp. 9–12, 10–13 Sept. 2006.
11. Pradeep S. Nair, Santosh Koppa, Eugene B. John, " A comparative Analysis of Coarse-grain and Fine-grain Power Gating for FPGA Lookup Tables", IEEE Int. Midwest Symposium on Circuits and Systems, pp. 507–510, 2–5 Aug 2007.
12. Chi-Yi Yeh, Hung-Ming Chen, Li-Da Huang, Wei-Ting Wei, Chao-Hung Lu, Chien-Nan Liu, "Using Power Gating Techniques in Area-Array SoC Floorplan Design", SOC Conference, 2007 IEEE International, pp. 233–236, 26–29 Sept. 2007.
13. Ehan Pakbaznia, Masoud Pedram, "Coarse-Grain MTCMOS Sleep Transistor Sizing using Delay Budgeting", Design, Automation and Test in Europe, pp. 385–390, 10–14 March 2008.
14. Szu-Pang Mu, Yi-Ming Wang, Hao-Yu Yang, Mango C.-T. Chao, " Testing Methods for Detecting Stuck-open Power Switches in Coarse-Grain MTCMOS Designs", IEEE/ACM International Conference on Computer-Aided Design, pp. 155–161, 7–11 Nov. 2010.
15. Yoshiki Saito, Shirai, T.; Nakamura, T.; Nishimura, T.; Hasegawa, Y.; Tsutsumi, S.; Kashima, T.; Nakata, M.;Takeda, S.; Usami, K.; Amano, H., " Leakage Power Reduction for Coarse Grained Dynamically Reconfigurable Processor Arrays with Fine Grained Power Gating Technique", Int. Conf. on ICECE Technology, pp. 329–332, 8–10 Dec. 2008.
16. Hailong Jiao, Volkan Kursum, "Threshold Voltage Tuning for Faster Activation with Lower Noise in Tri-Mode MTCMOS Circuits", IEEE Trans. on VLSI Systems, Vol. 20, No. 4, pp. 741–745, 3rd March 2011.
17. Zhiyu Liu, Volkan Kursun, "Charge Recycling Between Virtual Power and Ground Lines for Low Energy MTCMOS", IEEE Proc. of 8th International Symposium on Quality Electronic Design, pp. 239–244, 26–28 March 2007.
18. Ehsan Pakbaznia, Farzan Fallah, Massoud Pedram, "Charge Recycling in Power-Gated CMOS Circuits", IEEE Trans. On Computer-Aided Design of Integrated Circuits and Systems, Vol. 27, No. 10, pp. 1798–1811, 2008.
19. Suhwan Kim, Stephen V. Kosonocky, Daniel R. Knebel, Kevin Stawiasz, "Experimental Measurement of a Novel Power Gating Structure with Intermediate Power Saving Mode", IEEE International Symposium on Low Power Electronics and Design, pp. 20–25, 11th Aug. 2004.
20. Ehsan Pakbaznia, Massoud Pedram, "Design of a Tri-Modal Multi-Threshold CMOS Switch With Application to Data Retentive Power Gating", IEEE Tran. On VLSI Systems, Vol. 20, No. 2, pp. 380–385, February 2012.
21. Harish Gopalakrishnan, Wen-Tsong Shiue, "Leakage Power Reduction using Self-bias Transistor in VLSI Circuits", IEEE workshop on Microelectronics and Electron Devices, pp. 71–74, 2004.

22. Michael Powell, Se-Hyun Yang, Babak Falsafi, Kaushik Roy, T.N. Vijaykumar, "Gated-Vdd: A Circuit Technique to Reduce Leakage in Deep-Submicron Cache Memories", Proceedings of the 2000 International Symposium on Low Power Electronics and Design, pp. 90–95, 2000.
23. Afshin Nourivand, Asim J. Al-Khalili, Yvon Savaria, "Postsilicon Tuning of Standby Supply Voltage in SRAMs to Reduce Yield Losses Due to Parametric Data-Retention Failures", IEEE tran. On VLSI Systems, Vol. 20, No. 1, pp. 29–41, January 2012.
24. Joohee Kim, Marios C. Papaefthymiou, "Block Based Multiperiod Dynamic Memory Design for Low Data-Retention Power", IEEE Tran. On VLSI Systems, Vol. 11, No. 6, pp. 1006–1018, December 2003.

Phase Based Mel Frequency Cepstral Coefficients for Speaker Identification

Sumit Srivastava, Mahesh Chandra and G. Sahoo

Abstract In this paper new Phase based Mel frequency Cepstral Coefficient (PMFCC) are used for speaker identification. GMM with VQ are used as a classifier for classification of speakers. The identification performance of proposed features is compared with identification performance of MFCC features and phase features. The performance of PMFCC features has been found superior compared to MFCC features and phase features. Ten Hindi digits database of fifty speakers is used for simulation of results. This paper also explore the usefulness of phase information for speaker recognition.

Keywords PMFCC · Phase · Hindi digits · GMM

1 Introduction

In this modern world Automatic Speaker Recognition (ASR) is most developing area. ASR can be basically divided into two field as speaker identification (SI) and speaker verification (SV). The basic objective in speaker identification (SI) is to correctly find out, who is the speaker from a known set of speakers, with the help of utterance of the speaker [1–3]. On the other hand in speaker verification the claimed identity is tested by the machine. In ASR voice sample are used for testing. The use of the ASR system is increasing with the development of more advance technology, such as in banks for transaction authentication, money transfer, fraud detection.

S. Srivastava (✉) · G. Sahoo
Department of Computer Science & Enginnering, BIT Mesra, Ranchi, India
e-mail: sumit.srivs88@gmail.com

G. Sahoo
e-mail: gsahoo@bitmesra.ac.in

M. Chandra
Department of Electronics & Communication Enginnering, BIT Mesra, Ranchi, India
e-mail: shrotriya69@rediffmail.com

© Springer India 2016
S.C. Satapathy et al. (eds.), *Information Systems Design and Intelligent Applications*, Advances in Intelligent Systems and Computing 435,
DOI 10.1007/978-81-322-2757-1_31

Forensic department uses voice sample for security and monitoring such as development of smart home appliance, student and company attendance. Therefore proper ASR system is the need of modern world as it saves both human effort and time. This paper mainly deals with Automatic Speaker identification (ASI).

Many research scholars proposed different methods for speaker identification. There are many methods for features extraction of voice signals like LPC, LPCC [4–6], and MFCC etc. MFCC uses only magnitude of the Fourier transform in time domain of speech signal. Phase part is left out in MFCC. Several studies have shown direct implementation of phase in recognition system. The phase has its own importance because it carries the source information. The MFCC captures the speaker vocal tract knowledge. The parameters of features extracted from the source excitation characteristic are also useful for the speaker recognition. Some researchers [7–9] have used pitch in combination with other feature for better efficiency.

Phase is a sinusoidal functions. According to wave theory it has two different closely related meaning. First at its origin, it is an initial angle of a sinusoidal function which is considered as phase offset or phase difference. Second it is considered as a fraction of the wave cycle which has elapsed relative to its origin. The structure in phase spectrum of audio signals has been demonstrated and found useful in music processing e.g. in onset detection [10], beat tracking [11] or in speech watermarking in [12] where the idea was to embed the watermark signal into the phase of unvoiced speech segments. The phase spectrum has also proved useful in speech polarity determination [13] and detection of synthetic speech to avoid imposture in biometric system [10]. In speech coding and psychoacoustics, it shows that the capacity of human perception due to phase is higher than expected. It means that existing speech coders introduce certain distortions well perceived in particular for low-pitched voice. The more information on instantaneous higher order phase derivatives and the phase information embedded in speech [14]. The recent progress has been made towards estimation or incorporation of phase information and formant information [15].

Various types of the classifier have been used for speaker recognition. For speaker modeling generally, Gaussian Mixture Models (GMMs) [1, 2, 10] are often used as classifier due to their capability of representing a large class of sample distribution. GMMs are widely used for describing the feature distributions of individual speakers among a group of speakers [1]. Recognition decisions of speaker are usually considered based on the maximum log likelihoods by observing the features of the frames for a given speaker model. The GMM is used as a classifier for the model due to capability of the accepting the large class of sample distribution.

2 Proposed Feature

The main aim of feature extraction procedure is to convert the raw signal into an efficient and compact form which is more likely to be stable and distinct than the original signal.

2.1 Phase Based Mel Frequency Cepstral Coefficients (PMFCC)

In this work initially, work was started with Mel Frequency Cepstral Coefficients (MFCC) as a basic features. In course of extracting the features from the speech signal, first the speech signal is to be preprocessed. For this the speech signal was first go through a first order pre-emphasis filter [4] to spectrally flatten the speech signal and make the speech signal less susceptible to finite precision effects later in the speech signal processing approach, then normalization, mean subtraction and silence removal was performed. After preprocessing the speech signal was divided into frames of 30 ms with an overlap of 20 ms to avoid the effect of discontinuity. The speech signal is sampled at 16 kHz. Then each frame is passed through a hamming window. Then this way for each frame, MFCC features [4] are calculated.

Fast Fourier Transform The FFT is mainly used for conversion of the signal into the spatial domain from the time domain. Every frame has 480 samples which was converted into a frequency domain. Fast Fourier Transform is an algorithm that economically calculates the Discrete Fourier Transform (DFT). Basically the FFT and DFT are same but FFT and DFT differ in computational complexity. In the case of DFT, Fourier transform of each frame will be directly used as a sequence of N-M samples. The FFT computes the DFT then produces the exactly same result as by DFT directly. The FFT approach is much faster even in the presence of round-off error and sometimes it is more accurate then directly evaluating the DFT. The FFT is also a powerful tool because it will also calculate the DFT of an input speech signal in an efficient manner. It will also reduce the computational complexity and saving the processing power.

Mel Filter Bank The Mel scale is used to plot the calculated spectrum, which show the approximated value of the existing energy at each spot. In this filter bank, there are set of different band pass filters that have the spacing of the predefined steady bandwidth along Mel frequency time. Hence the Mel scale assists in the spacing of the given filter and calculate the appropriate width for the filter, so that as if there is an increase in the frequency filter width also keeps on growing. The given real frequency scale (Hz) and the perceived frequency scale (Mels) are mapped on the Mel scale. In the Mel scale mapping if the frequency is below 1000 Hz then the linear scale is used and for the frequency above 1000 Hz the spacing is logarithmic. Thus with the proper Mel scaled filter bank it is easy to get the estimated energies for each and every spot. When these energies are estimated for each spot, then log of these energy is calculated to get Mel spectrum. With this Mel spectrum we calculate 16 MFCC coefficients by taking the DCT of the Mel filter scale. The process of obtaining coefficients with the help of Mel-cepstral involves the use of a Mel-scale filter bank. The Mel-scale is a logarithmic scale resembling the way, the human can perceive the sound through ear. The filter bank is composed with the help of 20 triangular filters which are non-uniformly placed in frequency such that these filters will have a linear frequency response at low frequency (up to 1 kHz) and logarithmic at high frequencies. The Mel-scale is represented by the following equation:

$$\text{Mel (f)} = 2595 \log_{10}(1 + \text{f}/700) \quad (1)$$

where f is the frequency. The spectral coefficients of the frames are multiplied by the filter gain and result is obtained.

PMFCC features are proposed to get the better identification rate compared to MFCC features. Phase is speaker specific feature. The phase information is combined with MFCC. The phase carries the source information. The MFCC captures the speaker vocal tract knowledge. The Merger of two yields better result than that of any one alone. PMFCC has the advantage of both MFCC and phase. The next task is to find the phase of a speech signal. The spectrum of speech signal S(w, t) is obtained by DFT of input speech signal.

$$S(w,t) = X(w,t) + jY(w,t) \quad (2)$$

$$= \sqrt{X^2(w,t) + Y^2(w,t)} \times e^{j\theta(w,t)} \quad (3)$$

The phase θ(ω, t) changes with the clipping position of the given speech signal even for the same frequency ω. This problem can be solved by keeping the phase of a certain basic frequency same as ω. For example, setting the basis frequency ω to JI/4, we have

$$S'(\omega,t) = \sqrt{X^2(\omega,t) + Y^2(\omega,t)} \times e^{j\theta(\omega,t)} \times e^{j\theta(\frac{\pi}{4} - \theta(\omega,t))}, \quad (4)$$

where in the other frequency $\omega' = 2\pi f'$, the spectrum becomes

$$S'(\omega',t) = \sqrt{X^2(\omega',t) + Y^2(\omega',t)} \times e^{j\theta(\omega',t)} \times e^{j\theta(\frac{\pi}{4} - \theta(\omega',t))}, \quad (5)$$

$$= \tilde{X}(\omega',t) + j\tilde{Y}(\omega',t), \quad (6)$$

with this the phase can be normalized. Then the real and imaginary part of Eq. (7) becomes as

$$\tilde{X}(\omega',t) = \sqrt{X^2(\omega',t) + Y^2(\omega',t)} \times \cos\left\{\theta(\omega',t) + \frac{\omega'}{\omega}\left(\frac{\pi}{4} - \theta(\omega,t)\right)\right\} \quad (7)$$

$$\tilde{Y}(\omega',t) = \sqrt{X^2(\omega',t) + Y^2(\omega',t)} \times \sin\left\{\theta(\omega',t) + \frac{\omega'}{\omega}\left(\frac{\pi}{4} - \theta(\omega,t)\right)\right\}, \quad (8)$$

In similar way the phase features of all the frames of a speech sample are calculated. Three estimated phase features were concatenated with thirteen MFCC features to get sixteen PMFCC features as shown in Fig. 1.

Fig. 1 Feature extraction using MFCC and PMFCC

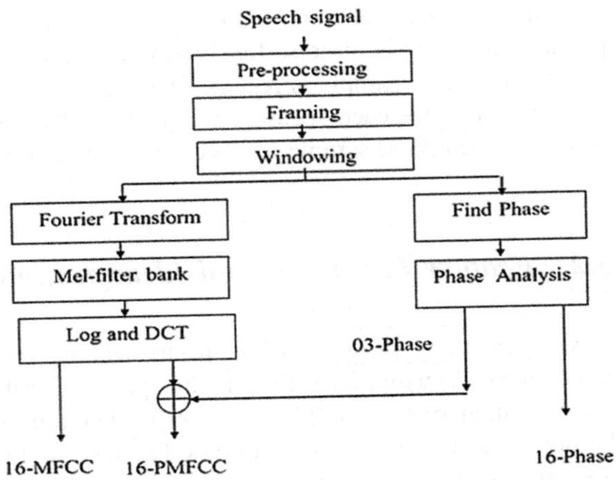

3 Simulation Set-Up

For speaker identification the simulation setup is given in Fig. 2. For speaker recognition process, the database preparation, then feature extraction and finally the classification are the basic building blocks.

3.1 Database

To perform the speaker identification experiments, a Hindi database of fifty speakers has been used. In this, database, ten Hindi digits from 0 to 9 have been

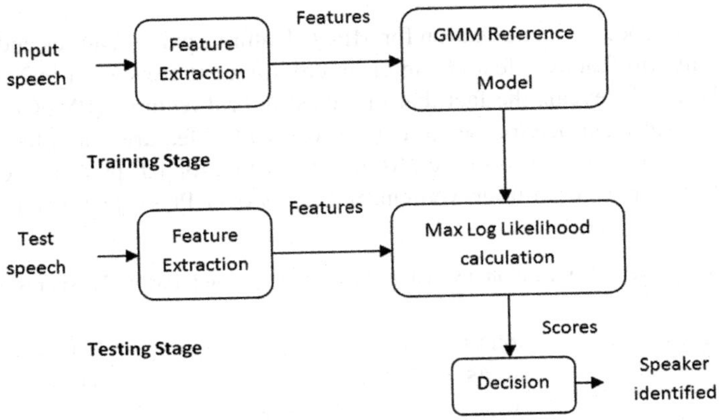

Fig. 2 Experimental set-up for speaker identification

spoken by each speaker in reading style manner. Goldwave software is used to prepare the database in noise free environment. The sampling frequency is 16 kHz mono. Each digit sample is repeated ten times to capture the different variation of speech sample for each speaker. So there are 100 sample for each speaker and in this way, total 5000 sample has been used for training purpose.

3.2 Feature Extraction and Identification

After performing pre-processing, normalization, mean subtraction and dc offset removal as discussed in Sect. 2.1, the speech signal is divided into a segment of 30 ms with an overlap of 20 ms. After this hamming window is used on each frame before extraction of MFCC and phase features. In this way sixteen MFCC features are calculated for each clean frame of Hindi digits samples for all speakers. The frame based extracted features were stored in fifty different files one for each speaker.

For Classification purpose GMM models were prepared for all the fifty speakers using the extracted MFCC features of respective speakers and these models were stored as a reference models to be used during testing. In this case training is done by using clean data base features. Decision about the speaker identification was made on the basis of maximum log likelihood for each tested feature vector with GMM models.

In case of PMFCC, first three features are extracted from phase and concatenated with thirteen frame based MFCC features. In this way total sixteen PMFCC features are obtained for each frame. Training and testing are done similarly as for MFCC.

4 Results and Discussion

The results for speaker identification for MFCC features, phase features and PMFCC (proposed hybrid features) features in clean environment are shown in Table 1 and Fig. 3. The result reveals the fact that proposed hybrid features (PMFCC) have an improvement in identification performance over MFCC features and phase features in clean environments. It is observed from Table 1 that for 10 speakers the PMFCC features have achieved an improvement of 12.1 % over Phase features and 0.75 %

Table 1 % speaker identification by using MFCC, phase and PMFCC features in a clean environment

No. of speakers	MFCC	Phase	PMFCC
10	98.35	87.00	99.10
25	97.44	85.60	99.02
50	96.04	84.67	98.42

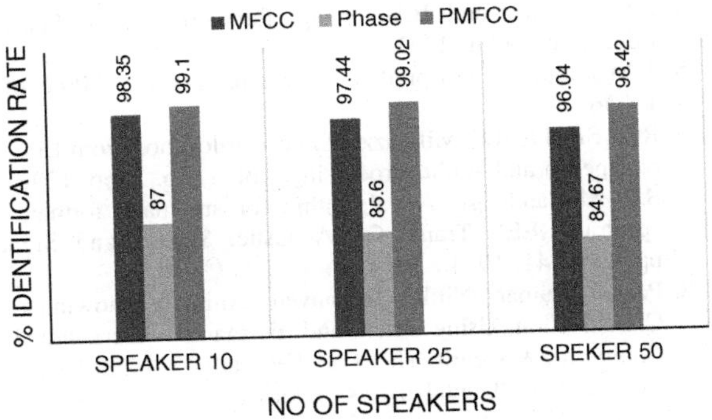

Fig. 3 % speaker identification in clean environment

over MFCC features. For 25 speakers PMFCC features have achieved an improvement of 13.42 % over phase features and 1.58 % over MFCC features. For 50 speakers PMFCC features have achieved an improvement of 13.75 % over phase features and 2.38 % over MFCC features in a clean environment.

5 Conclusion

The proposed PMFCC feature technique has shown an overall improvement in identification performance over phase and MFCC features in clean environments. The MFCC features are based on speech perception that how a human ear perceive a speech signal. The phase features are also useful to extract the information from signal. Through experimental results, it is also proved that short time phase spectrum also contribute in speaker recognition significantly. Phase features based on speaker production, vary with speaker to speaker. As phase features are more speaker specific thus they have helped in increasing the performance of speaker identification.

References

1. D.A. Reynolds, and R.C. Rose, "Robust Text-Independent Speaker Identification using Gaussian Mixture Speaker Models," IEEE Transactions on Speech and Audio Processing, vol. 3, no. 1, pp. 74–77, January 1995.
2. Md. Fozur Rahman Chowdhury, "Text independent distributed speaker identification and verification using GMM UBM speaker models for mobile communications," 10th International Conference on Information Science, Signal Processing and Their Application, 2010, pp 57–60.
3. Tomi Kinnunen, Evgeny Karpov and Pasi Franti "Real-time speaker identification and verification", IEEE Transaction on Audio, Speech and Language Processing, Vol. 14, No. 1, pp. 277–278, 2006.

4. L.R. Rabiner and B.H. Juang, Fundamentals of Speech Recognition, 1st ed., Pearson Education, Delhi, 2003.
5. J. Makhoul, "Linear prediction: A tutorial review," Proc. of IEEE, vol. 63, no. 4, pp. 561–580, 19756.
6. R.C. Snell and F. Milinazzo, "Formant location from LPC Analysis data," IEEE Transactions on Speech and Audio Processing, vol. 1, no. 2, pp. 129–134, Apr. 1993.
7. S.S. McCandless, "An algorithm for automatic formant extraction using linear prediction spectra," IEEE Trans. On Acoustic, Speech and Signal Processing, ASSP-22, No. 2, pp. 135–141, 1974.
8. Pawan Kumar, Nitika Jakhanwal, Anirban Bhowmick, and Mahesh Chandra, "Gender Classification Using Pitch and Formants" International Conference on Communication, Computing &Security (ICCCS), February 12–14, 2011, Rourkela, Odisha, India, pp. 319–324.
9. J.D. Markel, "Digital inverse filtering-A new tool for formant trajectory estimation," IEEE Trans. AU-20, pp. 129–1 37, 1972.
10. A. Holzapfel and Y. Stylianou, "Beat tracking using group delay based onset detection." in ISMIR, 2008, pp. 653–658.
11. M. E. P. Davies and M. Plumbley, "Context-dependent beat tracking of musical audio," IEEE Trans. on Audio, Speech, and Language Processing, vol. 15, no. 3, pp. 1009–1020, March 2007.
12. K. Hofbauer, G. Kubin, and W. Kleijn, "Speech watermarking for analog flat-fading bandpass channels," IEEE Trans. on Audio, Speech, and Language Processing, vol. 17, no. 8, pp. 1624–1637, Nov. 2009.
13. I. Saratxaga, D. Erro, I. Hernez, I. Sainz, and E. Navas, "Use of harmonic phase information for polarity detection in speech signals." in INTERSPEECH, 2009.
14. Munish Bhatia, Navpreet Singh, Amitpal Singh," Speaker Accent Recognition by MFCC Using KNearest Neighbour Algorithm: A Different Approach", in IJARCCE.2015.
15. Sumit Srivastava, Pratibha Nandi, G. Sahoo, Mahesh Chandra," Formant Based Linear Prediction Coefficients for Speaker Identification", SPIN 2014.

A New Block Least Mean Square Algorithm for Improved Active Noise Cancellation

Monalisha Ghosh, Monali Dhal, Pankaj Goel, Asutosh Kar, Shibalik Mohapatra and Mahesh Chandra

Abstract Acoustic noise is an undesired disturbance that is present in the information carrying signal in telecommunication systems. The communication process gets affected because noise degrades the quality of speech signal. Adaptive noise reduction is a method of approximating signals distorted by additive noise signals. With no prior estimates of input or noise signal, the levels of noise reduction are attainable that would be difficult or impossible to achieve by other noise cancelling algorithms, which is the advantage of adaptive technique. Adaptive filtering before subtraction allows the treatment of inputs that are deterministic or stochastic, stationary or time variable. This paper provides an analysis of various adaptive algorithms for noise cancellation and a comparison is made between them. The strengths, weaknesses and practical effectiveness of all the algorithms have been discussed. This paper deals with cancellation of noise on speech signal using three existing algorithms—Least Mean Square algorithm, Normalized Least Mean

---

The original version of this chapter was revised: The spelling of one of the author's name was corrected. The erratum to this chapter is available at DOI 10.1007/978-81-322-2757-1_66

M. Ghosh (✉) · M. Dhal
Electronics and Telecommunication Engineering, IIIT Bhubaneswar, Bhubaneswar, India
e-mail: b211030@iiit-bh.ac.in

M. Dhal
e-mail: b211031@iiit-bh.ac.in

P. Goel · M. Chandra
Electronics and Communication Engineering, BIT Mesra, Ranchi, India
e-mail: write2pankaj@rediffmail.com

A. Kar · S. Mohapatra
Electronics and Electrical Engineering, BITS Pilani, Hyderabad, India
e-mail: asutosh@hyderabad.bits-pilani.ac.in

S. Mohapatra
e-mail: shibalik.mohapatra4@gmail.com

© Springer India 2016
S.C. Satapathy et al. (eds.), *Information Systems Design and Intelligent Applications*, Advances in Intelligent Systems and Computing 435,
DOI 10.1007/978-81-322-2757-1_32

Square algorithm and Recursive Least Square algorithm and a proposed algorithm —advanced Block Least Mean Square algorithm. The algorithms are simulated in Simulink platform. Conclusions have been drawn by choosing the algorithms that provide efficient performance with less computational complexity.

Keywords Active noise cancellation · Adaptive filter · Least mean square · Normalized least mean square · Recursive least square · Block least mean square

1 Introduction

The speech is an information carrying signal that transfers information from one to other. With the emotion of a human voice speech signal carries the intended information. The speech signal is one-dimensional signal, time as its independent variable, random and non-stationary in nature. The audible frequency range for the human being is 20 Hz to 20 kHz. Again the human speech can carry significant frequency components only up to 8 kHz [1]. The reduction of noise components from desired speech is one of the focused research in speech signal analysis and processing.

The unwanted noise corrupts the signal and diminishes its information carrying capacities. The major sources of unwanted noise include traffic noise, electrical equipment noise and distortion products. Sometimes the sound system has large peaks in its frequency response which is not usual. In this case the speech signal may vary in amplitude and phase from the normal response [2–4]. One important relationship between the signal strength and unwanted noise present in the signal is called Signal-to-Noise Ratio (SNR). Ideally, if the SNR is more than 0 dB, it indicates that the speech signal is louder than the noise signal. The type of signal used is speech and it can in the form of music or in any form. The noise cancellation is a techniques that employ digital filter. Digital filters are driven by adaptive algorithms and the algorithms have their own concern like complexity, convergence, misadjustment etc. All these parameters should be taken care of when designing a new or improved ANC algorithm with the use of adaptive techniques.

2 Proposed Solution

This paper proposes an advanced version of the BLMS algorithm applied for noise cancellation. This advanced version of BLMS algorithm attempts to compensate the various drawbacks of the existing adaptive algorithms applied for ANC. Compared with the existing adaptive algorithms, the proposed algorithm has the following attributes i.e. it has a comparatively simple design and is more adaptable in the selection of the block length L, complex ratio is higher compared to the existing algorithms, it has a faster convergence rate and it has a good prospect for use in real-time processing environment. The basic idea between the proposed algorithms is that the weighting coefficients are modified for a block of input data rather than for each input sample. The proposed algorithm allows us to dynamically change the block length L.

Here, the input data sequence is divided into L-point blocks with the help of serial-to-parallel converter. These blocks of input data are applied to an FIR filter of length M, one block at a time. It is needed to maintain constant weights while each block of data is processed. Therefore, the weight vector is modified once per block of input data instead of one per block of input sample as in the existing algorithms [5–8]. The input signal vector at time 'n' is denoted as:

$$S(n) = [s(n),\ s(n-1),\ s(n-2),\ldots, s(n-M+1)]^T \tag{1}$$

The tap weight vector at time is 'n' denoted as:

$$\hat{W}(n) = [\hat{w}_0(n),\ \hat{w}_1(n),\ldots\ldots, \hat{w}_{M-1}(n)]^T \tag{2}$$

Block index 'p' and original sample time 'n' are related as:

$$n = pL + i,\ i = 0, 1, \ldots, L-1,\ p = 1,\ 2,\ 3, \ldots$$

The input data matrix for block 'p' is given as:

$$B^T(p) = [s(pL),\ s(pL+1),\ s(pL+2), \ldots, s(pL+L-1)] \tag{3}$$

The output produced by the filter:

$$y(pL+i) = \hat{W}^T(n)S(pL+i) = \sum_{j=0}^{M-1} \hat{w}_j(p)s(pL+i-j) \tag{4}$$

The desired response is denoted by $d(pL+i)$. The difference between the filter output and desired response is given by the error signal (Fig. 1):

$$e(pL+i) = d(pL+i) - y(pL+i) \tag{5}$$

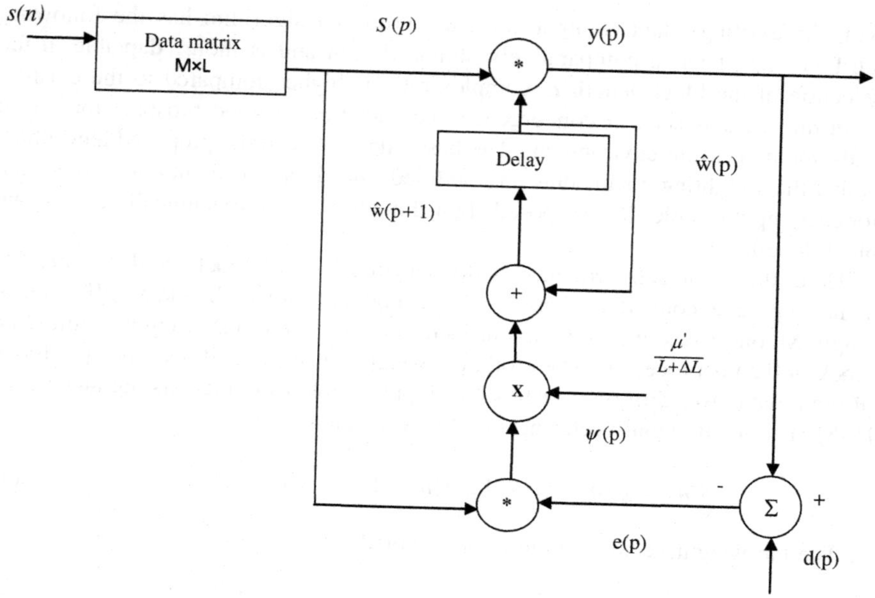

Fig. 1 Signal-flow graph representation of the advanced BLMS algorithm

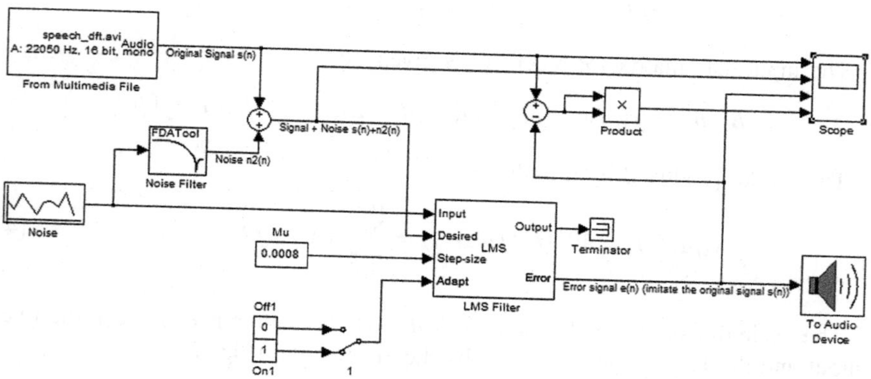

Fig. 2 Simulink model for advanced LMS algorithm

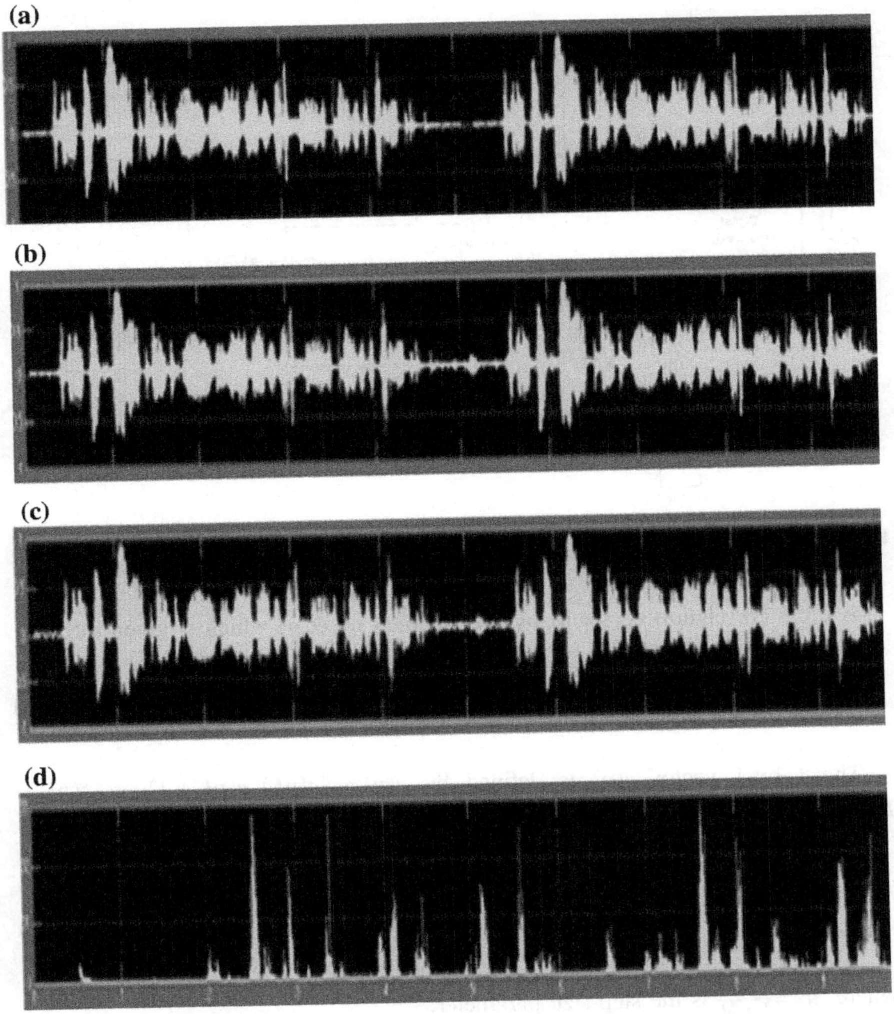

Fig. 3 Simulation results for advanced LMS algorithm. **a** Original signal. **b** Noisy signal. **c** Original + noisy signal. **d** Mean-squared error

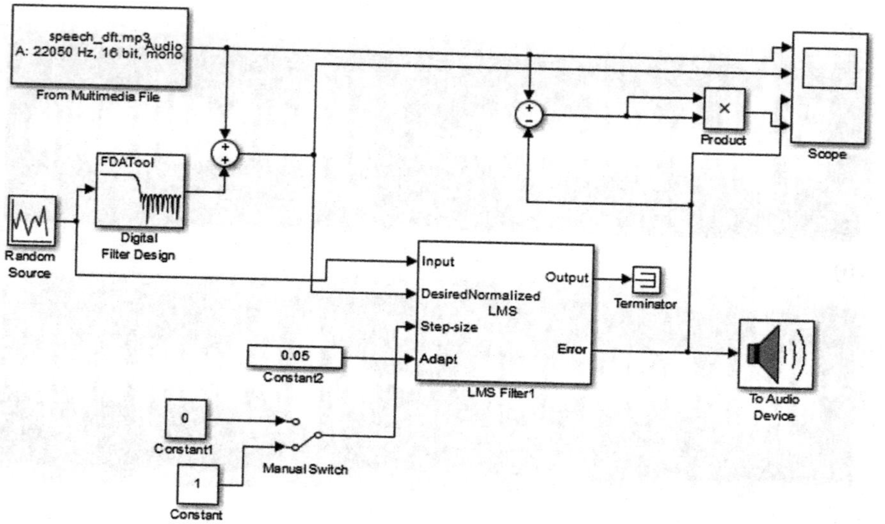

Fig. 4 Simulink model for NLMS algorithm

A cross correlation vector M by 1 is denoted by $\psi(p)$ which is given as:

$$\Psi(p) = \sum_{j=0}^{L-1} S(pL+i)e(pL+i) = B^T(p)e(p) \tag{6}$$

The L-by-1 vector $e(p)$ is defined by $e(p) = [e(pL), e(pL+1), \ldots, e(pL+L-1)]^T$

The weight updating equation is given by:

$$\hat{W}(p+1) = \hat{W}(p) + \frac{\mu'}{L+\Delta L}\Psi(p) \tag{7}$$

where, $\mu = \frac{\mu'}{L+\Delta L}$ is the step-size parameter.

Here, the incremental length ΔL is used to dynamically vary the block length L [9].

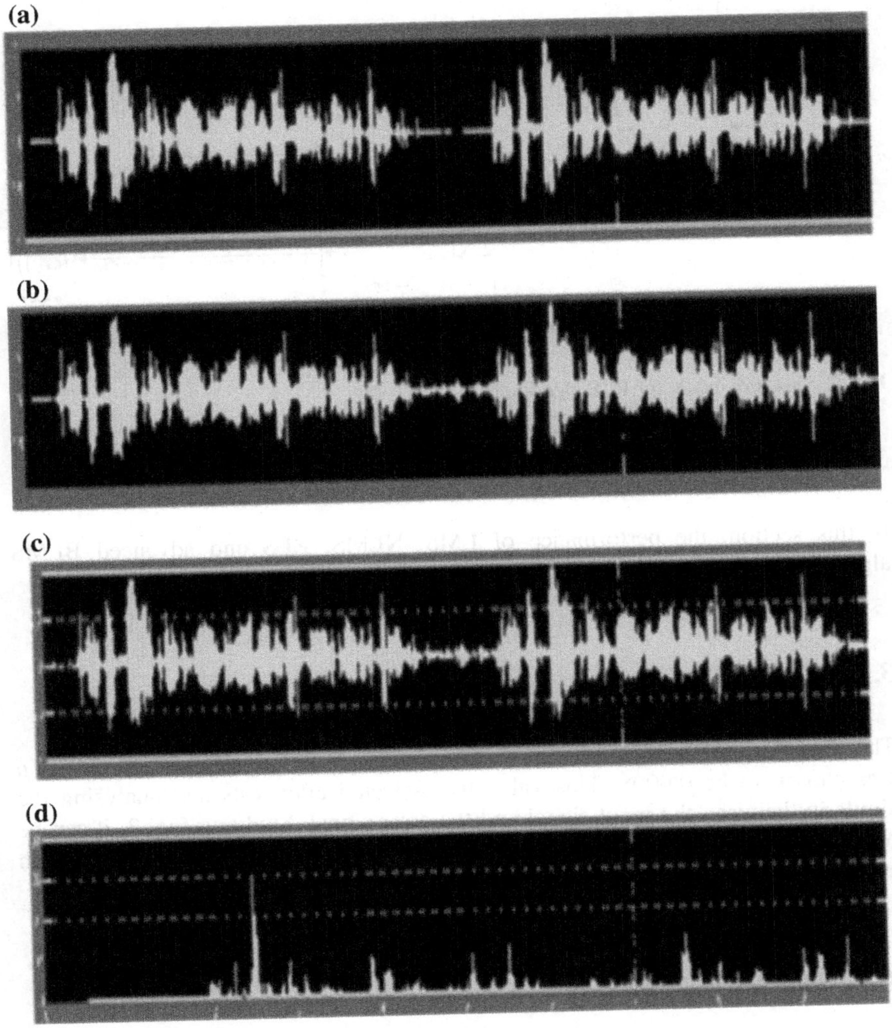

Fig. 5 Simulation results for NLMS algorithm. **a** Original signal. **b** Noisy signal. **c** Original + noisy signal. **d** Mean-squared error

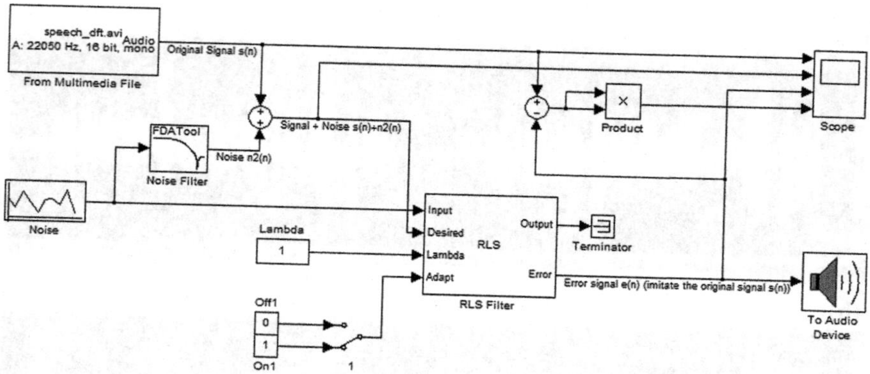

Fig. 6 Simulink model for RLS algorithm

3 Simulations and Results

In this section, the performance of LMS, NLMS, RLS and advanced BLMS algorithms in noise cancellation setup is evaluated (Fig. 2).

3.1 The LMS Solution

The filter has length 32 and the noise has variance 0.1. The step-size parameter μ was chosen to be 0.0008. This value was selected after tests and analysing the similarity between the input signal and the error signal. Studying Fig. 3, it can be observed that the algorithm takes about 6 s to present an acceptable conversion characteristic, having the squared error tending to zero (Fig. 4).

3.2 The NLMS Solution

The filter has length 32 and the noise has variance 0.1. The step-size parameter μ was chosen to be 0.0001. This value was selected after tests and analysing the similarity between the input signal and the error signal. Studying Fig. 5, it can be noticed that the algorithm takes about 5 s to present an acceptable conversion characteristic, having the squared error tending to zero (Fig. 6).

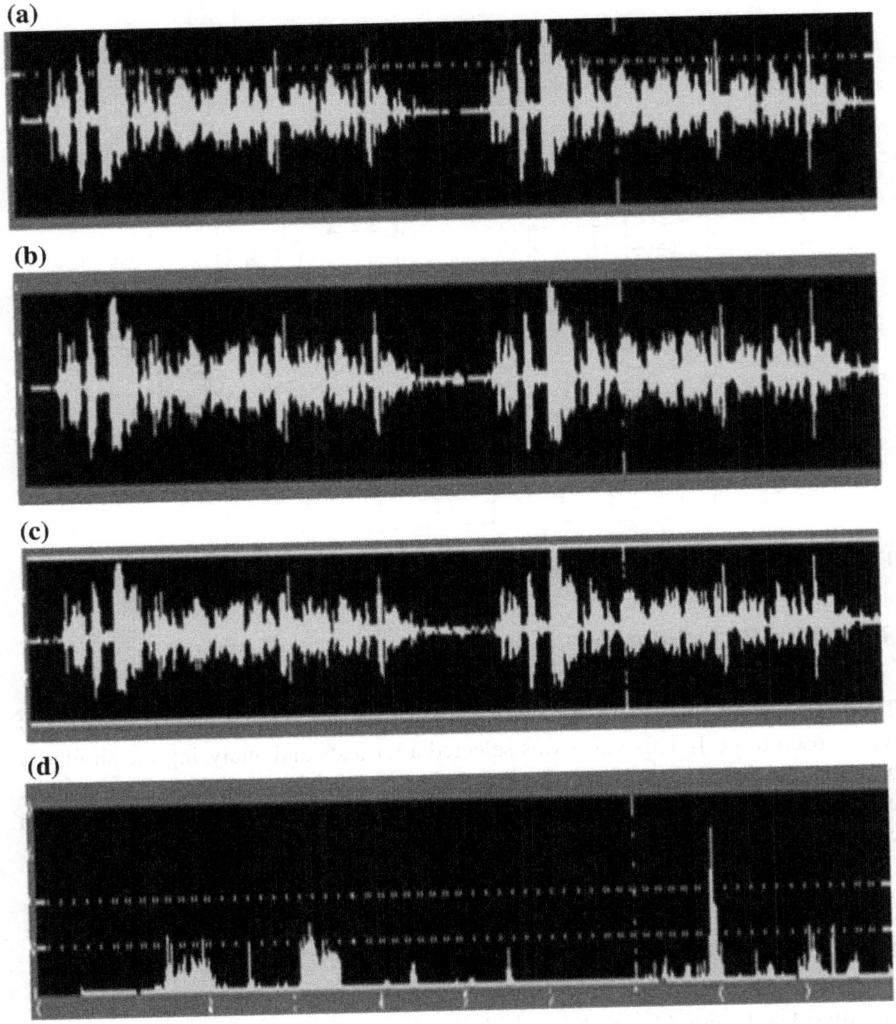

Fig. 7 Simulation results for RLS algorithm. **a** Original signal. **b** Noisy signal. **c** Original + noisy signal. **d** Mean-squared error

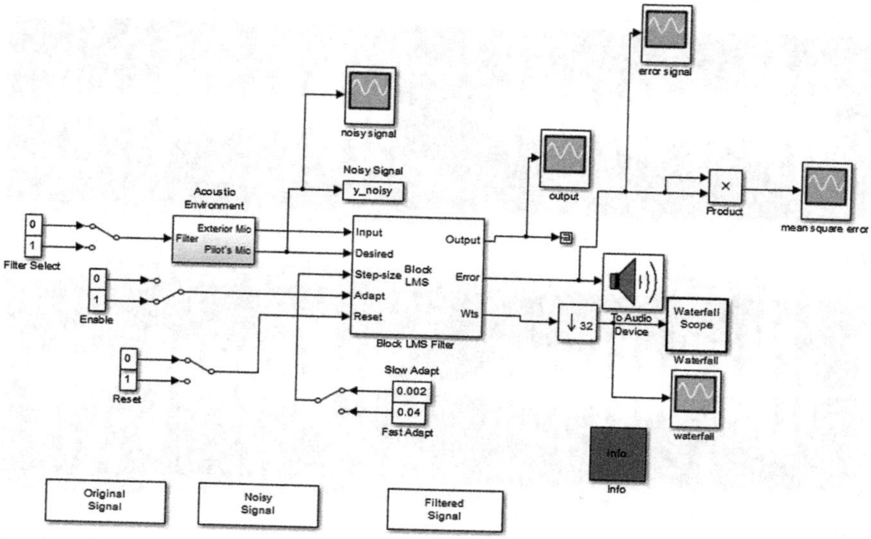

Fig. 8 Simulink model for BLMS algorithm

3.3 The RLS Solution

The filter has length 32 and the noise has variance 0.1. The step-size parameter λ was chosen to be 1. This value was selected after tests and analysing the similarity between the input signal and the error signal. Studying this Fig. 7, it can be observed that the algorithm takes about 3 s to present an acceptable conversion characteristic, having the squared error tending to zero (Fig. 8).

3.4 Advanced BLMS Solution

The filter has length 32 and the noise has variance 0.1. The step-size parameter μ was chosen to be 0.0001. This value was selected after tests and analysing the similarity between the input signal and the error signal. Studying Fig. 9, it can be seen that the algorithm takes about 1 s to present an acceptable conversion characteristic, having the squared error tending to zero.

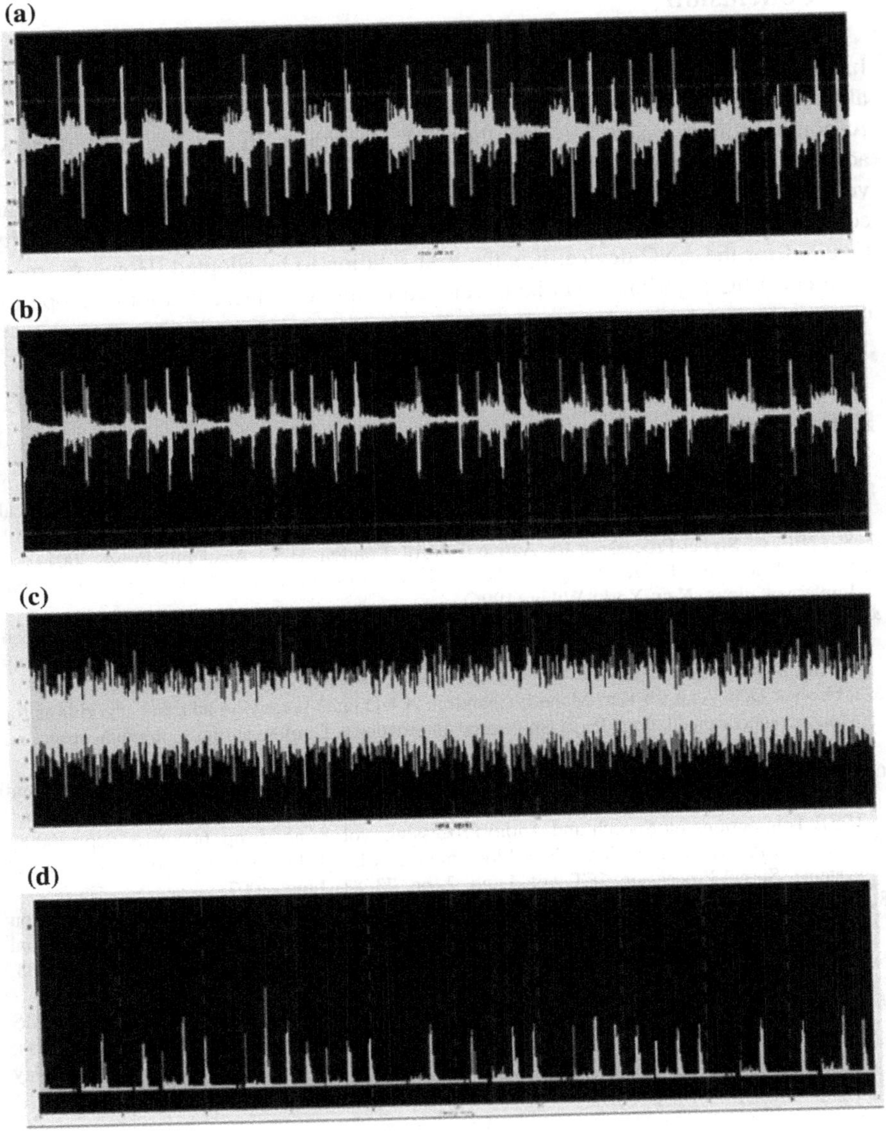

Fig. 9 Simulation Results of advanced BLMS algorithm. **a** Original signal. **b** Error signal. **c** Original + noisy signal. **d** Mean-squared error

4 Conclusion

In this paper, an ANC problem is described and the solutions of three existing algorithms along with the proposed algorithm are presented for the above problem. A comparison is made between the four algorithms. It is concluded that the advanced BLMS algorithm is more robust than the existing algorithms and has a very fast rate of convergence. The advanced BLMS algorithm has moderate computational complexity and computational cost, but depending on the quality required for the ANC device, it is the best solution to be adopted. However, more coherent ANC algorithms can be developed in future as there is a lot of scope for research on this topic.

References

1. Eberhard Hansler and Gerhard Schmidt.: Acoustic Echo and Noise Control: A Practical Approach", Wiley(2004).
2. S. J. Elliot.: Signal Processing for Active Control, London, U.K.: Academic Press (2001).
3. S. M. Kuo, and D. R. Morgan.: Active Noise Control Systems-Algorithms and DSP Implementations, New York: Wiley (1996).
4. Asutosh Kar, Ambika Prasad Chanda and Mahesh Chandra.: An improved filtered-x least mean square algorithm for acoustic noise suppression. vol. 1, pp. 25–32, Springer-Verlag, Germany (2014).
5. Abhishek Deb, Asutosh Kar, Mahesh Chandra.: A technical review on adaptive algorithms for acoustic echo cancellation. IEEE Sponsored International Conference on Communication and Signal Processing, pp. 638–642, Apr. 3–5, (2014).
6. Bouchard, M., Quednau, S.: Multichannel recursive-least-square algorithms and fast-transversal-filter algorithms for active noise control and sound reproduction systems. IEEE Transactions on Speech and Audio Processing, vol. 8, no. 5, pp. 606–618, Sep (2000).
7. Das, D.P., Panda, G., Kuo, S.M.: New block filtered-X LMS algorithms for active noise control systems. Signal Processing, IET, vol. 1, no. 2, pp. 73–81, June (2007).
8. Gnitecki, J., Moussavi, Z., Pasterkamp, H.: Recursive least squares adaptive noise cancellation filtering for heart sound reduction in lung sounds recordings. Proceedings of the 25th Annual International Conference of the IEEE on Engineering in Medicine and Biology Society, 2003, vol. 3, pp. 2416–2419, Sept. (2003).
9. Das, D.P., Panda, G., Nayak, D.K.: Development of Frequency Domain Block Filtered-s LMS Algorithm for Active Noise Control System. IEEE International Conference on Acoustics, Speech and Signal Processing, 2006, ICASSP 2006 Proceedings, vol. 5, pp. 14–19, May (2006).

An Improved Feedback Filtered-X NLMS Algorithm for Noise Cancellation

Bibhudatta Mishra, Ashok Behuria, Pankaj Goel, Asutosh Kar, Shibalik Mohapatra and Mahesh Chandra

Abstract The age of unmatchable technical expertise envisages noise cancellation as an acute concern, as noise is held responsible for creating hindrances in day to day communication. To overcome the noise present in the primary signal notable traditional methods surfaced over the passage of time being listed as noise barriers, noise absorbers, silencers, etc. The advanced modern day approach suppresses noise by continuous adaptation of filter weights of an adaptive filter. The change in approach was ground breaking that accredits its success to advent of adaptive filters which employs adaptive algorithms. The various premier noise cancellation algorithms include LMS, RLS etc. Further much coveted Normalized LMS, Fractional LMS, Differential Normalized LMS, Filtered-x LMS etc. ensued out of active framework in this field. The paper looks forward to provide an improved approach for noise cancellation in noisy environment using newly developed variants of Filtered x LMS (FxLMS) algorithm, Feedback FxLMS (FB-FxLMS). An initial detailed analysis of existing FXLMS algorithm and FB-FxLMS algorithm has been carried out along with the mathematics of the new proposed algorithm.

The original version of this chapter was revised: The spelling of one of the author's name was corrected. The erratum to this chapter is available at DOI 10.1007/978-81-322-2757-1_66

B. Mishra (✉) · A. Behuria
Electronics and Telecommunication Engineering, IIIT Bhubaneswar, Bhubaneswar, India
e-mail: b211013@iiit-bh.ac.in

A. Behuria
e-mail: b211010@iiit-bh.ac.in

P. Goel · M. Chandra
Electronics and Communication Engineering, BIT Mesra, Ranchi, India
e-mail: write2pankaj@rediffmail.com

A. Kar · S. Mohapatra
Electronics and Electrical Engineering, BITS Pilani, Hyderabad, India
e-mail: asutosh@hyderabad.bits-pilani.ac.in

S. Mohapatra
e-mail: shibalik.mohapatra4@gmail.com

© Springer India 2016
S.C. Satapathy et al. (eds.), *Information Systems Design and Intelligent Applications*, Advances in Intelligent Systems and Computing 435,
DOI 10.1007/978-81-322-2757-1_33

The proposed algorithm is applied to noise cancellation and the results for each individual process were produced to make a suitable comparison between the existing and proposed one.

Keywords Active noise control · Least mean square · Feed-forward ANC · Feedback ANC · Adaptive filter

1 Introduction

Acoustic Noise Cancellation (ANC) is the technique which brought revolutionary changes in the field of noise cancellation. In the mediaeval era a number traditional approach were instrumented in the field of acoustic noise control which emphasised upon the application of passive methods such as enclosures, barriers and silencers to attenuate unwanted sound waves. These approaches use either the concept of impendence change or the energy loss due to sound absorbing materials. The application of these silencers were largely preferred for their ability of attenuating noise globally; however, when queried about their feasibility they were found relatively bulky, over costly and ineffective for low frequency noise. In order to overcome these primary problems, Acoustic Noise Cancellation (ANC), has received reasonable scope of research due to the fact that it employs an electro-acoustic system which is responsible to create a local silence zone to attenuate the primary noise.

In this technology dominated era an important reason for the growing acceptance of ANC is due to the fact that it is more efficient than passive devices at low frequencies and ANC also uses digital signal processing system as its principal noise canceller. This system executes various complicated mathematical operations with high accuracy and precision in real time. With the use of above system, the amplitude and phase of both the primary and the secondary waves match closely. By virtue of it, maximum noise suppression can be achieved with enormous stability, and reliability. Generally ANC is widely regarded as one of the application of Active control of sound, which can be coined by the phenomenon by virtue of which the existing original sound field (primary field), due to some original (primary) sound sources, is modified to something else (desired sound field) by the help of controlled sound sources (secondary sources). Again Acoustic Noise Cancellation (ANC) can be implemented by using three broadly options: Active Noise Absorption (ANA), Active Noise Reflection (ANR) or Active Potential Energy Minimization. The active absorption (or Active Acoustic Attenuation (AAA)) utilises the secondary sources to absorb the original sound field which are being directed towards them, while in the active noise reflection (or active sound reflection), the original sound field is reflected back towards the primary sources by secondary sources. The last part which is the Active Potential Energy Minimization, the sound pressure can be minimized at selected points. While in active absorption the secondary sources acts like passive absorbents, the same secondary sources act as noise barriers in active noise reflection.

Table 1 Comparison of existing algorithms

Name of algorithm	Strengths	Weaknesses
LMS	Low computational complexity, stable	Slow speed of convergence and less robust
NLMS	Fast speed of convergence, stable	More computational complexity than LMS and less robust than RLS
FLMS	Faster speed of convergence, more stable	More computational complexity than NLMS
RLS	Fast speed of convergence, more robust	Very high computational complexity and unstable
FxLMS	Less computational complexity, simple real time realization	Slow speed of convergence
FxRLS	Fast convergence, Low steady state residual noise	Large computational complexity
FB-FxLMS	Better convergence than FxLMS, simple real time realization	More computational complexity than FxLMS and LMS

Generally the generation of the secondary signal is controlled by a suitable adaptive algorithm which adaptively changes the weight of the filter being used. ANC is mainly classified into two categories i.e. "feed-forward ANC" and "feedback ANC". In feed-forward ANC method a coherent reference noise input is sensed before it propagates past the secondary source while feedback ANC employs a method where the active noise controller attempts to cancel the noise without the benefit of an upstream reference input. In other words the first method uses a reference sensor for the noise cancellation while the latter one doesn't utilizes it all.

An active noise control system usually based upon four main parts. These include the plant or control path which acts as a physical system between the actuator and the sensor. The sensors generally perform the work of picking up the reference signal and the error signal respectively. The noise measured as the primary noise present in the system is cancelled out around the location of the error microphone by a process of generating and combining an anti-phase cancelling noise that is closely correlated to the spectral content of the unwanted noise. The actuators are also regarded as loudspeakers. The last part is the controller which controls the actuators using the signals picked up by the sensors. The comparative data of earlier existing algorithms are highlighted in Table 1 [1–7].

2 Proposed Algorithm and Implementation

The proposed algorithm is applied for both feed-forward as well as for feedback system. Hence we look forward to divide this section into two halves where in the first half the new proposed algorithm has been applied in feed-forward ANC and in the second half the proposed algorithm is applied for the feedback ANC. This section on a whole requires prior knowledge of both FxLMS and FB-FxLMS algorithm respectively.

2.1 Feedforward ANC

The above system primarily uses the FxLMS algorithm to cancel the noise source. The structure of the algorithm comprises of a primary plant through which noise propagates. This noise then reaches to the error sensor in the form of desired signal that is represented by $d(n)$. To cancel this approaching noise the controller generates a signal which has the same amplitude but opposite phase to the noise signal. The control signal $y(n)$ is then produced by filtering the reference signal $x(n)$ at the cancelling loudspeaker with the weight updating adaptive LMS algorithm at its disposal. The filter weight coefficients are defined by $W(n)$ with $y(n)$ being the convoluted signal between the reference signal $x(n)$ and filter coefficient. In the subsequent process $y(n)$ is filtered by the secondary path transfer function $S(n)$ that finally reaches the error sensor as output signal $y'(n)$. When this output signal is mixed with the noise signal, error signal $e(n)$ is obtained that is measured by error sensor. The residual error is obtained by acoustic combination of desired signal and control signal which is subsequently minimised by the adaptive algorithm that optimizes the filter coefficients. The wiener optimum filter for the filter to optimize is designated by W_{opt}. The weight update equation of FxLMS algorithm is given by:

$$W(n+1) = W(n) + \mu x'(n)e(n) \tag{1}$$

where the error signal is estimated by,

$$e(n) = d(n) - s(n) * \{W^T(n)x(n)\} \tag{2}$$

where

n Time index
$S(n)$ Impulse response of secondary path $S(z)$
$*$ Linear convolution

$$W(n) = [W_0(n), W_1(n), \ldots\ldots\ldots W_{L-1}(n)]^T \text{ Filter coefficient}$$
$$x(n) = [x(n), x(n-1), \ldots\ldots\ldots x(n-L+1)]^T \text{ Signal vectors}$$
$$x'(n) = \hat{S}(n) * x(n) \text{ Filtered response}$$

In order to improve the speed of convergence along with increasing the accuracy in system identification and cancelling the noise signal the following modifications are proposed to the existing Filtered x adaptive system. Here the existing LMS adaptive algorithm which appears to be the heart of noise cancellation has been suitably replaced by a new advanced version of normalised LMS adaptive structure. The weight adaptation coefficient of the new proposed advanced algorithm improves the existing one by,

$$\mu(n) = \frac{\alpha}{\beta + R} = \frac{\alpha}{\beta + x(n) * x^T(n)} \tag{3}$$

where the value scope of $\mu(n)$ is kept between the values ranging from 0 to unity or 1. The other parameter known as β is made quite small as possible to overcome the lacuna of SSM degradation with a conditional point of $r(n)$ being very small. This assumption increases the SSM performance of advanced NLMS by fore fold when compared to traditional LMS. The above relationship uses the concept of normalised value of LMS adaptation. The proposed advanced algorithm updates the step size by,

$$W(n+1) = W(n) + \left(\frac{\alpha}{\beta + R}\right) x'(n)e(n) \tag{4}$$

The proposed algorithm operates at larger step sizes as compared to the previous stated algorithm when the case of decreasing trend of residual error. Assuming the FIR filter weight to be $W(z)$ the filter length is taken L and the output results are computed likewise.

2.2 Feedback ANC

The above system primarily uses the FxLMS algorithm to cancel the noise source in the feedback path. This christens its name to FB-FxLMS. The FxLMS algorithm showed that anti-noise output to the loudspeaker not only cancels acoustic noise downstream, but unfortunately, it also radiates upstream to the input microphone, resulting in a contaminated reference input $x(n)$. This inherent problem was solved by the use of FB-FxLMS. Initially the system components are all the same parts as that same of FxLMS with slight modifications. Instead of the application of secondary sensors the linear predictor is used for generating the requisite reference signal. The linear predictor provides a robust, reliable and accurate way for estimating the parameters of secondary path.

From the diagram it can be figured out that $y(n)$ is the anti-noise signal produced by the controller $W(z)$. The output of $W(z)$ enters to the secondary path which goes on to get inverted and results in the generation of $y'(n)$. The signal is then added to the primary noise $d(n)$ and resulting into subsequent generation of error signal $e(n)$. It is also regarded as residual noise which is then fed back to the controller $W(z)$. By filtering the output of the controller and then adding it to the error signal, an estimate of the filtered primary noise $\hat{d}(n)$ is generated. This signal is then used as the reference signal $x(n)$, and is filtered through the model of the secondary path before being fed to the adaptive block. The expressions for the anti-noise $y(n)$, filtered x signal $x'(n)$, and the adaptation equation for the FB-FxLMS algorithm are the same as that for the FxLMS ANC system, except that $x(n)$ in FB-FxLMS algorithm is a feedback-free signal that can be expressed as:

$$x(n) \equiv \hat{d}(n) = e(n) + \sum_{m=0}^{M-1} \hat{s}_m y(n-m-1) \tag{5}$$

where $S_m(n)$, m $= 0,\ 1, \ldots..., \mathrm{M}-1$ are the coefficients of the *M*th order finite impulse (FIR) filter $\hat{S}(z)$ used to estimate the secondary path $S(z)$.

The weight update equation of existing FB-FxLMS algorithm is given by,

$$W(n+1) = W(n) + \mu x'(n)e(n) \tag{6}$$

In order to improve the existing algorithm the proposed adaptation is being applied in place of existing LMS and the Feedback Filtered x structure is optimized accordingly.

$$W(n+1) = W(n) + \left(\frac{\alpha}{\beta + R}\right) x'(n)e(n) \tag{7}$$

The important thing to underline that the phase error between $S(n)$ and $\hat{S}(n)$ should be less than orthogonal shift following rise of which leads to slowing of convergence. The performance of the above listed as well as new proposed algorithms depends widely upon the nature of physical plant, secondary path impulse response, noise bandwidth and the length of the traversal filter. It is clear from the above that these algorithms malfunction for long impulse response and large noise bandwidth while larger filter length leads to compromise in convergence. In the given paper various results for both types of systems are found out using different filter lengths (Fig. 1).

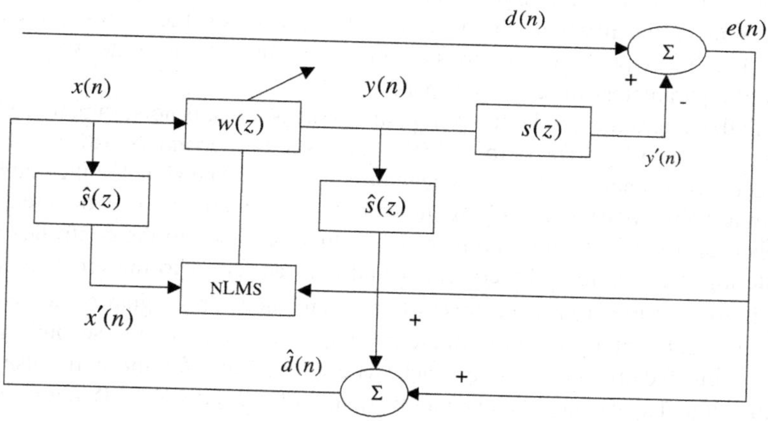

Fig. 1 Block diagram of proposed algorithm for feedback ANC

3 Simulation and Results

In the following section various results using FxLMS, FB-FxLMS and the proposed algorithm for noise cancellation are shown. For the above simulation process 100 Hz fundamental frequency is chosen along with a sampling frequency of 200 Hz. The filter length is specified to 16 along with a 1000 number of iterations being performed.

3.1 *Feedforward ANC*

Figure 2 visualises the application of FxLMS in identifying error while Fig. 3 shows the application of proposed FxNLMS for noise cancellation in feed-forward scenario. From the given set of figures we can see that when traditional algorithm is applied the convergence starts after 250 number of iterations while the value is close to 100 for the new proposed algorithm. Hence the proposed algorithm converges faster than FxLMS algorithm though it involves more computational complexity. The signal power distribution for the proposed system is also provided in Fig. 4.

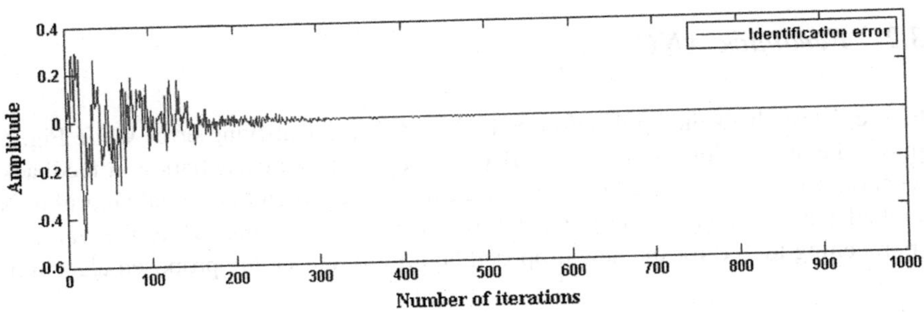

Fig. 2 Convergence of FxLMS

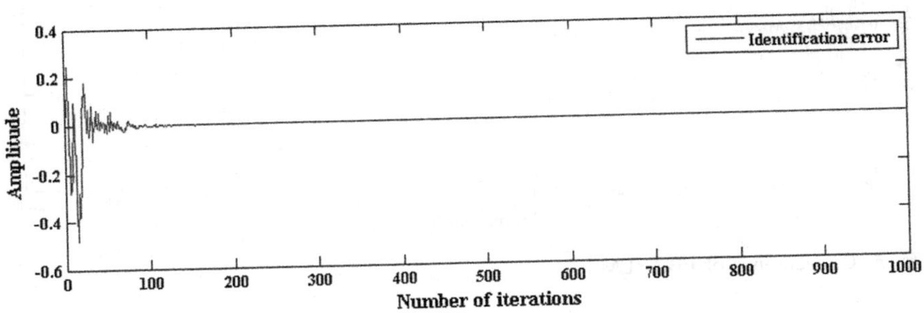

Fig. 3 Convergence of FxNLMS

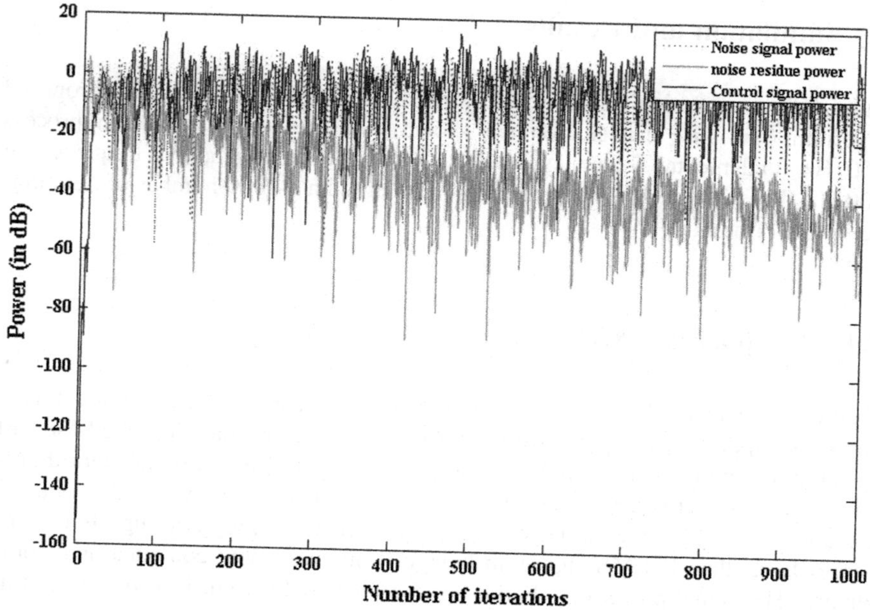

Fig. 4 Signal power distribution of FxNLMS

3.2 *Feedback ANC*

Figure 5 visualises the application of FB-FxLMS in identifying error while Fig. 6 shows the application of proposed FB-FxNLMS for noise cancellation in feedback scenario. From the given set of figures we can see that when traditional algorithm is applied the convergence starts after 150 number of iterations while the value is approaching to 100 for the new proposed algorithm. Hence the proposed algorithm

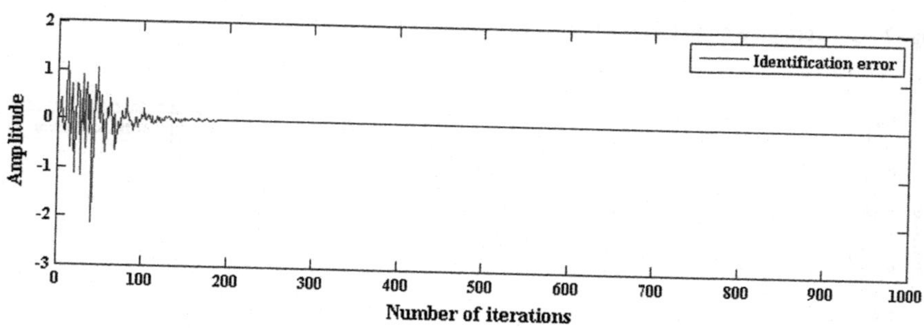

Fig. 5 Convergence of FB-FxLMS

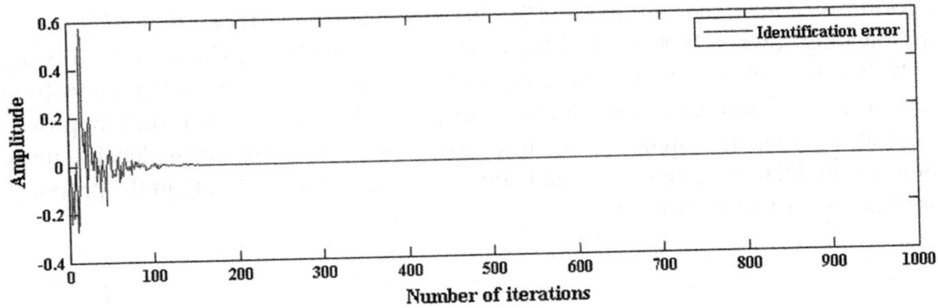

Fig. 6 Convergence of FB-FXNLMS

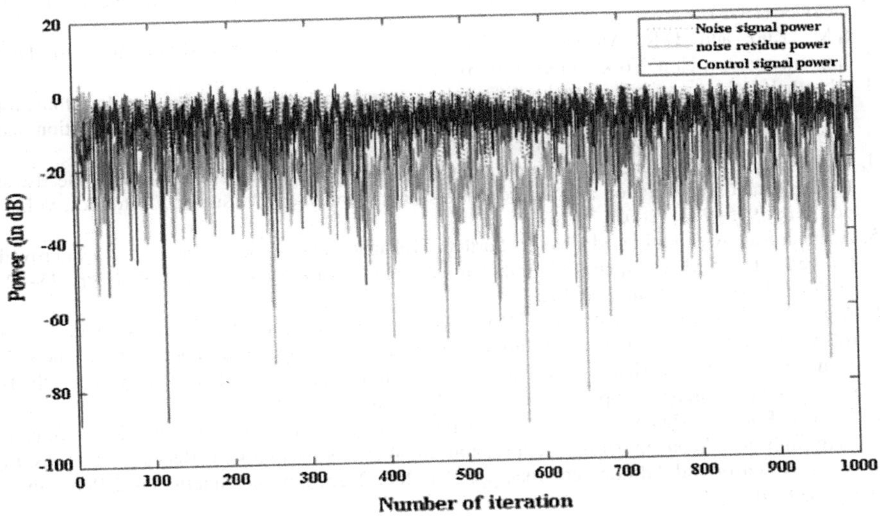

Fig. 7 Signal power distribution of FB-FxNLMS

converges faster than FB-FxLMS algorithm though it involves more computational complexity. The signal power distribution for the proposed system is also provided in Fig. 7.

4 Conclusion

In this paper the proposed algorithms were compared with the existing sets of algorithms like FxLMS and FB-FxLMS in their respective ANC systems. It is concluded from the above simulations that the results are much more acceptable for proposed algorithms as far as convergence is concerned. The proposed set of algorithms converged at about 100 iterations for feed-forward ANC and is approaching towards 100 for feedback ANC respectively. Again the feed-forward ANC involved in the attenuation of random noise at its disposal while the feedback ANC was restricted to narrowband noise respectively. This can be further improved by making requisite standard modifications to the system for wideband noise attenuation. Hence a scope for even better and accurate algorithm than the proposed one can be expected in future.

References

1. S.J. Elliott and P.A. Nelson Active noise control.: IEEE signal processing magazine, pp. 12–35, October (1993).
2. S.M. Kuo and D.R. Morgan.: Active Noise Control Systems-Algorithms and DSP Implementation, New York: Wiley, (1996).
3. Abhishek Deb, Asutosh Kar, Mahesh Chandra.: A technical review on adaptive algorithms for acoustic echo cancellation. IEEE Sponsored International Conference on Communication and Signal Processing, pp. 638–642, Apr. 3–5, (2014).
4. W.S. Gan, S. Mitra and S.M. Kuo.: Adaptive Feedback Active Noise Control Headsets: Implementation, Evaluation and its Extensions. IEEE Trans. on Consumer Electronics, vol. 5, no. 3, pp. 975–982, (2005).
5. Asutosh Kar, Ambika Prasad Chanda, Sarthak Mohapatra and Mahesh Chandra.: An Improved Filtered-x Least Mean Square Algorithm for Acoustic Noise Suppression. vol. 1, pp. 25–32, Springer-Verlag, Germany, (2014).
6. Ambika Prasad Chanda, Abhishek Deb, Asutosh Kar and Mahesh Chandra.: An improved filtered-x least mean square algorithm for cancellation of single-tone and multitone noise", Medical Imaging, m-Health and Emerging Communication Systems, MedCom, (2014) International Conference, pp. 358–363, 7–8 Nov. (2014).
7. Asutosh Kar and Mahesh Chandra, "An Optimized Structure Filtered-x Least Mean Square Algorithm for Acoustic Noise Suppression in Wireless Networks", Recent Advances in Wireless Sensor and Ad-hoc networks, pp. 191–206, 2015, Springer International Publishing, Switzerland.

Application of Internet of Things (IoT) for Smart Process Manufacturing in Indian Packaging Industry

Ravi Ramakrishnan and Loveleen Gaur

Abstract Smart Manufacturing is the need of the hour in India with growing concerns of environmental safety, energy conservation and need for agile and efficient practices to help Indian firms remain competitive against the low cost mass manufacturing and imports. The twelfth five year plan in India (2012–17) (Ministry of Commerce, 2012) has identified low technology intensity, inadequate costs and high transaction costs as major constraints. Smart Manufacturing can help companies gather and consolidate data on near real time basis at each step of their operations to get meaningful insights using existing instrumentation e.g. sensors in valves, motors, pressure and energy meters by connecting them to a digital network where data generated by them is constantly stored for proactive decision making. This paper critically examines the role of Internet of Things in taking flexible packaging manufacturing to the next level in India with relevant citations from a Indian company and alleviating manufacturing pitfalls due to infrastructural issues be it energy, transportation or machine efficiency.

Keywords Internet of Things · Flexible packaging · Smart manufacturing · Process manufacturing · Quality · Customer service · Inventory control · Asset management

1 Introduction

A rapid growth in Indian manufacturing can be through introduction of Internet of Things based sensor technology communicating and sending data over communication networks for further analytics and reports. Process manufacturing refers to using formulas and recipes to generate different outputs (main product, by product

R. Ramakrishnan (✉) · L. Gaur
Amity University, Noida, UP, India
e-mail: ravi.ramakrishnan@gmail.com

L. Gaur
e-mail: lgaur@amity.edu

© Springer India 2016
S.C. Satapathy et al. (eds.), *Information Systems Design and Intelligent Applications*, Advances in Intelligent Systems and Computing 435,
DOI 10.1007/978-81-322-2757-1_34

and wastage) from a basic set of outputs with the help of a process plant. These are specifically used in food and beverages industries, consumer goods packaging, petro chemicals and bio technology units.

Internet of Things (IoT) refers to a network of physical objects namely sensors, actuators and motors connected to a data network. It allows remote connection and control of objects integrating the physical world and computer systems [1–8].

1.1 The Flexible Packaging Process-Manufacturing Overview

The process of generating flexible packaging material involves a series of steps with following specialized plants

- **Biaxially Oriented Polypropylene (BOPP)** products made from low density and high density polypropylene, stretched in both machine directions i.e. length and width and primarily used in packaging products like candy wrappers and chips wrappers, print lamination films, Label Films and Industrial Films e.g. Tape Films
- **Biaxially-oriented polyethylene tetra phthalate (BOPET)** film which is a polyester film made from polyethylene tetra phthalate with high tensile strength, chemical stability and gas/aroma barrier properties used in print laminates/substrates used in food packaging like yogurt and covering over paper, clothes and photographic print

A complete Process in Manufacturing Plant for Flexible packaging is shown in Fig. 1.

1.2 Core Concern Areas in Flexible Packaging Manufacturing

The traditional shop floor in a flexible packaging manufacturing organization consists of a process plant which does a batch processing using inputs and

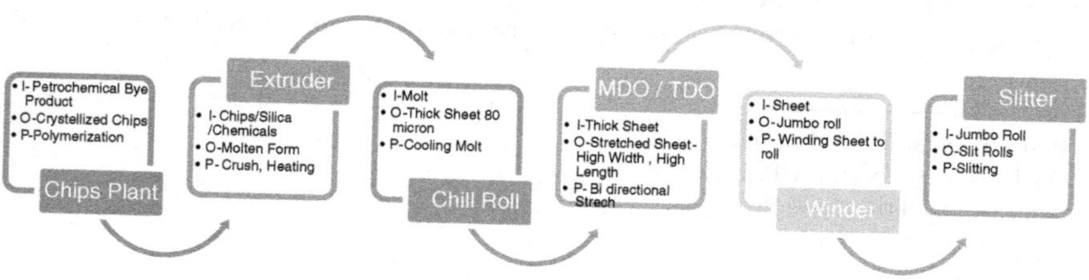

Fig. 1 Process plant—flexible packaging (*I* Input, *O* Output, *P* Process)

generating outputs. The following aspects of business operations becomes an important aspect to track in order to bring efficiency, prevent wastage and ensure compliance in any batch processing:

Inventory Tracking of Materials
In traditional Indian firms the manufacturing cost which includes direct materials cost, direct labor cost and manufacturing overheads accounts for over 75 % of the overall costs. Issues in tracking inventory will lead to plant downtimes for want of material, duplicate inventory, capital lock in and finally in-efficient process chain.

Energy Consumption Tracking
Energy is the main stay for any manufacturing operations with options in India to generate power or energy from state electric boards or captive consumptions. Intelligent energy monitoring and utilization system to ensure that energy demands and capacity of the generation units optimally factored on real time.

Ensuring Machine Efficiency
Higher machine efficiency is the ultimate means for higher unit of productions in mass process manufacturing. This can be achieved by running at optimal loads e.g. processing intake, motor capacity or speed, idling capacity or other parameters. It is possible that a machine set to operate at certain specifications gets altered by human behavior or control malfunctioning hence leading to a higher wear and tear or component break down. Tracking can be done using monitoring sensors which send out alerts using data networks to decision makers whenever a threshold level has been breached. In current scenario, all manufacturing systems which may be manned on unmanned display information only on local consoles not accessible outside the ecosystem of the machine. This has limited utility and also a limited retention life.

Tracking Environmental Variables
Environmental factors play a crucial role in manufacturing and include heat, light, sound, moisture content, corrosive air content like Sulphur or chemicals or gaseous substances which may be an undesirable but unavoidable by-product of the manufacturing process causing equipment failure, damage to inventory stocks and an environmental and safety hazard for workers and visitors. Specific sensors connected to data networks for detection of gas leaks, oil spills, pressure of outlet gas valves, pressure of cooling water and fluids, temperature of cold storages warehouse or normal packing areas, light intensity including direct sunlight, ultraviolet lights and others needs to be continuous monitored to ensure that they are not only minimally compliant but also fit for exports to outside countries with more stringent laws.

Ensuring Health and Safety
Safety detection based on sensors is very necessary to prevent loss of human life or property. While a human safety officer may skip a lot of things, a machine observer will do a round the clock monitoring and ensure safety. Watching over a data-network connected CCTV footage of the shop floor, access control systems,

safety gadget wearable's and early detection alarm systems for fire or structural safety are necessary for ensuring a round the clock safety measure in an operating premise.

Pollution Control
Most of the manufacturing units discharge effluents which are harmful waste products of the process cycle including effluent water and gases. While it is statutory to now process all effluents going outside a plant premise before it is discharged to sewage waste, to ensure that effluent treatment plants are working and waste processing stations are in good condition, firms should put sensors in place to detect water content and also gaseous discharge in different locations.

Asset Tracking and Preventive Maintenance
There are multiple assets which contribute to the manufacturing process like spares for machines including chain belts, spare motors, electrical equipment and electronic boards which need to be accounted and tracked not only for location traceability but also for their useful life and warranty periods.

Delivery Tracking and Customer Service
In-Transit visibility of goods and materials once it leaves factory premises is a concern area due to dependence on humans to capture the tracking information at specific landing points/stations and there is a time lag between data feeding and material movement.

Customer service in today's business environment in a key differentiator for manufacturing in the given quality focused demands for products and prompts after sales service. There is too much emphasis on the human aspect in customer service business process today which leads to inefficiency and delays.

Support Automation
Support automation would provide key benefits to an enterprise by reducing customer costs, reduced wait time, increase call resolution rates and thereby decrease customer negativity towards a product.

Automation today aims at reducing the "human element" in customer correspondences and communications. Automated systems persuade customers to self-diagnose problems based on a knowledge base.

1.3 Internet of Things Concepts for Industry 4.0

With the Advent of technology concepts and integration of Man and Machine, virtual-physical-cyber systems, Internet of Things and Services there will be a modular technology structure where virtual objects will monitor physical process and machines and transmit results over connected data networks on to Cloud Computing infrastructure or decision support systems for triggering alert and monitoring compliance for decentralized decisions.

Industry 4.0 refers to the computerization of manufacturing with human beings taking the role of a secondary observer while machines communicate with each other.

Indian firms to move towards Industry 4.0 will need to adapt the following measures.

- Inter connect between sensors, machines, humans, data networks, output alarms and decision support software's.
- Model for a virtualized smart factory where it will be possible to emulate and simulate using the actual shop floor data fine tune components and observe changes or improvements.
- Machine based decision making where components take their own actions e.g. a pressure valve releasing pressure in response to a meter reading.
- Modularity i.e. adapting legacy manufacturing components to add data networks and integrate with systems outside the scope of a standalone unit.

2 Areas Benefitting from IoT Implementation in Flexible Packaging Manufacturing Plants—Case Study of UFlex Ltd

Traditionally flexible manufacturing organizations like UFlex Ltd (A Global operations company with 6 countries manufacturing locations into Polyester Packaging Manufacturing) have automated systems with local console giving data and information on machine parameters and integrating to a back-office Enterprise Resource Planning software data relating to sales, production, logistics, asset management, energy consumption and inventory was tracked separately and analyzed as part of the management information systems. The manufacturing information system is separate from the digital network.

An Internet of Things implementation would read data from machines and integrate with other machines using sensors for tracking parameters.

Inventory Tracking of materials
With global manufacturing capacities traceability of inventory, is a big challenge in UFlex across huge captive warehouses and consigned inventory warehouses.

Inventory mismatch between enterprise data systems and physical stock happen due to wrong system entries, missed entries and substituted rolls. Inventory losses happen due to misplaced rolls, wrongly consigned rolls, perished rolls, damaged rolls during transit or storage. It is possible to minimize a lot of these issues using Internet of Things and RFID specifically to track inventories and monitor environmental conditions. UFlex has achieved a nearly 80 % accurate tracking with RFID/IoT.

Energy Conservation and Tracking
Energy and power requirements form a major cost component after raw material in the overall cost sheet. Energy consumption accounts for almost 5 % of the overall

cost of production in UFlex. Internet of Things technology can enable the organization to track power consumption and voltage fluctuation with analytics showing fluctuations which may also be a sign for equipment failure.

- Electrical Assets can be monitored including lighting and cooling equipment, circuit breakers and motors using Internet of Things sensors to track voltage and current and measure power drawn against rated capacity. This accounts for almost 20 % of the power consumption.
- Furnace Oil/Gas Based Captive Power Plants need to be operated under some specifications as per operating guidelines to ensure smooth running. Internet of Things sensors help monitor environmental conditions including heating of core components due to overloading and transmit data remotely for analysis. This accounts for another 25 % of the power consumption.
- Cooling/Chiller Plants are a major supporting asset for controlling the process manufacturing plant and Internet of Things can enable preventive maintenance programs by sending out details on meter reading.
- Air Conditioning Units can be upgraded to use Smart Meters to make sure power is utilized optimally and also to keep tab on temperature requirements.
- IT Infrastructure is used extensively be it networks, desktops, enterprise data centers and have their unique requirements for power and standard operating conditions. Using Internet of Things sensors for humidity and temperature in data centers reduces breakdown and malfunction for the Global IT in UFlex.

Machine Efficiency

In a packaging industry like UFlex Ltd there a number of machines which consist of hundreds of electrical and electronic circuits like Motors/Program Logic Units and Mechanical Units like Gears, Chains, Assemblies and Pressure valves. Each of these have their operating specifications and any deviations outside normal range can cause cascading failures and ultimately machine down time. Internet of Things sensors and data monitoring capabilities help engineers monitor data constantly and record any deviation and take preventive measures.

Environmental Variables

Using Internet of Things sensors at UFlex including plant locations and office locations, warehouses and storage rooms, utilities near machine areas or packaging areas it is now possible to identify the following measures and hence take corrective action where required.

- Temperature: Variation causing yeast or fungus contamination.
- Pressure: Causing film distortion or water droplets contamination.
- Light Intensity: Light specifically ultraviolet or infrared lights are harmful to the monomer and polymer molecules in a film and may cause stretching or reduce flexibility of a film roll.
- Effluents: Liquid effluents have harmful products in any process plants such as chemicals, molten melts of plastic. These are generally purified and recycled

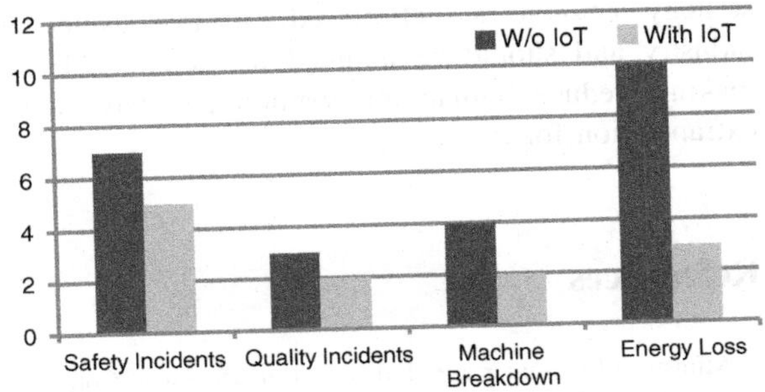

Fig. 2 Results of pilot run of IoT—UFlex Ltd

using a RO plant and a Effluent Treatment Plant. Internet of Things sensors help detect liquid water compositions and chemicals in these effluents.

- Gaseous Products: Gas based effluents, an undesired waste product generated in process manufacturing process including corrosive gases like CO_2 and SO_2 which can further corrode film surface and also harm electronic parts. Internet of Things sensors which can detect such gases when deployed help capture these data and can be used to retune the gas treatment and discharge systems.

Health and Safety

Human health and environmental safety are a major area where Internet of Things can be used to track composition of gases in the surrounding environment, reduce accidents by proximity sensors and Internet of Things enabled cameras, microphones and adequate lighting which can adjust as per requirement and complement natural sunlight in areas such as warehouse which are accident-prone (Fig. 2).

In a mini-Pilot run done in one of the plants for a period of 15 days in a global location for UFlex it was observed that with the implementation of IoT sensors the number of safety incidents went down from 7 to 4 (−43 %), the quality defects reported internally by the labs down from 3 to 2 (−33 %), machine breakdowns from 4 to 2 (−50 %) and energy loss (based on billed units) from 1000 to 300 kWh. This shows that there is a scope for significant improvements and benefits of IoT.

3 Conclusion

In traditional manufacturing like UFlex Ltd earlier almost 90 % of the data was not captured or had limited data life of a few days. Using IoT it is now possible to store data and bring process efficiency, eliminate downtimes, better compliance and traceability. The cost of sensors has gone down and the total cost of implementation of a basic Internet of Things infrastructure can see a very fast return on investment. Internet of Things can help in Preventive maintenance and proactive response, energy conservation, health and safety measures, machine efficiency, higher inventory traceability and customer satisfaction. Overall the Internet of Things strategy seems to be well placed with the business objectives of a manufacturing industry and should be adopted in incremental measures to automate decision making, reduce human intervention and have data visibility for analytics and extrapolation for foresights.

References

1. Ministry of Commerce & Industry, Department of Commerce. (2012). Report of the Working Group on 'Boosting India's Manufacturing Exports' Twelfth Five Year Plan.
2. Indian Council for Research on International Economic Relations (ICRIER). (2014).
3. Dennis Gerson. (2014). "Internet of Things, Smart Clothing, Smart Packaging and The Consumer Experience".
4. Content. (2015). Smart Sensors Improve Packaging Machinery Performance, Retrieved from http://iotevent.eu/big-data/smart-sensors-improve-packaging-machinery-performance/.
5. Ministry of Power, Government of India.(Jun 2015) Sourced from India Smart Grid Knowledge Portal.
6. US Department of Energy. (Jan 2013). Smart Meter Investments Benefit Rural Customers in Three Southern States.
7. Koen Kok., Jan Ringelstein. (Jun 2011)."Field Testing Smart Houses for a Smart Grid" 21st International Conference on Electricity Distribution.
8. Jim Cross key (2015). "Using the Internet of Things to detect asset failures before they occur" Retrieved from blogs by IBM Big Data and Analytics Hub.

Function Optimization Using Robust Simulated Annealing

Hari Mohan Pandey and Ahalya Gajendran

Abstract In today's world, researchers spend more time in fine-tuning of algorithms rather than designing and implementing them. This is very true when developing heuristics and metaheuristics, where the correct choice of values for search parameters has a considerable effect on the performance of the procedure. Determination of optimal parameters is continuous engineering task whose goals are to reduce the production costs and to achieve the desired product quality. In this research, simulated annealing algorithm is applied to solve function optimization. This paper presents the application and use of statistical analysis method Taguchi design method for optimizing the parameters are tuned for the optimum output. The outcomes for various combinations of inputs are analyzed and the best combination is found among them. From all the factors considered during experimentation, the factors and its values which show the significant effect on output are discovered.

Keywords Simulated annealing algorithm · Function optimization · Taguchi method · Robust design

1 Introduction

Optimization of a process is a critical issue during experimentation and important to improve the quality of output at the lowest time and cost [1, 2]. Every process constitutes inputs, its corresponding output, factors and noise affecting the process. The inputs, outputs and environmental noise cannot be controlled by experimenter [3, 4]. Optimization can be achieved by modifying the intermediate factors effecting

H.M. Pandey (✉) · Ahalya Gajendran
Department of Computer Science & Engineering, Amity University, Sector-125, Noida, Uttar Pradesh, India
e-mail: profharimohanpandey@gmail.com

Ahalya Gajendran
e-mail: ahalya.gajendran@gmail.com

© Springer India 2016
S.C. Satapathy et al. (eds.), *Information Systems Design and Intelligent Applications*, Advances in Intelligent Systems and Computing 435,
DOI 10.1007/978-81-322-2757-1_35

output [5]. For example, when NP-hard problems are solved, we do not know whether the obtained solution is optimum. So, different experimental design approaches are utilized to achieve global optimum and find the approach which results in a best solution within least polynomial time [6, 7]. The main objective of an experiment is to determine the most influencing factors and combination values of them for nominal output. This can be achieved by conducting the experiment involving change in values of the factors to notice the corresponding difference in output.

Simulated annealing is efficient for solving optimization problem and results in good solution [8]. The usage of simulated annealing is important because it does not get caught at local maxima which aren't the best solutions. It can be applied on complex problems and even deals with cost functions and random systems [9].

In this paper, the approach used for solving the problems is explained. The experimental design Taguchi design and the case studies considered for applying it like function optimization are discussed. It also gives the basic knowledge about the tool used to apply experimentation methods. These are followed by results of experiments and the observations. The results are analyzed and conclusions are made.

The rest of the paper is organized as: Sect. 2 represent the SA applied for Function optimization. The different types of experimental design methods with the discussion of each method are reported in Sect. 3. The discussion of experiment, parameter tuning and results are drawn in Sect. 4. Lastly, conclusion is of the study in given in Sect. 5.

2 Simulated Annealing for Function Optimization

The simulated annealing (SA) is an efficient method to find a good solution to an optimization problem as it does not stop at local maximum solutions [10, 11]. The temperature is slowed down, starting from an arbitrary high temperature gradually tending to pure greedy as it nears to zero. The algorithm starts with setting to initial winning solution and randomly determining an initial high temperature. The following steps are repeated until temperature is zero [10].

(a) While equilibrium stage is not gained, choose some arbitrary neighbor of present solution.
(b) Find the difference between the lengths of present solution and the neighbor.
   - If the difference is less than or equal to zero set the neighbor as the starting solution and If length of initial solution is less than is less than initial winning solution, set the winning solution to initial solution.
   - If the difference between their lengths is greater than zero, choose a random value between zero and one.

(c) Repeat till equilibrium is attained. Then reduce the initial temperature.
(d) End the algorithm execution when zero temperature is achieved.

The main aim of optimization problem is to minimize or maximize a real function by appropriately selecting input values [12] from an allowed set and function value is computed. The optimization theory generalization and methods to various establishments constitutes a major field in applied mathematics [13]. Generally, predicting apt accessible values of an objective function mentioned a domain which is predefined, also a diverse class of various types of objective functions and various domains. A representation of an optimization problem is given below:

Given: A function f from some set A to the real numbers (f: A → R)

Sought: An element x_0 which belongs to set A such that $f(x_0)$ is greater than or equal to f(x) for every x in set A called maximization or $f(x_0)$ is less than or equal to f(x) for every x in set A called minimization.

A function is said to be optimized when its maximum or minimum value is found by choosing the appropriate input values within the permeable set. The function used here for optimization is De Jong's function, which is generally a sphere function. The related function we considered is the weighted sphere function or hyper ellipsoid function which is given as follows in Eq. (1).

$$f(x) = \sum_{i=1}^{n} i x_i^2 \quad -5.12 \leq x_i \leq 5.12 \tag{1}$$

As SA uses random number generations, various different results are produced. There are many local minima where the algorithm might stop. But the chances of getting trapped at local minima are decreased because of wide search applied by SA.

3 Methods for Experiment Design

There are many traditional approaches for experiment designs were proposed like Build-test-fix and one-factor at a time. Build-test-fix is an intuitive and slow approach where it is difficult to find whether the optimum solution is attained or not and requires re-optimization [14]. In One-factor-at-a-time approach, the main principle involved is experiments are conducted by changing one-factor-at-a-time and keeping others constant [14]. Then, the other factors are varied simultaneously until their best values are achieved but prediction of effects of factors is less accurate and the interaction within the factors cannot be seen.

Experimental design approaches like Design of experiments (DOE) are used during the phase of development and starting levels of production for statistical analysis [15]. DOE utilizes statistical experimental methods to find the best factor and its appropriate level for design optimization or process optimization [2]. DOE constitutes various approaches like Taguchi design, factorial design, fractional factorial design etc. [15].

In full factorial design approach, each possible set is executed at least once. The resources however get exhausted when the factor count increases as it increases the number of runs [4]. For testing 5 factors at 2 levels, a full factorial approach would take 32 runs. The Taguchi method is one among the DOE approaches which is efficient, used to build systems of best quality [5]. The basic principle of Taguchi is to locate the factors and levels and to get the desired set of its values by DOE [15, 16]. To decrease the number of experiments, orthogonal arrays are used to study the parameter space in Taguchi approach. Here, a loss function is derived to find the deviation between the experimental value and expected value. The loss function might be changed into signal to noise ratio (S/N) [1]. Taguchi uses the concept of S/N ratio to calculate the quality characteristics deviating from actual values. In the analysis of S/N ratio, smaller is the best, larger is the best, nominal is the best are the three major groups [1, 4]. The larger S/N ratio shows the better quality irrespective of the category of S/N ratio. The S/N ratio can be determined by the following Eq. (2).

$$\frac{S}{N} = -10 \log\left(\frac{1}{n}\sum_{i=1}^{n} yi^2\right) \tag{2}$$

where, y_i is measured output and n is the number of replication. The effect of each control factor can easily be represented in response graphs. One of the crucial steps of Taguchi design is parameter design which reduces the experimental cost.

4 Simulation Model

The tool used for statistical analysis to find the factors which have significant effect is Minitab that helps in easy design creation, storage, interpretation of results and statistical analysis. The configuration of hardware used for experimentation is Intel Core i5 processor, 4 GB RAM, 500 GB hard disk. To solve constrained and unconstrained function optimization, SA is one of the best methods that can be used. The approach involves the method of heating a material and then gradually decreasing the temperature to reduce defects, in return reducing the system energy [10]. At every iteration, arbitrarily a new point is initiated by SA. The distance between the present point and the new point is based on probability distribution where the temperature is proportional to scale. The points that lower the objective are accepted by the algorithm and also points that elevate the aim which has definite probability. This helps the algorithm to avoid being caught at local minimum and paves a way to explore many possible global solutions [8]. As the algorithm execution continues, to reduce the temperature an annealing schedule is chosen in a systematic way. The simulated annealing algorithm reduces the search converges to

Table 1 Factors and levels selected for the simulation

Factors	Level 1	Level 2
Initial temperature (I_Temp)	50	100
Re-annealing temperature (Re_ann)	50	100
Annealing function (ann_fun)	Fast	Boltzman
Temperature function (Temp_fun)	Exponential	Logarithmic

minimum as there is a reduction in temperature. The factors and their levels taken here are mentioned in Table 1.

We implement the problem varying these four factors at two variety levels. As we need to find the minimum value for the given function, the option chosen is smaller is better. This will show the significant factors which effect the result in finding out least value of the function.

For each given combination the process runs and the results are stored. As it is combinatorial problem, the mean of different results gathered by running the experiment several times. These results are used to analyze the Taguchi design and gain the knowledge about the factors that have significant effect on the output. It is graphically represented which is shown above in Fig. 1a, b. We can learn that the factor annealing function have a significant effect in finding the least possible value for function optimization. The other factors initial temperature, re-annealing temperature and temperature function do not show much effect on the output.

The residual plots in Figs. 2 and 3 represent the normal distribution of the residuals explain that there is no normality between the residuals. This can be seen from the uneven distribution of the residuals around the reference line. This may be because of small samples taken but can vary when large data samples are considered. The assumption taken, stating that the variance is constant all over experimentation is achieved as we see that the residuals are spread along two sides

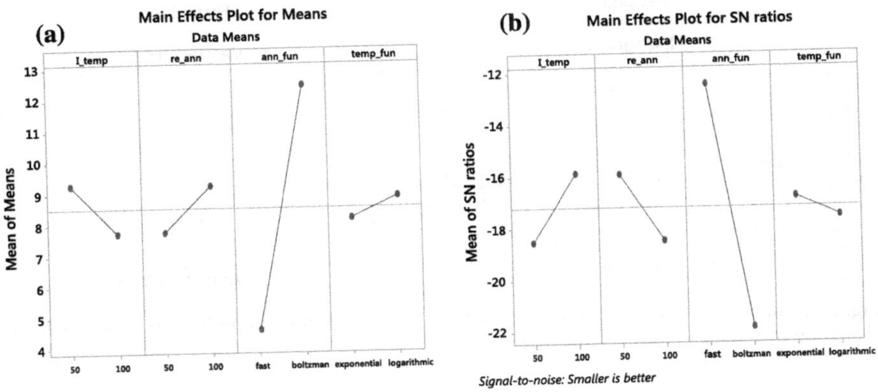

Fig. 1 **a** Main effects plot for means. **b** Main effect plot for SN ratio

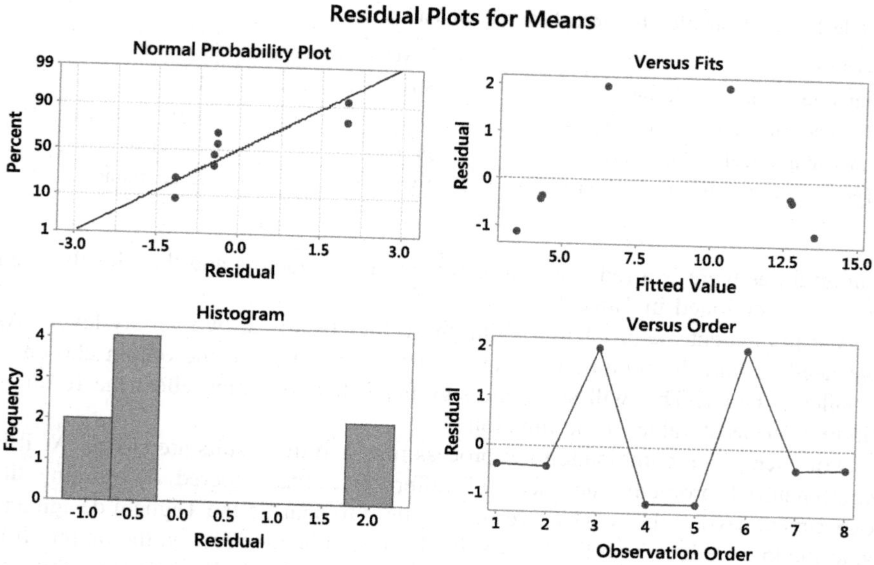

Fig. 2 Residual plot for means graph showing four graph (normal probability plot, versus fit, histogram and versus order) in one plot

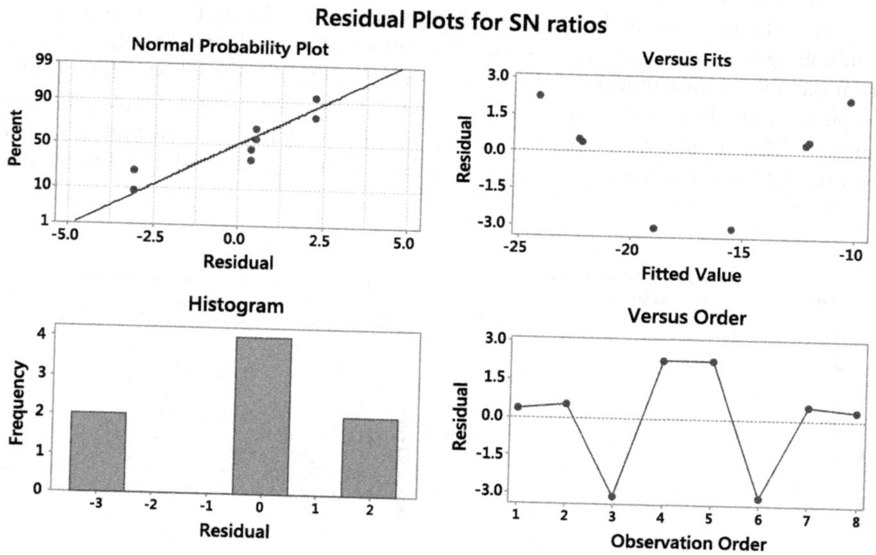

Fig. 3 Residual plot for SN ratio graph showing four graph (normal probability plot, versus fit, histogram and versus order) in one plot

Table 2 Response table for means

Level	I_TEMP	RE_ANN	ANN_FUN	TEMP_FUNC
1	9.282	7.780	4.594	8.174
2	7.750	9.252	12.439	8.859
Delta	1.532	1.472	7.845	0.685
Rank	2	3	1	4

Table 3 Response table for SN ratio

Level	I_TEMP	RE_ANN	ANN_FUN	TEMP_FUNC
1	−18.53	−15.91	−12.48	−16.82
2	−15.85	−18.46	−21.90	−17.56
Delta	2.68	2.55	9.42	0.75
Rank	2	3	1	4

of zero without any recognizable pattern from the residual versus fitted value plot. The histogram plot shows the general physical characteristics of residuals like typical values and shape. There are outliers in this. The residual versus observation order also helps to find the time related effects only when they are randomly distributed along two sides of zero i.e. reference line.

The values of the means and the SN ratios are also stored in the response table given in Tables 2 and 3 respectively. The tables give the information about the values means at two different levels for each factor. The variable delta represents the difference between the means at two different levels. Ranks are given based on delta values; rank is high when delta is high that in turn shows the highest significance. In support of this, the analysis report for means and signal-to-noise ratio is also displayed below.

The values of signal to noise ratios and other particulars for minimizing a function using simulated annealing are given in Table 4.

At the end, the results can be predicted which gives the best association of factor values that results in the optimized output. For this experiment conducted, it gives best solution when initial temperature is 100, re-annealing temperature is 50, annealing function is fast function and temperature function is logarithmic. The mean result for this combination is 3.43375 and signal to noise ratio is −10.2376.

Table 4 Table represents various parameters level combination employed in robust experiment design

I_TEMP	RE_ANN	ANN_FUN	TEMP_FUN	SNRA1	STDE1	MEAN1
50	50	Fast	Exponential	−11.8114	0.091924	3.895
50	50	Boltzman	Logarithmic	−21.8499	0.424264	12.37
50	100	Fast	Logarithmic	−18.5555	1.131371	8.43
50	100	Boltzman	Exponential	−21.893	0.091924	12.435
100	50	Fast	Logarithmic	−7.99189	1.513209	2.27
100	50	Boltzman	Exponential	−21.9973	0.120208	12.585
100	100	Fast	Exponential	−11.5597	0.254558	3.78

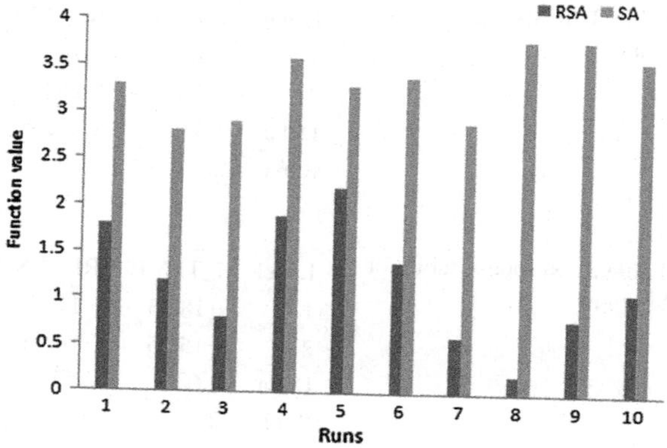

Fig. 4 Comparison chart for function optimization uses RSA (robust simulated annealing) and SA (simple)

A comparative study is done between the performance of simple simulated annealing algorithm and robust simulated annealing which is represented using bar graph in Fig. 4. For the function considered, it is clear from the graph that the function is minimized effectively by Robust Simulated annealing than simple annealing algorithm. Irrespective of the number of runs, the minimized function value produced by RSA is less than SA and there is a huge difference between the results generated by both approaches.

5 Conclusions

The objective of this paper is to analyze the experimental approaches which are used in finding the appropriate parameter values which effect the performance of an algorithm to find optimum results was achieved. A case study on function optimization was solved using SA. SA statistically guarantees in finding an optimal solution due to its gradually cooling process and is easy to code even for complex problems. The cooling Experimental design method called Taguchi is applied on the problem considered using Minitab tool for analyzing the results. The analysis report was delivered in the form of graph plots and tabular forms. The analysis report was thoroughly studied to find the factors which have significant effect on the outcome and the values at which it produces optimum results. It is seen that Taguchi is good to find the optimum result and parameters influencing it. The comparison is done—shows that tuning process improves the performance of an algorithm.

References

1. Bérubé, J and C.F.J. Wu, "Signal-to-noise ratio and related measures in parameter design optimization: an overview." *Sankhyā: The Indian Journal of Statistics, Series B*, 2000, pp. 417–432.
2. H Akbaripour and E Masehian. "Efficient and robust parameter tuning for heuristic algorithms." *Int. J. Ind. Eng* 24, no. 2, 2013, pp. 143–150.
3. Sekulić, Milenko, et al, "Optimization of cutting parameters based on tool-chip interface temperature in turning process using Taguchi's method", *Trends in the Development of Machinery and Associated Technology*, 2011.
4. Hari Singh and Pradeep Kumar, "Optimizing feed force for turned parts through the Taguchi Technique." *Sadhana*, 31, no. 6, 2006, pp. 671–681.
5. K Yew Wong. "Parameter tuning for ant colony optimization: a review." *Computer and Communication Engineering, International Conference on*, IEEE, 2008, pp. 542–545.
6. Adenso-Diaz et al, "Fine-tuning of algorithms using fractional experimental designs and local search," *Operations Research*, no. 6, 2006, pp. 099–114.
7. GS Tewolde et al. "Enhancing performance of PSO with automatic parameter tuning technique." *Swarm Intelligence Symposium*, IEEE, 2009, pp. 67–73.
8. Kirkpatrick, Scottet al. "Optimization by simulated annealing." *science* 220, no. 459, 1983, pp. 671–680.
9. G Ye and X Rui. "An improved simulated annealing and genetic algorithm for TSP." *Broadband Network & Multimedia Technology, Fifth IEEE International Conference*, IEEE, 2013.
10. D Bookstabe, "Simulated Annealing for Traveling Salesman Problem.", 1997.
11. Xu, Qiaoling, et al. "A robust adaptive hybrid genetic simulated annealing algorithm for the global optimization of multimodal functions." *Control and Decision Conference*. IEEE, 2011.
12. Rosen Bruce. "Function optimization based on advanced simulated annealing." *IEEE Workshop on Physics and Computation-PhysComp*, vol. 92, 1992, pp. 289–293.
13. Patalia, T.P., and G.R. Kulkarni. "Behavioral analysis of genetic algorithm for function optimization." *Computational Intelligence and Computing Research, IEEE International Conference*, 2010.
14. P Mach and S Barto, "Comparison of different approaches to manufacturing process optimization." *Design and Technology in Electronic Packaging, 16th International Symposium*, IEEE, 2010.
15. Z Wahid and N Nadir, "Improvement of one factor at a time through design of experiments." *World Applied Sciences Journal* 21, no. 1, 2013, pp. 56–61.
16. LS Shu et al."Tuning the structure and parameters of a neural network using an orthogonal simulated annealing algorithm." *Pervasive Computing, Joint Conferences on*. IEEE, 2009, pp. 789–792.

Analysis and Optimization of Feature Extraction Techniques for Content Based Image Retrieval

Kavita Chauhan and Shanu Sharma

Abstract The requirement of improved image processing methods to index increasing image database that results in an alarming need of content based image retrieval systems, which are search engines for images and also is an indexing technique for large collection of image databases. In this paper, an approach to improve the accuracy of content based image retrieval is proposed that uses the genetic algorithm, a novel and powerful global exploration approach. The classification techniques—Neural Network and Nearest Neighbor have been compared in the absence and presence of Genetic Algorithm. The computational results obtained shows the significant increase in the accuracy by incorporating genetic algorithm for both the classification techniques implemented.

Keywords CBIR systems · Feature extraction · Genetic algorithm · Neural network · Nearest neighbor classification

1 Introduction

The utilization of images for human communication is not recent—the ancestors used to paint their cave walls, and also at the time of pre-Romans the maps and building plans were used to convey information. The invention of various technologies like photography, television and not the least the computer had marked the revolution of image usage in various fields like capturing, processing, transmission, storage and others, thus realizes the importance of creating a collection of all the

Kavita Chauhan (✉) · S. Sharma
Department of Computer Science & Engineering, ASET, Amity University, Noida, Uttar Pradesh, India
e-mail: kavita.chauhanmolife@gmail.com

S. Sharma
e-mail: shanu.sharma16@gmail.com

© Springer India 2016
S.C. Satapathy et al. (eds.), *Information Systems Design and Intelligent Applications*, Advances in Intelligent Systems and Computing 435,
DOI 10.1007/978-81-322-2757-1_36

available images in the electronic form [1]. Huge amount of multimedia data is being produced by various computer technologies related to the image data.

The collection of images known as image database integrates and stores the images. In small databases, a simple search query can identify an image, but in a huge database, the user encounters the problem of image retrieval. To resolve this problem, two techniques known as content-based and text-based were adopted to investigate and retrieval of image from an image database. In content based image retrieval (CBIR), the image content is used for searching and retrieving images and in text based image retrieval (TBIR), the keywords and the subject headings are used for image retrieval. The TBIR system focuses only on the text and keywords which are described by humans and thus limits the range of image retrieval. To overcome the drawbacks of this technique, CBIR systems came into use where the actual image contents, i.e., the visual features like color, texture and shape are used to explore images. Even though CBIR systems have turn out to be an active research field, it is still young. The underlying motivation is to contribute towards this area by recognizing latest problems or issues and also enhancing the retrieval performance and accuracy. For example, in many current applications which are using large image databases, the traditional methods of indexing images have proven to be insufficient. A huge amount of work had been done on this area in the last two decades; however, a lot more is to be done to overcome the recognized challenges. More efforts are required to be made on CBIR as access to appropriate information is the fundamental necessity in the modern society.

The purpose of this work is to improve the accuracy of the image retrieval and compare the results with the existing techniques. The specific objective of this paper includes conducting research on two major parts of the paper, which are (1) improving the accuracy of the CBIR system using the genetic algorithm which helps in minimizing the semantic gap for accurate results and (2) comparing the implemented results of the Neural Network and Nearest Neighbor Classifier in the absence and presence of genetic algorithm.

In this paper, Sect. 1 introduces the brief overview of CBIR systems, its applications, motivation and objective of the paper. Section 2 provides a systematic literature review. Section 3 focuses on the implementation part of the paper describing the various used features and Sect. 4 explains the neural network, nearest neighbor and genetic algorithm along with their used configuration. Section 5 presents the experimental results and Sect. 6 concludes the paper.

2 Background and Related Work

A thorough and systematic literature has been done on various CBIR systems which is summarized in this section.

Agarwal et al. [2] worked on CBIR systems and proposed a feature descriptor which includes the features of both Color Histogram and Co-occurrence of Haar like Wavelet Filter. The technique extracts the properties of images by considering its various perspective and thus improving the CBIR system efficiency. Jeyabharathi et al. [3] studied various techniques of feature extraction and classification and worked on them which include Independent Component Analysis (ICA), Principal Component Analysis (PCA), Linear Discriminant Analysis (LDA), Nearest Neighbor and Support Vector Machine (SVM). Ezekiel et al. [1] developed a multi-scale feature extraction technique for CBIR which was based on Pulse Coupled Neural Network which includes the integration of Contourlet transformation coefficients applied on Rescaled Range (R/S) analysis technique and fast wavelet transformation. Syam et al. [4] considers the color, texture and shape features for feature extraction using GA for Medical Image Retrieval. For efficient image retrieval they uses Squared Euclidean Distance for similarity measurement. Ligade et al. [5] proposed an efficient image retrieval technique using GA in CBIR systems based on multi-feature similarity synthesis. The method includes the relevance feedback in terms of both implicit and explicit feedback. Rashedi et al. [6] worked on Binary Gravitational Search Algorithm (BGSA) for feature extraction for better precision. Thy also compared the working of Binary Particle Swarm Optimization and GA and found BGSA better than the other two algorithms. Pighetti et al. [7] came up with an innovative architecture integrating Support Vector Machine inheriting its property of user evaluation learning and multi-objective IGA with its ability of global optima. Their work shows the efficient improvement as compared with the previous work. Ligade et al. [8] reviews various techniques of feature extraction including Neural Network, Relevance Feedback and Interactive Genetic Algorithm for image retrieval. These techniques were compared with respect to their characteristics and also discussed their achievements, merits and demerits along with their applications.

Having a review, it can be considered that no single method can be resulted as best or very good for all types of images or all the methods uniformly good for a specific image type. So to improve the working of CBIR systems, this paper combines the Genetic Algorithm with the Neural Network and Nearest Neighbor to increase its accuracy and provide better results.

3 CBIR System Implementation

The dataset considered for the image retrieval is Wang dataset [9]. It consists of 1000 images subdivided into 10 categories having 100 images in each category. It is divided into two parts: training and testing dataset. The size of each image is same, which is 256×384 or 384×256 pixels. The image dataset, which is assumed to be the training data, used for the comparison purpose in the form of query image is performed. The query image is picked from the test dataset (Wang dataset) [9, 10]. The objective of passing this image is to retrieve the similar images of the same class to which this query image belongs. Feature extraction is a mean of

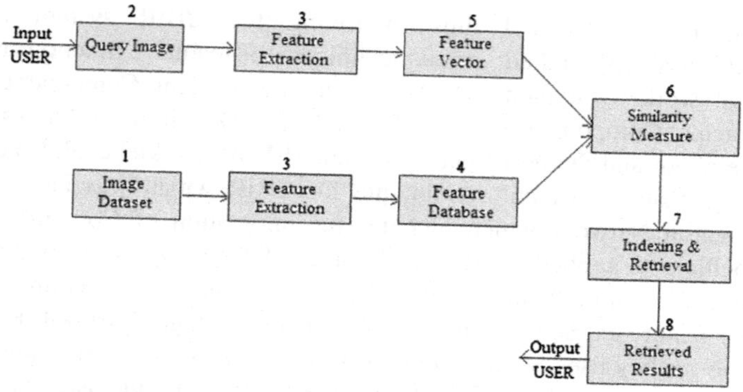

Fig. 1 CBIR system architecture [3]

extracting valuable information from images, used as a signature for the image. The general flowchart of CBIR system is shown in Fig. 1 [3].

The image database containing 900 training images having 407 features stored in a form of a matrix. This matrix is of size 407 × 900. This database is used by the system for comparing the query image features and finding the respective class to which it belongs. The features of the query images are stored in a feature vector which is of the size 1 × 407. This feature vector is used by the system to compare the feature values with the database values.

The similarity measure of the project is the main module where the performance and the accuracy of the system depend. It measures the similarity between the query and the training dataset using either neural network or nearest neighbor classifier. Now from the above stage, class of the query image is given as the output. The images of those classes can be indexed according to their minimum distance and the retrieval process is started. The images from the previous stage are collected and shown to the user as the result of their query image. The outline of the features used in this work is shown in Table 1.

Table 1 Summary of used features

Feature type	Feature name	No. of features
Texture	Gabor features	225
	GLCM features	8
Shape	Hu-7 moment	7
Color	Color moment	6
	Histogram of oriented gradient (HOG)	81
	Edge oriented histogram (EOH)	80
Total	6	407

4 Classification Techniques

Two types of classifiers are used in this project to identify the class of the query image which are: Neural Network and Nearest Neighbor classifier. These two classifiers are implemented in the absence and presence of the Genetic algorithm (GA) and the accuracy of Neural Network and Nearest Neighbor are compared with and without the GA algorithm.

4.1 Neural Network

In the paper, this technique is used as a pattern recognition tool [1]. Neural network is fed in the input layer a matrix of 407 × 900 size which contains all the features of the training dataset. The output layer is given in the form of 407 × 900 matrix consisting of 0 and 1 in specified order. The network returns the class of the query image to which it belongs as the output. Thus the accuracy of the system can be calculated by checking for the correct number of class for the testing image. The accuracy of the Neural Network without using Genetic Algorithm is coming to be 72 %.

4.2 Nearest Neighbor Classifier

It is a classification technique which classifies the element based on their distance from the other elements. The distance between the elements is calculated by using the Euclidean Distance [11]. In this project, the Nearest Neighbor classification is applied on the feature vector of the query image and the feature database of the training images. The distance between them is calculated using the Euclidean distance. Thus input to this technique is the features of the query and the training database, which returns the minimum distance of the query image from the database. Thus by evaluating the distance range, the class of the query image can be obtained. In this way, the accuracy of the system can also be evaluated. The accuracy of the Nearest Neighbor without Genetic Algorithm is calculated to be 34 %.

4.3 Genetic Algorithm

It is one of the heuristic technique which tries to simulate the behavior of biological reproduction and its capability to solve a given problem [1]. This technique is assumed to be robust, computational and stochastic search procedures modelled on the natural genetic system mechanism [12]. General procedure of genetic algorithm is presented in Fig. 2 [4].

- **Generate initial chromosome**

The population size is fixed to be 30 in this experimentation. Thus the number of chromosomes to be generated is 30. The size of one chromosome is 407 as the system contains these many features. The first chromosome contains all 1's. The remaining chromosomes contain random order of 1 and 0 and these are randomly generated. These chromosomes are then passed to the tournament selection for the selection of parent chromosomes.

- **Tournament selection**

In this step two parent chromosomes are selected where randomly two numbers are chosen and the chromosome of the number with less value will be elected as the first parent chromosome. Similarly the second chromosome is selected. These selected parent chromosomes are passed to the net phase for the generation of the offspring.

Fig. 2 Genetic algorithm process [4]

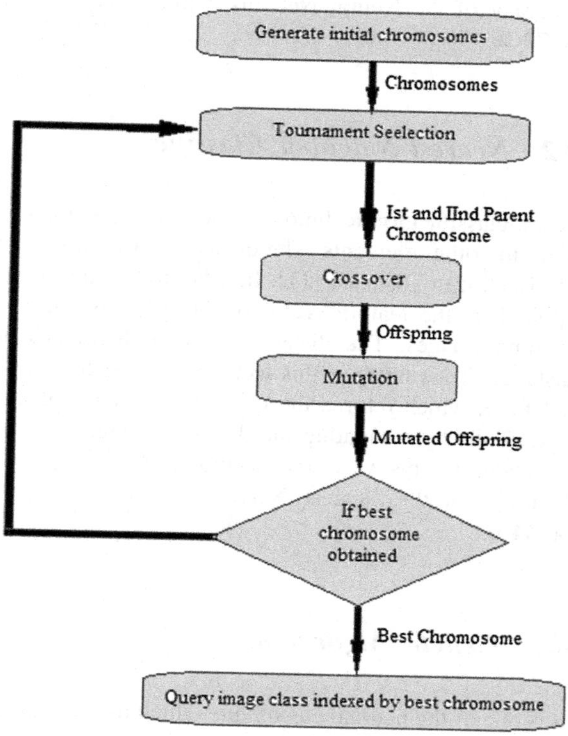

- **Crossover**

In this stage, two randomly crossover sites are chosen and the two point crossover if performed. Here two offspring's are generated. The one which is the best among the two is selected as the offspring for further process and the other is rejected.

- **Mutation**

Mutation process is applied on the offspring. The offspring is tested to provide the required results. If it gives the best result then it is assumed to be the output, but if it's not good then previous results, then again the complete process of selection is repeated till the best chromosome is achieved. Once the output is obtained, the class of the query image can be determined and also the accuracy of the system.

The accuracy of the system using Neural Network with Genetic Algorithm is calculated to be between 85 and 95 % and also, the accuracy of the system using the Nearest Neighbor Classifier with Genetic Algorithm is calculated to be between 70 and 90 %.

5 Results

The results and their analysis are mainly focused on two main program code: feature extraction and class recognition. The feature extraction code calculates the different types of color, texture and shape features. For class recognition, the Neural Network and Nearest Neighbor are used which determines the class to which the query image belongs to. Also comparison of the accuracy of the two techniques is performed in presence and absence of the Genetic Algorithm.

When the accuracy of the Neural Network in presence of Genetic Algorithm was calculated, it has been observed that the accuracy increased from 72 to 91 % as shown in Table 2. The accuracy with Genetic Algorithm is not coming to be exact because of the random generation of chromosomes and also the tournament

Table 2 Table of neural network in the absence of genetic algorithm

Image class	Images retrieved	Relevant images	Accuracy
Class 1	11	6	54.5
Class 2	12	8	66.2
Class 3	6	2	33.3
Class 4	7	6	85.7
Class 5	12	10	83.3
Class 6	10	6	60
Class 7	10	10	100
Class 8	12	10	83.3
Class 9	10	6	60
Class 10	10	8	80

Table 3 Table of nearest neighbor in the absence of genetic algorithm

Image class	Images retrieved	Relevant images	Accuracy
Class 1	13	2	15.3
Class 2	8	2	25
Class 3	11	2	18
Class 4	5	1	20
Class 5	12	10	83.3
Class 6	12	5	41.6
Class 7	5	2	40
Class 8	14	6	42.8
Class 9	12	3	25
Class 10	8	2	25

selection which selects the parent chromosomes randomly from the initial population.

Now, when the accuracy of the system was calculated for Nearest Neighbor in presence of Genetic Algorithm, it had been resulted to 87 % as shown in Table 3. Thus the accuracy increased from 70 to 90 %.

6 Conclusion

In this work, we have designed and implemented a CBIR system that evaluates the comparison of query image with each of the images of image dataset in terms of color, texture and shape characteristics and returns the query image class for determining the system accuracy. For texture analysis Gabor and GLCM features are extracted, for shape analysis, Hu-7 Moment feature is extracted and lastly for color analysis, color moment, Histogram of Oriented Gradient (HOG) and Edge Oriented Histogram (EOH) are used. In the absence of Genetic Algorithm, Neural Network provides the accuracy of 72 % and Nearest Neighbor Classifier provides the accuracy of 34 %. In the presence of Genetic Algorithm, Neural Network provides the accuracy in the range of 85–95 % and the Nearest Neighbor Classifier provides the accuracy in the range of 70–90 %.

References

1. Ezekiel, S., Alford, M.G., Ferris, D., Jones, E., Bubalo, A., Gorniak, M., Blasch, E.: Multi-Scale Decomposition Tool for Content Based Image Retrieval. In: Applied Imagery Pattern Recognition Workshop (AIPR): Sensing of Control and Augmentation, IEEE, pp. 1–5. (2013).

2. Agarwal, M.: Integrated Features of Haar-like Wavelet Filters. In: 7th International Conference on Contemporary Computing (IC3), IEEE, pp. 370–375. (2014).
3. Jeyabharathi, D., Suruliandi, A.: Performance Analysis of Feature Extraction and Classification Techniques in CBIR. In: International Conference on Circuits, Power and Computing Technologies (ICCPCT), IEEE, pp. 1211–1214. (2013).
4. Syam, B., Victor, J.S.R., Rao, Y.S.: Efficient Similarity Measure via Genetic Algorithm for Content Based Medical Image Retrieval with Extensive Features. In: International Multi-Conference on Automation, Computing, Communication, Control and Compressed Sensing (iMac4 s), IEEE, pp. 704–711. (2013).
5. Ligade, A.N., Patil, M.R.: Optimized Content Based Image Retrieval Using Genetic Algorithm with Relevance Feedback Technique. In: International Journal of Computer Science Engineering and Information Technology Research (IJCSEITR), pp. 49–54, TJPRC, (2013).
6. Rashedi, E., Nezamabadi-pour, H.: Improving the Precision of CBIR Systems by Feature Selection Using Binary Gravitational Search Algorithm. In: 16th CSI International Symposium on Artificial Intelligence and Signal Processing (AISP), IEEE, pp. 039–042, (2012).
7. Pighetti, R., Pallez, D., Precioso, F.: Hybrid Content Based Image Retrieval combining Multi-objective Interactive Genetic Algorithm and SVM. In: 21st International Conference on Pattern Recognition (ICPR), pp. 2849–2852, (2012).
8. Ligade, A.N., Patil, M.R.: Content Based Image Retrieval Using Interactive Genetic Algorithm with Relevance Feedback Technique—Survey. In: International Journal of Computer Science and Information Technologies, IJCSIT, pp. 5610–5613, (2014).
9. [Online] “Wang Database”, (2015) Available: http://wang.ist.psu.edu/docs/related/
10. Jia Li, James Z. Wang: Automatic linguistic indexing of pictures by a statistical modeling approach. In: IEEE Transactions on Pattern Analysis and Machine Intelligence, Vol.25 No.9. pp. 1075–1088. (2003).
11. Giorgio, G.: A Nearest Neighbor Approach to Relevance Feedback in Content Based Image Retrieval. In: 6th ACM International Conference on Image and Video Retrieval, CIVR’07, pp. 456–563, (2007).
12. Gali, R., Dewal, M.L., Anand, R.S.: Genetic Algorithm for Content Based Image Retrieval. In: 4th International Conference on Computational Intelligence, Communication Systems and Networks (CICSyN), IEEE, pp. 243–247, (2012).

Parametric Curve Based Human Gait Recognition

Parul Arora, Smriti Srivastava, Rachit Jain and Prerna Tomar

Abstract In this paper we institute a baseline technique for human identification based on their body structure and gait. This paper presents a unique human identification system based on self-extracted gait biometric features. Recurring gait analysis is done to deduce key frames from the gait sequence. The gait features extracted are height, hip, neck and knee trajectories of the human silhouette from the body structure Here, we propose two new parametric curves Beizer curve and hermite curve, based on gait pattern. The projected approach has been applied on the SOTON covariate database, which comprises eleven subjects. The testing samples are compared to training samples using normalized correlation, and subject classification is performed by nearest neighbor matching among correlation scores. From the conducted experimental results, it can be accomplished that the stated approach is successful in human identification.

Keywords Gait · Model based · Beizer curve · Hermite curve · Nearest neighbor

1 Introduction

Most image-based human identification techniques use proximity-based analysis such as fingerprint sensing or iris biometric systems. Human gait is the pattern of movement of the limbs constituting the locomotion as an individual. Many studies

P. Arora (✉) · S. Srivastava · R. Jain · P. Tomar
Netaji Subhas Institute of Technology, New Delhi, India
e-mail: parul.narula@gmail.com

S. Srivastava
e-mail: smriti.nsit@gmail.com

R. Jain
e-mail: jain.rachit11@gmail.com

P. Tomar
e-mail: prernatomar23@gmail.com

© Springer India 2016
S.C. Satapathy et al. (eds.), *Information Systems Design and Intelligent Applications*, Advances in Intelligent Systems and Computing 435,
DOI 10.1007/978-81-322-2757-1_37

have now shown that it is possible to recognize people by the way they walk. Gait is one such feature of human entity which eliminates the need of close-kinship for exact results [1]. Hence, gait analysis is a better option when human identification is of higher concern than individual identification. Gait analysis has undergone repetitive exploration after regular intervals due to ever-increasing input of people, thereby increasing the need for security and surveillance [2–4]. This gives the research a new domain.

Gait recognition has been classified into two types depending on silhouette consideration—model based and model free [5]. Model based methods capture the structure or shape of the human body in motion. This approach generally models the human body structure or motion and extracts the features to match them to the model components. It incorporates knowledge of the human shape and dynamics of human gait into an extraction process. This approach helps to derive dynamic gait features directly from model parameters. It is not affected from background noise as well as change in viewpoint and cloth variation [6]. However, it includes lots of mathematical relations and variables, which results in a complex model. So due to the complexity involved in analysis of model based approaches, model free approaches are more widely used. Model free approach does not need any pre-defined model to satiate structure; it only works on binary silhouettes [7, 8].

Previously used models [9] include stick and volumetric models, which are most commonly used- a model consisting of six segments comprising of two arms, two legs, torso and head represented the human body structure in the silhouette by a stick figure model which had ten sticks articulated with six joints.

Because of the higher advantages and better results, we have combined model free and model based approach to extract the subsequent gait features. Model free features include adaptive height, neck joint, hip and knee joints. Model based features include area under the legs during the gait cycle and Beizer and hermite implementation of the extracted joints.

Nearest neighbor classifier is used to classify the subjects based on these evaluated features.

The paper is organized as follows. Section 2 describes the preprocessing of gait images. Section 3 proposes the model based and model free feature extraction process. Section 4 is devoted to evaluations and Sect. 5 gives the conclusion.

2 Data Preprocessing

In this paper, we have used the SOTON small dataset. It consists of gait silhouette sequences of 11 persons from side view with speed, shoulder bag (accessory) and clothing variations respectively. Every frame has been selectively used such that the start of the gait cycle for each individual is made similar. The complete binary image shown in Fig. 1 has been used without the need for defining any region of interest or image cropping [10].

Fig. 1 Sample picture of Soton data

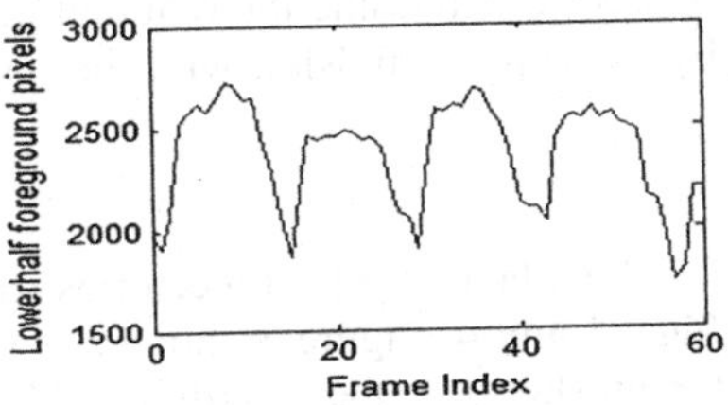

Fig. 2 Variations of foreground pixels over time

2.1 *Gait Period Extraction*

The gait period can be defined as the number of images in one gait cycle. Due to the speed variations for each sequence the number of frames per gait cycle may not be same. This is one of the bottlenecks of our study, where neither the start of the sequence of the cycle nor the number of frames for each is same. The gait period can be calculated by white pixels in the silhouette image. Since the lower part of the body is more dynamic than the upper part, we use lower half of the silhouette image in order to find the gait period. When both legs are far apart, the number of pixels would be more as compared to when legs overlap each other. Thus, by counting the number of pixels in every frame, we get a periodic waveform, which is used to calculate gait period as shown in Fig. 2.

3 Gait Feature Extraction

3.1 *Model Free Features*

Gait recognition can either use the individual frames or several frames over a gait cycle. We have aimed at the individual frame approach, even though it increases complexity, but it enhances accuracy rates by computing the variation in heights for the person with each step taken during the entire gait cycle.

A normal human body is a framework where the set ratios with respect to height define the approximate location of various other body joints like neck, shoulder, waist, elbows, pelvis, knees, ankles, etc. These have a proven scientific backing and are represented for an average human body. Since our derived heights are adaptive, i.e. continuously varying for each person from frame to frame as he walks, so are our various other locations of the body joints.

This completes the vertical detection of frames. For horizontal detection, a bidirectional scanning consisting of calculating the white pixel intensity is used. This gives an accurate location of the desired point. Thus the first pixel (x_s) and last pixel (x_l) on the scan line give the initial and final point. The total number of pixels is used to determine the required thickness and hence the exact point location [11] by subsequent division with the integer 2 or 4 as shown in Eq. 1.

$$x_{center} = x_s + (x_l - x_s)/2 \tag{1}$$

The adaptive joint trajectories for each person in a gait sequence are plotted in Figs. 3 and 4. Figure 3 shows the variation of height feature for different persons. Graph shows that each individual has different way of walking. Figure 4 shows the intra-similarity between different samples of a single person.

3.2 Model Based Features

Beizer Curve We use a cubic Bezier curve model described by four control points to give a vivid picture of motion statistics in a gait cycle. The control points are supposed to lie on the centroid of the knee outer contour and the hip shown in Fig. 5. The equation defining a parametric cubic curve [12] is shown in Eq. 2.

$$P(r) = (1 - r)^3 P_0 + 3r(1 - r)^2 P_1 + 3(1 - r)r^2 P_2 + r^3 P_3 \tag{2}$$

P(r) is any point on the curve described by control points P_0, P_1, P_2, P_3 which denote the centroids of the silhouette (hip), and left and right knee respectively in a

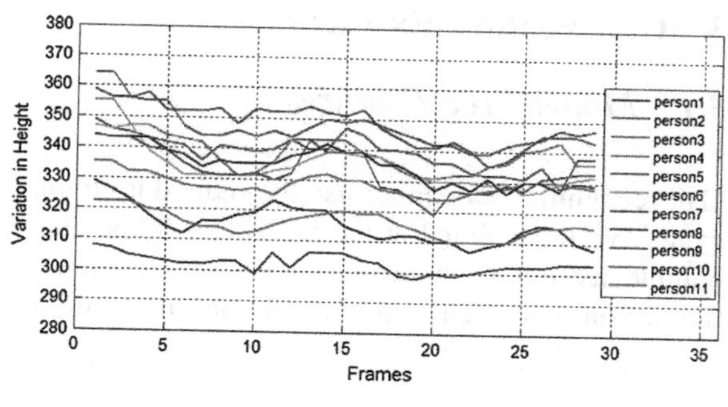

Fig. 3 Variation of heights with frame number of 11 subjects

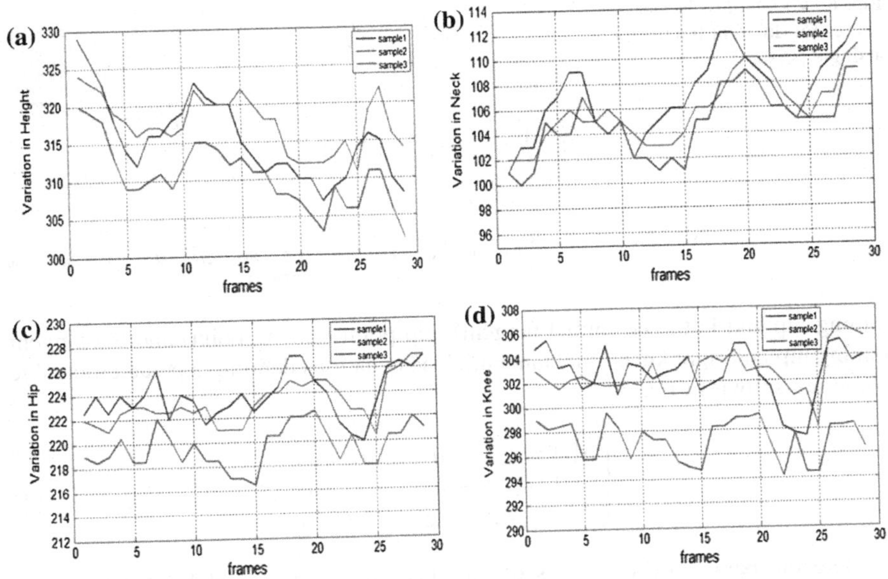

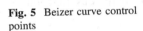

Fig. 4 Plot of the height, neck, hip and knee location variation with respect to the frame number for the three samples of same person

Fig. 5 Beizer curve control points

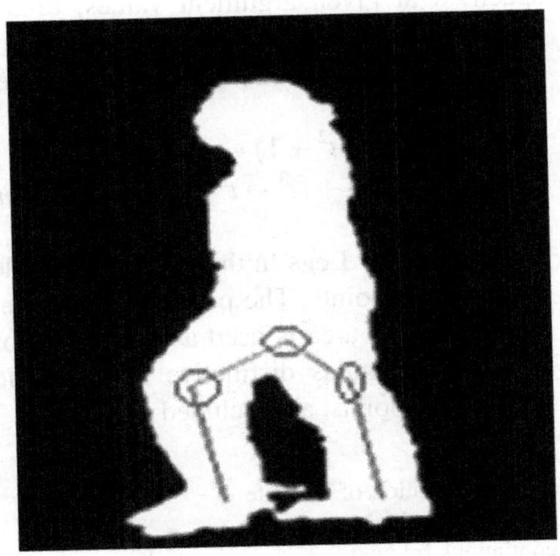

gait frame. This method corresponds to the first model based features. Figure 6 shows the variation of Beizer curves in continuous frames for a single person.

Hermite Curve We have used a cubic Hermite curve model described by four control points to give a vivid picture of motion statistics in a gait cycle. For a cubic curve, the control points are supposed to be hip point (starting point), tangent

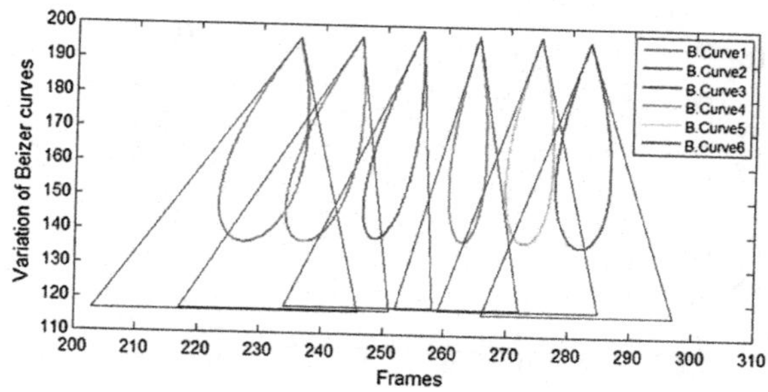

Fig. 6 Variation of Beizer curves for different frames for a single person

joining hip and left knee (initial tangent), hip point (ending point) and the tangent joining hip and right knee (final tangent). The equations defining a parametric cubic curve are shown in Eq. 3.

$$\begin{cases} x(t) = a_3t^3 + a_2t^2 + a_1t + a_0 \\ y(t) = b_3t^3 + b_2t^2 + b_1t + b_0 \\ z(t) = c_3t^3 + c_2t^2 + c_1t + c_0 \end{cases} \quad (3)$$

Here the parametric equations of the curve is given separately as $x(t)$, $y(t)$, $z(t)$ and 200 points are taken to form the curve. This method corresponds to the second model based features

Cubic Hermite splines are typically used for interpolation of numeric data specified at given argument values, to obtain a smooth continuous function as shown in Fig. 7 for continuous frames.

The resultant equation of Hermite curve is:

$$x(t) = (2t^3 - 3t^2 + 1) * xhip - (2t^3 - 3t^2) * xhip + (t^3 - 2t^2 + t) * (xleftknee - xhip) + (t^3 - t^2) * (xhip - xrightknee) \quad (4)$$

Area under Legs In the area under the legs, the hip and the two knees are taken as the major points. The points are joined and then the area is calculated as shown in Fig. 8. This area is used as a unique tool to determine the way a person walks. The walk contains distinctive features and these features, with the help of the determined points, are included in the area. Thus, the continuously changing area in

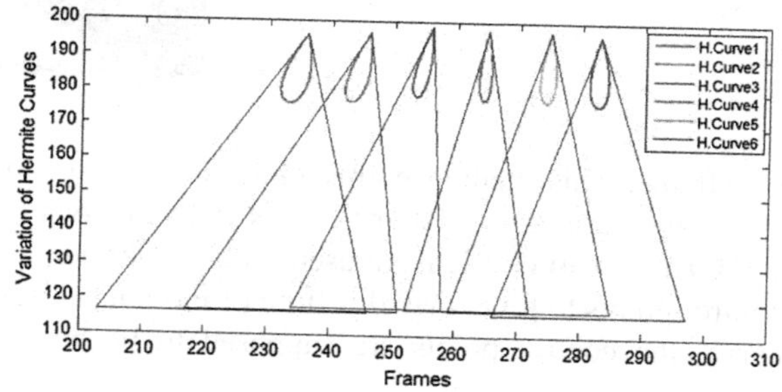

Fig. 7 Variation of Hermite curves for a single person in continuous frames

Fig. 8 Area under the legs

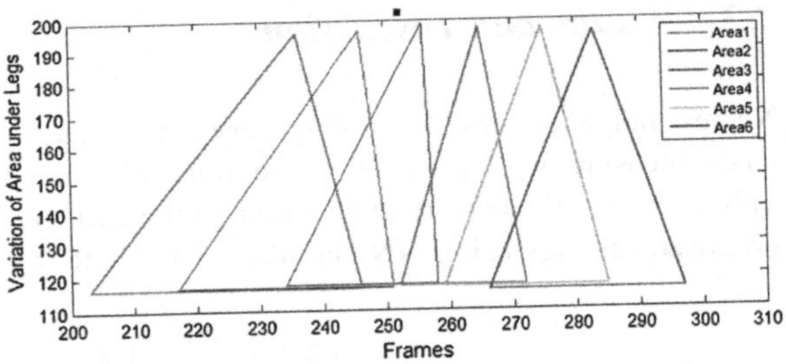

Fig. 9 Variation of area curves for different frames for a single person

every frame during the gait cycle is considered to be a distinctive feature of an individual as shown in Fig. 9. This method corresponds to the third model based features.

4 Experimental Results

4.1 Dataset

All the study is done on SOTON small dataset [13] provided by Southampton academy. This database consists of 11 persons with variation in carrying and clothing conditions. We considered only the normal gait sequences, which has 3 samples. We used 2 samples for training and 1 sample for testing.

Table 1 Recognition results from proposed features

Feature	Correct classification rate (%)
Height	81.82
Neck	81.82
Hip	63.64
Knee	72.73
Beizer	63.64
Hermite	72.73
Area under the leg	72.73
Average	72.73

Table 2 Comparative results

Features	Dynamically coupled oscillator [14]	PCA on spatial silhouette [15]	Hierarchical [16]	Proposed method
Recognition results (%)	55	70	64	72.73

4.2 Results and Discussion

Nearest neighbor classifier, being very simple uses the normalized Euclidean distance measure to evaluate the similarity between training and testing samples. Table 1 shows the performance based on the correct classification rate (CCR) of the proposed features using NN classifier. The CCR is calculated as follows:

$$CCR = \frac{N_c}{N} \times 100\,(\%) \tag{5}$$

where N_c is the total number of samples, which are correctly recognized. N represents total number of gait samples.

From Table 1, it is observed that static features as well as dynamic features contributed to average recognition of 72.73 %, which is quite good as compared to the previous techniques. Comparison of proposed features with existing features [14–16] is done in Table 2.

Comparison with the previous technique shows that the proposed features perform far better.

5 Conclusions

We have described a new method for extracting the gait signatures for analyzing and identifying the gait motion, guided by known anatomy. An adaptive method for calculating the heights of each frame is considered. Subsequently, all other power points are obtained along with the joint trajectories. The resultant body power points are used to create the Beizer model and Hermite model for gait analysis. The gait signature is extracted from the human body contour by determining the scan position using known body topology. A comparative study of model based, used for the extraction of a feature, viz., the area traversed between the lower limbs in a frame of the silhouette has been done. This area feature captures the dynamics of the structural characteristics over the gait cycles of different individuals thus possessing the discrimination power.

References

1. Dawson, M.R.: Gait Recognition. Final Thesis Report, Department of Computing, Imperial College of Science, Technology & Medicine, London (2002).
2. Perry, J., Davids, J.R.: Gait analysis: normal and pathological function. Journal of Pediatric Orthopaedics, 12(6), 815 (1992).
3. Wang, L., Hu, W., & Tan, T.: Recent developments in human motion analysis. Pattern recognition, 36(3), 585–601 (2003).
4. Guo, Y., Xu, G., Tsuji, S.: Understanding human motion patterns. In: 12th IAPR International Conference on Pattern Recognition, pp.325–329. IEEE Press, Jerusalem (1994).
5. Murray, M.P., Drought, A.B., Kory, R.C.: Walking Patterns of Normal Men. The Journal of Bone & Joint Surgery, 46(2), 335–360 (1964).
6. Attwells, R.L., Birrell, S.A., Hooper, R.H., Mansfield, N.J.: Influence of Carrying Heavy Loads on Soldiers' Posture, Movements and Gait. Ergonomics, *49*(14), 1527–1537(2006).
7. Arora, P., Srivastava, S.: Gait Recognition Using Gait Gaussian Image. In: 2nd International Conference on Signal Processing and Integrated Networks, pp. 915–918. IEEE Press, India (2015).
8. Arora, P., Hanmandlu, M., Srivastava, S.: Gait based authentication using gait information image features. Pattern Recognition Letters (2015).
9. Boyd, J.E., Little, J.J.: Biometric gait recognition. In: Advanced Studies in Biometrics, LNCS, vol. 3161, pp. 19–42, Springer, Heidelberg (2005).
10. Bhanu, B., Han, J.: Human recognition on combining kinematic and stationary features. In: Audio-and video-based biometric person authentication, pp. 600–608, Springer, Heidelberg (2003).
11. Yoo, J.H., Nixon, M.S., Harris, C.J.: Extracting Human Gait Signatures by Body Segment Properties. In: 5th IEEE Southwest Symposium on Image Analysis and Interpretation, pp. 35–39. IEEE Press, Sante Fe, New Mexico (2002).
12. Kochhar, A., Gupta, D., Hanmandlu, M., Vasikarla, S.: Silhouette based gait recognition based on the area features using both model free and model based approaches. In: IEEE International Conference on Technologies for Homeland Security, pp. 547–551. IEEE Press, Massachusetts (2013).
13. Shutler, J. D., Grant, M. G., Nixon, M. S., Carter, J. N.: On a large sequence-based human gait database. In: Applications and Science in Soft Computing, LNCS, Vol.24, pp.339–346, Springer Berlin Heidelberg (2004).
14. Yam, C., Nixon, M. S., Carter, J. N.: Gait recognition by walking and running: a model-based approach. In: 5th Asian Conference on Computer Vision, pp. 1–6, Australia (2002).
15. Wang, L., Tan, T., Ning, H. Hu, W.: Silhoutte analysis based gait recognition for human identification, IEEE Trans Pattern Analysis and Machine Intelligence (PAMI), 25(12), 1505–1518 (2003).
16. Wagg, D.K., Nixon, M.S.: On automated model-based extraction and analysis of gait. In: 6th IEEE International Conference on Automatic Face and Gesture Recognition. pp. 11–16, IEEE Press (2004).

An Energy Efficient Proposed Framework for Time Synchronization Problem of Wireless Sensor Network

Divya Upadhyay and P. Banerjee

Abstract Now a day's Wireless Sensor Network are used in various applications where full or partial time synchronization plays a vital role. Basic aim of Time synchronization is to achieve equalization of local time between all the nodes within the network. This paper proposes a new framework for time synchronization problem of wireless sensor network. Time synchronization protocols are very popular and widely used in WSN now days. An analysis has been performed utilizing the proposed frame work, which leads to a conclusion that it consumes less energy than the traditional time synchronization protocols: Reference Broadcast Time Synchronization and Time-Sync Protocol for Sensor Network. It has been observed that the proposed frame work do not require a Global Positioning System or any other external system to coordinate with time, as a typical Network Time Protocol for wireless sensor system uses. The proposed time synchronization protocol is categorized as peer to peer, clock-correcting sender- receiver network-wide synchronization protocol depending upon the characteristics of WSN. The maximum probability theory is used in order to analyze the clock offset. It was observed that resynchronization interval is required to achieve a specific level of synchronization correctness. Results are obtained by simulating the WSN on NS2 to demonstrate the energy efficient feature of the proposed protocol.

Keywords Clock offset · Time synchronization · Wireless sensor network

Divya Upadhyay (✉)
Computer Science & Engineering, Amity School of Engineering & Technology, Amity University, Noida, India
e-mail: dupadhyay@amity.edu

P. Banerjee
Electronics & Communication Engineering, Amity School of Engineering & Technology, Amity University, Noida, India
e-mail: pbanerjee@amity.edu

© Springer India 2016
S.C. Satapathy et al. (eds.), *Information Systems Design and Intelligent Applications*, Advances in Intelligent Systems and Computing 435,
DOI 10.1007/978-81-322-2757-1_38

1 Introduction

The advancement in research & technology have made possible the development of very small, low energy operated devices competent in performing intelligent sensing and communication operations. These technologies are called Sensors and they emerged the wireless sensor networks and therefore very popular among several researcher. These are a special category of ad hoc networking, where small wireless sensor or nodes are connected together to form an infrastructureless network. These nodes communicate with each other by forwarding information packets for delivery from one source to another destination.

Similar to other distributed systems, a sensor network is also highly depends on time synchronization. The definition of Time synchronization with respect to computer network is providing a common clock time-scale between local clocks of the sensor nodes within the network [1]. Due to difference in manufacturing of hardware clocks, they are imperfect. Because of this local clock of sensor nodes might drift away with each other nodes in time, therefore the observation time or intervals of time duration may vary for each and every sensor node of the wireless network. Hence the problem of time synchronization for the local clocks of wireless sensor nodes in the network is an important and popular topic of research over the last two decades [2]. Still there is a lot of scope and opportunity available as there is no precise scheme for synchronization with high degree of accuracy and scalability is achieved. The complex and collaborative character of sensor nodes causes a trivial problem for itself. Most of the wireless sensor network applications depend upon the time to be coordinated inside the network itself. Some of these applications include data fusion, Environment monitoring, movable target tracking system, radio scheduling [3].

The work presented in this paper is a new proposed protocol for time synchronization using maximum probability theory for WSN and is simulated and analyzed using NS2.

The rest of the paper is structured as follows: In Sect. 2 the working and operation of sensor clocks is described along with the time synchronization problem on networks. Section 3 presents related work and motivations for studying problem of time synchronization in wireless sensor networks. Section 4 is devoted towards the details of the proposed time synchronization protocols. In Sect. 5 discusses and analyzes the results of proposed synchronization methods for sensor networks along with its comparison with some already implemented protocols. Last section i.e. Section 6 concludes the paper with some remarks and future research directions.

2 Sensor Clock and Time Synchronization Problem

Each and every sensor of the network possesses its own clock. The counter inside the sensor clock is incremented in accordance with the edges of the periodic output signal of the local oscillator [4]. As soon as the counter approaches the defined threshold value, the interrupt is generated and stored in the memory. The frequency of the oscillator and the threshold value determine the resolution of the clock [4]. The ideal clock of the sensor node must be configured as [4]

$$C(t) = t \tag{1}$$

where t is the ideal reference time.

On the other hand, because of the imperfections of the clock's oscillator, the clock function of the ith sensor node is modeled as [4]

$$Ci(t) = \varphi + \omega t + \varepsilon \tag{2}$$

The factors φ is the clock offset and ω is called the clock skew and ε is for random noise.

The issue of time synchronization for WSNs is a difficult and complicated problem because of the following reasons:

1. Every oscillator of sensor clock has its different clock factors despite of its kind.
2. As it is known that clock skew ω is a time dependent random variable (TDRV). The factors that affect the TDRV of clock parameters are short-term and long-term stabilities. These instabilities are primarily because of environmental factors, such as high temperature difference or instabilities resulting due to subtle effects, like oscillator aging, respectively [5]. Therefore, in a generalized way, the relative clock offset factor keeps on changing with time. Hence, the wireless sensor networks have to execute periodically time-resynchronization in order to adjust the clock factors/parameters [6].

3 Related Work

In last few decades, lot of work has been done in designing many protocols for maintaining time synchronization between physical clock and sensor clock. They all are simple connectionless protocol for messaging and exchange information regarding clock between different nodes or between clients and servers. All these protocol consist of an algorithm within the nodes for the correction and updating the local clocks depending upon the information received from different sources [7]. These protocols differ in certain respect: either the network can be kept consistent internally or it can be synchronized with some external standard; whether the server has to be considered for a canonical clock, or an arbiter clock of client, and so on.

Based on the clock synchronization state space model [8] Wang Ting et al. discussed in his paper a model for synchronizing error variance and its design issues. For his analysis the sensor nodes are used to exchange information regarding the clock within the network depending upon packet loss. He analyzed that minimum information of the clock is directly proportional to arrival rate of the packet in order to assure the synchronization accuracy at synchronizing nodes.

As discussed in [9] by Elson and Deborah, notified that wireless sensor networks need some of the improvements to get the better results and to compete with the latest requirements of the network. They proposed their own low-power synchronization, post facto synchronization. By the use of less energy they implemented the short-lived and localized network for higher precision synchronization. Performance of post-facto synchronization analyzed here was good, where NTP [10] was used for maintaining the clock and precision got was about 1μs whereas when implemented with NTP alone and precision got was 100μs.

According to Ranganathan [11], study on many of the conservative time synchronization protocols like RBS, TPSN, FTSP and many more like node based approach, diffusion based approaches and a hybrid approaches for getting the network time exists.. They tried for the implementation of new method of secured time synchronization protocol having many of the features like: scalability, independent of topology, provides better efficient energy and which is less application dependence. Many of metrics are discussed in [10, 11] by using them the researchers can compare new methods with existing methods to get accurate new method.

As discussed in [12, 13], with the growing progress in the field of wireless sensor network, the need of simulators is also increased which can simulate the network and give the better results. Musznicki et al. [12] and Jevti´c 14] have shown many of the simulators for different type of scenario. They have divided the simulators in many of the broad areas: like emulators and code level, topology control, environment and wireless medium, network and application level, cross level, NS-2 based, OMNeT++ based, Ptolemy II based simulators. Singh et al. [15] and Egea-López et al. [16] they have compiled all the important simulators for better understanding and select the best one out as per their requirement and also to keep our researchers up to date.

TPSN based on the simplest approach sender-receiver approach which takes 2 messages to be sent and 2 to receive for the time synchronization between nodes. This protocol is best solution for all the defined problems in the time synchronization. In many of the context the time synchronization is been emerged like in: data fusion, TDMA scheduling, synchronized sleep time. As discussed in [17], there solution focused for the deterministic time synchronization methods getting minimum complexity, minimum storage so to achieve good accuracy. Message was sent after every second for almost 83 min which resulted in almost 5000 data-points.

4 Proposed Framework for Time Synchronization Protocol

This section discusses the proposed framework divided into three phases outlined as.

For the first phase an experiment is conducted is to find the local clock offset as a reference offset. For this purpose one or more time offset of different sensor clocks at the same time is recorded with in the network. Analysis is done using maximum probability theory. Maximum Probability theory is used to estimate the statistic that the outcome of the experiment performed is due to the assumed set of inputs depending upon the constraints of the experiment. The result of the estimation consist of a set of statistic which includes mean and variance in order to estimate the best time to be used as local clock offset.

Second phase consist of flooding phase. After the selection of the reference offset, the timing packets containing the reference clock offset are forwarded to neighboring nodes at (TO) in order to synchronize. The receiving node will analyze the time packet (TO). The third phase is Synchronization stage. In this phase, the node synchronizes its clock offset with clock of the referenced offset.

This process of synchronization should be repeated after a short span of time, in order to resynchronize and update the sensor nodes within the network. If during the process of synchronization some nodes are in sleep mode then these node should be synchronized when they are in active mode. The duration after which the resynchronization process is repeated should be equal to the time a node in sleep node.

4.1 *Proposed Algorithm for Time Synchronization*

Phase 1 To select a reference clock offset:

I. Calculate the offset or time difference between the clocks of the nodes on the network as

   a. No. of clocks(nodes) participated = 4
   b. Select a largest integer in (4/2) + 1 = 3 (minimum majority subset)
   c. Select the majority subset i.e. represented as a selection of 3 out of 4 possible combinations equal to C(4,3)
      As the value for No. of clocks (nodes) goes high the no. of computation required goes high. So here the maximum value of No. of clocks (nodes) is assumed as 4 (Table 1).

   So as shown in Table 2 the value for majority subset is 4.

Table 1 Difference of time of 4 sensor clock from universal standard time

S. no.	Sensor clock	DIFF of time from UST (s)	Accuracy of synch (s)
1	S1	10.8	0.053
2	S2	0.8	0.018
3	S3	1.4	0.087
4	S4	0.7	0.036

Table 2 Majority subset for x = 4

C(4,3)	Mean	Variance
10.8, 0.8, 1.4	4.3	21.0
10.8, 0.8,0.7	4.1	22.4
10.8, 1.4, 0.7	4.3	21.2
0.8, 1.4, 0.7	1.0	0.1

II. Table 2 also represents the value for calculated mean and variance of the associated majority subset.
III. Select the smallest variance i.e. 0.1 and return the associated mean i.e. 1.0 as the most probable estimate for the local clock offset.

Phase 2 Flooding Phase: to circulate the calculated offset to the neighboring nodes

I. Select a node to act as reference node. For this purpose the selection of reference is random depends totally upon the energy level of node.
II. Synchronize its clock by using the value of local clock offset as calculated in phase 1.
III. Broadcast time_packet with local clock offset value to neighboring nodes using simple user datagram protocol (UDP).

Phase 3 Synchronization Phase:

I. Analyze the time_packet by all the nodes(n1, n2, …, ni) on the network
II. Fetch the value of local clock offset.
III. Synchronize their clock by using this fetched value

Sensor Nodes Clock is now synchronized.

5 Results and Discussion

As compared with the classical time synchronization protocol which are already implemented and are being used it was observed that the proposed protocol is energy efficient and it can be used at the places where optimization of energy is required. Two most popular protocols RBS and TPSN are used in this paper to analyses energy conservation along with the proposed protocol. Assuming various criteria, the proposed time synchronization protocol is categorized as peer to peer, clock-correcting sender- receiver network-wide synchronization protocol.

Although, time synchronization protocol should be able to work optimally in terms of all the design requirements imposed on time synchronization in WSNs, [4] which include energy efficiency, scalability, precision, security, reliability, and robustness to network dynamics. However, the complex nature of WSNs makes it very difficult to optimize the protocol with respect to all these requirements simultaneously. Table 2 describes the comparison survey between RBS, TPSN and

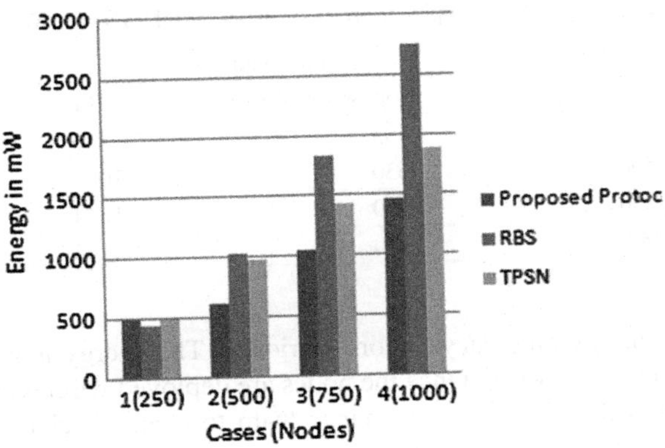

Fig. 1 Comparing energy factor of proposed protocol, RBS and TPSN for 4 scenarios

the proposed protocol depending upon the energy consumed. To simulate the work for this paper in order to implement the flooding and Synchronization phase four different scenarios have been considered in NS2. For first case no. of nodes have been restricted to 250 nodes. For case 2 the no. nodes set is 500 approx. similarly for case 3 and case 4 the no. of nodes considered are 750 and 1000 respectively. The energy consumed by RBS and TPSN is discussed by Robert, and Yanos in their research is takes for reference.

Case 1 When the no. of nodes are restricted 250, the energy consumed by all the three protocol is approximately equal. But RBS wins the race.

Case 2 As the numbers of nodes are increased to 500, the energy consumed by proposed protocol is 66.03 and 56.03 % less then RBS and TPSN respectively.

Case 3 Again as the numbers of nodes are increased to 750, the energy consumed by proposed protocol is approx 75.6 and 36.5 % less then RBS and TPSN respectively

Case 4 And finally when the no of nodes is set to 1000 the consumed energy is approx. 87.89 % less than energy consumed by RBS and approx. 28.23 % less than energy consumed by TPSN.

As depicted by Fig. 1 it can be concluded that on an average proposed protocol consumes less energy than both RBS and TPSN as the no. of nodes increases.

6 Conclusion

The complex character of Wireless sensor network creates a very difficult environment to optimize the proposed protocol with respect to all the requirements which includes scalability, energy efficiency, security, robustness and reliability to network dynamics simultaneously In this paper the proposed protocol is optimized

Table 3 Energy consumption by proposed protocol, RBS and TPSN in mW

Cases	No. of nodes	Energy consumed in mW		
		Proposed protocol	RBS	TPSN
1	250	505	446	511
2	500	630	1046	983
3	750	1050	1844	1434
4	1000	1470	2762	1885

considering the energy efficiency factor as priority. The energy is major factor for every wireless sensor network. Once the nodes are deployed wirelessly on field then it may not be possible for the maintenance team to change/recharge the batteries using external sources.

The principle of maximum probability density function are well recognized and broadly used in the area of networking and communication. For the paper this principles is applied depending upon the majority subset procedure for conditions where small numbers of sensor nodes are involved. The proposed protocol was simulated on raw data collected from different sensor clocks over internet. It was observed that by using the principle of maximum probability density function the clock offset can be selected without using any external source. Hence it accumulates energy of the sensor nodes.

As observed in the paper the proposed protocol is better than RBS and TPSN in terms of optimization of energy consumption in WSN. For analysis the proposed work is simulated with four different scenarios. In all the four cases the energy consumed by our proposed protocol is always less than the RBS and TPSN as shown in Table 3.

It can also be stated that resynchronization interval is required to achieve a specific level of synchronization correctness.

References

1. F.Y. Ren, Time Synchronization in Wireless Sensor Networks, Tsinghua university, 2005.
2. Amulya Ratna Swain and R.C. Hansdah. 2015. A model for the classification and survey of clock synchronization protocols in WSNs. Ad Hoc Netw. 27, C 219–241, April 2015.
3. Gautam, G.C.; Sharma, T.P.; Katiyar, V.; Kumar, A., "Time synchronization protocol for wireless sensor networks using clustering," Recent Trends in Information Technology (ICRTIT), 2011 International Conference on, pp. 417, 422, 3–5 June 2011.
4. Serpedin, Erchin and Qasim M. Chaudhari. Synchronization in Wireless Sensor Networks. 1st ed. Cambridge: Cambridge University Press, 2009. Cambridge Books Online. Web. 21 June 2015.
5. J. Elson, L. Girod, and D. Estrin, Fine-grained network time synchronization using reference broadcasts, in Proceedings of the 5th Symposium on Operating System Design and Implementation, Boston, MA, December 2002, pp. 147–163. ACM, 2002.
6. J. R. Vig, Introduction to quartz frequency standards, Technical Report SLCETTR-92-1, Army Research Laboratory Electronics and Power Sources Directorate, Oct 1998.

7. Jeremy Elson and Kay Römer. 2003. Wireless sensor networks: a new regime for time synchronization. SIGCOMM Comput. Commun. Rev. 33, 1 (January 2003), 149–154.
8. Wang Ting, Guo Di, Cai Chun-yang, Tang Xiao-ming, and Wang Heng, "Clock Synchronization in Wireless Sensor Networks: Analysis and Design of Error Precision Based on Lossy Networked Control Perspective," Mathematical Problems in Engineering, vol. 2015, Article ID 346521, 17 pages, 2015.
9. Elson, Jeremy Eric, and Deborah Estrin. Time synchronization in wireless sensor networks. Diss. University of California, Los Angeles, 2003.
10. G. Werner-Allen, G. Tewari, A. Patel, M. Welsh, and R. Nagpal, "Firefly inspired sensor network synchronicity with realistic radio effects," in Proceedings of the 3rd international conference on Embedded networked sensor systems, pp. 142–153, ACM, 2005.
11. Ranganathan Prakash, and Kendall Nygard. "Time synchronization in wireless sensor networks: a survey." International journal of UbiComp (IJU) 1.2, 2010.
12. Musznicki, Bartosz, and Piotr Zwierzykowski. "Survey of simulators for wireless sensor networks." International Journal of Grid and Distributed Computing 5.3, pp. 23–50, 2012.
13. Chen, Gilbert, et al. "SENSE: a wireless sensor network simulator." Advances in pervasive computing and networking. Springer US, 2005.
14. M. Jevti´c, N. Zogovi´c, and G. Dimi´c, "Evaluation of Wireless Sensor Network Simulators," in Proceedings of TELFOR 2009, 17th Telecommunications forum, (Belgrade, Serbia), 24–16 November 2009.
15. C. P. Singh, O. P. Vyas, and M. K. Tiwari, "A Survey of Simulation in Sensor Networks," in Proceedings of CIMCA'08, 2008 International Conference on Computational Intelligence for Modeling Control and Automation, (Vienna, Austria), 10–12 December 2008.
16. E. Egea-López, J. Vales-Alonso, A. S. Martínez-Sala, P. Pavón-Marino, and J. García-Haro, "Simulation tools for wireless sensor networks," in Proceedings of SPECTS 2005, Summer Simulation Multiconference, (Philadelphia, USA), 24–28 July 2005.
17. Sichitiu, Mihail L., and Chanchai Veerarittiphan. "Simple, accurate time synchronization for wireless sensor networks." Wireless Communications and Networking, 2003. WCNC 2003. 2003 IEEE. Vol. 2. IEEE, 2003.

Epigenetic and Hybrid Intelligence in Mining Patterns

Malik Shamita and Singh Richa

Abstract The term Epigenetics science is an element of a 'postgenomic' analysis paradigm that has more and more place in the theoretical model of a unidirectional causative link from DNA → polymer → supermolecule → constitution. Epigenetics virtually means that "above" or "on high of" biological science. It refers to explicitly modifications to deoxyribonucleic acid that flip genes "on" or "off." These changes don't amendment the deoxyribonucleic acid sequence, however instead, they have an effect on however cells "read" genes. Epigenetic changes alter the natural object of DNA. One example of associate degree epigenetic amendment is DNA methylation—the addition of a alkyl group, or a "chemical cap," to a part of the DNA molecule, that prevents sure genes from being expressed. In this paper, an algorithm i-DNA-M has been proposed which would improve the result of the mining intelligent patters in dataset. Patterns further helps to reconstruct phylogenetic network. The idea behind i-DNA-M is rearranging the input sequences in a way that the new arrangement gives a better tree, since the patterns or motifs affects the outcomes of phylogenetic network.

Keywords Epigenetics · Genetic algorithms · Phylogenetic · Pattern mining · Weighted motifs

1 Introduction

The functional naming of the term "Epigenetics" refers to the study of heritable changes in the gene activities, which are not caused by the changes in DNA sequence. The word "Epigenetics" has been derived from the word 'epi' which

M. Shamita (✉)
Amity School of Engineering Technology, Amity University, Noida, India
e-mail: smalik@amity.edu

S. Richa
Amity Institute of Information Technology, Amity University, Noida, India
e-mail: rsingh10@amity.edu

© Springer India 2016
S.C. Satapathy et al. (eds.), *Information Systems Design and Intelligent Applications*, Advances in Intelligent Systems and Computing 435,
DOI 10.1007/978-81-322-2757-1_39

means 'around, over, outside of' the gene. Epigenetic changes can modify the activation/inactivation status of certain genes, but they do not cause any alteration in nucleotide sequences. Additionally, there may be changes in the DNA binding proteins called as 'chromatin' and this may affect the functional activation of a gene, leading to an epigenetic change. Epigenetic changes are the molecular basis of differentiation in multi-cellular organisms, where only certain 'essential' genes are expressed in a differentiated tissue. Epigenetic changes are preserved when cells divide.

Some examples of epigenetic changes include paramutation, gene silencing, reprogramming, maternal effects, effects of teratogens and carcinogens and histone modifications. DNA damage can also include epigenetics changes. DNA methylation, chromatin remodeling, micro RNA mediated RNA silencing are most common epigenetic changes occurring in cells. Disturbances in epigenetic regulation has pathological consequences, leading to disease development and in some cases even microbial pathogens, such as human immunodeficiency virus (HIV) may dysregulate epigenetic mechanisms in host cells.

The idea behind this paper is rearranging the input sequences on the basis of mining important patterns. These patterns carry important information like some disease, disability, viruses etc. Input sequences are arranged on the Sequence patterns. Since patterns are becoming increasingly important in the analysis of gene regulation. In this way the new arrangement gives a better tree, since the order of input sequences affects the outcomes of phylogenetic network.

2 Epigenetics and HIV Sequences

In epigenetic inheritance it may be possible to pass down epigenetic changes to future generations. But these changes are possible if and only if, the changes occur in spermatozoan or egg cells. Most epigenetic changes that occur in spermatozoan and egg cells get erased once the 2 mix to create a fauna, in an exceedingly method known as "reprogramming." Reprogramming helps vertebrate cells to begin again and develop their own epigenetic. However scientists assume a number of the epigenetic changes in parents' spermatozoan and egg cells could avoid the reprogramming method, and build it through to consequent generation. If this is assumed correct, for example, what a person eats before they conceive could influence the future born child. But, it has not been confirmed. HIV was the first human pathogen implicated in induction of epigenetic changes in host cells [1–4].

It was observed that HIV infection leads to DNA methylation in anti-viral gene, Interferon-gamma's (IFNg) promoter, thus diminishing the anti-viral immune response of the host cells. Apart from this, HIV infection also leads to epigenetic changes in tumor suppressor gene, thereby altering the immune regulation in the host. Epigenetic mechanisms in HIV-host interaction also mark the basis of clinical drug resistance in HIV infected cells. Drug induced silencing of cellular thymidine-kinase gene (TK) leads to formation of epigenetically altered cells that are highly resistant to cytotoxic effects of AZT, a highly popular and widely used anti-retroviral drug [5]. The epigenetic effects of AZT are observed long after the perinatal exposure in the infants. Although the epigenetics of HIV promoter silencing has been studied extensively, high throughput characterization of the underlying mechanisms in HIV infected cells is still a promising research area. Characterization of the epigenomes of the primary targets of HIV in host immune system such as CD4+ T-cells, monocytes and macrophages before and after HIV infection may have pathogenetic and therapeutic implications. At present epigenetic therapies targeting the reactivation of latent HIV infected CD4+ T-cells are under major screening and offer ample hope for control of HIV epidemic. The assessment of epigenetic modifications in HIV-host interaction by latest bioinformatics tools such as target-site prediction will enable biologists to tackle the viral epigenetic modifications in a more defined and specific manner [4, 5].

The paper is organized as follows. In Sector 3, the new improved methodology for finding most frequent occurring patterns. The experimental results are given in Sect. 4, Sect. 5 concludes report and appendix.

3 Methodology

We accessed database of genome sequences downloaded from National Centre for Biotechnology Information. In this paper we have taken DNA sequences of HIV-1 virus. A set of multiple DNA sequences is represented as D = {s1, s2, s3, …, sn}. Firstly we focus on sequence alignments. These alignments are required because it is the way of arranging the sequences of DNA, RNA or proteins. It helps us to identify the regions of similarity that may be a consequence of functional, structural or evolutionary relationships between the sequences [6]. In this paper we have used ClustalW2 for alignment of HIV-1 virus sequences. Every gene contains a regulatory region (RR). This regulatory region typically get stretched to 100–1000 bp upstream of the transcriptional start site. Locating RR is tedious process. To get the set of most patterns, we run PMS5 [7]. We further applies weights to get the most optimized patterns [8].

Definition 1 A database of virus

$$D = \{S_i\}_{i=1}^{m}$$

where S is sequences in dataset and m is number of sequences.

Definition 2 Calculate the occurrence of patters by comparing data sequences.

$$P(I_1, I_2) = \sum_{j=1}^{n} w_j' \sqrt{\left(f_{I_1}^{j} - f_{I_2}^{j'}\right)^2}$$

[1	2	3	4	5	6	7	8	9	10	11]
[1]										
[2]	0.00									
[3]	0.00	0.00								
[4]	0.00	0.00	0.00							
[5]	99.00	99.00	99.00	99.00						
[6]	100.00	100.00	100.00	100.00	1.00					
[7]	100.00	100.00	100.00	100.00	1.00	0.00				
[8]	100.00	100.00	100.00	100.00	1.00	0.00	0.00			
[9]	107.00	107.00	107.00	107.00	106.00	107.00	107.00	107.00		
[10]	109.00	109.00	109.00	109.00	111.00	111.00	110.00	111.00	31.00	
[11]	107.00	107.00	107.00	107.00	106.00	107.00	107.00	107.00	0.00	31.00

Above matrix show number of difference between the sequences. Apply weights from the above matrix to the sequences

$$w_j^{'} = \frac{w_j}{\sum_{j=1}^{n} w_j}$$

where w_j is weight assigned to the sequences of dataset.

Definition 3 Let B be set of best matched pattern set

$$\mathrm{B} = \{\mathrm{B}_1, \mathrm{B}_2, \ldots, \mathrm{B}_\mathrm{k}\}$$

Definition 4 The performance of best patterns can be evaluated for precision as

$$T = \frac{a}{k}$$

$$Q = \frac{a}{z}$$

$$a = \sum_{i=1}^{z} v_i$$

where T is ratio of retrieval of same pattern number, Q is ratio of same pattern with maximum of 1 bit change, a is retrieved relevant patterns, k is mean of best patterns, z is pattern number.

$$\mathrm{V_i} = \{\mathrm{A}, \mathrm{G}, \mathrm{T}, \mathrm{C}\}$$

Definition 5

$$y(W) = \sum_{i=1}^{num} \sqrt{\frac{T^2 + Q^2}{2}}$$

where num is the best suitable pattern.

Table 1 Comparison between current and new methodology for tree reconstruction

Algorithms	Current methodology	i-DNA-M algorithm
Neighboring joining	Maximum bootstrap score given is 100 and 71.	Maximum bootstrap score given is 100 and 72.
Minimum evolution	Maximum bootstrap score given is 100 and 69.	Maximum bootstrap score given is 100 and 71.
Maximum parsimony	Maximum bootstrap score given is 99, 69, 33 and 28.	Maximum bootstrap score given is 99, 67, 33 and 27
UPGMA	Maximum bootstrap score given is 100 and 70.	Maximum bootstrap score given is 100 and 72.

After rearranging the sequences on the basic of i-DNA-M algorithm, we run phylogenetic reconstruction program MEGA followed by bootstrap process see Appendix, after using i-DNA-M for reconstruction of phylogenetic and bootstrap process.

4 Result and Analysis

By implementing i-DNA-M algorithm, we can analysis some improved results in reconstructing phylogenetic network reconstruction. The analysis of algorithm is done on Molecular Evolutionary Genetics Analysis Version 6.0. Sequences of HIV dataset is given for reconstruction of phylogenetic network. We have taken 10 strands of HIV virus for testing our algorithm. As the algorithm adds filters to increase the occurrence of weighted pattern called motifs. Sequences arranged on the basis of weighted patters shows improved bootstrap values. Table 1 summarize the results. In case of neighboring joining the bootstrap value is increased by 1 %. In case of minimum evolution and UPGMA it is increased by 2 % while in case of maximum parsimony it is decreased by 1 % on two taxa elements.

5 Conclusion

This paper is basically experimental paper which tries to show the impact of sequence rearrangement. By implementing i-DNA-M algorithm, we can rearrange database on the basics of scores of patterns, which shows improved results in some cases. This approach is robust to various kinds of databases, it can get the best features combination with best patterns to further reconstruct phylogenetic network. This algorithm helps us to trace the evolution of the disease with the help of weighted motifs and to calculate the duplex stability of the datasets.

Acknowledgments The authors thank Dr. Ashok K. Chauhan, Founder President, Amity University, for his support and motivation along with providing us with the necessary infrastructure for research.

Appendix

Appendix, shows graphical representation of phylogenetic network after rearranging sequences through i-DNA-M algorithm. See Figures 1, 2, 3, 4, 5, 6, 7 and 8.

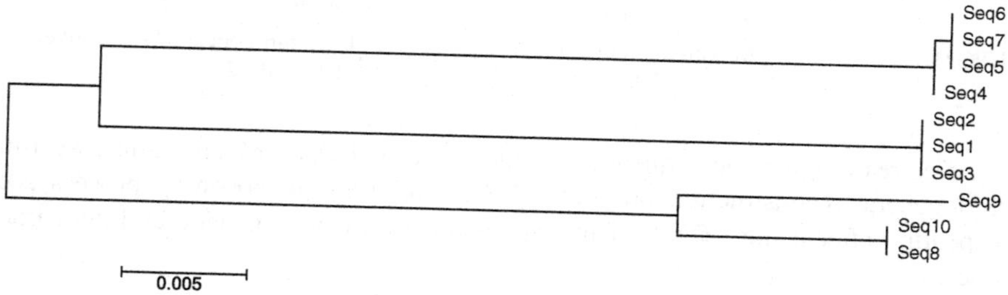

Fig. 1 Phylogenetic network construction by neighbor joining algorithm after i-DNA-M algorithm

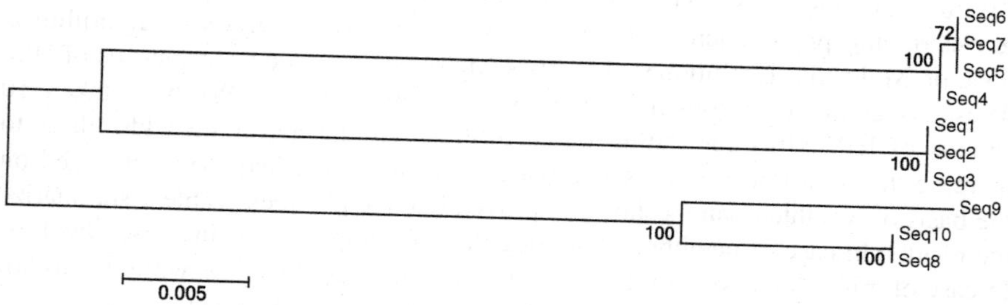

Fig. 2 Result after bootstrap test of phylogenetic network construction by neighbor joining algorithm after i-DNA-M algorithm

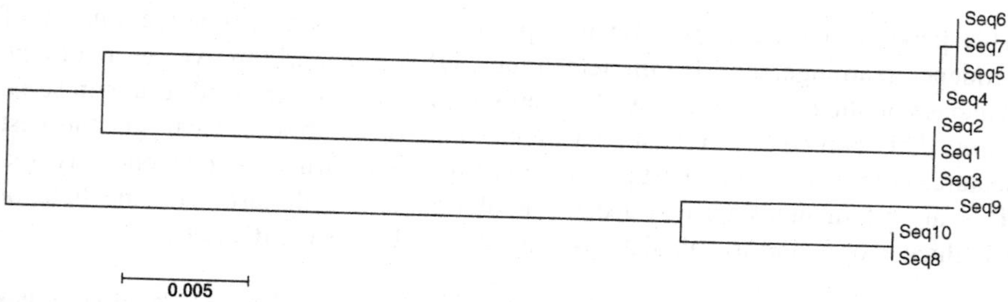

Fig. 3 Phylogenetic network construction by minimum evolution algorithm after i-DNA-M algorithm

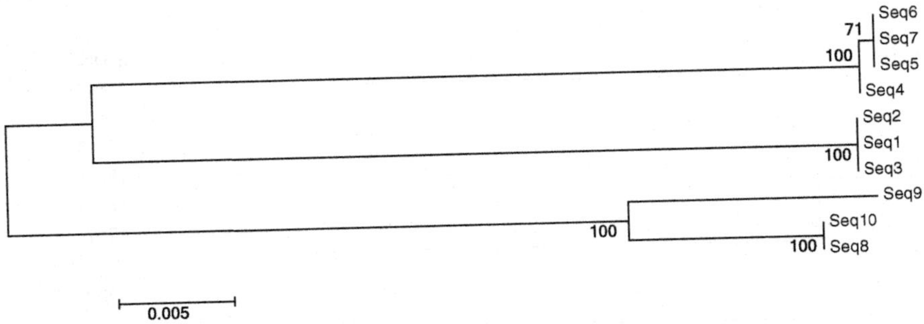

Fig. 4 Result after bootstrap test of phylogenetic network construction by minimum evolution algorithm after i-DNA-M algorithm

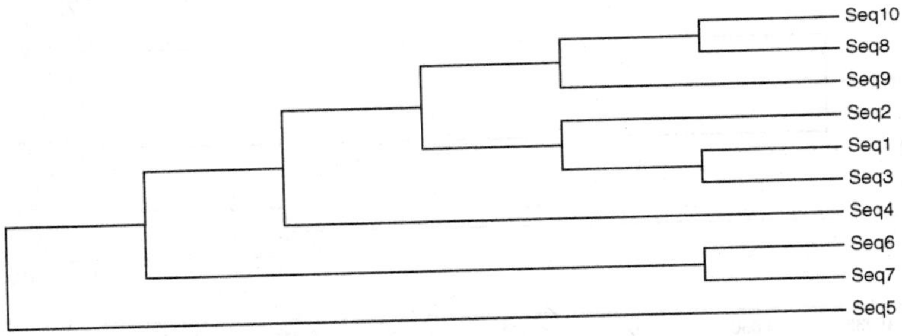

Fig. 5 Phylogenetic network construction by maximum parsimony algorithm after i-DNA-M algorithm

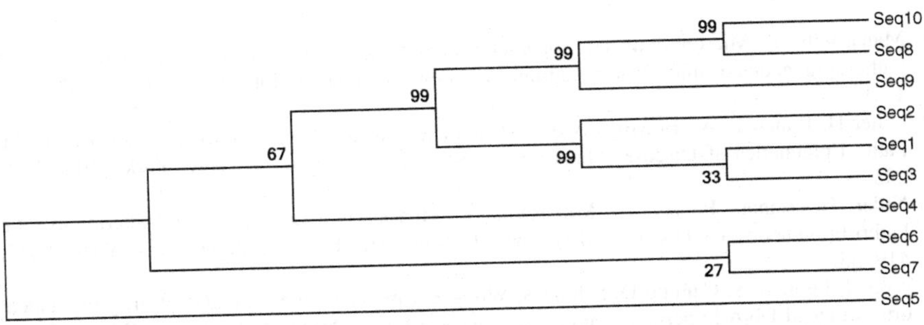

Fig. 6 Result after bootstrap test of phylogenetic network construction by maximum parsimony algorithm after i-DNA-M algorithm

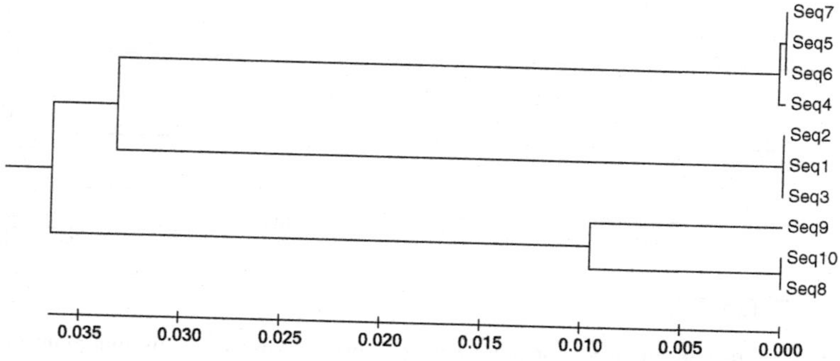

Fig. 7 Phylogenetic network construction by UPGMA algorithm after i-DNA-M algorithm

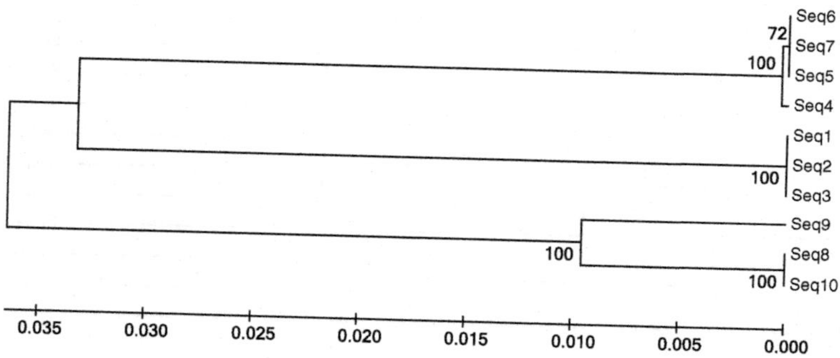

Fig. 8 Result after bootstrap test of phylogenetic network construction by UPGMA algorithm after i-DNA-M algorithm

References

1. Manarovits J. Microbe-induced epigenetic alterations in host cells: the coming era of patho-epigenetics of microbial infections. A review. Acta Microbiol Immunol Hung. 2009; 56: 1–19.
2. Niller H, Banati F, Ay E, Minarovits J. Epigenetic changes in virus-associated neoplasms. In: Patho-Epigenetics of Disease; Minarovits J, Niller H(Eds); Springer: New York, 2012; 179–225.
3. Niller H, Banati F, Ay E, Minarovits J. Microbe-induced epigenetic alterations. In: Patho-Epigenetics of Disease; Minarovits J, Niller H, (Eds); Springer: New York, 2012; 419–55.
4. Nyce J, Leonard S, Canupp D, Schulz S, Wong S. Epigenetic mecha nisms of drug resistance: drug-induced DNA hypermethylation and drug resistance. Proc Natl Acad Sci USA. 1993; 90: 2960–4.
5. K. Tamura, D. Peterson, N. Peterson, G. Stecher, M. Nei and S. Kumar, "MEGA5: Molecular Evolutionary Genetics Analysis using Maximum Likelihood, Evolutionary Distance and Maximum Parsiomny methods", Molecular Biology and Evolution, 2011.
6. B. Efron and R. J. Tibshirani. "An Introduction to the Bootstrap." CRC Press, Boca Raton, FL, 1998.

7. Dinh, Hieu, Sanguthevar Rajasekaran, and Vamsi K. Kundeti. "PMS5: an efficient exact algorithm for the (ℓ, d)-motif finding problem." BMC bioinformatics 12.1 (2011): 410.
8. Zhang, Zhao, et al. "Evolutionary Optimization of Transcription Factor Binding Motif Detection." Advance in Structural Bioinformatics. Springer Netherlands, 2015. 261–274.

A Technical Review on LVRT of DFIG Systems

Pretty Mary Tom, J. Belwin Edward, Avagaddi Prasad, A.V. Soumya and K. Ravi

Abstract The most important issue with doubly fed induction generator (DFIG) wind turbine is low voltage ride-through performance. To solve this problem, several techniques have been introduced. The paper discusses some of the most commonly used solutions for Low Voltage Ride Through (LVRT) of wind turbine generators which is the most important feature to be attained according to grid codes. A technical survey is presented with comparison between these techniques.

Keywords DFIG · LVRT · Flexible AC transmission system (FACTS) · Dynamic voltage regulator (DVR) · Static compensator (STATCOM) · Fault current limiter (FCL)

1 Introduction

In the recent years, there has been a tremendous increase in the electricity demand. The conventional sources of energy like coal and oil takes millions of years to form, moreover they have a negative impact on the environment causing pollution when they are used for power generation. This resulted in the growth of power generation from renewable energy resources particularly wind energy.

P.M. Tom (✉) · J. Belwin Edward · A. Prasad · A.V. Soumya · K. Ravi
School of Electrical Engineering, VIT University, Vellore, Tamil Nadu, India
e-mail: prettytom8@gmail.com

J. Belwin Edward
e-mail: jbelwinedward@vit.ac.in

A. Prasad
e-mail: prasadavagaddi@gmail.com

A.V. Soumya
e-mail: soumyanirudh@gmail.com

K. Ravi
e-mail: k.ravi@vit.ac.in

© Springer India 2016
S.C. Satapathy et al. (eds.), *Information Systems Design and Intelligent Applications*, Advances in Intelligent Systems and Computing 435,
DOI 10.1007/978-81-322-2757-1_40

There is a vast increase in the exploitation of wind energy which needs research and development of higher efficiency wind energy conversion systems. Figure 1 shows the increase in growth of installed wind capacity in Tamil Nadu. There are some issues which should be concerned when wind farms are connected to grid. Firstly, the power qualities of fluctuating wind power. Secondly, during grid faults a voltage dip occurs at the point of connection of wind turbine, which causes an increase in rotor circuit current. The capability of a wind turbine to endure a short duration voltage dip without tripping is referred to as Low Voltage Ride Throughability. Figure 2 depicts the LVRT curve. Fault ride through capability is the biggest issue related to wind farms and if improved will reduce the adverse effects on power system stability.

This paper investigates and presents a review on different techniques that have been used to enhance the operational ability of Wind Energy Conversion Systems during fault. Several methods introduced to improve the LVRT capability is evaluated with their pros and cons.

1.1 System Configuration

The variable speed Doubly Fed Induction Generator improves the voltage stability and works in combination with wind turbine, gearbox, converters and their control components. Figure 3 depicts the system configuration of grid connected DFIG based wind energy conversion system.

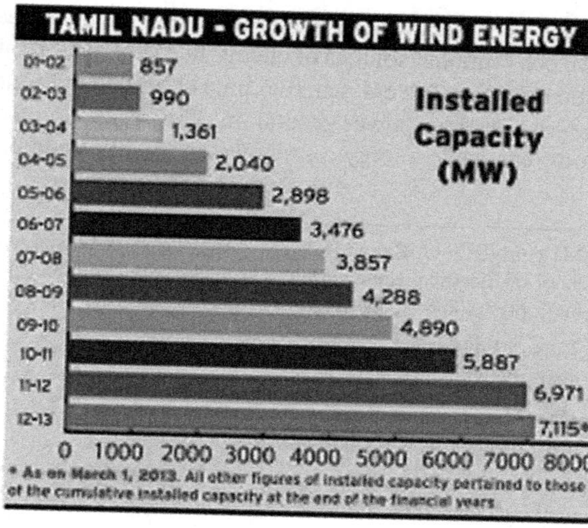

Fig. 1 Growth of wind energy in Tamil Nadu

Fig. 2 LVRT curve

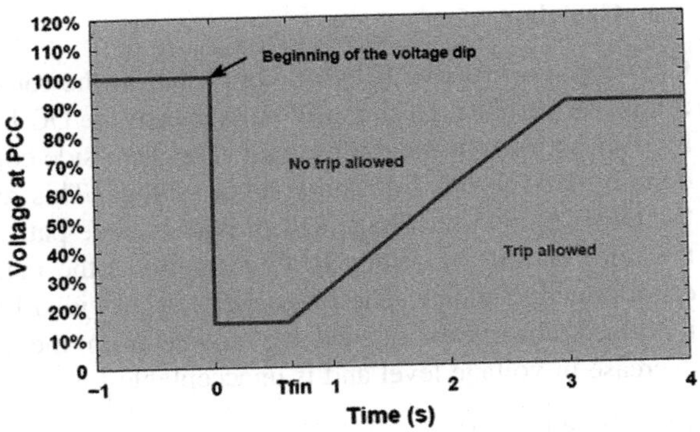

2 LVRT Techniques

Initially the wind energy conversion systems were allowed to disconnect from the grid during faults/disturbances. But now since the wind farms contribute a significant portion of load and also it can lead to a chain reaction leading the other generators to trip, generators are required to be connected to grid and contribute to voltage control for a defined fault clearing time. Several techniques have been proposed to enhance the LVRT capability of DFIG based wind energy conversion systems.

Wind farm disconnection during faults leads to

- Decrease in reactive power consumption
- High reactive consumption at reconnection
- Power generation loss increases the voltage dip

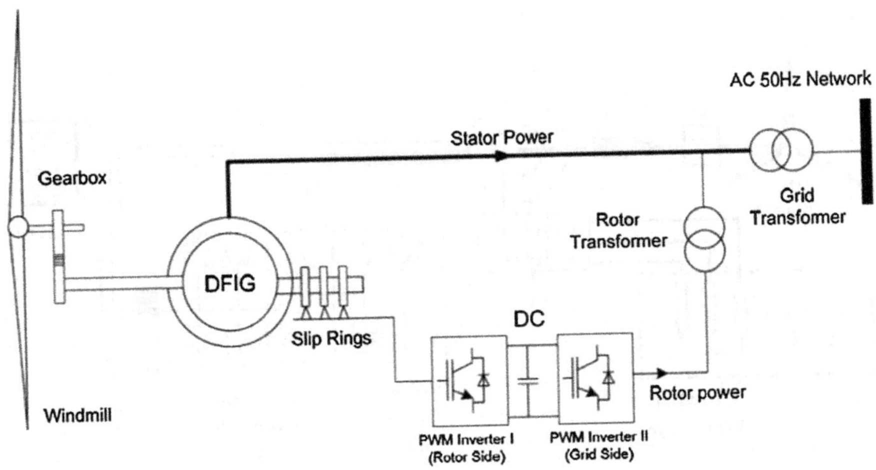

Fig. 3 Grid connected DFIG based wind energy conversion system

(a) **Crowbar**

Crowbar is a resistive network connected to the rotor circuit for safety in case of rotor over currents. During grid voltage dips the DC link voltage increases followed by over currents in the rotor circuit. The rotor side converter protection can also be done by feed forward of faulty stator voltage. This resistive network connected to the DFIG limits the voltage and provides a safe path for the currents by bypassing the rotor circuit [1]. When RSC is bypassed the machine behaves like a squirrel cage induction motor. The magnetization provided by RSC is lost which leads to increased absorption of reactive power from the grid, which leads to further decrease in voltage level and is unacceptable for grid connected operation (Fig. 4).

(b) **DC chopper**

A power resistor is incorporated in parallel with the DC link as shown in Fig. 5. in order to protect the increase in DC link voltage beyond the threshold value. During voltage dip the RSC is tripped and the huge increase in DC link voltage caused the triggering of chopper circuit [2]. This prevents a dangerous increase in DC link voltage. Once the fault is cleared the rotor side converters begin to work normally. The DC link can be protected only by proper designing of the chopper circuit, otherwise it will damage the DC link capacitor.

(c) **DVR**

Dynamic Voltage Restorer (DVR) is a voltage source converter connected in series between the grid and the generator. To maintain a constant stator voltage the DVR output voltage is fed to the grid, so the machine is not affected directly by grid disturbances. Unbalanced voltage dip ineffectiveness together with cost and complexity of DVR, limits its efficiency as fault ride through technique with DFIG wind energy systems [3] (Fig. 6).

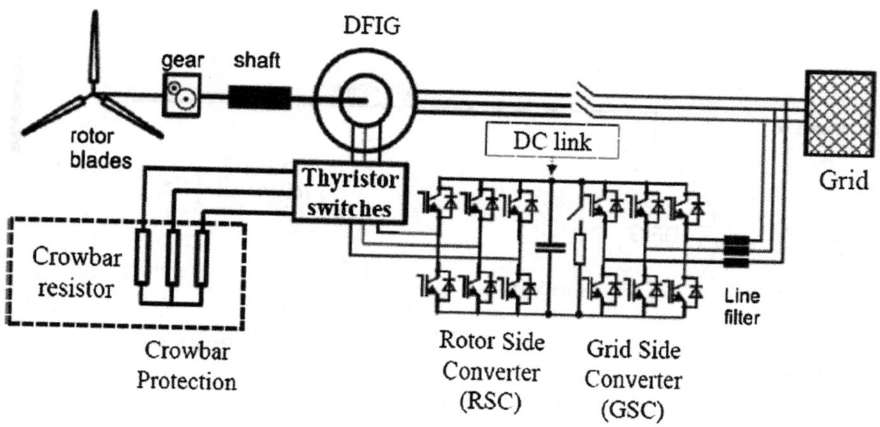

Fig. 4 Crowbar circuit with DFIG wind energy system

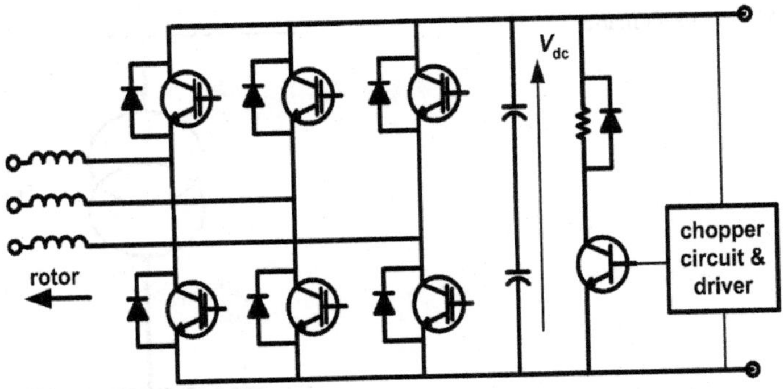

Fig. 5 Rotor side converter with DC chopper

(d) STATCOM

STATCOM is a shunt connected FACTS device capable of delivering and absorbing reactive power. If the voltage of the system is less than the STATCOM terminal voltage then reactive power is fed to the system so that voltage level is maintained, if it is high then reactive power is absorbed [4]. Figure 7 shows the schematic of STATCOM, which consists of voltage sourced converter, a capacitor and a coupling transformer [5].

(e) VSC-HVDC System

The power generated from wind energy can be connected to grid in two ways, either AC connection or VSC-HVDC connection. The system performance during a three phase short circuit is analyzed [6] and it was found that when VSC-HVDC connection is employed the capacitor in the DC link stores energy during fault and so LVRT requirements can be met, provided the capacitor selection is accurate (Fig. 8).

(f) UPFC

Unified Power Flow controller (UPFC) is connected at PCC to increase the damping of wind turbine generator and also to provide support during faults. It improves the LVRT by controlling the active and reactive powers of the bus to which it is connected [7]. UPFC can control both the active and reactive powers

Fig. 6 Single line diagram of DVR

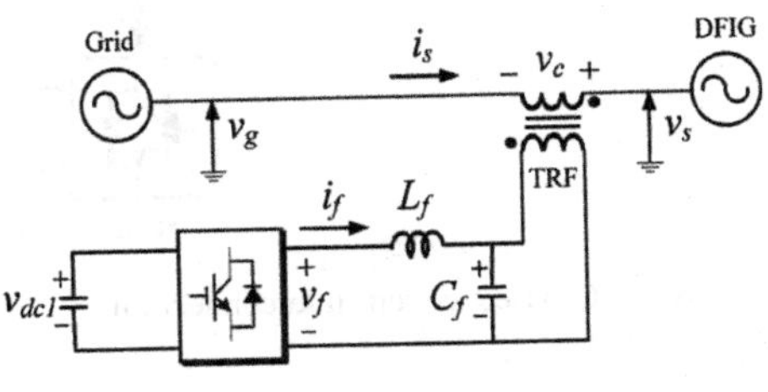

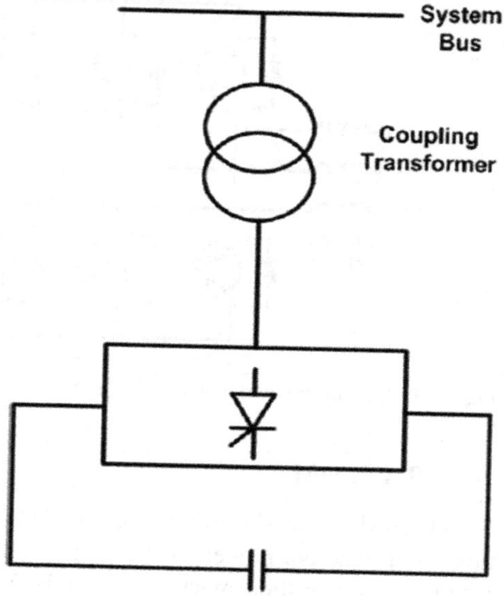

Fig. 7 STATCOM schematic

easily because it consists of both shunt and series converters. When fault is cleared the UPFC changes to idle condition and there is no exchange of reactive power (Fig. 9).

(g) SMES-Fault Current Limiter

During normal operation superconducting magnetic energy storage with fault current limiting function (SMES FCL) is used to smoothen the power fluctuation of DFIG system. Figure 10 shows DFIG system with SMES FCL. During grid faults the super conducting magnetic energy storage limits the fault current and allows the wind turbine generator to be remained connected to the system without failure [8].

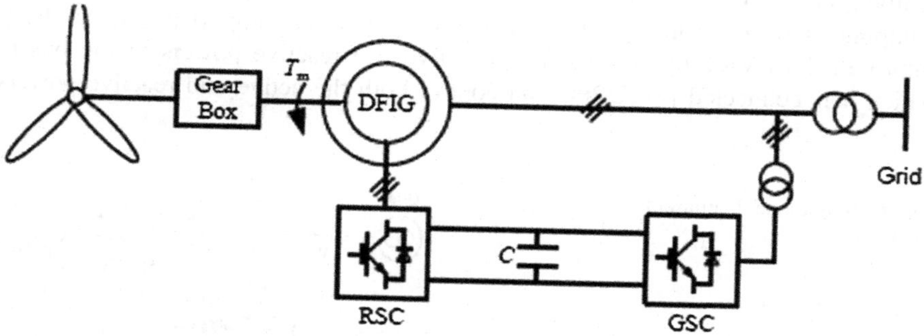

Fig. 8 VSC-based system interconnection

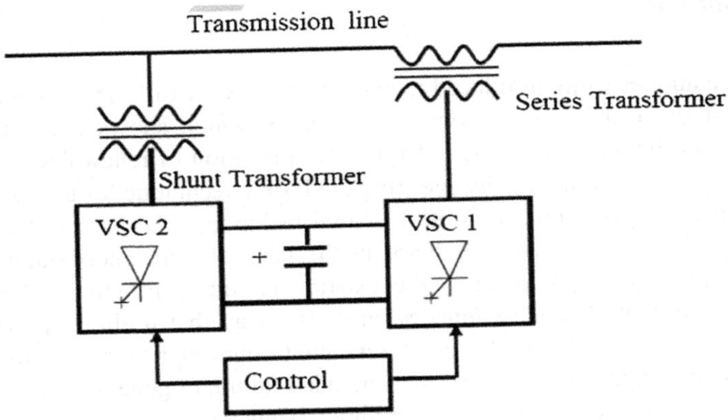

Fig. 9 UPFC connected to WTG

(h) Bridge Type Fault Current Limiter

A bridge type fault current limiter (BFCL) was introduced to enhance the LVRT during asymmetrical faults [9]. Here the BFCL has two parts, a bridge part and a shunt part. During normal operation the current flows through the bridge part and whenever there is a fault it flows through shunt part. Bridge has no impact on steady state. The excess energy in the fault current is absorbed by the shunt part. The comparison of system operation was done with series dynamic braking resistor and was found that BFCL provides stabilized control during a fault.

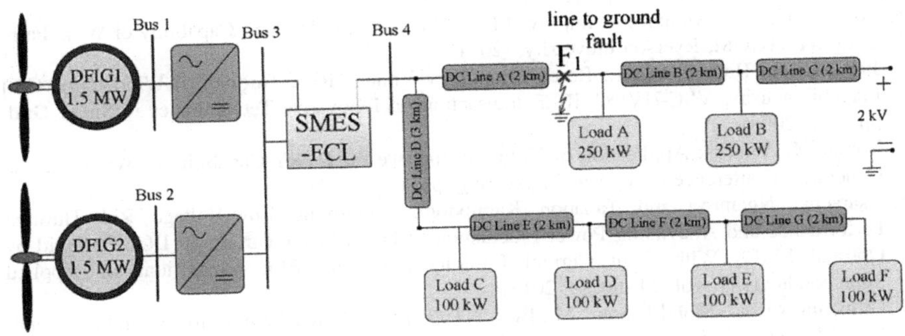

Fig. 10 SMES FCL with DFIG system

3 Conclusion

Various techniques were introduced to enhance the LVRT of WTGs. A comparison of some of the techniques is done in this paper which shows that with a few limitations all the techniques aids LVRT. The utilisation of renewable energy can be done in an efficient manner by selecting the LVRT technique for the system. The paper proposes a co-ordinated control of two techniques. DVRs can be used for improving LVRT performance but not in the case of unbalanced fault. A fault current limiter can be included in the existing circuit to provide the remaining functions. The control scheme must be in such a way that it allows protection of converters and allows the DFIG to be remained connected to the circuit without tripping, which might otherwise cause a cascaded failure of generators connected to the grid. Co-ordinated control of two methods would provide a suitable technique which will enhance the LVRT of wind energy conversion system.

References

1. Christian Wessels.: Fault Ride through of DFIG Wind Turbines during symmetrical voltage dip with Crowbar or Stator Current Feedback Solution", IEEE transaction on Energy Conversion Congress and Exposition. pp 2771–2777, (2010).
2. Graham Pannell, Bashar Zahawi,.: Evaluation of the Performance of a DC-Link Brake Chopper as a DFIG Low-Voltage Fault-Ride-Through Device, IEEE transaction, on Energy conversion, vol. 28, no. 3, pp 535–542, (2013).
3. Ahmad Osman Ibrahim.: A Fault Ride-Through Technique of DFIG Wind Turbine Systems Using Dynamic Voltage Restorers, IEEE transactions on energy conversion, vol. 26, no. 3, pp 871–882, (2011).
4. A F Abdou.: Application of STATCOM to improve the LVRT of DFIG during RSC Fire-through Fault, AUPEC proceedings, pp 1–6, (2012).
5. Miad Mohaghegh Montazeri.: Improved Low Voltage Ride Through Capability of Wind Farm using STATCOM, Ryerson university, (2011).
6. Jian Liu and Haifeng Liang.: Research on Low Voltage Ride Through of Wind farms Grid Integration using VSC-HVDC, IEEE transaction on Innovative Technologies in Smart Grid, pp 1–6, (2008).
7. Yasser M. Alharbi.: Application of UPFC to Improve the LVRT Capability of Wind Turbine Generator, Conference on Power Engineering, pp1–4, (2012).
8. Issarachai Ngamroo and Tanapon Karaipoom.: Improving Low-Voltage Ride-Through Performance and Alleviating Power Fluctuation of DFIG Wind Turbine in DC Microgrid by Optimal SMES With Fault Current Limiting Function, IEEE transactions on Applied Superconductivity, vol. 24, no. 5, (2014).
9. Gilmanur Rashid, Mohd Hasan.: Ali, Bridge-Type Fault Current Limiter for Asymmetric Fault Ride-Through Capacity Enhancement of Doubly Fed Induction Machine Based Wind Generator, IEEE Conference on Energy Conversion Congress and Exposition, pp 1903–1910, (2014).

An AIS Based Approach for Extraction of PV Module Parameters

R. Sarjila, K. Ravi, J. Belwin Edward, Avagaddi Prasad and K. Sathish Kumar

Abstract This article presents the calculation of the parameter extraction of photovoltaic (PV) panel using artificial immune system (AIS) and compared with genetic algorithm (GA) using MATLAB, Simulink at different environmental conditions (for different irradiations (200–1000 w/m^2)). The proposed method showing I_{pv} versus V_{pv} curves and to compare the obtained curve to the ideal values in order to obtain the absolute error curve. For extracting parameters of a PV cell, the proposed method useful because it can handle nonlinear functions. The proposed method compared with manual data and are validated by three different types of PV modules named as, Multi-crystalline (SHELL S36), Mono-crystalline (SHELL SP70) and Thin-film (SHELL ST40). Data derived from these calculations beneficial to decide the suitable computational technique to build an accurate and efficient simulators for a PV system.

Keywords PV cell · GA · AIS · Parameter extraction

1 Introduction

Engineers need to apply new methods to optimize the PV system performance due to its high initial cost. Selecting the modelling approach is a big challenge for PV model, among all approaches single diode model and two diode model [1, 2], are

R. Sarjila (✉) · K. Ravi · J. Belwin Edward · A. Prasad · K. Sathish Kumar
School of Electrical Engineering, VIT University, Vellore, Tamil Nadu, India
e-mail: r.sarjila@yahoo.com

K. Ravi
e-mail: k.ravi@vit.ac.in

J. Belwin Edward
e-mail: jbelwinedward@vit.ac.in

A. Prasad
e-mail: prasadavagaddi@gmail.com

K. Sathish Kumar
e-mail: kansathh21@yahoo.co.in

© Springer India 2016
S.C. Satapathy et al. (eds.), *Information Systems Design and Intelligent Applications*, Advances in Intelligent Systems and Computing 435,
DOI 10.1007/978-81-322-2757-1_41

the most popular methods [3–5]. System performance can be monitored by using parameters of PV cell. So the parameter extraction is very important. The current–voltage characteristics plays a major role in parameter extraction of PV cell. Earlier researchers developed many techniques named as analytical methods [6], Simple parameter extraction [7], EA computational method [8], GA [9, 10], nonlinear regression analysis [11], particle swarm optimization [12, 13] and other methods [14–16] for parameter extraction. But due to drawbacks in conventional methods, research is still going on this area for better techniques for parameter extraction. This article presents a new technique to evaluate parameter extraction of two diode PV model using AIS. In this paper, the proposed technique tested with three different PV modules and obtained results also compared with GA results.

This article is structured as follows. Section 2 explains two diode model of PV cell. The GA and AIS approaches explained in Sect. 3. Formulation and computation of both the methods explained in Sect. 4. This is followed by results and comparisons in Sect. 5. The final Sect. 6 gives the conclusion.

List of symbols

I	Output current (A)
V	Output voltage (V)
R_s	Series resistance (Ω)
R_p	Shunt resistance (Ω)
I_{pv}	Current generated by the incidence of light (A)
I_{ol}	Reverse saturation currents of diode 1 (A)
I_{o2}	Reverse saturation currents of diode 2 (A)
$VT_{1,2}$	Thermal voltage of the PV module (V)
Ns	Number of cells connected in series
q	Electro charge ($1.60217646 \times 10^{-19}$ C)
k	Boltzmann constant ($1.3806503 \times 10^{-23}$ J/K)
T	Temperature of the p-n junction in K
a_1, a_2	Diode ideality constants
I_m, V_m	Data pair of I–V characteristics curve
N	Number of data
J	Nonlinear function
ΔT	Temperature difference
V_{oc}	Open circuit voltage (V)
I_{sc}	Short-circuit current (A)
K_v	Open circuit voltage coefficient
K_t	Short circuit current coefficient
STC	Standard test conditions
I_m, V_m	Data pair of I–V characteristics curve

2 Two Diode Model of PV Cell

The equivalent two diode circuit model of PV cell shown in Fig. 1, two diode model gives better results than a single diode model. It needs calculation of parameters (I_{pv}, I_{o1}, I_{o2}, R_p, R_s, a_1, and a_2). To reduce the complexity in calculations, so many researchers fixed diode ideality constants i.e. $a_1 = 1$ and $a_2 = 2$ [8].

The required equations [3, 8], for extraction of model parameters are given below,

The output current equation is,

$$I = I_{pv} - I_{O1}\left[\exp\left(\frac{V+IR_S}{a_1 V_{T1}}\right) - 1\right] - I_{O2}\left[\exp\left(\frac{V+IR_S}{a_2 V_{T2}}\right) - 1\right] - \left(\frac{V+IR_S}{R_P}\right) \tag{1}$$

where, VT 1, 2 = (NskT)/q.

The parameters in Eq. (1) are updated because the curve fitting process is applied using both the algorithms GA or AIS. So the Eq. (1) can be written as:

$$y(I,\, V, \Phi) = I_{pv} - I_{O1}\left[\exp\left(\frac{V+IR_S}{a_1 V_{T1}}\right) - 1\right] - I_{O2}\left[\exp\left(\frac{V+IR_S}{a_2 V_{T2}}\right) - 1\right] - \left(\frac{V+IR_S}{R_P}\right) - I \tag{2}$$

where, $\Phi = [I_{PV}, I_{O1}, I_{O2}, R_S, R_P, a_1, a_2]$. For every iteration the parameters in Φ will be changed. The process stops whenever it reaches the maximum number of iterations.

The objective function is,

$$J = \sqrt{\frac{1}{N}\sum_{m=1}^{N} y(I_m, V_m, \Phi)^2} \tag{3}$$

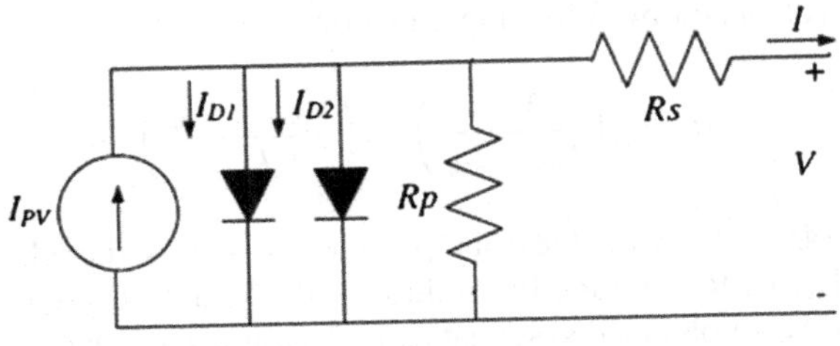

Fig. 1 Equivalent two diode circuit model of PV cell [3]

The PV current equation is,

$$I_{pv} = (I_{PV\_STC} + K_i \Delta T)\frac{G}{G_{STC}} \tag{4}$$

The reverse saturation current equation for a single diode model, in terms of temperature difference can be written as,

$$I_o = \frac{(I_{PV\_STC} + K_i \Delta T)}{exp\left[\left(V_{oc,STC} + K_v \Delta T\right)/\alpha V_T\right] - 1} \tag{5}$$

For the equivalent two diode circuit model, I_{o1} and I_{o2} can be calculated using iteration approach. The computation time increases due to iterations. A new equation is derived to calculate both saturation currents, i.e.

$$I_{o1} = I_{o2} = I_o = \frac{(I_{PV\_STC} + K_i \Delta T)}{exp\left[\left(V_{oc,STC} + K_v \Delta T\right)/\{(a_1 + a_2)/p\}V_T\right] - 1} \tag{6}$$

According to Shockley's diffusion theory and simulation results, $a_1 = 1$ and $a_2 \geq 1.2$. Since $(a_1 + a_2)/p = 1$, gives p value ≥ 2.2. Generalization can remove the vagueness in picking the values of a_1 and a_2. Equation (1) can be written in terms of p as,

$$I = I_{pv} - I_O\left[\exp\left(\frac{V + IR_S}{V_T}\right) + \exp\left(\frac{V + IR_S}{(p-1)V_T}\right) - 2\right] - \left(\frac{V + IR_S}{R_P}\right) \tag{7}$$

From Eq. (7), the equation for R_p for a single diode model at maximum power point (M_{PP}) is,

$$R_P = \frac{V_{mp} + I_{mp}R_s}{\left\{I_{PV} - I_o\left[\exp\left(\frac{V_{mp} + I_{mp}R_s}{V_T}\right) + \exp\left(\frac{V + IR_S}{(p-1)V_T}\right) - 2\right] - \frac{P_{\max,E}}{V_{mp}}\right\}} \tag{8}$$

To start the iteration process for equivalent two diode circuit model, the appropriate initial values for R_p and R_s are allocated. Hence, R_s is assumed to be zero and R_p is calculated by following equation,

$$R_{po} = \left(\frac{V_{mp}}{I_{scn} - I_{mp}}\right) - \left(\frac{V_{ocn} - V_{mp}}{I_{mp}}\right) \tag{9}$$

Using Eqs. (4) and (6) all the four parameters of this model can be readily extracted, i.e., I_{pv}, I_o, R_p, and R_s. From these, only R_p and R_s are calculated using iteration. I_{pv} and I_o are obtained systematically. Variable p can be selected to be any value larger than 2.2. These calculation shave developed the computational speed of the simulator.

3 Brief Review of Optimization Algorithms

3.1 Genetic Algorithm

The GA belongs to the theory of biological evolution. Selection, mutation and crossover are most important operators used in GA. To choose the chromosomes that contribute in the reproduction procedure to give birth to the next generation, the selection method is useful. To explore the new solutions by introducing changes in parameters of the population, the mutation method is needed. To obtain individuals for new generation, the crossover process is used (Fig. 2).

3.2 Artificial Immune System

AIS has attained worldwide attention because of its ability to solve complex optimization problems. AIS imitates immune system of the human body by fighting bacteria, viruses and other foreign particles. The antigen antibody reaction is the basic mechanism behind AIS which yields better performance in the field of optimization. The immune cells present in the human body are called antibodies which fight against the antigen, the foreign particles in the human body.

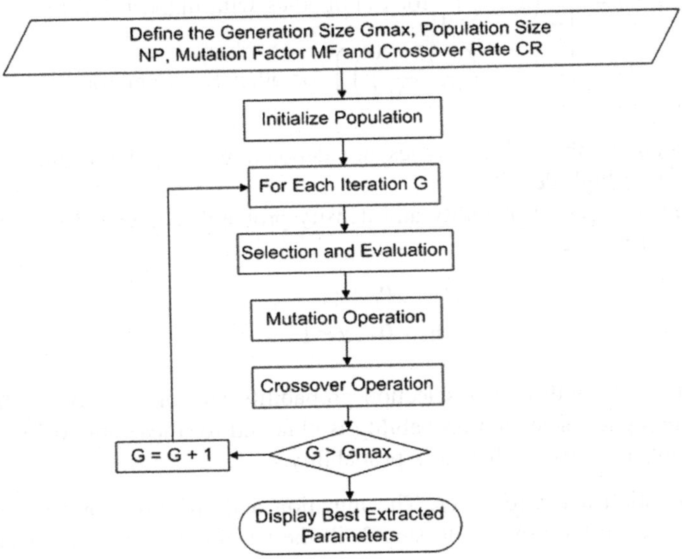

Fig. 2 Flowchart for GA

3.2.1 Various Steps Involved in AIS

Step 1. Identification of antigen: The objective function to be minimized and the constraints associated with that is considered as the antigen.

Step 2. Creation of initial antibody population: The initial population of antibody is generated randomly.

Step 3. Analysis of objective function: Each antibody is allowed to interact with the antigen and the fitness value is evaluated.

Step 4. Affinity calculation: The affinity between the antibody and antigen is determined. The antibody with higher affinity towards antigen, i.e. the 'healthy' antibody which has the ability to annihilate the antigen are added to the memory cells.

Step 5. Calculation of selection probability: The selection probability is evaluated based on probabilities of density and fitness and is given by

Fitness probability can be written as:

$$P_f = f(x_i) / \sum_{j=1}^{S} f(x_j) \tag{10}$$

The density of antibody is the proportion of antibodies with same affinity to the total number of antibodies is,

$$\begin{aligned} P_d &= \left\{ \frac{1}{S}\left(1 - \frac{t}{S}\right) \right\} \text{ for t antibodies with highest density} \\ P_d &= \left\{ \frac{1}{S}\left(1 + \left(\frac{t^2}{S^2 - St}\right)\right) \right\} \text{ for other S} - \text{t antibodies.} \end{aligned} \tag{11}$$

Here, total number of antibodies are denoted by S and t is the number of antibodies with high density.

The sum of fitness probability and density probability is called selection probability, which is given by

$$\begin{aligned} P &= \alpha P_f + (1 - \alpha) P_d \\ &\text{where } 0 < \alpha < 1 \end{aligned} \tag{12}$$

The antibodies with higher selection probability are selected for the next generation. The highest selection probability is obtained for those antibodies with low density probability and high fitness probability.

Step 6. Mutation and Crossover: The mutation and crossover process in AIS is similar to the GA. Go to step 3 till the termination criteria is reached.

The following AIS parameters are used in this work:

Size of population = 100, Maximum generations = 250, Rate of crossover = 0.8, Rate of mutation = 0.3, α = 0.4 (Fig. 3).

Fig. 3 Flowchart for AIS

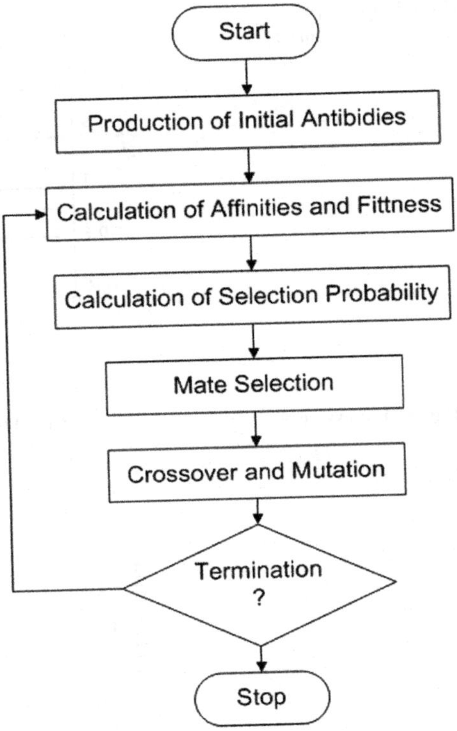

4 Formulation and Computation Using GA and AIS Algorithms

Here the computation of I versus V curves for both GA and AIS algorithms will be shown for three types of PV-SHELLS along with the IDEAL curves which are based on Experimental values while the executed curves will be based on the computed values for I_{pv} and V_{pv}. While for the significance of Objective function The FITNESS graphs will be shown for each PV-SHELL boards for 1000 w/m^2 Irradiance.

4.1 For GA

(a) For SHELL-S36:
See Figs. 4 and 5.
(b) For SIEMENS-SP70:
See Figs. 6 and 7.
(c) For SHELL-ST40:
See Figs. 8 and 9.

Fig. 4 Ipv–Vpv characteristic curve

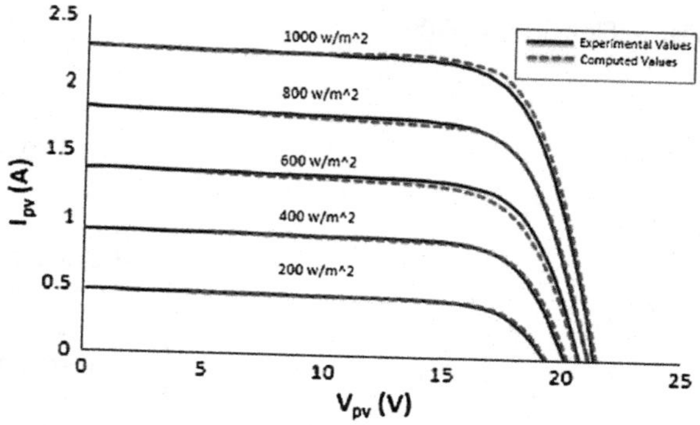

Fig. 5 Fitness curve

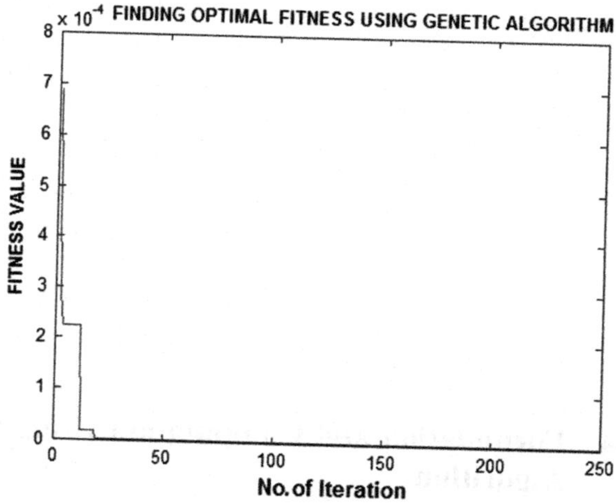

Fig. 6 Ipv–Vpv characteristic curve

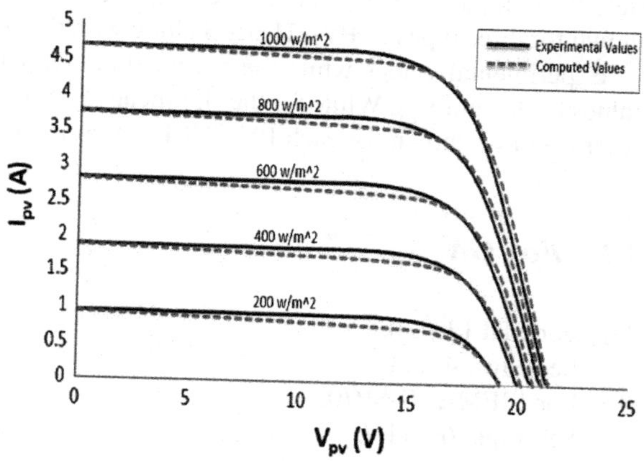

Fig. 7 Fitness curve

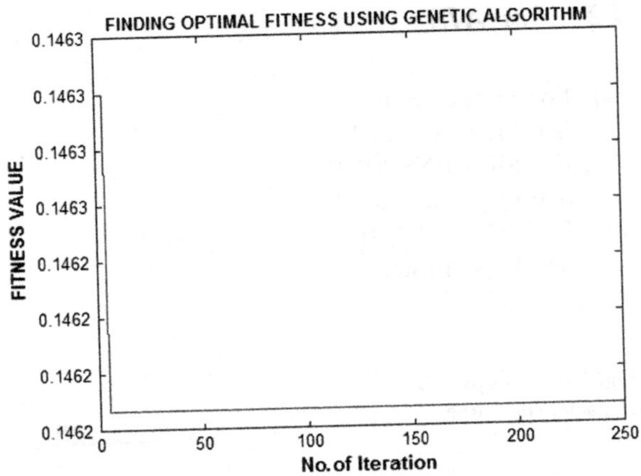

Fig. 8 Ipv–Vpv characteristic curve

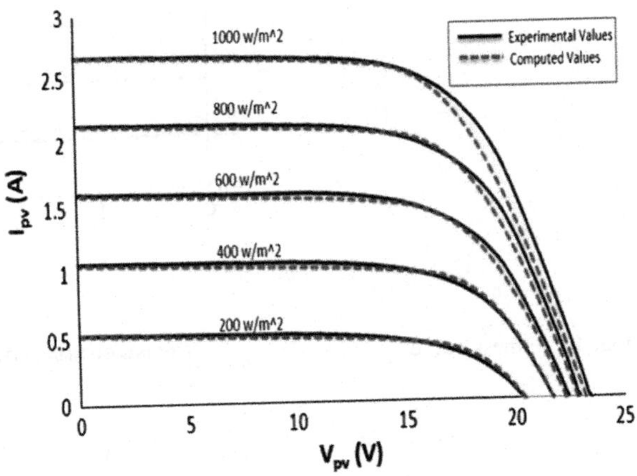

Fig. 9 Fitness curve

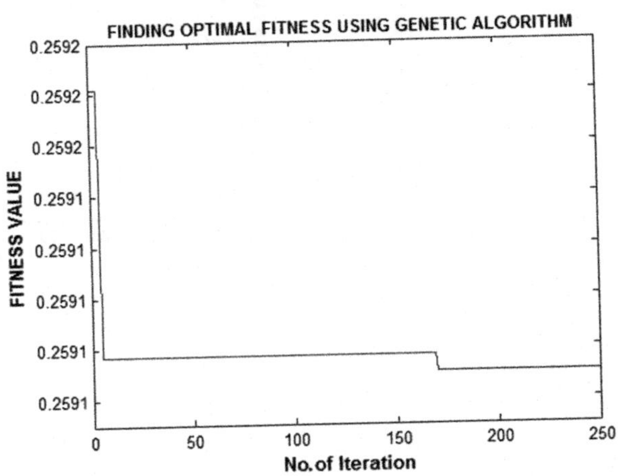

4.2 For AIS

(a) For SHELL-S36:
See Figs. 10 and 11.
(b) For SIEMENS-SP70:
See Figs. 12 and 13.
(c) For SHELL-ST40:
See Figs. 14 and 15.

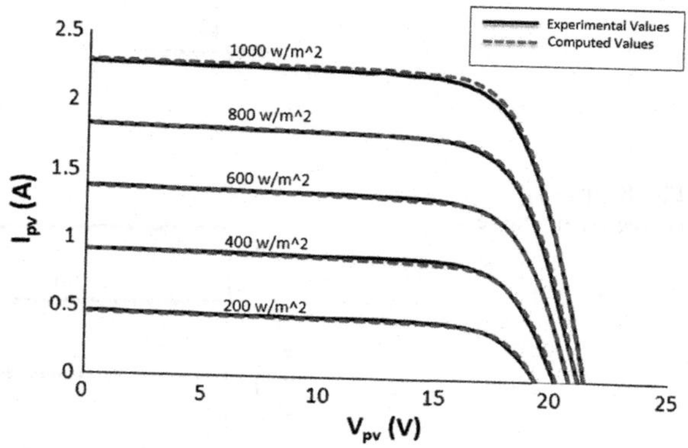

Fig. 10 Ipv–Vpv characteristic curve

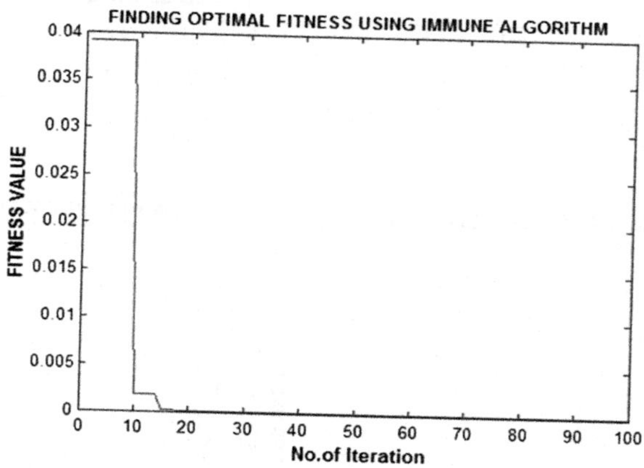

Fig. 11 Fitness curve

Fig. 12 Ipv–Vpv characteristic curve

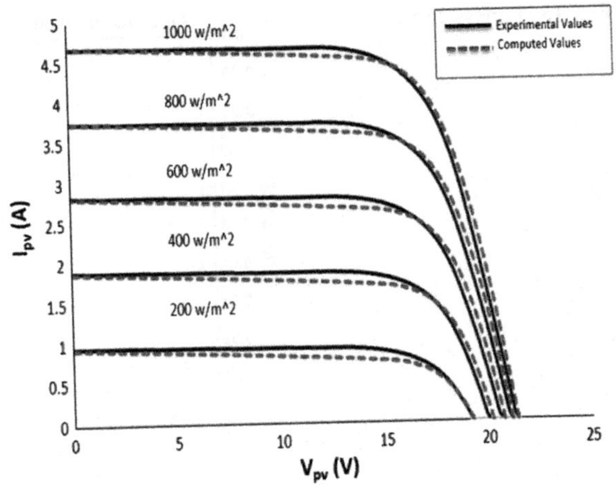

Fig. 13 Fitness curve

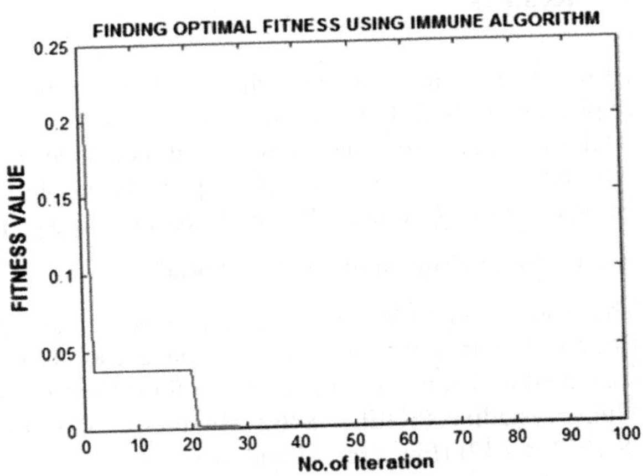

Fig. 14 Ipv–Vpv characteristic curve

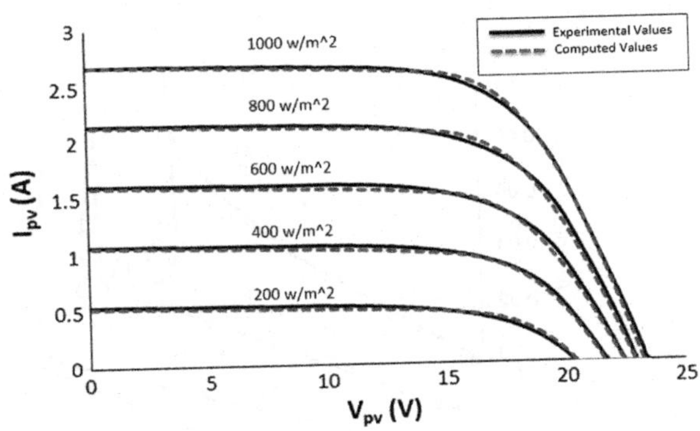

Fig. 15 Fitness curve

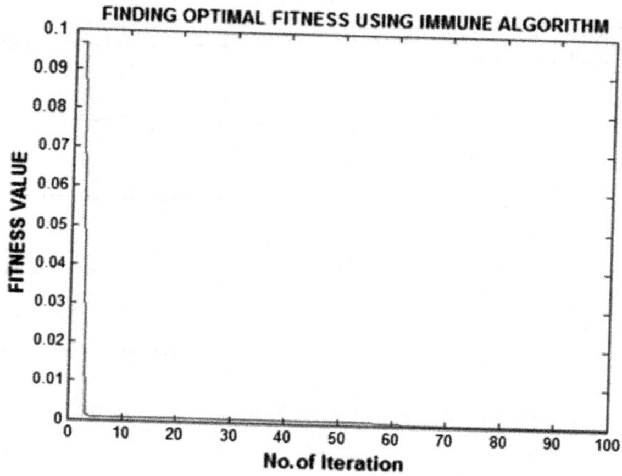

5 Results

Now, since we have obtained the 3-D Error graphs for all types of PV-SHELLS implementing both GA and AIS, now we will be comparing both the algorithms taking Error as a reference and we will conclude that which algorithm has a less Absolute error, so now we will display the results for comparison between 3-D Error graphs of GA and AIS for all the three types of PV-SHELLS.

Error comparison between GA Andais

Therefore, from all the above graph plots we can easily observe that amount of Error in GA is more for every Irradiance 200, 400, 600, 800 and 1000 w/m^2 as compared to that in the AIS algorithm for different types of three PV-SHELLS, i.e. Multi-crystalline (SHELL S36), Mono-crystalline (SHELL SP70) and Thin-film (SHELL ST40) (Figs. 16, 17 and 18).

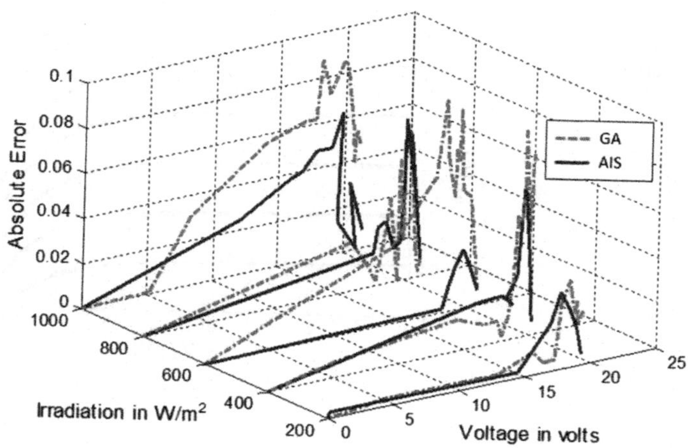

Fig. 16 A comparison of absolute errors between GA and AIS for SHELL S36

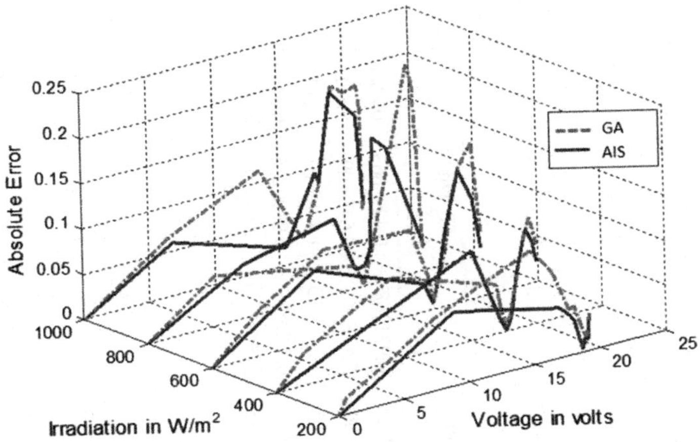

Fig. 17 A comparison of absolute errors between GA and AIS for SIEMENS SP70

5.1 *Extracted Parameters*

In this section we will be displaying all the parameters extracted implementing algorithms GA followed by AIS. Here the parameters a_1 and a_2 have been taken as fixed quantities whose values are $a_1 = 1$ and $a_2 = 1.2$.

Here the parameters are displayed for all the three types of PV-SHELLS named as, Multi-crystalline (SHELL S36), Mono-crystalline (SHELL SP70) and Thin-film (SHELL ST40) for different Irradiance (200–1000 w/m^2) (Tables 1, 2 and 3).

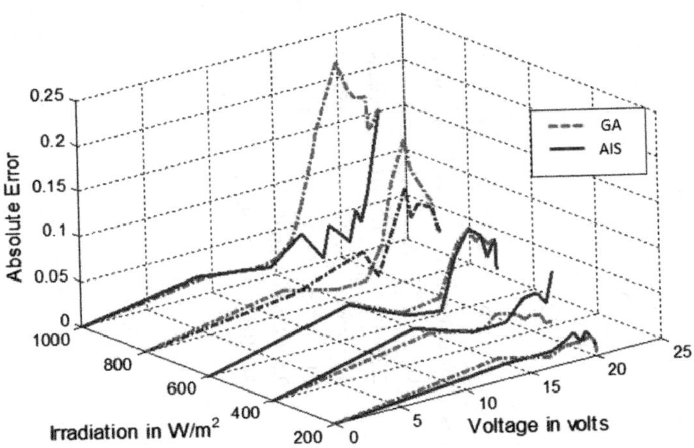

Fig. 18 A comparison of absolute errors between GA and AIS for SHELL ST40

Table 1 Comparison of extracted parameters for multi-crystalline (S 36)

Parameters	1000 w/m²		800 w/m²		600 w/m²		400 w/m²		200 w/m²	
	GA	AIS	GA	AIS	GA	AIS	GA	AIS	GA	AIS
I_{pv} (A)	2.2844	2.3031716	1.799465	1.8425373	1.3495988	1.381903	0.8997325	0.9212687	0.4498663	0.4606343
I_{01} (A)	4.73E−09	4.95E−09	4.85E−09	5.23E−09	1.04E−07	5.23E−09	5.23E−09	5.47E−09	5.64E−09	5.89E−09
I_{02} (A)	8.67E−11	9.15E−11	9.33E−11	1.08E−10	1.03E−10	1.08E−10	1.17E−10	1.24E−10	1.47E−10	1.56E−10
R_s	0.5962854	0.6060606	0.5434995	0.7214076	0.7526882	0.7292278	0.8797654	0.742913	0.9990225	0.344086
R_p	343.1085	321.60313	351.90616	228.739	402.73705	264.90714	477.02835	245.35679	247.31183	354.83871

Table 2 Comparison of extracted parameters for mono-crystalline (SP 70)

Parameters	1000 w/m^2		800 w/m^2		600 w/m^2		400 w/m^2		200 w/m^2	
	GA	AIS	GA	AIS	GA	AIS	GA	AIS	GA	AIS
I_{pv} (A)	4.7000654	4.7048512	3.7588559	3.763881	2.8216345	2.8229107	1.8844132	1.8819405	0.9471918	0.9409702
I_{01} (A)	1.01E−08	1.01E−08	9.88E−09	1.04E−08	9.57E−09	1.07E−08	9.16E−09	1.12E−08	8.50E−09	1.20E−08
I_{02} (A)	1.87E−10	1.87E−10	1.74E−10	2.01E−10	1.58E−10	2.21E−10	1.38E−10	2.53E−10	1.10E−10	3.18E−10
R_s	0.0999022	0.2248289	0.0997067	0.1915934	0.0997067	0.1075269	0.0999022	0.0175953	0.0999022	0.0078201
R_p	402.73705	473.11828	145.65005	300.09775	94.819159	391.98436	247.31183	386.11926	107.52688	488.75855

Table 3 Comparison of extracted parameters for thin film (ST 40)

Parameters	1000 w/m^2		800 w/m^2		600 w/m^2		400 w/m^2		200 w/m^2	
	GA	AIS	GA	AIS	GA	AIS	GA	AIS	GA	AIS
I_{pv} (A)	2.6810513	2.6800543	2.1432458	2.1440434	1.6074343	1.6080326	1.0716229	1.0720217	0.5358114	0.5360109
I_{01} (A)	2.31E−08	2.30E−08	2.43E−08	2.42E−08	2.59E−08	2.58E−08	2.84E−08	7.53E−07	3.31E−08	8.79E−07
I_{02} (A)	5.63E−10	5.61E−10	6.25E−10	6.23E−10	7.15E−10	7.13E−10	8.65E−10	8.62E−10	1.20E−09	1.19E−09
R_s	0.9110459	1.2785924	1.0478983	1.4780059	1.2512219	1.7478006	1.5601173	1.6304985	1.998045	1.4780059
R_p	185.92375	230.69404	241.05572	280.54741	265.68915	258.06452	294.42815	244.37928	242.81525	296.18768

Observation

For both the algorithms 'GA' and 'AIS' for all the three PV-SHELLS we may observe that when there is a decrease in 'Irradiation' also there is a decrease in 'I_{pv}' current value in a considerable amount and similarly in case of increase, so 'Irradiation is directly proportional to 'I_{pv}'.

6 Conclusion

In this work, PV model parameters are extracted from current-voltage characteristics by using AIS method. The AIS method evaluated based on convergence speed, computational efficiency, consistency, accuracy and the essential number of control parameters. Results indicate that the AIS technique gives better results than a GA approach. The feasibility of the AIS approach has been validated by manual and experimental I–V dataset.

References

1. Kashif Ishaque, Zainal Salam, Hamed Taheri.: Simple, fast and accurate two-diode model for photovoltaic modules. pp. 586–594, Solar Energy Materials and Solar Cells 95 (2), pp. 384–391, (2011).
2. Paramjit Saha, Saurabh Kumar, Sisir kr. Nayak, Himanshu Sekhar Sahu.: Parameter estimation of double diode photo voltaic-module. 1st conference on Power, Dielectric and Energy Management at NERIST, pp. 1–4, (2015).
3. Kashif Ishaque, Zainal Salam, Syafaruddin.: A comprehensive MATLAB, Simulink PV system simulator with the partial shading capability based on two-diode model, Solar Energy, pp. 2217–2227, Elsevier, (2011).
4. Hiren Patel and Vivek Agarwal.: MATLAB-based modeling to study the effects of partial shading on PV array characteristics. IEEE Transactions on Energy Conversion, 23 (1), pp. 302–310, (2008).
5. Marcelo Gradella Villalva, Jonas Rafael Gazoli, and Ernesto Ruppert Filho.: Comprehensive approach to modeling and simulation of photovoltaic arrays. IEEE Transactions on Power Electronics, 24 (5), pp. 1198–1208, (2009).
6. Chan, D.S.H., Phang, J.C.H.: Analytical methods for the extraction of solar-cell single- and double-diode model parameters from I–V characteristics, IEEE Transactions on Electron Devices, pp. 286–293, (1987).
7. Chegaar, M., Azzouzi, G., Mialhe, p.: Simple parameter extraction method for illuminated solar cells. Solid-State Electronics, pp. 1234–1237, (2006).
8. Kashif Ishaque, Zainal Salam, Hamed Taheri, Amir Shamsudin.: A critical evaluation of EA computational methods for Photovoltaic cell parameter extraction based on two diode model, Solar Energy, pp. 1768–1779, Elsevier, (2011).
9. Jervase, J.A., Boudocen, H., Ali M Al-lawati.: Solar cell parameter extraction using genetic algorithms. Measurement Science and Technology 12 (11), (2001).
10. Moldovan, N., Picos, R., et al.: Parameter extraction of a solar cell compact model using genetic algorithms. Spanish Conference on Electron Devices, (2009).

11. Liu, C.-C., Chen, C.-Y., Weng C.-Y., Wang, C.-C., Jeng, F.-L., Cheng, P.-J., Wang, Y.-H., Houng M.-P.: Physical parameter extraction from current-voltage characteristic for diodes using multiple nonlinear regression analysis. Solid-State Electronics 52 (6), pp. 839–843, (2008).
12. Ye, M., Wang, X., Xu, Y.: Parameter extraction of solar cells using particle swarm optimization. Journal of Applied Physics 105 (9), pp. 094502–094508, (2009).
13. Zwe-Lee, G.: A particle swarm optimization approach for optimum design of PID controller in AVR system. IEEE Transactions on Energy Conversion 19 (2), (2004).
14. Ortiz-Conde, A., García Sánchez, Fransico, J., Juan Muci.: New method to extract the model parameters of solar cells from the explicit analytic solutions of their illuminated I–V characteristics. Solar Energy Materials and Solar Cells 90 (3), pp. 352–361, (2006).
15. Javier Cubas, Santiago Pindado, Assal Farrahi.: New method for analytical photovoltaic parameter extraction. International Conference on Renewable Energy Research and Applications, pp. 873–877, (2013).
16. Xiaofang Yuan, Yongzhong Xiang, Yuking He.: Parameter extraction of solar cell models using mutative-scale parallel chaos optimization algorithm. Solar Energy, pp. 238–251, Elsevier, (2014).

Word Sense Disambiguation in Bengali: An Auto-updated Learning Set Increases the Accuracy of the Result

Alok Ranjan Pal and Diganta Saha

Abstract This work is implemented using the Naïve Bayes probabilistic model. The whole task is implemented in two phases. First, the algorithm was tested on a dataset from the Bengali corpus, which was developed in the TDIL (Technology Development for the Indian Languages) project of the Govt. of India. In the first execution of the algorithm, the accuracy of result was nearly 80 %. In addition to the disambiguation task, the sense evaluated sentences were inserted into the related learning sets to take part in the next executions. In the second phase, after a small manipulation over the learning sets, a new input data set was tested using the same algorithm, and in this second execution, the algorithm produced a better result, around 83 %. The results were verified with the help of a standard Bengali dictionary.

Keywords Natural language processing · Bengali word sense disambiguation · Naïve Bayes probabilistic model

1 Introduction

A word which carries different senses in its different field of uses, is called an ambiguous word and the methodology to resolute the actual sense of that particular word in a particular context is called Word Sense Disambiguation (WSD) [1–6]. As an example, the English word "Bank" has several senses in several contexts as "Financial institution", "River side", "Reservoir" etc.

A.R. Pal (✉)
Department of Computer Science and Engineering, College of Engineering and Management, Kolaghat, India
e-mail: chhaandasik@gmail.com

D. Saha
Department of Computer Science and Engineering, Jadavpur University, Kolkata, India
e-mail: neruda0101@yahoo.com

© Springer India 2016
S.C. Satapathy et al. (eds.), *Information Systems Design and Intelligent Applications*, Advances in Intelligent Systems and Computing 435,
DOI 10.1007/978-81-322-2757-1_42

The WSD algorithms are categorized into three major classes as (i) Knowledge based methods, (ii) Supervised methods and (iii) Unsupervised methods. The Knowledge based methods disambiguate the sense by the help of some machine readable dictionary, sense inventory, thesauri etc. The Supervised approaches depend on some previously annotated tagged data and the Unsupervised approaches depend on some external sources for sense disambiguation. The unsupervised approaches do not tag a particular sense to an ambiguous word; rather they classify the instances into different clusters.

The Knowledge based approaches are classified into four categories; (1) Lesk algorithm, (2) Semantic Similarity based algorithms, (3) Selectional Preference based algorithms, (4) Heuristic method based algorithms.

The Supervised approaches are further categorized into seven classes; (1) Decision List based algorithms, (2) Decision Tree based algorithms, (3) Naïve Bayes algorithms, (4) Neural Network based algorithms, (5) Exemplar-based or Instance-based algorithms, (6) Support Vector Machine based algorithms, and (7) Ensemble Methods.

The un-Supervised approaches base on (1) Context Clustering, (2) Word Clustering, (3) Spanning tree, and (4) Co-occurrence Graph.

The proposed algorithm is designed on the Naive Bayes [7] probability distribution formula. First, the algorithm was tested on a data set, collected from the Bengali corpus and the algorithm produced nearly 80 % accurate result. In addition to the disambiguation task, the sense evaluated sentences were inserted into the related learning sets to take part in the next executions. In the second phase, after a small manipulation over the learning sets, a new input data set was tested using the same algorithm, and the algorithm produced nearly 83 % accurate result.

All the results were verified by the standard Bengali dictionary.

2 A Brief Survey

Research in WSD was started notably in 1949 with the work of Zipf, who proposed that the words which are used more frequently, have more number of senses than the less frequently used words. Next, in 1950 Kaplan proposed his theory that the adjacent two words residing at the both sides of an ambiguous word in a sentence, have an equivalent meaning. There were lot of works in this field in 1980. Lesk proposed his work in 1986, which is based on overlaps among the glosses (Dictionary definitions) of an ambiguous word and its collocating words in a particular sentence. In 1990, the invention of WordNet [8] started a new era in this field of research.

As the datasets, machine readable Dictionaries, corpuses in different languages around the world are different, there were no such benchmark standard in this field of research. In 1997, Resnik and Yarowsky started Senseval evaluation procedure, which brought all the researches in this domain under a single umbrella. Today, the

Senseval exercises have brought all the researchers around the world under a same umbrella to share their thoughts and ideas in this domain.

The Asian languages, as Bengali, Hindi, Tamil, Telugu, etc. are morphologically so strong that with comparison to English language, lesser amount of work has been implemented till now.

We have gone through a detail survey [9] on the works in this domain for different Indian languages. But in Bengali, no such bench mark wok has been published except a work by Ayan Das and Sudeshna Sarkar [10].

3 Overview of Bengali Morphology

In English, there are less number of morphological varieties, compared to Asian languages. As for example, the English word "go" has five morphological varieties, as "go", "went", "gone", "going", and "goes". But in Asian languages, a word can have wide set of morphological varieties. For example, the Bengali verb খাওয়া (khāoyā: Eat) has around 165 morphological deformities in *chalit* and *sādhu* format in total.

In comparison with verb forms, the sets of nominal and adjectival deformities in Bengali are quite smaller. Generally, Bengali nouns have their inflections based on their seven grammatical categorization, two numbers as singular and plural, several determiners and a few emphatic markers etc. The Bengali adjectives have different inflections based on several primary and secondary adjectival suffixes. This large scale of inflections present in the Bengali language is the main obstacle in the research works in this language [11–21].

4 The Present Bengali Text Corpus

The present Bengali text corpus was developed under the project of the Government of India named Technology Development for the Indian Languages. This text corpus is consisted of 85 categories of text samples, as Accountancy, Banking, Chemistry, Drawing, Economics, Folk Lores, Games and Sport, History and Arts and Journalism etc.

As a whole, the corpus contains 11,300 (Eleven Thousand, Three Hundred) A4 pages. There are 271,102 (Two Lakh, Seventy One Thousand, One Hundred and Two) sentences consisted of 3,589,220 (Thirty Five Lakh, Eighty Nine Thousand, Two Hundred and Twenty) words in their inflected and non-inflected forms. This Corpus contains 199,245 (One Lakh, Ninety Nine Thousand, Two Hundred and Forty Five) unique words with their different frequency of occurrence.

5 Proposed Approach

First, the proposed model disambiguates the actual sense of an ambiguous word in a particular context using Naïve Byes probability distribution. Next, it populates the learning sets with the sense evaluated sentences, as they can be used in further execution. This auto increment property of the learning sets, to increase the accuracy of the result is the key feature of the algorithm.

5.1 *Text Annotation*

The texts, collected from the corpus were un-normalized in nature. So, all the documents were transformed into a similar format by the following steps, as (a) the font size of all the sentences are made similar, (b) the uneven spaces, new lines are removed, (c) the punctuation symbols as parenthesis, colon, semi-colon, slash, hyphen, underscore etc. are removed, (d) the regular sentence termination symbols (as "!", ".", "?") and special Bengali sentence termination symbol "|" (dāri) are taken into account.

5.2 *Flowchart of the Proposed Approach*

The flowchart of the proposed approach is given in Fig. 1.

The four commonly used Bengali ambiguous words have been used for evaluation of the algorithm, "মাথা" (māthā), "তোলা" (tolā), "হাত" (hāt) and "দিন" (din) and the performance of the algorithm is detailed in the next section.

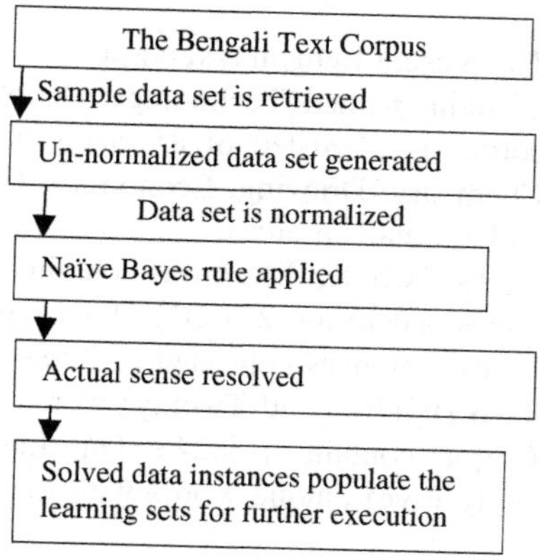

Fig. 1 The flowchart of the proposed approach

6 Results and Discussion

Total 200 instance sentences, containing the above mentioned four Bengali ambiguous words were taken for evaluation. In the first execution, 100 sentences were considered and after the sense evaluation those sentences were inserted in the relevant learning sets to enrich the learning sets (Table 1).

The steps for evaluation of output are given below:

6.1 *Preparing the Input Data*

Figure 2 depicts a part of the input data in tagged format.

Table 1 The input data

Words	Sentences taken	Different senses
"মাথা" (māthā)	30	মস্তক (mastak: head), চিন্তা (chintā: thought), প্রান্ত (prānta: end)
"তোলা" (tolā)	30	ওঠানো (othāno: pick), সৃষ্টিকরা (sreestikarā: create), সংগ্রহকরা (sangrahakarā: collect), উত্থাপনকরা (utthāpankarā: propose),প্রত্যাহারকরা (pratyāhārkarā: withdraw), অর্পণকরা (arpankarā: give)
"হাত" (hāt)	20	অবদান (abadān: contribution), হাতপাতা (hātpātā: beg) হাতবদল (hātbadal: exchange), হস্ত (hasta: hand),
"দিন" (din)	20	প্রতিদিন (pratidin: everyday), দিবস (dibas: day), দিনকাটা (dinkātā: live), দেওয়া (deyā: give)

<Sentence 1> মার্কিন যুক্তরাষ্ট্রে ফিল্ম তৈরির সঙ্গে সঙ্গেই এদেশে হোটেল ব্যবসাতেও <wsd id=1, pos=noun> হাত </wsd> লাগিয়েছেন বিজয় অমৃত রাজ।<Sentence 2> কোথাও সহানুভূতি নেই সহযোগিতার <wsd id=1, pos=noun> হাত </wsd> এতটুকু প্রসারিত নেই একটি ভাল কথা কারও মুখ থেকে শোনা যায় না।<Sentence 3> এই সব কয়টি দেশেরই জাতীয় মুক্তি ও সাম্রাজ্যবাদ বিরোধিতার আন্দোলনে ভারত একনিষ্ঠভাবে পাশে দাঁড়াইয়াছে সাহায্য ও সহযোগিতার <wsd id=1, pos=noun> হাত </wsd> প্রসারিত করিয়াছে।<Sentence 4> সঠিক ও বাস্তব মূল্যায়নের অভাব খাপছাড়াভাবে

Fig. 2 A sample tagged input

6.2 Preparing a Standard Output

The standard output file was generated previously using a Bengali lexical dictionary. The system generated output was compared with this desired output using another program (Fig. 3).

One close observation about this proposed approach is that, the disambiguated sentences in the first execution are inserted into the learning sets for further use. So, the auto-update procedure is mandatory to be 100 % accurate; otherwise the results in the second phase must be affected. But, the percentage of accuracy in the first phase is around 80 %. So, a little manipulation is done to prepare an accurate learning set.

6.3 Performances in the Two Consecutive Executions

The results of the Sense Disambiguation tasks in consecutive two executions are given in Table 2.

<Sentence 1> মার্কিন যুক্তরাষ্ট্রে ফিল্ম তৈরির সঙ্গে সঙ্গেই এদেশে হোটেল ব্যবসাতেও <wsd_id=1, pos=noun, sense=abadaan> হাত </wsd> লাগিয়েছেন বিজয় অমৃত রাজ।<Sentence 2> কোথাও সহানুভূতি নেই সহযোগিতার <wsd_id=1, pos=noun, sense=abadaan> হাত </wsd> এতটুকু প্রসারিত নেই একটি ভাল কথা কারও মুখ থেকে শোনা যায় না।<Sentence 3> এই সব কয়টি দেশেরই জাতীয় মুক্তি ও সাম্রাজ্যবাদ বিরোধিতার আন্দোলনে ভারত একনিষ্ঠভাবে পাশে দাঁড়াইয়াছে সাহায্য ও সহযোগিতার <wsd_id=1, pos=noun, sense=abadaan> হাত </wsd> প্রসারিত করিয়াছে।<Sentence 4> সঠিক ও বাস্তব মূল্যায়নের অভাব খাপছাড়াভাবে কলোনি উন্নয়নের কাজে <wsd_id=1, pos=noun, sense=abadaan> হাত </wsd> দেওয়া যথাযথ ব্যবস্থা না করে শিবির বন্ধ

Fig. 3 A sample desired output

Table 2 Comparison between two consecutive executions

Word	Sentences taken	Accuracy in the first execution (%)	Accuracy in the second execution (%)
"মাথা" (māthā)	30	83	84
"তোলা" (tolā)	30	82	84
"হাত" (hāt)	20	80	81
"দিন" (din)	20	80	82

It is observed that after the first execution the sense evaluated sentences are inserted into the learning sets, which are used in the next executions. As the learning sets are enriched in the first execution, it produced a better output in the next phase.

7 Conclusion and Future Works

In this work, initially we have considered four ambiguous words and we observed the betterment in the accuracy of results due to the auto-increment property of the learning sets. Although, we accept that a well-structured input data set and a highly populated learning set could generate a better result. In near future, we plan to compare the percentage of accuracy among more number of executions as well as more number of ambiguous words.

References

1. Ide, N., and Véronis, J.: Word Sense Disambiguation: The State of the Art. Computational Linguistics, Vol. 24, No. 1, Pp. 1–40 (1998).
2. Cucerzan, R.S., Schafer, C., and Yarowsky, D.: Combining classifiers for word sense disambiguation. In: Natural Language Engineering, Vol. 8, No. 4, Cambridge University Press, Pp. 327–341 (2002).
3. Nameh, M. S., Fakhrahmad, M., Jahromi, M.Z: A New Approach to Word Sense Disambiguation Based on Context Similarity. In: Proceedings of the World Congress on Engineering, Vol. I. (2011).
4. Xiaojie, W., Matsumot, Y.: Chinese word sense disambiguation by combining pseudo training data. In: Proceedings of The International Conference on Natural Language Processing and Knowledge Engineering, Pp. 138–143 (2003).
5. Navigli, R.: Word Sense Disambiguation: a Survey. In: ACM Computing Surveys, Vol. 41, No. 2, ACM Press, Pp. 1–69 (2009).
6. Gaizauskas, R.: Gold Standard Datasets for Evaluating Word Sense Disambiguation Programs. In: Computer Speech and Language, Vol. 12, No. 3, Special Issue on Evaluation of Speech and Language Technology, pp. 453–472 (1997).
7. http://en.wikipedia.org/wiki/Naive_bayes Dated: 27/02/2015.
8. Miller, G.A.: WordNet: A Lexical Database. In: Comm. ACM, Vol. 38, No. 11, Pp. 39–41 (1993).
9. http://arxiv.org/ftp/arxiv/papers/1508/1508.01346.pdf.
10. http://cse.iitkgp.ac.in/~ayand/ICON-2013_submission_36.pdf date: 14/05/2015.
11. Dash, N.S.: Bangla pronouns-a corpus based study. In: Literary and Linguistic Computing. 15(4): 433–444 (2000).
12. Dash, N.S.: Language Corpora: Present Indian Need, Indian Statistical Institute, Kolkata, (2004). http://www.elda.org/en/proj/scalla/SCALLA2004/dash.pdf.
13. Dash. N.S.: Methods in Madness of Bengali Spelling. In: A Corpus-based Investigation, South Asian Language Review, Vol. XV, No. 2 (2005).
14. Dash, N.S.: From KCIE to LDC-IL: Some Milestones in NLP Journey in Indian Multilingual Panorama. Indian Linguistics. 73(1–4): 129-146 (2012).

15. Dash, N.S. and Chaudhuri, B.B.: A corpus based study of the Bangla language. Indian Journal of Linguistics. 20: 19–40 (2001).
16. Dash, N.S., and Chaudhuri, B.B.: Corpus-based empirical analysis of form, function and frequency of characters used in Bangla. In: Rayson, P., Wilson, A., McEnery, T., Hardie, A., and Khoja, S., (eds.) Special issue of the Proceedings of the Corpus Linguistics 2001 Conference, Lancaster: Lancaster University Press. UK. 13: 144–157 (2001).
17. Dash, N.S. and Chaudhuri, B.B.: Corpus generation and text processing. In: International Journal of Dravidian Linguistics. 31(1): 25–44 (2002).
18. Dash, N.S. and Chaudhuri, B.B.: Using Text Corpora for Understanding Polysemy in Bangla. In: Proceedings of the Language Engineering Conference (LEC'02) IEEE (2002).
19. Dolamic, L. and Savoy, J.: Comparative Study of Indexing and Search Strategies for the Hindi, Marathi and Bengali Languages. In: ACM Transactions on Asian Language Information Processing, 9(3): 1–24 (2010).
20. Dash, N.S.: Indian scenario in language corpus generation. In: Dash, Ni S., P. Dasgupta and P. Sarkar (Eds.) Rainbow of Linguistics: Vol. I. Kolkata: T. Media Publication. Pp. 129–162 (2007).
21. Dash, N.S.: Corpus oriented Bangla language processing. In: Jadavpur Journal of Philosophy. 11(1): 1–28 (1999).

Efficient Methods to Generate Inverted Indexes for IR

Arun Kumar Yadav, Divakar Yadav and Deepak Rai

Abstract Information retrieval systems developed during last 2–3 decades have marked the existence of web search engines. These search engines have become an important role player in the field of information seeking. This increasing importance of search engines in the field of information retrieval has compelled the search engine companies to put their best for the improvement of the search results. Therefore the measurement of the search efficiency has become an important issue. Information retrieval is basically used for identifying the activities which makes us capable to extract the required documents from a document repository. Information retrieval today is done on the basis of numerous textual and geographical queries having both the textual and spatial components. The textual queries of any IRS are resolved using indexes and an inversion list. This paper mainly concentrates on the indexing part and the analysis of the algorithm. Several structures are in existence for implementing these indexes. Hash tables, B-trees, sorted arrays, wavelet trees are few to name. For an efficient data structure there are different deciding parameters that are to be taken into account. Some important parameters considered in this paper are index creation time, storage required by inverted file and retrieval time. This paper provides a detailed comparative study of different data structures for the implementation of inverted files.

Keywords Geographical information system · Information retrieval · Wavelet tree · Hash function

---

A.K. Yadav (✉) · D. Rai
Ajay Kumar Garg Engineering College, Ghaziabad, India
e-mail: ak_yadav@yahoo.com

D. Rai
e-mail: ramdeepakniwash@gmail.com

D. Yadav
Jaypee Institute of Technology, Noida, India
e-mail: dsy99@rediffmail.com

© Springer India 2016
S.C. Satapathy et al. (eds.), *Information Systems Design and Intelligent Applications*, Advances in Intelligent Systems and Computing 435,
DOI 10.1007/978-81-322-2757-1_43

1 Introduction

There is no right or wrong way to carry out search over the web. We need to do experiment by adding and taking away words and trying the different combinations. No search engine has indexed the entire web, and no two are the same. Search engines stores the documents in a storehouse and maintains an index same as in any database systems. These indexes are processed in order to evaluate the queries for identifying the matches which are then returned to the user.

Indexing is basically the process of building a data structure that which will make the complete process of text searching very fast [1]. There are many existing retrieval approaches based on which different classes of indexes are made. These indexes are either based on tree or on hashing. In the complete process of retrieval filtering always precedes indexing [2]. Filtering simplifies the entire process of searching by reducing the size of text [3, 4]. Filtering involves removal of some common words using stop words, transformation of uppercase letters to lowercase letters, removal of special symbols, reduction of multiple spaces to one space, transformation of dates and numbers to standard format [5].

There are three parts of inverted file namely, inverted list, dictionary and vocabulary [6]. All the distinct items present in the database are listed by vocabulary. And, the information about the document in which the item is present is contained in the inverted list. Example 1 explains the working of inverted file.

Suppose we have a list of documents as shown in Table 1 with their document ids. Now the task is to identify potential index-able elements in the documents. In designing such a system, we need to define the word 'term'. We need to further clarify whether the 'term' or the 'keyword' is an alphanumeric character or punctuation. The next step would be to delete the 'stop' words from the given documents such as articles, conjunctions, prepositions, pronouns etc. Having completed the above stated steps, we get a resultant document file. Using this file, we create a global dictionary where weights are assigned to terms in the index file. These weights can be made more sophisticated by measuring the number of times the term has occurred in the document. Finally, inverted file is created which stores the index information. The same is searched for each and every query. Table 2 shows the inverted file which consists of all the keywords in a set of documents and their corresponding document number.

Table 1 List of documents

Documents (id)	Text
1	Hot and cold milk bowl
2	Milk bowl in the pot
3	New date of old
4	I like it hot and cold, some like it cold
5	Some like it in the pot

Table 2 Inversion list

Number	Text	Documents
1	Bowl	1, 2
2	Cold	1, 4
3	Date	3
4	Hot	1, 4
5	I	4
6	In	2, 5
7	It	4, 5
8	Like	4, 5
9	Milk	1, 2
10	Old	3
11	Pot	2, 5
12	Some	4, 5
13	The	5

For implementing inverted files there are many different structures available. These structures are used both for storing text in secondary memory and for searching algorithms. Arrays have not been considered in this paper as because they are very popular and are well known by almost all. This paper discusses a new wavelet tree based index structure in addition to B-trees and Hash trees. This new structure keeps a proper balance between the search efficiency and the storage space required. Wavelet tree implements inverted index with in inverted file. B-tree, hash tree and wavelet tree differs mainly in the way of searching. In search trees the complete value of a key is used for searching. In hashing the original order of the data is randomized which makes the searching faster. But, hashing has a disadvantage that we cannot scan sequentially.

Further the paper is divided in several sections. Section 2 describes the related concepts. Section 3 present different data structures for implementing inverted file. Section 4 present results and analysis. Finally, we conclude in Sect. 5.

2 Literature Review

From early 1994 users started using appropriate tools for accessing documents over internet. Before that, it was only possible to index and manage the title [7]. After that much advancement have been done in this field. These advancements were one of the biggest events in the history of information retrieval system. Among this Web Crawler, was the first tool which was developed at the University of Washington (USA). A web crawler facilitates complete text searching of the entire web documents. It is in work since April 1994. In July 1994 a different search engine named Lycos was developed at Carnegie Mellon University (USA) [8].

With these developments it became possible for the users to have different web tools with effective functionalities for information retrieval.

Based on numerous data structures the inverted file has been till date implemented in a number of ways. From time to time the efficiency of one or the other data structure used to implement an inverted file has been computed. *Guillaume Latu* in his paper in 2010 has given an efficient mechanism of creating hash tables for better usage in an inverted file [9]. He clearly stated the principle of the hash table with the specification of one of the drawbacks of hashing. Hash function should have a unique slot index corresponding to every possible key. But, ideally this is not achievable. In most of the hash table designs occurrence of hash collisions are considered to be normal. But, there is an approach to deal with this. We can have a linked list of keys that are hashed to the same location. Searching in this is done by hashing its key to find the index of a bucket, and the linked list stored into the bucket is scanned to find the exact key. These types of data structures that use linked lists are called chained hash tables.

When the drawbacks of the hash table were overcame and B-Tree came into existence, the graph structure proved to be very instrumental [10, 11].

Their works clearly reflected the computational mechanism of a simple B-Tree and how the searching and traversing algorithms can be effectively implemented in a B-Tree. *Phadke* on the other hand showed the implementation of B-Tree over an inverted file in his project analysis report [12]. He showed by example that how a B-Tree divides the indexes of an inverted file over the node structure giving complete analysis of the B-Tree.

This paper hence, tries to provide a complete comparative analysis of the various data structures that have been used in the past and the ones that are being used in the present to implement an inverted file for search engine indexing. It tries to set a benchmark and aim to aid further help in the field of search engine indexing.

3 Data Structure for Inverted File

3.1 Inverted File Using Hash Table

Hash table based inverted file can be created using the Java class library [13]. It basically maps keys and their corresponding values. Here, the keys are the keywords and the lists of URLs are the values. Whenever collision searching is done sequentially hash table remains open. The quantity up to which the hash table is to be populated before increasing its capacity is decided by a parameter known as load factor. For the creation of inverted index sequential visit of URL's are done and the words contained in them are read. Then hashing of the key is performed. The structure of hash table based inverted file index is shown in Fig. 1.

One of the advantages of the hash table is that the required creation time is less in comparison to sorted array. Also it requires only an elementary hash function for

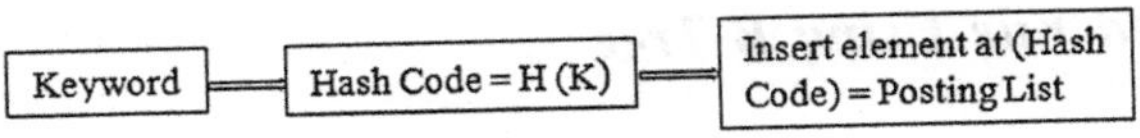

Fig. 1 Structure of Inverted File

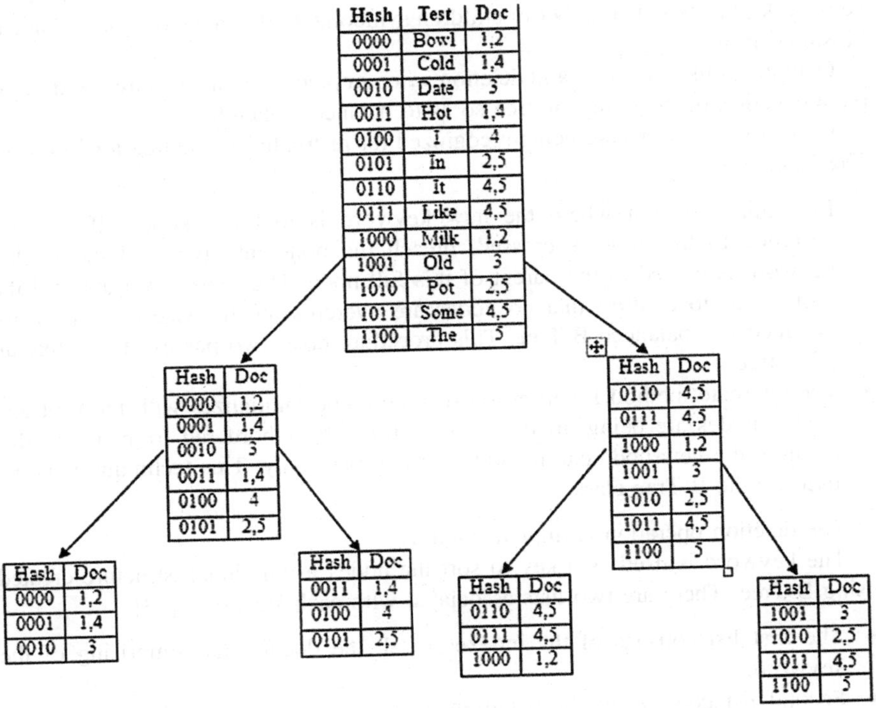

Fig. 2 Hash tree

the computation of the hash. This computed hash works as an index of the array where the posting list is stored. Hash table is very efficient data structure in the absence of the binary search for the creation of the inverted index. However maintaining the multiple hash tables is a consequential memory overhead. Hence a different memory efficient data structure can also be used.

For, implementing the inverted file using hash tree, the example explained in Sect. 1 has been considered. The constructed hash-tree is shown in Fig. 2.

Now, for the hash tree shown in Fig. 2, if we keep only one node at the leaf in each level then the total number of levels would equal to 5.

3.2 Inverted File Using B-Tree

A balanced tree having all their leaf node at the same level is referred to as B-Tree. In such a tree, a key is used to sort all the nodes. Every node can store 'n' number of keys as its upper limit and the minimum holds at 'n/2'. A node having number of keys as 'k', has 'k + 1' number of children symbolizing the range of values that can be stored in it.

Operations like insertion and deletion when implemented in such tree lead to the reconstruction of the complete tree so as to balance its height.

Basically, two ways are being recognized for restructuring the tree for insertion. They are:

- Recognize the slot where the new keyword is to be reckoned. If the size specified to the node is crossed, breach the node into two and the medial keyword is passed to the parent of this full node. This step is repeated till the root, so as to confirm that no nodes are overcrossing the size and the result obtained is a balanced B-Tree. This procedure takes two passes of tree like an AVL tree.
- The alternate method for insertion of a new node initializes with the root and further nodes are being divided whose size is over bordered from 'n'. In this method, the new node can be added in one pass only. This technique is being used by the B-Tree class.

The deletion operation is implemented as:

The keyword is used as a key to sort the nodes in this index structure created using B Tree. There are two lists associated with each keyword, [13]

- The first list consists of the indexes to all the documents comprising of the keyword.
- Second list consists of the occurrence list i.e., frequency of keyword in the document.

Each node has a minimal of 'n/2' and maximum 'n' children (n > 1) and 'n' is a fixed integer. The root has at least two children while the leaf nodes have no child. Complexity of insertion of a keyword in B Tree is O (nlogn i), 'n' is the order of tree and 'I' is the total number of keywords indexed [13].

One of the chief advantages of B tree is the balanced height. As well as, B Trees are compatible for on-disk storage and searching since we can synchronize the node size to the cylinder size. This helps in storing more data members in each node and thus the tree is oblate/flattened, and henceforth less node to node transitions are being required [14]. The inverted file implemented using B Tree is explained in Sect. 1. Constructed B Tree is shown in Fig. 3.

Analyzing the most probable B Tree implementation, it is observed that the level is 4. It is espied that the number of levels when Hash tree implementation is used are more than when we use B Tree implementation. Hence forward, B Tree index implementation is more competent in terms of complexity of implementation.

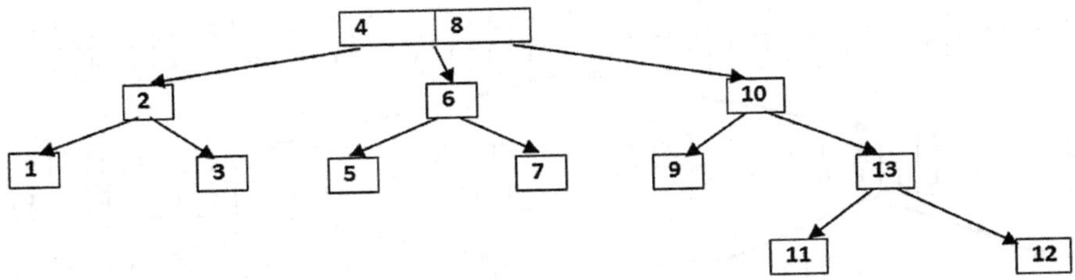

Fig. 3 B-tree

3.3 *Inverted File Using Wavelet-Tree*

Wavelet tree refers to a data structure that maintains a sequence of symbols over an alphabet. The wavelet tree is a balanced binary tree.

We identify the two children of a node as left and right. Each node represents a range R of the alphabet. The inverted file for any document consists of a list of keywords of the document along with the occurrence positions of the respective keywords. Based on their relative positions in the document a wavelet tree can be constructed wherein the nodes represent the key terms' positions and a range of values from 0 to 1 are assigned to each position. The tree is subdivided into its child nodes by placing all the 0 numbered indexes at the left sub node and the right node consists all those with value assigned 1. However once the positions are subdivided into sub nodes they are again allotted values from 0 or 1 based on the criterion that lower numbered positions get value 0 and consequently the higher numbered positions get value 1 as shown in the Fig. 4.

Fig. 4 Wavelet tree

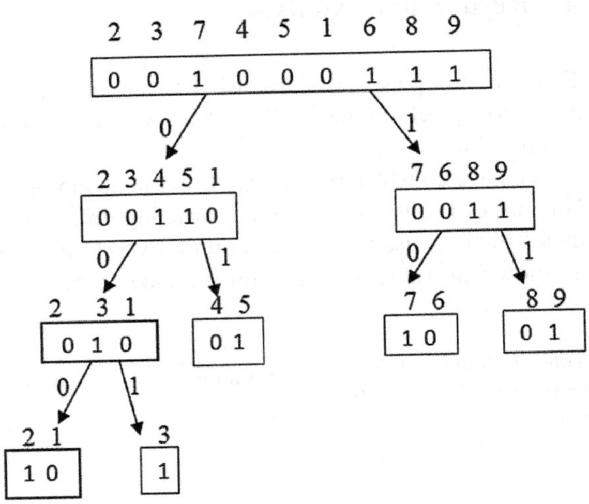

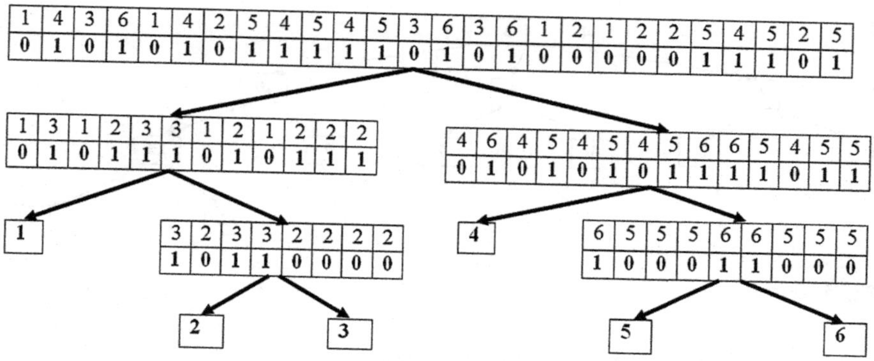

Fig. 5 Wavelet tree for the example in Sect. 1

Hence, from the tree in Fig. 4 the keyword can be found out in the document which gives the URL of the document so needed.

The biggest advantage of using a wavelet tree is that it gives maximal performance for breadth-first traversal. Many fields are there to support the efficiency of wavelet trees. In the field of computer science it has proved to be a reliable technique, especially in computer graphics where it has been used for multi resolution representation.

Although the Hash tree and B-tree implementation have been the most popular ones to be used in creating an inverted index but the wavelet tree is gaining quick importance these days. For implementing inverted file using wavelet tree, the example explained in Sect. 1 has been considered. The constructed wavelet tree is shown in Fig. 5.

4 Results and Analysis

Through this paper we have tried to provide a comparison as to which implementation would be most efficient when implementing the inverted index using an inverted file.

We have considered an example explained in Sect. 1 for the purpose of analysis. Number of levels required in constructing different trees corresponding to different data structures used in implementing inverted file has been considered for analysis as shown in Table 1. Based on the data in Table 3, we have plotted a graph as

Table 3 Number of levels required by different data structures

Hash tree	B-tree	Wavelet tree
5	4	3

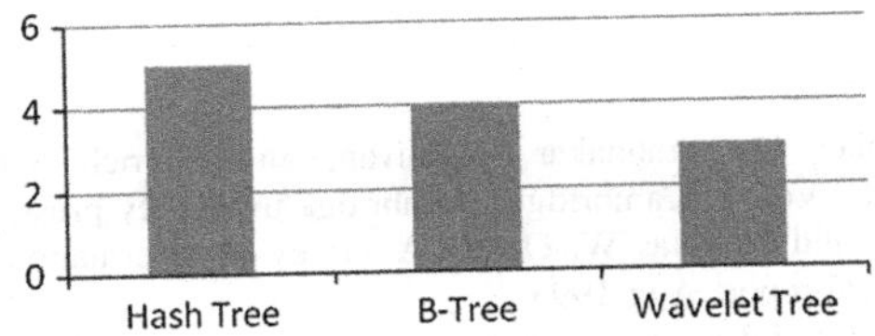

Fig. 6 Graph showing number of levels

shown in Fig. 6, which shows the performance measure of the different data structures of the inverted file in the form of the number of levels needed to represent one particular implementation where the Y axis represent the number of levels needed for each data structure.

It is clear from the graph that the number of levels required is minimum for wavelet tree in comparison to hash tree and B-tree. Apart from using the number of levels metric for differentiating among the various data structures of the inverted file we have also considered time complexity factor for differentiation among them. It is found that, Hash tree have the searching time of order 1, whereas B-tree have $O(m\log_m n)$ search time. Here, m is the order of the B-tree.

5 Conclusion and Future Scope

This paper tried to study the world of inverted file based indexing techniques. It basically provides the researchers with the theoretical results of indexing algorithms. Here, we examined the available data structures like Hash tree, B-tree and Wavelet tree for implementing inverted file through an example. We considered the number of levels and time complexity as the criteria for the analysis. The overall performance analysis of the proposed data structures and the inverted index file concludes that for having efficient search engines the most important thing required is to have efficient algorithm and data structures. And it has been found from this paper that wavelet tree based inverted file is most efficient among all available data structures for the inverted file. It is found to be efficient in terms of time complexity, space complexity and the complexity of implementation.

This paper summarizes the works done in the field of inverted file based indexing till now. These findings provide better understanding and motivations for further research. In future wavelet tree can be efficiently used for both textual and spatial indexing.

References

1. Manning, Christopher D., Prabhakar Raghavan, and Hinrich Schütze,.*Introduction to information retrieval*. Vol. 1. Cambridge: Cambridge university press, 2008.
2. Faloutsos, Christos, and Douglas W. Oard. "A survey of information retrieval and filtering methods." A technical report Aug 1995.
3. Singhal, Amit. "Modern information retrieval: A brief overview." *IEEE Data Eng. Bull.* 24.4 (2001): 35–43.
4. Belkin, Nicholas J., and W. Bruce Croft. "Information filtering and information retrieval: two sides of the same coin?." *Communications of the ACM* 35.12 (1992): 29–38.
5. Amble, Ole, and Donald Ervin Knuth. "Ordered hash tables." *The Computer Journal* 17.2 (1974): 135–142.
6. Callan, James P., W. Bruce Croft, and Stephen M. Harding. "The INQUERY retrieval system." *Database and expert systems applications*. Springer Vienna, 1992.
7. Agosti, Maristella, and Massimo Melucci. "Information retrieval on the web."*Lectures on information retrieval*. Springer Berlin Heidelberg, 2001. 242–285.
8. S. Davis Herring. *The value of interdisciplinarity: A study based on the design of Internet search engines*. Journal of the American Society for Information Science, 50(4):358–365, 1999.
9. Latu, Guillaume. "Sparse data structure design for wavelet-based methods."*ESAIM: Proceedings*. Vol. 34. EDP Sciences, 2011.
10. Broder, Andrei Z., et al. "Indexing shared content in information retrieval systems." *Advances in Database Technology-EDBT 2006*. Springer Berlin Heidelberg, 2006. 313–330.
11. Broder, R. Kumar, F. Maghoul, P. Raghavan, S. Rajagopalan, R. Stata, A. Tomkins, and J. Wiener. *Graph Structure in the Web*. In Proceedings of WWW9 Conference, 2000.
12. Biswas, Ingrid, and Vikram Phadke. "Project Report Comparative Analysis of Data Structures for Inverted File Indexing in Web Search Engines."
13. Ingrid Biswas and Vikram Phadke, "Comparative Analysis of Data Structures for Inverted File Indexing in Web Search Engines".
14. Waisman, Amnon, and Andrew M. Weiss. "B-tree structured data base using sparse array bit maps to store inverted lists." U.S. Patent No. 4,606,002. 12 Aug 1986.

Intelligent Mail Box

Hitesh Sethi, Ayushi Sirohi and Manish K. Thakur

Abstract In 21st century, email is one of the most effective ways of written communication due to its easy and quick access. But now days with each individual receiving large number of emails, mostly promotional and unnecessary mails, organization of emails in individual's inbox is a tedious task to do. In last decade, researchers and scientific community have contributed lot for organization of individual's inbox by classifying the emails into different categories. In this paper, we propose an intelligent mail box where email classification has been carried out on the basis of labels created by users and needs few training mails for future classification. Hence it provides more personalized mail box to the user. The proposed system has been tested with various classifiers, viz. Support Vector Machine, Naïve Bayes, etc. and obtained the highest classification accuracy (66–100 %) with Naïve Bayes.

Keywords Personalized mail box · Email classification · Email organization · Stemming · Support vector machine · Naïve bayes

1 Introduction

With the advent of internet based technologies, 21st century has seen emergence of emails as one of the most prominent forms of written communication which is evident from the fact that "Gmail" (Google email service provider) itself has about 425 million active users in the year 2012.

H. Sethi (✉) · A. Sirohi · M.K. Thakur
Jaypee Institute of Information Technology, Noida 201307, India
e-mail: hitesh.28jan@gmail.com

A. Sirohi
e-mail: ayushi160793@gmail.com

M.K. Thakur
e-mail: mthakur.jiit@gmail.com

© Springer India 2016
S.C. Satapathy et al. (eds.), *Information Systems Design and Intelligent Applications*, Advances in Intelligent Systems and Computing 435,
DOI 10.1007/978-81-322-2757-1_44

E-mail is one of the most common ways of communicating, sharing data, promoting etc. in today's time. Emails have made human lives easier as one can send and receive information in moments. This information can belong to different categories viz. personal messages, project related or job related information, advertisements etc. Now a days, this has increased both the number and the diversity of emails received by a person, resulting into flooded inbox (of individual's email account) with lot of emails which is not so urgent or unwanted along with important emails. With increase in number of emails a user receives, it becomes more and more difficult for user to find out the important and urgent mails.

Hence, it has been everyone's need to arrange the flooded inbox such that emails with immediate attention or specific emails can be easily identifiable. Manually classifying the emails is one of the ways to arrange the inbox, but it is a time consuming process and hence auto classification of emails becomes a need for the end users now a days. Using this one can either auto prioritize the emails or emails can be auto classified under different categories.

Further, there are several email clients who provide the facility to apply filters using which one can move the filtered mail to an assigned folder or attach a specified label. But these filters are not always effective as they seek user's attention and are tedious to use. Hence it is not a very common practice among users [1].

In addition, as users receive diversified emails so there can be no custom defined folders for all users as user's needs are so varied and large in number. Hence, it is required to shift the focus from customization (which is provided by many email clients like Gmail) to personalization i.e. providing controls to the user to auto classify his/her inbox according to the needs [2]. Personalization lead to an increase in user satisfaction because here the machine adapts to the whims and need of the user rather than user adapting to the functionality provided by the machine.

Further, personalization can be achieved when the machine is in constant supervision and receives feedback from the user. Here, the application has to be constantly analyzed by the user viz. what folders the user uses to classify mails, what are the keywords which distinguish the emails from one category to another and the user response in wrong classification of email. In brief, the classifier has to adapt to the user needs and correctly classify the emails [3].

Considering these issues, we present in this paper an Intelligent Mail Box to increase the user experience by providing personal assistance to users. It is an application developed to enhance user experience and having simple user interface for classifying emails quickly and effortlessly.

Besides introduction in Sect. 1, remaining of the paper is organized as follows: Sect. 2 discusses the background study; Sect. 3 presents the proposed approach; implementation and results have been discussed in Sect. 4 followed by conclusion and references.

2 Background

In this section, we present the recent contributions for developing the auto classification systems for emails. These contributions are majorly based on the supervised learning, where an input x is mapped to desired output y using some classifiers for predicting the output of new unseen data. These supervised learning approaches include, Rule Based Approach [4], SVM (Support Vector Machine) [5], Decision Tree Approach [6], Naïve Bayes Approach [7], etc.

Further, the background study also involves new ideas like how one can classify mails in folders and clusters using the NGram [8]. Authors in [8] presented that classification of documents and emails in particular require many techniques of natural language processing like stemming, parsing, etc. Further, "Glabel" [2], is an interface agent helping to improve the user experience of existing mail client by automatic labeling of new mails with the knowledge of the labels of existing emails. Also for improving research in email classification Enron Corpus [9] was made available which consists of large number of emails on which folder prediction can be analyzed. In addition, document classification plays an important role for email classification, viz. multi-labeling of documents [10], which is applied in real for Czech News Agency.

Subsequently, we present some of the classification strategies viz. Stemming, TF-IDF (Term Frequency-Inverse Document Frequency), SVM, etc. for email classification along with the briefing of our approach which utilizes these concepts.

2.1 Stemming

Natural language contains many variants of same root word like Find, Finds, and Finding etc. Stemming is a very common pre-processing step which helps in reducing the morphological forms i.e. different grammatical form of word (noun, verb, adverb, adjective, etc.) to the common base word. Stemming is necessary as it helps in identifying inflectional and sometimes derivational form of the word [11]. While converting a word to its stem it is assumed that all morphological forms of the stem have same semantic meaning. Thus after application of stemming the key terms in the mail are reduced to their root form and so minimizing the distinct terms which in turn help in minimizing the processing time. Stemming is rule based and most common algorithms used for stemming are Potter's Stemmer [12] and Snowball Stemmer [13].

2.2 TF-IDF

TF-IDF is used in determination of words which belong or are more favorable to be present in a given document [14]. The algorithm determines the relative frequency of a specific word in a mail as compared to the inverse proportion of that word over the entire emails received. Mathematically it can be represented as follows:

$$w_m = f_{w,m} * \log(|M| f_{w,M})$$

where, M is the collection of emails, w represents a word in a mail, m stands for a mail (m ϵ M) and $f_{w,m}$ is the frequency of word w in a mail m, $|M|$ stands for number of emails and $f_{w,M}$ stands for number of mails in which word w appear in mails.

Words which have a high value of TF-IDF for a given email have a strong relation with that mail and hence helps in classifying emails more effectively [15, 16].

2.3 Support Vector Machines (SVM)

SVM is a set of related supervised learning methods for classification and regression [17] and mostly used as a framework of binary classification. In case of multiple classes, the process of classification is achieved by decomposing it to multiple binary classes and observing one class versus others. In general, SVM model represent examples as points in space which are mapped such that examples/points of different classes are separated by a gap which should be as wide as possible [17, 18].

In our approach, we applied SVM to classify mails according to the classifier or folder created by the user. We have used LIBSVM which is a library supporting SVM [19] and supports most of the SVM classification, distribution, regression, multi-class classification, cross validation for model selection, probability estimates, various kernels, weighted SVM for unbalanced data, etc.

2.4 Naïve Bayes

Naïve Bayes classifier is the simplest of the Bayesian Classification models and is applied with an assumption that all the attributes of an example are independent of each other given the context of a class [20, 21]. Naïve Bayes gives a base line in classification and its mathematical model is as follows:

$$y_0 = \arg\max_{y=1,..,m} P(c_y|d_i, \theta) = \arg\max_{y=1,...,m} P(c_y|\theta)P(d_i|c_y, \theta)$$

where, d_i stands for the document which is assigned to a category c_{y_o} and the parameter θ is estimated from the training set which generally use multinomial Bernoulli Model and maximum likelihood $P(d_i|c_y, \theta)$ is calculated as

$$P(d_i|c_y, \theta) = \prod_{k=1}^{|d_i|} P(w_k|c_y, \theta)$$

where $w_1, \ldots, w_{|d_i|}$ represents words in a given document d_i.

2.5 *KNN*

K-Nearest Neighbor (KNN) share similarities with Nearest Neighbor algorithm with the exception that in KNN algorithm only the K instances/objects are looked into for classifying the unclassified instance/object. Distance between the neighbors is calculated by computing Euclidean Distance between them as follows:

$$D = \sqrt{\sum_{i=1}^{k} (x_i - y_i)^2}$$

The value of K is obtained by hit and trail method and for our approach (email classification) we have chosen K = 5.

3 Proposed Approach

This section presents the design of our proposed Intelligent Mail Box, where we discuss the back end and front end algorithms of our proposed system. Here user can make his/her own folders and the new receiving mails will go to the respective folders. We have proposed to classify our new incoming mails by using Naïve Bayes (NB) classification [21]. For classification, a user needs to create his/her label based on the needs (viz. Alumni, Project, etc.) followed by putting some seed emails in that corresponding label which will serve as training data set. The incoming new mails will be the test data set and if user finds that the classification is wrong, the proposed feedback mechanism will run and will re-train the machine making it adaptive to the users' need. This process will automate the whole process of user classification without any need of filters.

Subsequently we present the classes or labels, training set, and algorithms for back end (training of the system) and front end (working of client i.e. user level).

Classes or labels Classes in our application are user defined labels whose names serves as the name of the folder/classifier/class they represent and accordingly mails are classified. A user needs to create his/her class/label, if does not exist. The classifier needs to be trained/re-trained (based on user's feedback) for the newly created classes/labels using some of the seed mails belonging to these classes/labels.

Training set for each user

- Dataset: Our dataset consists of classified mails received by the mail ids of the user which he initially puts in the folders he has created.
- Classes: Can be any labels especially the categories from which the user receives maximum mails, viz. Inbox/Primary, Jobs, Promotions, etc. are the classes that relate more personally to graduate students who are looking out for new ventures. Project may be an important class for students studying in school, whereas Alumni can serve as a class for professors who want to in touch with their old students.
- Seed Mails: We used at least 10 mails in each folder for email classifications.
- Retraining: Retraining of the classifier is done if the user indicates that the mail has been classified in the wrong folder.

As stated earlier, initially user has to map the folder with the mails he/she wants to classify in that folder and from here our machine starts to learn. We have also parsed the HTML present in some of the incoming mails.

Algorithm Here we present the Front End and Back End algorithms used for the proposed system.

Front End – Working of client

Step 1: User opens the application
Step 2: Enter Credentials
Step 3: If user wants to use existing labels from Gmail then go to Step 7
Step 4: Download the mails
Step 5: User to make new labels
Step 6: Each folder is given enough seed mails (10 to 50) to train the machine
Step 7: Classified view of the mails
Step 8: Ask user for feedback. If the mails are correctly classified go to Step 10
Step 9: Repeat Step 6 to Step 8
Step 10: Show classified mails

Back End – Training of the system

Step 1: Row mails (String text, Class) // A function which have the content of Mail as String, Class stands for the classifier.
Step 2: Remove Stop words
Step 3: Apply Potter's Algorithm (Using Snowball)
Step 4: Apply String to Word Vector Transform
Step 5: Apply TF-IDF transform
Step 6: Train the system using Naïve Bayes (On reduced feature set)#
Step 7: Test the new mails and put in suitable folder.
#SVN and KNN may also be applied for training the system.

4 Implementation and Results

This section presents the implementation of the proposed algorithms along with the obtained results followed by discussion on the validation of obtained results.

Figure 1 presents the snapshot of the user interface (front end) of the inbox of a client, implemented using the front end algorithm presented in previous section. Here, labels can be created by clicking on the add category button, where one can write the label or classifier name, viz. Alumni, Personal, Job, etc. One can create as many labels or classifiers as possible. For training the new classifiers created one just have to right click on the mail and select move to from the drop down menu (Fig. 2), after selecting the classifier, selected mails will be moved to the desired folder. Here, we have also provided an option for those users who already have labels in their existing mail account so that they don't have to waste time in making those labels again.

After this, selected emails (ranging from 10 to 90 % of total emails of specific class viz. Alumni) are used as training data set for training the classifiers (Naïve Bayes, SVM, and KNN) and remaining emails (90–10 %) have been used for testing. Later, the trained classifier classifies the remaining or incoming (additional) mails to be placed into respective labels or classes. Table 1 presents the obtained classification accuracies (for the class Alumni) while implementing the email classification scheme using the classifiers, Naïve Bayes, SVM, and KNN. As seen from Table 1, we obtain the best email classification using Naïve Bayes approach.

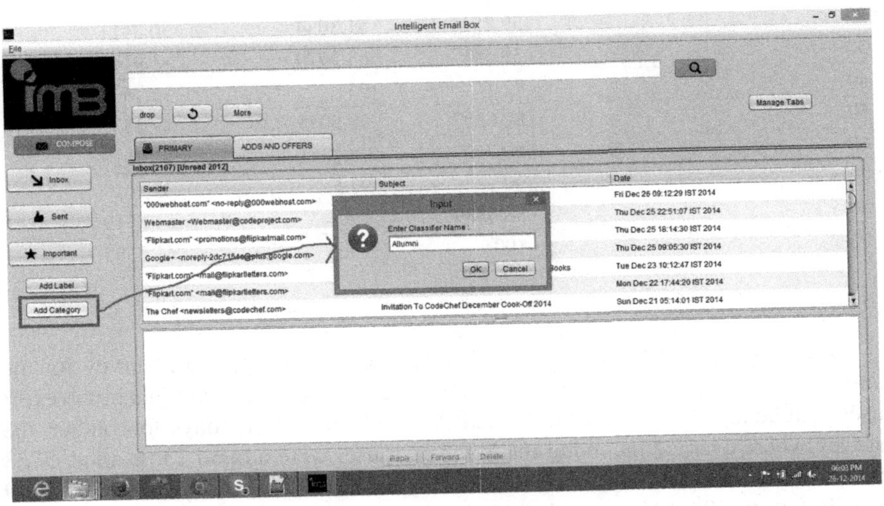

Fig. 1 Snapshot of the user interface of our application depicting the creation of labels/classes

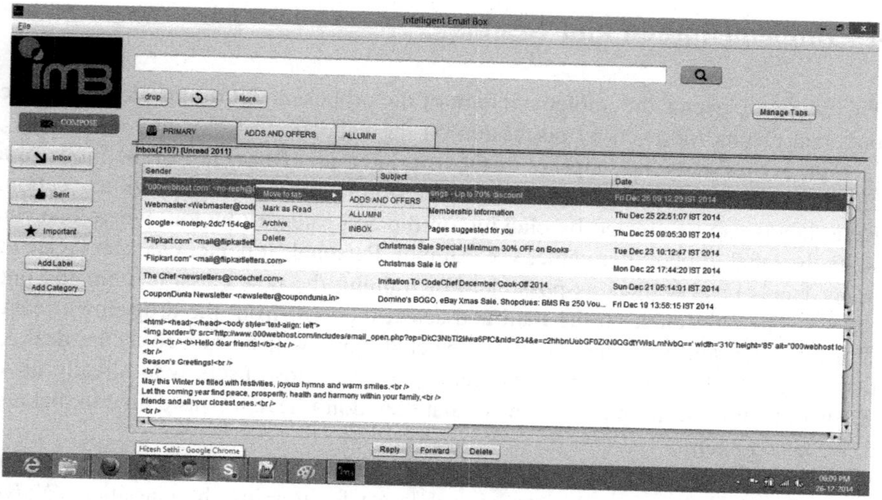

Fig. 2 Snapshot of the user interface depicting emails selection for training the classifier

Table 1 Achieved classification accuracies using different classifiers to classify the remaining emails of the Alumni class, i.e. not used for the training of the classifiers

Percentage of emails used for training of classifiers (%)	Percentage classification accuracy obtained by different classifiers		
	SVM	Naïve Bayes	KNN (k = 5)
10	28.0992	66.1157	23.1405
20	39.2523	81.3084	30.8411
30	39.3617	88.2979	31.9149
40	72.5	95	33.75
50	88.0597	98.5075	49.2537
60	90.7407	98.1481	51.8519
70	90	97.5	45
80	88.8889	100	51.8519
90	100	100	61.5385

Besides testing of the classification accuracies, we conducted a survey for the end users who experienced our application for email classification. This involves 17 users (students) aging between 17 and 23 years. It took 15 days to conduct the survey where average incoming mails for each user were around 23. In-total, 5738 emails were received, out of which our system could correctly classified 5249 emails (around 92.8 % correctly identified emails) into different labels or classes created by 17 users.

5 Conclusion

In this paper, we presented an intelligent email classification system where we targeted to arrange emails by their meaning rather than who send it or on the basis of message filtering. Unlike, filtering which is tedious task and prone to errors, our application will be a great user experience for people who receive large number of emails and do not know how to apply filters. Our application provides the automation to the users where they are free to choose labels describing the general feature of the folder to which that label belongs and putting a few training mails and then testing the new mails and retraining the machine in case of an error. Furthermore as the time grows machine becomes more intelligent and thus helps user to have a well organized inbox. The proposed system performed best with the Naïve Bayes classification where, we observed the accuracy ranging between 66 and 100 % for different amount of test data sets.

References

1. Manco, G., Masciari, E., and Tagarelli, A.: A Framework for Adaptive Mail Classification. In Proceedings of the 14th IEEE International Conference on Tools with Artificial Intelligence, ICTAI '02, USA, 387–392 (2002).
2. Armentano, M. G. and Amandi, A. A.: Enhancing the experience of users regarding the email classification task using labels. In Knowledge-Based Systems, Volume 71, 227–237 (2014).
3. LeeTiernan, S., Cutrell, E., Czerwinski, M., and Scott, H. H.: Effective Notification Systems Depend on User Trust. In Proceedings of Human-Computer Interaction-Interact '01, IOS Press, Tokyo, Japan, 684–685 (2001).
4. Cohen, W. W.: Learning rules that classify e-mail. In Proceedings of AAAI Spring Symposium on Machine Learning in Information Access, (1996).
5. Drucker, H., Wu, D., and Vapnik, V. N.: Support vector machines for spam categorization. In IEEE Transactions on Neural Networks, 10(5), 1048–1054 (1999).
6. Youn, S. and McLeod, D.: Spam email classification using an adaptive ontology. Journal of Software, 2(3), 43–55 (2007).
7. Sahami, M., Dumais, S., Heckerman, D., Horvitz, E.: A Bayesian approach to filtering junk e-mail. In Proceedings of AAAI Workshop on Learning for Text Categorization, 55–62 (1998).
8. Alsmadi, I. and Alhami, I.: Clustering and classification of email contents. Journal of King Saud University—Computer and Information Sciences, Vol. 27, Issue 1, 46–57 (2015).
9. Kossinets, G., Kleinberg, J., and Watts, D.: The structure of information pathways in a social communication network. In Proceedings of the 14th ACM SIGKDD International Conference on Knowledge Discovery and Data Mining, 435–443 (2008).
10. Pavel, K. and Lenc, L.: Confidence Measure for Czech Document Classification. In Computational Linguistics and Intelligent Text Processing, Springer International Publishing, 525–534 (2015).
11. Han, P., Shen, S., Wang, D., Liu, Y.: The influence of word normalization in English document clustering. In Proceedings of IEEE International Conference on Computer Science and Automation Engineering (CSAE 2012), 116–120 (2012).
12. The Porter stemmer Algorithm. http://tartarus.org/~martin/PorterStemmer/.

13. The English (Porter2) stemming algorithm. http://snowball.tartarus.org/algorithms/english/stemmer.html.
14. Ramos, J.: Using TF-IDF to Determine Word Relevance in Document Queries. In Proceedings of the First Instructional Conference on Machine Learning, (2003).
15. Berger, A. and Lafferty, J.: Information retrieval as statistical translation. In Proceedings of the 22nd Annual International ACM SIGIR Conference on Research and Development in Information Retrieval, 222–229 (1999).
16. Oren, Nir.: Reexamining tf. idf based information retrieval with genetic programming. In Proceedings of SAICSIT, 1–10 (2002).
17. Cortes, C. and Vapnik, V.: Support Vector Networks. Machine Learning, 20 (2), 273–297 (1995).
18. Christianini, N. and Shawe-Taylor, J.: An Introduction to Support Vector Machines. Cambridge University Press, (2000), http://www.support-vector.net.
19. Chang, C. C. and Lin, C. J.: LIBSVM: a Library for Support Vector Machines. ACM Transactions on Intelligent Systems and Technology (TIST), Vol. 2, No. 3, Article 27 (2011).
20. Murphy, K. P.: Naive Bayes Classifiers. University of British Columbia (2006).
21. McCallum, A. and Nigam, K,: A Comparison of Event Models for Naive Bayes Text Classification. In AAAI Workshop on Learning for Text Categorization (1998).

Learning Based No Reference Algorithm for Dropped Frame Identification in Uncompressed Video

Manish K. Thakur, Vikas Saxena and J.P. Gupta

Abstract For last many years video authentication and detection of tampering in a video are major challenges in the domain of digital video forensics. This paper presents detection of one of the temporal tampering (frame drop) under no reference mode of tampering detection. Inspirit of the scheme presented by Upadhyay and Singh, this paper extends the features to train the SVM classifier and accordingly classify frames of given video as tampered or non-tampered frames, i.e. detects the tampering of frame drop. Subsequently given video is classified as tampered or non-tampered video. The obtained results with enhanced features show significant improvement in classification accuracy.

Keywords Video forensics · Tampered videos · Frame drop · Temporal tampering · Support vector machine · Mean squared error · Entropy

1 Introduction

The continuous advancements in visual technologies (videos and images) have facilitated the society at several domains and are becoming modern need [1]. Besides many applications in diversified domains, there are some darker sides of visual (video) information, viz. misuse or the wrong projection of information through videos. One of them is video tampering, where a forger can intentionally

---

M.K. Thakur (✉) · V. Saxena
Jaypee Institute of Information Technology, Noida 201307, India
e-mail: mthakur.jiit@gmail.com

V. Saxena
e-mail: vikas.saxena@jiit.ac.in

J.P. Gupta
Lingaya's University, Faridabad 121002, India
e-mail: jaip.gupta@gmail.com

© Springer India 2016
S.C. Satapathy et al. (eds.), *Information Systems Design and Intelligent Applications*, Advances in Intelligent Systems and Computing 435,
DOI 10.1007/978-81-322-2757-1_45

manipulate actual or original videos to create tampered or doctored videos for malpractice [2, 3].

Depending on the domain in which manipulation is done, there can be following types of video tampering: Spatial Tampering, Temporal Tampering, and Spatio-Temporal Tampering. A forger can tamper source videos spatially by manipulating pixel bits within a video frame or in adjacent video frames, whereas forger can tamper source videos with respect to time (i.e. temporal tampering) by disturbing the frame sequence through frames sequence reordering, frames addition, and by the removal of video frames or frame drop [4–7].

With the help of forensic experts and tools, forensic laboratories play major role to examine these videos against tampering and ensure the authenticity. Depending upon availability of reference, forensic experts generally examine videos in three different modes viz. full reference (FR), reduced reference (RR), and no reference (NR). While tampering detection, forensic experts may have prior information (i.e. FR or RR) about video contents or may not have any information about video contents or original video, i.e. NR mode of tampering detection [8].

In this paper we present one of the temporal tampering, frame drop or frame deletion and its detection (i.e. location of dropped frame) under NR mode. The paper is inspired from the work presented in [5] by Upadhyay and Singh, where they used Support Vector Machines (SVM) to classify the frames in given video as tampered or non-tampered frames and accordingly the video.

Besides introduction presented in Sect. 1, remaining paper is organized as follows; Sect. 2 presents the problem statement and related work; Sect. 3 describes different features used for detection of frame drop; training and test modules have been presented in Sects. 4 and 5 respectively followed by conclusion and references.

2 Problem Statement and Related Work

Let us consider an original video, $\boldsymbol{O}$ with m video frames which is being tampered by dropping some video frames, thus creates a temporally tampered video, $\boldsymbol{T}$ with n video frames, where, $m \geq n$, as presented in Fig. 1.

Objectives are for given video, $\boldsymbol{T}$, if, $\boldsymbol{O}$ is not available (i.e. NR tampering detection):

(a) Examine the authenticity of video, $\boldsymbol{T}$ (i.e. classify the video, $\boldsymbol{T}$ as either tampered or non tampered video), and
(b) If, $\boldsymbol{T}$ is classified as tampered video, then identify the location of tampering in video, $\boldsymbol{T}$, i.e. frame indices of $\boldsymbol{T}$ after which frames were dropped.

As presented in Fig. 1, drop of k frames between ith and $(i + k)$th frames in original video, $\boldsymbol{O}$ will make these frames as adjacent to each other in the tampered video, $\boldsymbol{T}$, i.e. the frame $\boldsymbol{O}(i+k)$ will now be the $(i + 1)$th frame in video, $\boldsymbol{T}$. Usually adjacent frames in a video are redundant, therefore the tampering of frame drop may

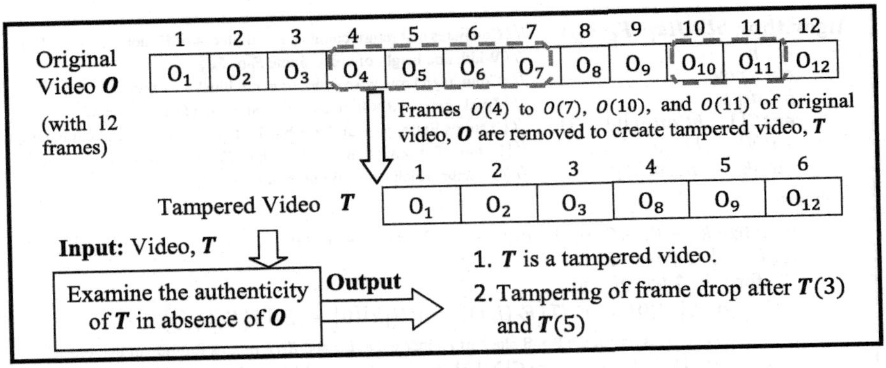

Fig. 1 An example (original video with 12 frames and count of deleted frames is 6, i.e. tampered video is created with 6 frames) depicting the problem statement

disturb the temporal redundancy among the frames in video, ***T*** resulting into inconsistent adjacent frames in video, ***T***. Tracing such inconsistency in a given video, ***T*** is the key idea behind the schemes developed under no reference (NR) mode of tampering detection [5, 7].

One of such scheme is proposed by Upadhyay and Singh in [5], where they trained the kernel for SVM learning involving training datasets of 1100 tampered and 1100 non tampered frames. They defined *Tampered Frame* as a frame in a video, V after which frames have been dropped. In other words, while creation of the tampered video, V from the original video, V_O, if, k frames between ith and $(i + k)$th frames in V_O are deleted, then they defined, $V(i)$ as *Tampered Frame*, otherwise non *Tampered Frame*. Further, they extracted the features, viz. Average Object Area (AOA) and Entropy from the binary difference frame and created the training dataset with two feature vectors involving 2200 data points. Similar procedure has been adopted while classification of test/input video, V, and classified frames in V as tampered or non-tampered frames by the trained SVM. If, none of the frames are classified as *Tampered Frame*, then they call the video, V as non-tampered otherwise tampered.

In this paper, we extend above scheme with additional features defined/used by us. Next section presents the extended set of features to train the SVM classifier.

3 Proposed Features

In addition to the features (Entropy and AOA) used by authors in [5], we used a feature Mean Squared Error, MSE [9] and proposed a feature, count of displaced block (DISP) for the stated problem. Subsequently we discuss the feature, DISP.

Count of displaced blocks (DISP) Inspirit of the schemes to identify the block displacement [10] between two frames (mostly used in video compression, with

```
Algorithm: blkDisp(F_C, F_N)      //{Computes and return count of 8 × 8 blocks of F_C not present in F_N}
    w , h                         //{Width and height of video frame F_C or F_N}
    col , row                     //{Column and row wise count of 8 × 8 block in a frame}
    bCount                        //{Total count of 8 × 8 blocks in a frame i.e. bCount = row × col}
    entC[1 .. bCount][1 .. 3]     //{It stores Entropy of an 8 × 8 block of F_C and its row and col index}
    entN[1 .. bCount][1 .. 3]     //{It stores Entropy of an 8 × 8 block of F_N and its row and col index}
    mark[1 .. bCount]             //{A vector which stores Boolean value}
  begin
    initialize k ← 0, mC ← 0, row ← h/8 , col ← w/8
    for i ← 1 to row do
        for j ← 1 to col do
            entC[k][0] ← ENT(F_C(i,j)) , entN[k][0] ← ENT(F_N(i,j))
            //{F_C(i,j) is one of the 8 × 8 block of F_C indexed at i, j and ENT returns Entropy of that block}
            entC[k][1] ← i , entC[k][2] ← j , entN[k][1] ← i , entN[k][2] ← j
            k ← k + 1
    for i ← 1 to bCount do
        mark[ i ] ← 0
    for i ← 1 to bCount do
        for j ← 1 to bCount do
            v_1 ← m_1 × entN[j][0]
            v_2 ← m_2 × entN[j][0]               //{m_1 and m_2 are the margins and need to be tuned}
            if v_1 ≥ entC[i][0] and v_2 ≤ entC[i][0] then
                check ← MSE(F_C(entC[i][1], entC[i][2]), F_N(entN[j][1], entN[j][2]))
                //{MSE returns Mean Squared Error between 8 × 8 block of F_C and 8 × 8 block of F_N}
                if check ≤ bThr and mark[i] = 0 then        //{bThr is a threshold}
                    mC ← mC + 1 , mark[i] ← 1
    return (bCount − mC)
  end   {End of the algorithm blkDisp}
```

Fig. 2 The scheme *blkDisp*() computes the count of of 8 × 8 blocks of a video frame not present in another video frame

block size of 8 × 8 or 16 × 16), the proposed feature, DISP, represents the count of 8 × 8 blocks in a frame not present in another frame. To compute such counts in two frames, viz. F_C and F_N, a scheme, *blkDisp* is presented in Fig. 2.

In the scheme, *blkDisp*, we split the frames (height, *h* and width, *w*) for which DISP needs to be computed, viz. F_C and F_N into blocks of size 8 × 8 as follows

$$F_C = \{B_C(1,1), B_C(1,2), B_C(\text{row}, col), \text{and}$$
$$F_N = \{B_N(1,1), B_N(1,2), B_N(\text{row}, col)\}$$

where, row $= h/8$; $col = w/8$; $B_C(a, b)$is an 8 × 8 blocks in F_C; and $B_N(a, b)$ is an 8 × 8 blocks in F_N.

Further, we computed the Entropy for each block (of size 8 × 8) in F_C and F_N. Based on the Entropy of a block, $B_C(a, b)$ in frame F_C we selected blocks in F_N to which $B_C(a, b)$ may probably matched. Two margins/thresholds, viz. m_1 and m_2 have been used to decide the probable matching blocks, e.g. if $E_C(a, b)$ is the

Entropy of block $B_C(a, b)$ then it's probable matching blocks in F_N have been selected as the blocks of F_N having Entropy in between $m_1 \times E_C(a, b)$ and $m_2 \times E_C(a, b)$. If, there is no such block available in F_N to which $B_C(a, b)$ may probably matched, then we considered it as a block which is not present in F_N. If, such blocks are available, then we compared the block, $B_C(a, b)$ with each probable matching block of F_N using MSE. Using a threshold, ***bThr*** (represents the maximum MSE between two blocks to call them almost identical) we decided the single matching block from the set of probable matching blocks.

Based on above discussion to compute the count of displaced blocks, we computed the DISP, $D(i)$ for each frame with its successive frames in a video, V as follows

$$D(i) = blkDisp(\mathrm{V}(i+1), \mathrm{V}(i))$$

where, the method *blkDisp* returns the approximate count of blocks (of size 8 × 8) of a video frame $\mathrm{V}(i)$ which are not present in successive video frame $\mathrm{V}(i+1)$.

4 Training Module

This section presents our approach to train the kernel of SVM. Subsequently we present the training dataset, feature extraction and cross-validation.

As shown in Fig. 3, we used 15 videos to create the database of training videos includes tampered and non-tampered videos having *Tampered Frames* and non *Tampered Frames*. These videos are available at http://media.xiph.org/video/derf/ in compressed format; we uncompressed these videos and called them original or non-tampered videos. We created 100 copies of each original video and randomly

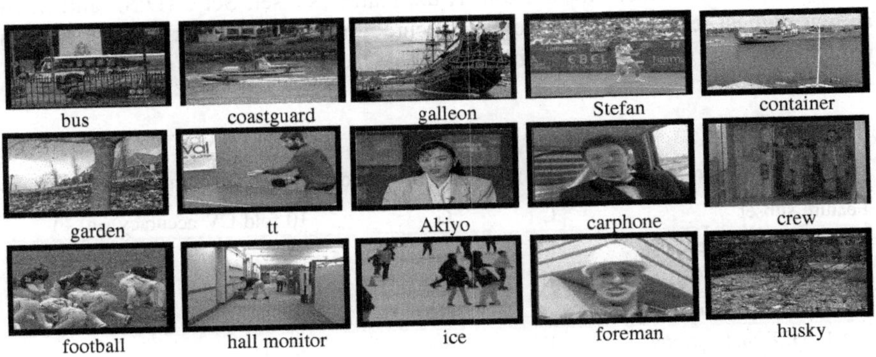

Fig. 3 Videos (http://media.xiph.org/video/derf/) used to create training videos dataset

dropped 5 to 20 frames in each copy of original videos. Hence, there are 1500 tampered videos and 15 original or non-tampered videos in our database of training videos, resulting into 1500 *Tampered Frames* and around 5000 non *Tampered Frames*.

For each original video in our training video database, we extracted the features viz. Entropy, etc. between each frame and its successive frame in a video. We stored the extracted features into the training dataset, where each extracted feature stored in the dataset constitutes a feature vector and all the 4 features corresponding to a non *Tampered Frame* represent a data point. Further, for each tampered video in our training video database, we obtained the *Tampered Frames* and extracted the features viz. Entropy, etc. involving each *Tampered Frame* and its successive frame in the respective tampered video and accordingly stored into the training dataset.

We manually labeled the data points corresponding to non *Tampered Frames* as class "0" and data points corresponding to *Tampered Frames* as class "1". In order to have equal count of data points of each class in the training dataset, we randomly deleted some data points of class "0" and retained 1600 such data points and hence, there are in total 3100 data points of class "0" and class "1" in the training dataset.

Further, tenfold cross validation is applied with following feature subset: Set 1 (Entropy and AOA), Set 2 (MSE and DISP), Set 3 (Entropy and DISP), and Set 4 (Entropy and MSE). Entropy and AOA are the features used in [5], whereas other subsets shown comparatively lesser overlapping data points of both classes, *Tampered* and *Non-tampered frames* than the other combinations of feature subset. Besides, these we used one more subset having all the four features, i.e. Set 5 (Entropy, AOA, MSE, and DISP) to train the RBF kernel of SVM.

Further, the RBF kernel has been separately trained for each selected feature subsets involves respective feature vector in training dataset and the label (class "0" or class "1"). While training, we used SVM classifier of LibSVM library [11].

Table 1 presents the tenfold cross validation accuracy of the trained SVM kernels with different feature subsets and kernel parameter, gamma (g) and penalty parameter, *C*. As evident from Table 1, the feature subset, Set 2 (DISP and MSE) shown highest tenfold cross validation accuracy (in %), which is an improvement over the feature subset, Set 1 (Entropy and AOA).

Table 1 10 fold cross validation (CV) accuracy (in %) for different feature subsets

Feature subset	C	G	10 fold CV accuracy (in %)
Set 1 : AOA and entropy	1	32	72.19
Set 2: DISP and MSE	16	6	86.57
Set 3: DISP and entropy	2	0.0312	79.35
Set 4: MSE and entropy	10	8	82.21
Set 5: All four features	1	13	78.63

5 Test Module

This section presents the test module discussing the conducted experiments with trained SVM classifier to identify tampered or non tampered frames in a video and accordingly classify the video either as tampered or authentic. We conducted the experiments under following categories involving different sets of test video data:

E_A We randomly selected 100 tampered videos from the training video database, where each video has exactly one tampered frame. Thus, under this category of experiments, in total we have 100 *Tampered Frames* and 100 tampered videos, where E_A includes the same set of tampered videos which were involved in the training.

E_B We randomly selected 5 original videos from the training video database and created 5 copies of each original video. Further, we randomly deleted 5–20 frames at 4 different locations in the copy of each original video, viz. we randomly deleted 5–20 frames after *i*th, *j*th, *k*th, and *l*th frames in the copies of original videos, where *i*, *j*, *k* and *l* are randomly selected index for each video. Thus, under this category of experiments, we have 25 tampered videos, each with 4 *Tampered Frames*; in total we have 100 *Tampered Frames*, where E_B includes the set of tampered videos which were not involved in training but created from same set of original videos which were involved in the training.

E_C In this experiment category we used 4 new or unknown videos [12–15] as original videos, i.e. frames in these videos were neither involved as *Tampered Frames* nor as non *Tampered Frames* while training the kernel of SVM. We uncompressed these videos and for uniformity in frame size of each video, scaled to 352 × 288 pixels. Further, we created 5 copies of each original/new video and dropped 5–20 frames at 5 different locations in each copy. Under this category of experiments we created 20 tampered videos, each with 5 *Tampered Frames*; totaling 100 *Tampered Frames*, where E_C includes the set of tampered videos which were neither involved in training nor created from the original videos which were involved in the training.

Further, for each experiment category, we created the test datasets. While creation of test datasets, all frames in the test video database of respective experiment categories have been obtained and features were extracted from these frames similar to the process elaborated in training module including the labeling of data points. We independently input different subset of feature vectors in unlabelled test dataset to the respective SVM. Data points corresponding to respective feature vectors have been projected into the respective hyper-plane of trained SVM. If, the output of the SVM is 0 for each of the data point in the input unlabelled test dataset, then we reported no frame drop and accordingly classified the videos as authentic videos. If, the output of the SVM is 1 for any of the data point, then we obtain the respective data point in the labeled test dataset (viz. T1_Label) and if found the label as 1 then reported the respective video as tampered video and location of frame drop in the tampered video as the respective frame index in the labeled test dataset.

Further, performance of these feature subsets to detect the location of frame drop (i.e. classify a frame as *Tampered Frame*) and verify the authenticity of a video (i.e. classify a video as tampered video) have been analyzed under following categories.

(a) Classification with known videos (i.e. E_A + E_B)—There are in total 200 *Tampered Frames* and 125 tampered videos under this category.
(b) Classification with unknown videos (i.e. E_C)—There are in total 100 *Tampered Frames* and 20 tampered videos under this category.

Percentage accuracy to detect the location of frame drop and classify a video as tampered or non tampered video have been computed respectively as follows (where *TF* stand for *Tampered Frames* and *TV* stands for *Tampered Videos*)

$$\text{Accuracy (in \%)} = (\text{Count of correctly detected } TF)/(\text{Total count of of } TF) \times 100$$
$$\text{Accuracy (in \%)} = (\text{Count of correctly identified } TV)/(\text{Total count of of } TV) \times 100$$

Table 2 presents the average classification accuracies (in %) achieved by each feature subset with known and unknown test videos. As observed from Table 2, the highest average classification accuracy (in %) to detect the location of tampering of frame drop has been achieved by feature subset, Set 2 (MSE and DISP), i.e. 80 %, if the test videos are known to the SVM classifier. The least average classification accuracy, i.e. 69 % has been shown by feature subset, Set 1 (Entropy and AOA).

Similar observations have been made for the average classification accuracy to verify the authenticity of a video; it is around 83 %, achieved with Set 2, whereas it is around 72 %, achieved with Set 1. This shows an improvement in classification accuracy with the feature subset, Set 2 over the feature subset, Set 1.

Further, none of the feature subsets shows satisfactory classification accuracy, to detect the location of tampering of frame drop as well as verification of the authenticity of a video, if test videos are unknown. However, the SVM trained with feature subset, Set 2 (MSE and DISP) shown comparatively better average accuracy to detect the location of tampering of frame drop. Similar observations have been made for the average classification accuracy to verify the authenticity of a video, if test videos are unknown to the SVM classifier.

Table 2 Average classification accuracies (in %) to detect the location of frame drop (TF) and authenticity of a video (TV), achieved by each set with known and unknown videos

Videos	Feature subset				
	Set 1 (entropy and AOA)	Set 2 (MSE and DISP)	Set 3 (entropy and DISP)	Set 4 (entropy and MSE)	Set 5 (entropy, AOA, MSE and DISP)
Known Videos (E_A + E_B)	69 % (TF) 71.2 % (TV)	80 % (TF) 82.4 % (TV)	75.5 % (TF) 78.4 % (TV)	78.5 % (TF) 80.8 % (TV)	73.5 % (TF) 75.2 % (TV)
Unknown Videos (E_C)	29 % (TF) 50 % (TV)	37 % (TF) 70 % (TV)	34 % (TF) 60 % (TV)	36 % (TF) 65 % (TV)	31 % (TF) 60 % (TV)

6 Conclusion

In this paper we proposed a learning based NR scheme to verify the authenticity of a video as well as location of the tampering of frame drop. We extended the work proposed in [5] by exploring two additional features, MSE and DISP. We trained the RBF kernel of SVM and found significant improvement with proposed feature subset (DISP and MSE) over the feature subset (Entropy and AOA) to verify the authenticity of a video as well as detection of location of tampering of frame drop, if videos are known. However, none of the feature subsets had shown satisfactory classification accuracy, if videos were unknown to the SVM classifier.

References

1. Madden, M.: The Audience for Online Video Sharing Sites Shoots Up. http://fe01.pewinternet.org/Reports/2009/13–The-Audience-for-Online-VideoSharing-Sites-Shoots-Up.aspx. Accessed 30 Jun 2015
2. Rocha, A., Scheirer, W., Boult, T., and Goldenstein, S.: Vision of the unseen: Current trends and challenges in digital image and video forensics. ACM Computing Surveys, Vol. 43, No. 4, Article 26, 1–42 (2011)
3. Farid, H.: Image Forgery Detection. IEEE Signal Proc. Magazine, Vol. 26, Issue 2, 16–25 (2009)
4. Atrey, P.K., Yan, W.Q., and Kankanhalli, M.S.: A scalable signature scheme for video authentication. Multimed Tools Appl, Vol. 34, 107—135 (2007)
5. Upadhyay, S. and Singh, S.K.: Learning Based Video Authentication using Statistical Local Information. In: Proc. International Conference on Image Information Processing, Nov 3-5, 1–6 (2011)
6. Chen, S. and Leung, H.: Chaotic Watermarking for Video Authentication in Surveillance Applications. IEEE Trans. Circuits and Systems for Video Technology, Vol. 18, No. 5, 704–709 (2008)
7. Shanableh, T.: Detection of Frame Deletion for Digital Video Forensics. Digital Investigation, 10, 350–360 (2013)
8. Thakur, M.K., Saxena, V., Gupta, J. P.: Data-parallel full reference algorithm for dropped frame identification in uncompressed video using genetic algorithm. In: Proc. 6th International Conference on Contemporary Computing (IC3 2013), 467–471 (2013)
9. Wang, Z., Sheikh, H.R., and Bovik, A. C.: Objective Video Quality Assessment. Chapter 41 in The Handbook of Video Databases: Design and Applications, B. Furht and O. Marqure, ed., CRC Press, 1041–1078 (2003)
10. Bovik, A.: The Essential Guide to Video Processing. Second Edition, Academic Press, Elsevier, (2009)
11. Chih-Chung Chang and Chih-Jen Lin. LIBSVM: a library for support vector machines. Software available at http://www.csie.ntu.edu.tw/~cjlin/libsvm. Accessed 30 Jan 2015
12. 2013 Volvo FH Series on the Road, at www.youtube.com/watch?v=VZX-o9jzX0k. Accessed 30 Jan 2015
13. The New Volvo FH, at www.youtube.com/watch?v=bQmmlIXS0fc. Accessed 30 Jan 2014
14. Volvo Trucks—How Volvo FMX was tested, http://www.youtube.com/watch?v=QokdT75uFf4, Accessed 30 Jan 2015
15. All new Volvo FH16 750, at http://www.youtube.com/watch?v=XiK-qd8iNwY. Accessed 30 Jan 2015

Multi-view Video Summarization

Chintureena Thingom and Guydeuk Yeon

Abstract Video summarization is the most important video content service which gives us a short and condensed representation of the whole video content. It also ensures the browsing, mining, and storage of the original videos. The multi- view video summaries will produce only the most vital events with more detailed information than those of less salient ones. As such, it allows the interface user to get only the important information or the video from different perspectives of the multi-view videos without watching the whole video. In our research paper, we are focusing on a series of approaches to summarize the video content and to get a compact and succinct visual summary that encapsulates the key components of the video. Its main advantage is that the video summarization can turn numbers of hours long video into a short summary that an individual viewer can see in just few seconds.

Keywords Video summarization · Euclidean distance · Key frames · Spatio temporal · Threshold values

1 Introduction

Multi-view videos are of a particular activity taken from different perspectives i.e. different views of an activity. Such videos are used in surveillance systems and sports broadcast where multiple cameras are used to record an event. The rapidly growing amount of digital video in today's network gives rise to problem for efficient browsing and management of video data. The plethora of the data stored in a video when compared with other forms of media such as word or audio file effects

C. Thingom (✉) · G. Yeon
Centre for Digital Innovation, Christ University, Bangalore, India
e-mail: chintureena.thingom@christuniversity.in

G. Yeon
e-mail: yeon@christuniversity.in

© Springer India 2016
S.C. Satapathy et al. (eds.), *Information Systems Design and Intelligent Applications*, Advances in Intelligent Systems and Computing 435,
DOI 10.1007/978-81-322-2757-1_46

the effective management very difficult. To reduce the amount of storage requirement, most of the video clips are compressed and reduced to a lesser size using a compression standard such as MPEG. Even after going through compression, the videos are still too big and large to transfer over into any storage device in a general networks.

1.1 Image Processing

In the world of computer science, the image processing is a system where in the input is an image or a video frame; and the output of image processing may be either itself as an image or, a set of parameters related to the image. Image processing usually concerns with the digital image processing, but optical and analog image processing also are referred. Digital image processing is with the help of computer algorithms to process a digital image. In the area of digital signal processing, digital image processing has advantages over other analog image processing. Image processing inculcates different scopes to eliminate the problems such the noise of the image and distortion during image processing. The images are defined in most of case in 2Ds, the digital image processing can put the image in the form of Multi-dimensional systems.

- Key Frame: a single frame from the original video
- 24–30 frames/s
- Scene: collection of shots forming a semantic unity
- Theoretically, a single time and place

2 Related Work

Herranz et al. [1] the analysis of the algorithm which is used with the help of the iterative ranking procedure is an extension of the previously done research work. The outcome of the algorithmic research is the ranked list, which is a representation of the sequence used for video summarization. The summarized video is a compiled from all the important frames or bits of the original video using bit extractions method. Ren et al. [2], the authors propose a method that relies on the machine learning. The method first detects what relies on machine learning. The method first sees what they can focus on as the principal features; these features are based on their pixel value, edges and histograms. Then these factors are converted into their specific system to predict video transitions. For each video transition, if the video have undergone the techniques of neural network and finds out that the video exceeds a threshold limit, then a key frame is selected and is marked to be included in the video summarization. According to their research, this method has been found as a robust method while dealing with full fledge movies where many video effects are incorporated such as Fade-in, Fade-out or Cut/Dissolve are presented.

Coelho et al. [3], a new algorithm is proposed which has more simplicity and also ensures a good performance but is not as robust as [2]. The method proposed takes as input the desired temporal rate, i.e. the total number of frames Tf that the final summary will have. Then the method adds the frame by default to the summary. And then, the method finds out the distortions of the current frames. If the distortion is very large and we have not reached expected Tf, then the identified video frame will be added to the video summary. This greedy algorithm performs fast but according to their results the final summary is not robust enough. Li et al. [4] had introduced an algorithm which is based on dynamic programming where a MINMAX optimization is used. The proposed method also based on the distortions of the frames with the help of the dynamic programming approach to eliminate the distortion of the video summary under the assumption that this will provide a better video summary for the interface user.

3 Problem Definition

If a system can automatically identify an image or a set of important images which can give a sense of content or a summary then it will be useful in a number of environments. These selected video frames can summarize a large collection of media in novel ways, giving a gist of the whole video rather than a precise linguistic summary. However, such important visual frames are often found to be very beneficial as indicators of the gestalt of the media than more traditional language summaries.

The main goal of the project is to view the shot summary of videos from many views of a single event instead of viewing the whole videos one after the other.

4 Research Work

Almost of all the research methods proposed by various scholars in the literature represents a general scheme. Further, our research schemes can be categories into two distinct steps. First, we can identify and extract few important frames from the original video which are relevant. Second, after the identification we can incorporate all these relevant frames and form a final video summary. We have focused to implement the method proposed by [1]. To explain the research model, please refer Fig. 1 which explains the systematic process of the model (Fig. 2).

In this proposed model, first we will load a video. Second, we will identify its corresponding frames. Each of the video frames extracted, we need to determine spatio temporal feature detection using Hessian Matrix. Then, we will compute the level of activity in the video frames. With a benchmark limit, if any activity level is too high, we can mark that frame as a key frame with the help of Euclidean Distance Formula. We can continue throughout the whole of the video. We will

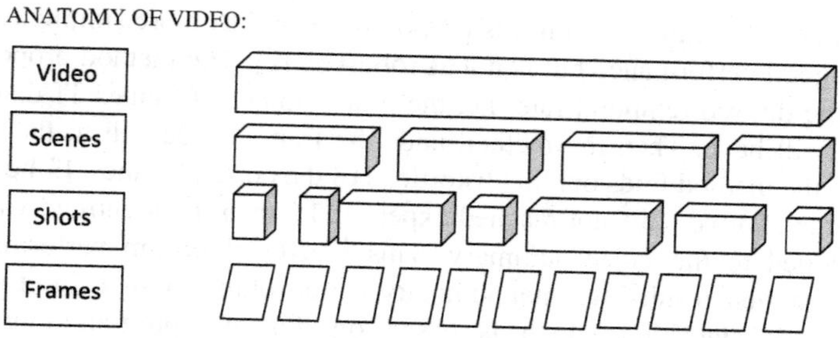

Fig. 1 Anatomy of video

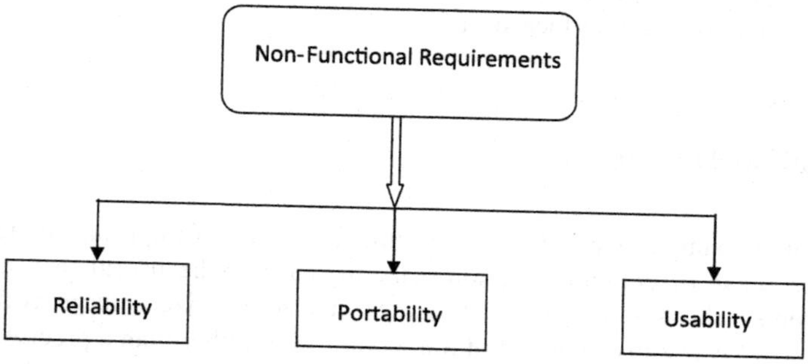

Fig. 2 Non-functional requirements to retrieve important frames

have a set of important key frames from the original video. After processing all the frames, next we can filter (by using hyper graph) to find out the most relevant key frame. In the end, once we have identified all the relevant ones, we can construct a video sequence of all the selected key frames. The outcome of this model will be a summarized video.

In the Fig. 3 diagram, it represents the different videos arrived from different cameras to the computing system. The image processing take these image and sends the results to it's as summary of the video. We have used Spatio Temporal Feature Detection to apply the Hessian matrix, which is known to produce stable features and is also used in object recognition and feature matching applications. After extracting the frames from a view, the next step is to detect important frame. To detect the important frames Euclidean's Distance formula is used. These difference values is store in a matrix then find the maximum variation in the difference values to get the locations of local maxima. Using the local maxima we retrieve the values at these points to get important frames. The maximum and minimum difference value is calculated. Taking the average as the threshold, only those frames are selected which exceed this value. These frames are the key frames. The key frames from different views are represented in the hyper graph. These selected key frames from multiple videos will be used to give the final outcome for the video summarized in short duration which the user can see to get a gist of the entire activity.

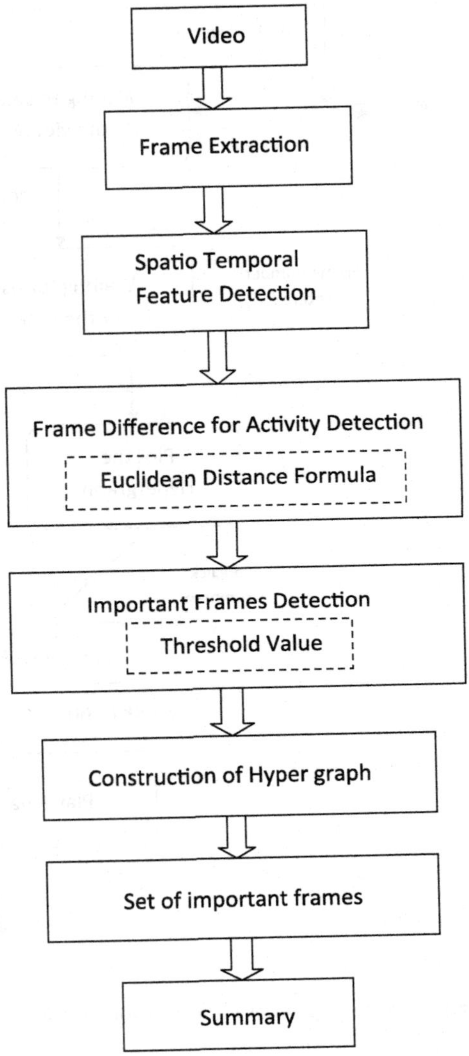

Fig. 3 Flow diagram depicting the steps for summarization from euclidean distance

Here the three suggested parameters are presented in three different modules.
Module 1: Video Parsing and Spatio Temporal Feature Detection
Module 2: Important Frame Detection Using Euclidean Distance Formula
Module 3: Construction of Hyper graph (Fig. 4).

Module 1: Video Parsing and Spatio Temporal Feature Detection

After loading the video, the most important task is to include parsing process of all the multi-view videos i.e. videos of a specific event taken from different angle into a different frame perspective. The original videos can be from any format. Normally,

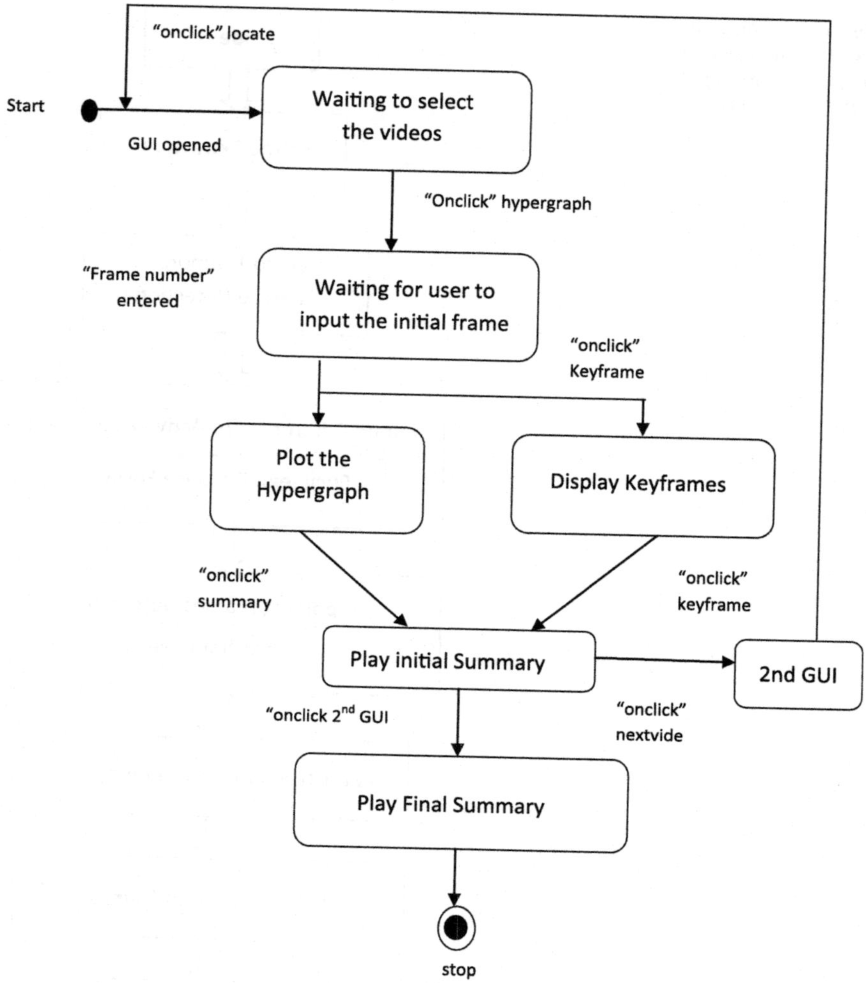

Fig. 4 State diagram of summarization of videos

these videos from different perspective of the event will be converted first into .avi file format using a video converter application.

(1) Spatio-Temporal Feature Detection: We need to follow the method of Hessian matrix to ensure stable feature detection and to use in the reorganization of the object or of an activity with those of feature matching applications. In our research, we will incorporate 3D scenes with space and time details as an input. As such, the information will be represented in a frame as I(x; y; z), where x; y is the spatial information and t is the temporal information of I frame.

Our Hessian matrix is defined as:

$$H(I) = \begin{bmatrix} \frac{\partial^2 I}{\partial x^2} & \frac{\partial^2 I}{\partial x \partial y} & \frac{\partial^2 I}{\partial x \partial t} \\ \frac{\partial^2 I}{\partial x \partial y} & \frac{\partial^2 I}{\partial y^2} & \frac{\partial^2 I}{\partial y \partial t} \\ \frac{\partial^2 I}{\partial x \partial t} & \frac{\partial^2 I}{\partial y \partial t} & \frac{\partial^2 I}{\partial t^2} \end{bmatrix} \quad (1)$$

We work with frame I as an image, so for each pixel in I we use Eq. 1 to compute its corresponding Hessian Matrix H. We are using the following masks to compute the second derivation:

$$mx = \begin{bmatrix} 1 & -2 & 1 \end{bmatrix}$$

$$my = \begin{bmatrix} 1 \\ -2 \\ 1 \end{bmatrix}$$

$$mxy = \frac{1}{4} \times \begin{bmatrix} 1 & 0 & -1 \\ 0 & 0 & 0 \\ -1 & 0 & 1 \end{bmatrix}$$

(2) Frame Activity Level: We have defined a threshold value (th). As we have already calculated H, we will now compute the determinant of the matrix H. If det(H) > th then, the pixel is marked as a feature in the image. We store all the features detected in another matrix 10 of the same size of I. once all the pixels have been processed, we count the number of features in matrix 10 and we will store this value in a structure n, so for each frame I we will have its correspondent n (I). We will consider as key frames the ones with the most salient activity levels. In order to extract the key frames we use the concept of local maxima. While choosing the algorithm for scheduling of resources, there scheduling algorithm choices are given.
As seen in Fig. 5, a local maxima occurs at A. we can detect the local maxima of a function using:

Fig. 5 Local maxima of a function

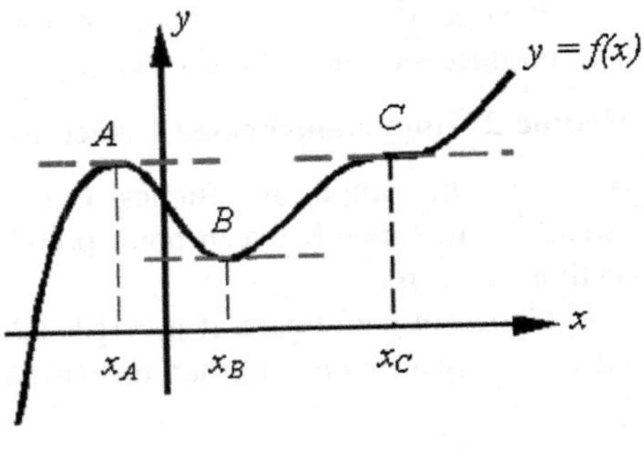

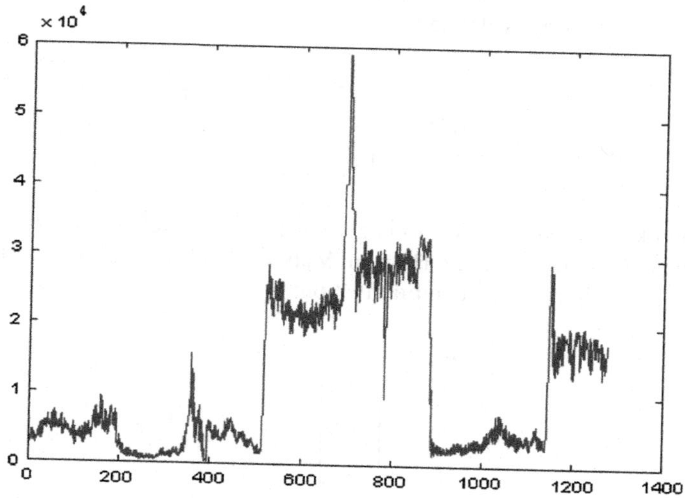

Fig. 6 Activity levels of a movie

$$\frac{\partial^2 f}{\partial x^2} < 0$$

In Fig. 6, we will like to show the level of activity of a certain video. As we can see, if we were to detect the local maxima in that signal, a lot of key frames would be selected making our summary useless. The idea of filtering the signal so that we can detect only the ones with the most salient levels is useful here.

(3) Key frames Filtering: In this stage, we have got all the important activity levels of all the frames, which are the ones with the most salient activity levels of the frames. These frames will be our key frames. In this process, we have used Median Filter technique to filter the signal. The median filter is best tool which can identify the high frequencies while smoothing the signal. As it is well beneficial tool, we are applying the median filter allows us to attenuate these "fake" high levels of activity. So after the filtering only the real and most salient frames are selected. In Fig. 7, we can see the final result after filtering the signal of Fig. 6. As we can see, filtering the signal of levels of activity produces a much better summary.

Module 2: Important Frame Detection Using Euclidean Distance Formula

To detect the important frames Euclidean's Distance formula is used. The **Euclidean distance** between point p and q is the length of the line segment connecting them **(pq)**.

In Cartesian coordinate, if p = (p1, p2, …, pn) and q = (q1, q2, …, qn) which will be the two points and then the distance from p to q or vice versa is given as:

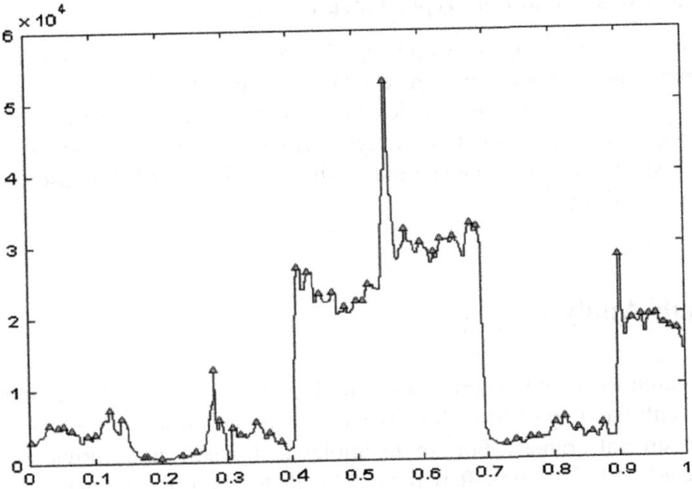

Fig. 7 Filtered signal of the activity levels

$$\mathrm{d(p,q)} = \mathrm{d(q,p)} = \sqrt{(q1 - p1)^2 + (q2 - p2)b2 + \cdots + (qn - pn)2}$$
$$= \sqrt{\sum_{i=1}^{n} (qi - pi)2}$$

This concept is applied to determine the difference between frames. A sequence of three frames is taken at every iteration level. The frame is converted into grayscale image and the data field is used to compute the difference.

If the sequence is i, i + 1, i + 2 representing the data information of the first, second, third frame respectively, then Euclidean formula to compute the difference in frames is

$$De = \sqrt{(((i+2) - (i-)) - ((i-1) - (i)))^2}$$

These difference values is store in a matrix then find the maximum variation in the difference values to get the locations of local maxima. Using the local maxima we retrieve the values at these points to get important frames. The maximum and minimum difference value is calculated. Taking the average as the threshold, only those frames are selected which exceed this value. These frames are the key frames. The key frames from different views are represented in the hyper graph. These key frames from multiple views are used to give the final summary which can be used by the user to get a gist of the entire activity.

Module 3: Construction of Hyper Graph

As a hyper graph is a graph in which an edge connects more than two vertices, thus the computation for frame difference takes into account three frames each. Each difference value represents the relation between three frames. The key frames from different views are represented in the hyper graph. These key frames from multiple views are used to give the final summary which can be used by the user to get a gist of the entire activity.

5 Result Analysis

In our research work, the experiments of multi view video summarization have been executed with the help of Matlab software. As we are aware, a movie is composed by many several frames. For an example, a 1 (one) min movie have about approximately 1200 frames. It will be difficult to show the whole sequence of a video in this research paper, therefore the experiment is executed with an extraction of a video. The outcome of this research is shown at Fig. 8. The model is shown with the help of the combined sequence of frames. With the logical sequencing, the proposed model has detected 60 key frames with the most salient level of activity. The remaining of the frames is disposed and only 60 key frames are considered for the final video summarization.

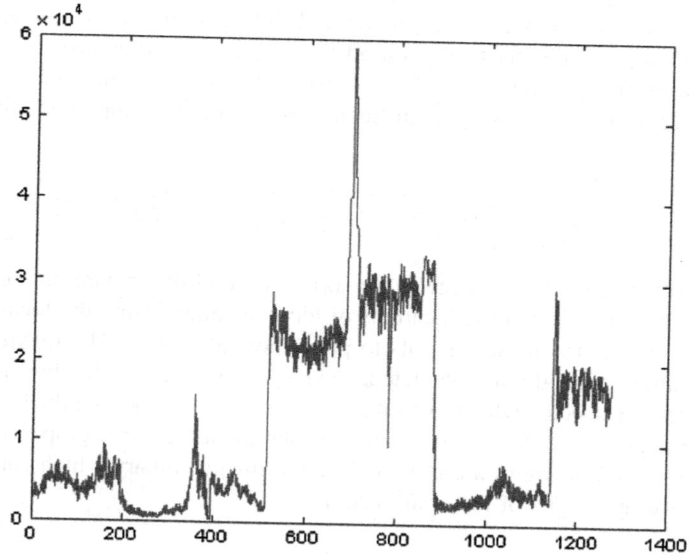

Fig. 8 Key frames detection in hyper graph

Frame 10 Frame 13 Frame14

Frame 18 Frame 50 Frame 55

Frame 82 Frame 83 Frame 84

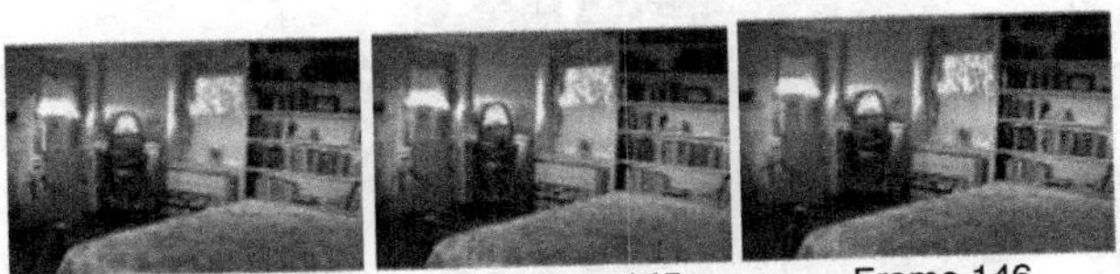

Frame 144 Frame 145 Frame 146

Frame 211 Frame 212 Frame 213

Frame 214 Frame 251 Frame 252

In our test experiment, we have totally 60 key frames which are detected in the video using the parameter set in the research. With the help of the constant, alpha we have eliminated many other frames after looking at the benchmarked threshold level because excess key frames will be produced if smaller alpha value is used. Some key frames produced by our proposed method for the .avi file are as follows:

6 Conclusion

In our research paper, we have presented important key frames generated using Euclidean's Distance Formula. The primary objective of our method is to generate relevant key frames from the original video with relatively at a faster speed. According to result of research, all the important video key frames are generated to give the gist of the video so that the interface user need not see the whole video but only the video summary which will have all the important details. Furthermore, our proposed method holds our research objectives. As for our future works, we will continue our research on this field, especially in feature extraction and pattern recognitions from a video.

References

1. Luis Herran Zandv Jose M. Martinez, "A Frame for Scalable Summarization of Video" IEEE Transactions on Circuits and Systems For Video Technology, Vol. 20, Issue 9, September 2010.
2. W Ren and Y Zhu, "A video summarization approach based on machine learning" International Conference on Intelligent Information Hiding and Multimedia Signal Processing, IEEE, 2008.
3. L. Coelho, L. A. D. S. Cruz, L. Ferreira, and P. A. Assunção, "An improved sub-optimal video summarization algorithm" International Symposium ELMAR (Electronics in Marine)—2010, IEEE, 2010
4. Z. Li, G. M. Schuster, and A. K. Katsaggelos, "Minmax optimal video summarization" IEEE Transactions on Circuits and Systems for Video Technology, 2005

Comprehensive Review of Video Enhancement Algorithms for Low Lighting Conditions

G.R. Vishalakshi, M.T. Gopalakrishna and M.C. Hanumantharaju

Abstract Video enhancement becomes a very challenging problem under low lighting conditions. Numerous techniques for enhancing visual quality of videos/images captured under different environmental situations are proposed by number of researchers especially in dark or night time, foggy situations, rainy and so on. This paper discusses brief review of existing algorithms related to video enhancement techniques under various lighting condition such as De-hazing based enhancement algorithm, a novel integrated algorithm, gradient based fusion algorithm and dark channel prior and in addition it also presents advantages and disadvantages of these algorithms.

Keywords Video enhancement · De-hazing · Integrated algorithm · Gradient fusion · Dark channel prior · Review

1 Introduction

Video enhancement is an effective and difficult field of video processing. Pre-processing includes a poor quality video as an input and post-processing includes a high quality video as an output. Video enhancement has become very interesting field of research in modern days and chosen as an interesting area of research. The main objective of video enhancement is to enhance fine details which

G.R. Vishalakshi
Department of Electronics & Communication Engineering,
Dayananda Sagar College of Engineering, Bangalore, India

M.T. Gopalakrishna
Department of Computer Science Engineering, Kammavar Sangam School
of Engineering & Management, Bangalore, India

M.C. Hanumantharaju (✉)
Department of Electronics & Communication Engineering,
BMS Institute of Technology & Management, Bangalore, India
e-mail: mchanumantharaju@gmail.com

© Springer India 2016
S.C. Satapathy et al. (eds.), *Information Systems Design and Intelligent Applications*, Advances in Intelligent Systems and Computing 435,
DOI 10.1007/978-81-322-2757-1_47

is hidden in a video and to provide "better transform" representation for future automated processing. It provides analysis of background information to study different characteristics of object without requiring much human visual inspection. Digital video processing includes numerous applications in the field of industry and medicine.

Most of the digital video cameras are not designed for capturing videos under different environmental conditions, therefore they may fail to capture videos practically for different illumination conditions [1]. Videos captured become unusable for several real time applications. Therefore, for various real time applications involves employment of such cameras becomes inefficient. Several far and near infrared based techniques [1–4] are used in numerous systems. Conventional enhancement techniques based on contrast enhancement technique, Histogram equalization (HE), gamma correction are widely employed to recover image details in low lighting conditions by reducing noise [5, 6]. However, this result in good enhancement of a video but the implementation of these techniques increases computational complexity and hence not suitable for real time videos.

Our paper focuses on recent existing enhancement methods for low lighting conditions. Low lighting conditions such as night time videos, foggy situations, rainy and so on, are very difficult for human perception and therefore many researchers provided different methods to improve such videos.

2 Video Enhancement Algorithms

This section presents some of the existing approaches to enhance the visual qualities of videos.

Various researchers have been proposed their techniques to improve the poor quality videos that capture under various illuminated conditions.

2.1 De-Hazing Algorithm

Kaiming et al. [7] proposed an algorithm to enhance videos in hazing conditions. This algorithm applies an inversion over the background pixels of high intensities and main objects of low intensity pixels. The result of this technique is as shown below in Fig. 1.

The main drawback of this is given by comparing both the figures it is observed that inverted and haze videos are not much comparable and are almost the same similar.

Based on the above observation, a novel integrated algorithm [1] was proposed by Xuan et al. for enhancing a video under low lighting conditions such as haze, night time and foggy videos. This method has low computational complexity and outputs good visual quality video. This algorithm applies the inversion technique on

Fig. 1 *First row* Videos under low illumination. *Second row* Resultant video frame obtained due to inversion. *Third row* Haze videos

low-contrast video frames and removes haze of the inverted video. Thus this technique removes noise from the hazy videos. The main advantage of this algorithm is to increase the speed by 4 times than when enhancement is done in frame-wise.

Let I be the Low-lighting input frame, which can be inverted using mathematical relation as:

$$R^c = 255 - I_c(x) \quad (1)$$

where, c indicates RGB color channel. Ic indicates color channel intensity of the low-lighting video input I. R^c is the inverted image intensity.

The hazy image obtained after inversion is given by:

$$R(x) = J(x)t(x) + A(1 - t(x)) \quad (2)$$

where, A is a constant indicating. R(x) is pixel intensity that the camera catches pixel. J(x) is the intensity of the original objects or scene. t(x) gives the amount of illuminated light from the objects. Evaluation of A and t(x) is very difficult in haze removal techniques.

$$T(x) = e^{-\beta d(x)} \quad \text{for homogeneous atmosphere} \quad (3)$$

where β is a constant called as coefficient of scattering of the atmosphere d(x) is the scene's path. t(x) can be determined using the distance between object and the camera, d(x).

$$t(x) = 1 - \omega \min_{c\in\{r,g,b)} \left(\min_{C\in\Omega(x)} \left(\frac{R^C}{A^C} \right) \right) \tag{4}$$

where ω is 0.8, Ω{x} is the local block centered at x.

From Eq. (2) we can estimate J(x) by using equation:

$$J(x) = \frac{R(x) - A}{t(x)} + A \tag{5}$$

Equation (5) results under-enhancement for low-lighting regions. Thus, this equation is further modified in algorithm to get smoother videos by introducing a multiplier p(x) into (5).

$$P(x) = \begin{cases} 2t(x), & 0 < t(x) < 0.5 \\ 1, & 0.5 < t(x) < 1 \end{cases} \tag{6}$$

Then J(x), the intensity of recovered scene is given by:

$$J(x) = \frac{R(x) - A}{P(x)t(x)} + A \tag{7}$$

If suppose t(x) < 0.5 indicates that corresponding pixel requires boosting. If p(x) is small then p(x) t(x) becomes much lower value and therefore increases intensities of RGB (Fig. 2).

2.2 *Video Fusion Based on Gradient*

The main aim of image fusion is to fuse details obtained from different images of same video, which is more applicable to human vision, perception and interpretation. Image fusion can be done mainly for image segmentation, feature extraction and object recognition. Various methods have been proposed for image fusion by numerous authors.

Stathaki [8] suggested a pixel–based approach to enhance details of night time video which can be explained by using following equation:

$$F(x, y) = A_1(x, y)\, L_n(x, y) + A_2(x, y)\, L_{db}(x, y) \tag{8}$$

where F(x, y), $L_n(x, y)$, $L_{db}(x, y)$ are resultant illumination component of enhanced video frame, illumination of night time video frame, and day time video

Fig. 2 Examples of processing steps involved in low-lighting enhancement algorithms: input image I (*top left*), inverted image of the input R (*top right*), haze removal result J of the image R (*bottom left*) and output image (*bottom right*)

frame respectively. A1(x, y), A2(x, y) are having values in the intervals [0, 1]. This operation results in enhancement of background regions. The enhancement of the foreground regions are achieved by fusion of night time and day time videos. But the result will be a ghost pattern.

"Delighting" algorithm was given by Yamasaki [9]. This approach can be explained with the following expression:

$$L(x, y) = \frac{L_{db}(x, y)}{L_{nb}(x, y)} L_n(x, y) \tag{9}$$

where L(x, y) indicates enhanced illumination component of enhanced video frame, $L_{nb}(x, y)$ represents illumination component of the enhanced night time background images, and Ln denotes illumination component of the night time background images. The main disadvantage of this method is the ratio of illumination component of background corresponds to day time video frames to the illumination component of night time video frame is smaller than 1 then enhanced video frame will lose static illumination.

Another efficient method based on gradient fusion was proposed by Rasker et al. [10] to enhance details hidden in various video frames by extracting

Fig. 3 *First row* High illuminated daytime image, a night time image, and with a foreground person. *Second row* A simple binary mask obtained by subtracting reference from foreground, the resultant image obtained after processing the binary mask, and resultant image obtained by gradient based fusion algorithm

illumination–dependent properties. This technique become an efficient tool to develop surrealist create images and videos for artists. This method fuses high illuminated background of the daytime with low illuminated foreground of night time video frame. This can be given by expression:

$$\mathrm{G(x,y)} = N_i(x,y) * w_i(x,y) + D(x,y) * (1 - w_i(x,y)) \tag{10}$$

where G(x, y) indicates mixed gradient field, N_i denotes gradient of the night time video, D(x, y) is the gradient of daytime background and w_i denotes the importance of the image. This algorithm eliminates visual artifacts as ghosting, aliasing and haloing. The main limitations of this algorithm are computational complexity and color shifting problem (Fig. 3).

To overcome the limitations of above mentioned methods an efficient method based on gradient fusion enhancement was proposed by Yumbo et al. In this technique, gradient of video frames corresponds to background of day time are fused with gradient corresponds to foreground of night time video frames. As a result of this a high quality video results. Sobel operators are used to compute gradient of the video frame in the horizontal and vertical directions. This algorithm includes mainly two conditions to obtain enhanced video frame: one is fusing of gradient corresponds to background of day time and the gradient corresponds to night time video frames along horizontal and vertical directions whish can be expressed as:

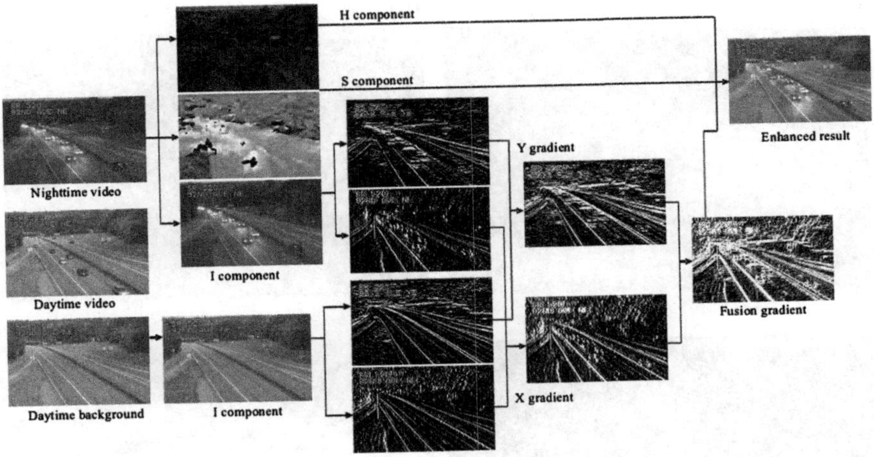

Fig. 4 Block diagram of gradient based fusion algorithm

$$G_{x/y} = \begin{cases} G_{x-db} + G_{x-nvideo}, \\ G_{y-db} + G_{y-nvideo}, \end{cases} \tag{11}$$

where G_x and G_y represents gradient of enhanced night time video frame in the horizontal and vertical directions respectively.

Second, is the fusing of gradient of G_x and G_y components.

$$G_{enhancement} = \begin{cases} \alpha.G_x + (1-\alpha).G_y, & if\ G_x \geq G_y \\ (1-\alpha).G_x + \alpha.G_x, & if\ G_x < G_y \end{cases} \tag{12}$$

where $G_{enhancement}$ is resultant gradient enhanced video, α is the weighting factor combining both daytime and night time gradient (Figs. 4, 5 and 6).

2.3 Dark Channel Prior

Haze removal technique is widely used for photography and in computer vision. Kaiming et al. [7] provided an efficient algorithm for removal of haze from the single input image using Dark channel prior. Various techniques have been developed by making use of multiple images or additional information. One such approach is based on polarization [13, 14] used to remove haze in different images of different degrees of polarization.

The dark channel prior is a kind of statistics of the haze-free outdoor images [7]. Since the dark channel prior is based on statistic, therefore it may not be an efficient method for the scene objects which are inherently similar to the atmospheric light

Fig. 5 **a** Night time video frame, **b** I component extracted, **c** H component extracted, and **d** S component extracted

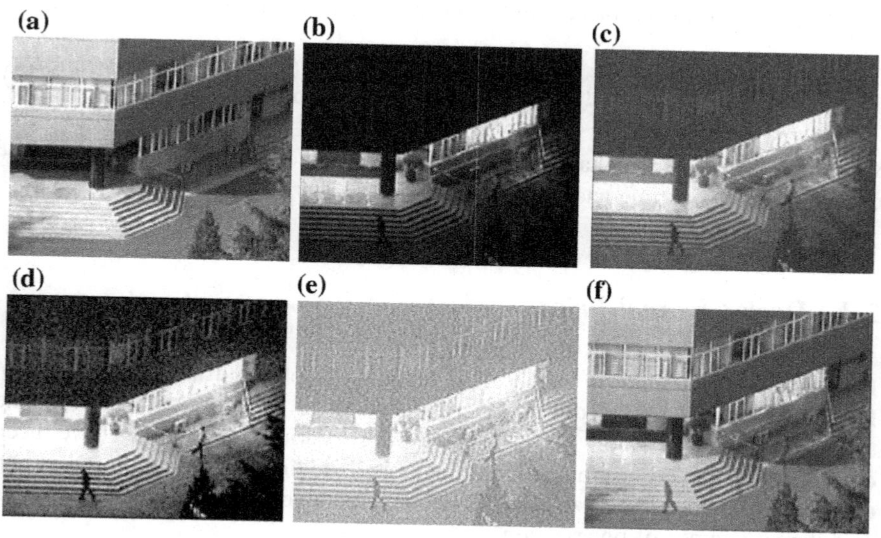

Fig. 6 **a** Day time video frame with high illuminated background, **b** night time video frame with dark scene, **c** enhanced night time video sequence using power-law transform in [11], **d** histogram equalization, **e** enhanced night time video sequence using tone mapping function in [12], and **f** gradient based fusion approach

and no shadow is cast on them; for such situations the dark channel prior fails. Main limitations of this technique is that it may be inefficient for various visible objects. More complicated algorithms are needed to implement sun's influence on the sky region, and the bluish hue near the horizon [15] (Figs. 7 and 8).

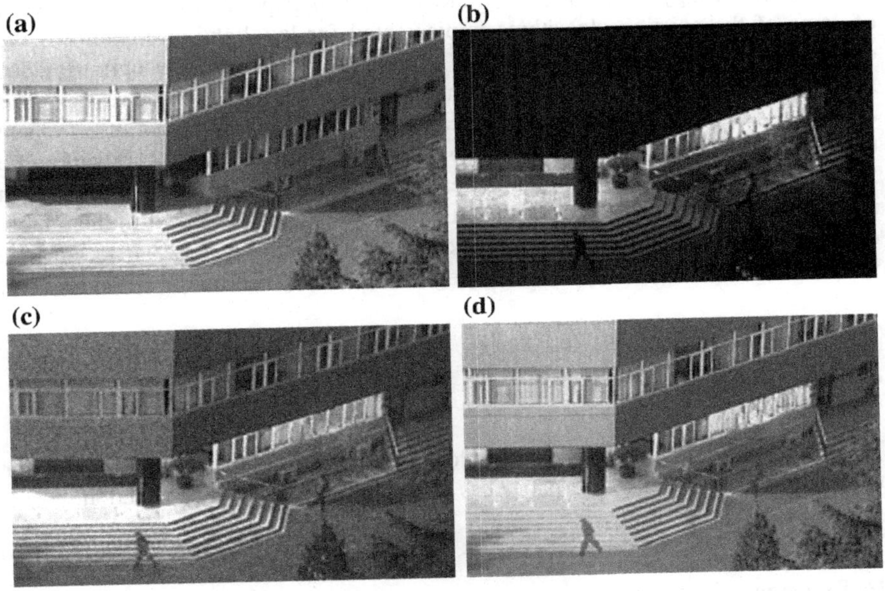

Fig. 7 Enhanced night time video frame: **a** using pixel-based method in [8], **b** using "denighting" method in [9], **c** using gradient domain in [10], and **d** gradient based fusion approach

Fig. 8 Haze removal results. *First column* Input images. *Second column* restored haze free images. *Third column* indicate depth maps

3 Conclusions

Video enhancement becomes a very challenging problem under low lighting conditions. Various algorithms are proposed for enhancing visual quality of videos/images captured under different environmental conditions especially in dark or night time, foggy situations, rainy and so on. This paper presented an overview of some of the existing algorithms implemented for low-lighting conditions. We discussed Hazy removal algorithm that applies inversion over low lighting video frames the result obtained is of noisy similar to hazy videos. From the result of this approach is as shown in the Fig. 1, we can concluded that there is no improvement to enhance poor quality videos and also we presented some of the Gradient based fusion algorithms proposed by Yumbo et al. that fuses background of daytime with night time video frames. As a result of this, there is an improvement in the quality of videos obtained and also we observed that this approach may fail to enhance videos like sun's influence on the sky region, and the bluish hue near the horizon.

4 Declaration

The images used in this paper are downloaded from the research papers selected for review and as a courtesy, the paper has been cited in the reference section.

References

1. Xuan dong, Guan wang, yi(Amy) pang, Weixin Li, Jiangtao(Gene) Wen "Fast Efficient algorithm for enhancement of low-lighting video.
2. M. Blanco, H. M. Jonathan, and T. A. Dingus. "Evaluating New Technologies to Enhance Night Vision by Looking at Detection and Recognition Distances of Non-Motorists and Objects," in Proc. Human Factors and Ergonomics Society, Minneapolis, MN, vol. 5, pp. 1612–1616 Jan. 2001.
3. O. Tsimhoni, J. Bargman, T. Minoda, and M. J. Flannagan. "Pedestrian ¨Detection with Near and Far Infrared Night Vision Enhancement," Tech. rep., The University of Michigan, 2004.
4. L. Tao, H. Ngo, M. Zhang, A. Livingston, and V. Asari. "A Multi-sensor Image Fusion and Enhancement System for Assisting Drivers in Poor Lighting Conditions," in Proc. IEEE Conf. Applied Imagery and Pattern Recognition Workshop, Washington, DC, pp. 106–113, Dec. 2005.
5. H. Ngo, L. Tao, M. Zhang, A. Livingston, and V. Asari. "A Visibility Improvement System for Low Vision Drivers by Nonlinear Enhancement of Fused Visible and Infrared Video," in Proc. IEEE Conf. ComputerVision and Pattern Recognition, San Diego, CA,, pp. 25 enhancement of Low lighting video". Jun 2005.
6. H. Malm, M. Oskarsson, E. Warrant, P. Clarberg, J. Hasselgren, and C. Lejdfors. "Adaptive Enhancement and Noise Reduction in Very Low Light-Level Video," in Proc. IEEE Int. Conf. Computer Vision, Rio de Janeiro, Brazil, Oct. 2007, pp. 1–8.
7. K. He, J. Sun, and X. Tang. "Single Image Haze Removal Using Dark Channel Prior," in Proc. IEEE Conf. Computer Vision and Pattern Recognition, Miami, FL, pp. 1956–1963, Jun. 2009.

8. T. Stathaki, Image Fusion: Algorithms and Applications, Academic Press, 2008.
9. A. Yamasaki et al., "Denighting: Enhancement of Nighttime Image for a Surveillance Camera," 19th Int. Conf. Pattern Recog., 2008.
10. R. Raskar, A. Ilie, and J.Y. Yu, "Image Fusion for Context Enhancement and Video Surrealism," Proc. SIGGRAPH 2005, ACM New York, NY, USA, 2005.
11. R.C. Gonzalez and R.E. Woods, Digital Image Processing, 3rd ed., NJ: Prentice Hall, 2007.
12. F. Durand and J. Dorsey, "Fast Bilateral Filtering for the Display of High-Dynamic Range Images," ACM Trans. Graphics, vol. 21, no. 3, pp. 257–266., 2002.
13. Y. Y. Schechner, S. G. Narasimhan, and S. K. Nayar. Instant dehazing of images using polarization. CVPR, 1:325, 2001.
14. S. Shwartz, E. Namer, and Y. Y. Schechner. Blind haze separation. CVPR, 2:1984–1991, 2006.
15. A. J. Preetham, P. Shirley, and B. Smits. A practical analytic model for daylight. In SIGGRAPH, pages 91–100, 1999.

A Detailed Review of Color Image Contrast Enhancement Techniques for Real Time Applications

P. Agalya, M.C. Hanumantharaju and M.T. Gopalakrishna

Abstract Real-time video surveillance, medical imaging, industrial automation and oceanography applications use image enhancement as a preprocessing technique for the analysis of images. Contrast enhancement is one of a method to enhance low contrast images obtained under poor lighting and fog conditions. In this paper, various variants of histogram equalisation, Homomorphic filtering and dark channel prior techniques used for image enhancement are reviewed and presented. Real-time processing of images is implemented on Field Programmable Gate Array (FPGA) to increase the computing speed. Further this paper focus on the review of contrast enhancement techniques implemented on FPGA in terms of device utilization and processing time.

Keywords Histogram equalisation · CLAHE · Homomorphic filtering · Dark channel prior · Fog images · Poor lighting image

---

P. Agalya (✉)
Department of Electronics & Communication Engineering,
Sapthagiri College of Engineering, Bangalore, India
e-mail: agal_s@yahoo.co.in

M.C. Hanumantharaju
Department of Electronics & Communication Engineering,
BMS Institute of Technology & Management, Bangalore, India
e-mail: mchanumantharaju@gmail.com

M.T. Gopalakrishna
Department of Computer Science Engineering, Kammavar Sangam School
of Engineering & Management, Bangalore, India
e-mail: gopalmtm@gmail.com

© Springer India 2016
S.C. Satapathy et al. (eds.), *Information Systems Design and Intelligent Applications*, Advances in Intelligent Systems and Computing 435,
DOI 10.1007/978-81-322-2757-1_48

1 Introduction

Image processing is an essential and promising research area in various real time application fields such as medical imaging, video surveillance, industrial X ray imaging, oceanography etc. [1]. Image enhancement is a preprocessing technique in many image processing applications that can produce an improved quality image than the original image so that the output image is more suitable for analysis by human or machine in specific applications. In general image enhancement techniques are classified into two broad categories such as spatial and frequency domain techniques. In spatial domain techniques, the pixels itself are directly modified to enhance an image. In frequency domain method, modification is done on the Fourier transformed image and inverse Fourier transform is applied to the modified image to get the enhanced image.

Quality of the image gets affected by uneven or poor illumination, external atmospheric condition such as fog or haze, wrong lens aperture setting of the camera, noise etc. So these degraded quality images are improved by increasing the brightness and contrast, by de-noising the image through various enhancement techniques. Researchers have developed numerous enhancement techniques that are good in enhancing the contrast of an image, while some are good for de-noising the images. In real time applications an enhancement technique should be capable of enhancing real color images in lesser time with lesser computational cost by reducing (i) the effect of haze or fog, (ii) poor or uneven illumination effect on an image and (iii) noise introduced in an image.

This research review work focus on various color image enhancement techniques that improves the contrast of real time images. Requirements of real time image enhancement techniques [2] are (i) It should be adaptive in nature (i.e.) should be able to enhance any type of images for a specific application, (ii) Should enhance a image in less processing time, (iii) It should utilize less computational resources.

1.1 Histogram Equalisation Based Enhancement Techniques

Histogram equalisation is a simplest technique to enhance the contrast and luminance of an image. It enhances the edges of different objects in a image, but could not preserve the local details in the object, and also it enhances brighter parts of the image heavily, so that the natural look and color of the image is lost. So to overcome these effects researchers have come out with various histogram techniques as local and adaptive histogram techniques. Local histogram techniques like Bi histogram equalisation techniques such as Brightness BI Histogram Equalisation Method (BBHE) proposed by Kim et al. [3], Dualistic Sub Image Histogram Equalisation Method (DSIHE) by Wang et al. [4], Minimum Mean Brightness Error BI Histogram Equalisation Method (MMBEBHE) by Hossain et al. [5] divide the image into two sub images based on mean, median and mean brightness difference, and apply histogram equalisation locally on each sub image. These techniques enhance the overall visual appearance of an image without distorting the local details. MMBEBHE technique preserves the brightness of the image better than BBHE and DSIHE.

Multi histogram equalisation techniques like Multi Variance Multi Histogram Equalisation (MVMHE) and Optimal Thresholding Multi Histogram Equalisation (OTMHE) [6] divide the original histogram into two sub images based on mean brightness value. All sub images are then divided into several sub histograms based on the minimum variance of each sub image. Histogram equalisation is applied on each sub image histograms and the resultant output image preserves the brightness by improving the contrast much more effectively with less mean square error comparatively with a view to bi histogram techniques. It also introduces unnecessary artifacts for images in poor lighting condition.

A Modified Histogram Equalisation (MHE) approach proposed by Padmaja et al. [7] explores the image in YIQ color space. In this work RGB image is converted into YIQ color space and then Y component is normalized using a sigmoid function given in Eq. (1) and then histogram equalisation is applied to the normalized Y component.

$$S_n = 1/1 + \sqrt{\frac{1 - In}{In}} \tag{1}$$

S is the sigmoid function normalised output, I is the input to be normalized.

Experimental results reveal that this technique can enhance the color images under fog and dim region by increasing the contrast and brightness of the darker region, but dilutes the uniform brighter background region in foggy image and reduce the color effect of the image in dim images. Since this technique involves less computation time and load, it may be well suited in real time applications if it improves to preserve the color effect (Fig. 1).

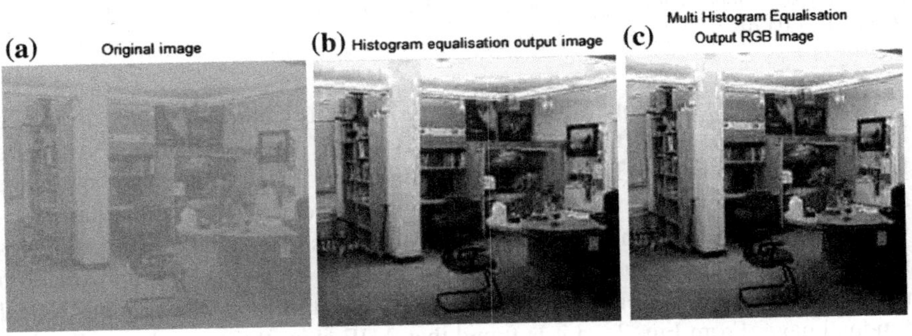

Fig. 1 Experimental results of histogram equalisation based techniques. **a** Original image of size 256 × 256. **b** Histogram equalised image with peak signal to noise ratio (PSNR) = 10.18 dB, root mean square error (rmse) = 79.00. **c** Modified histogram equalised image with PSNR = 3.11 dB, RMSE = 178.56

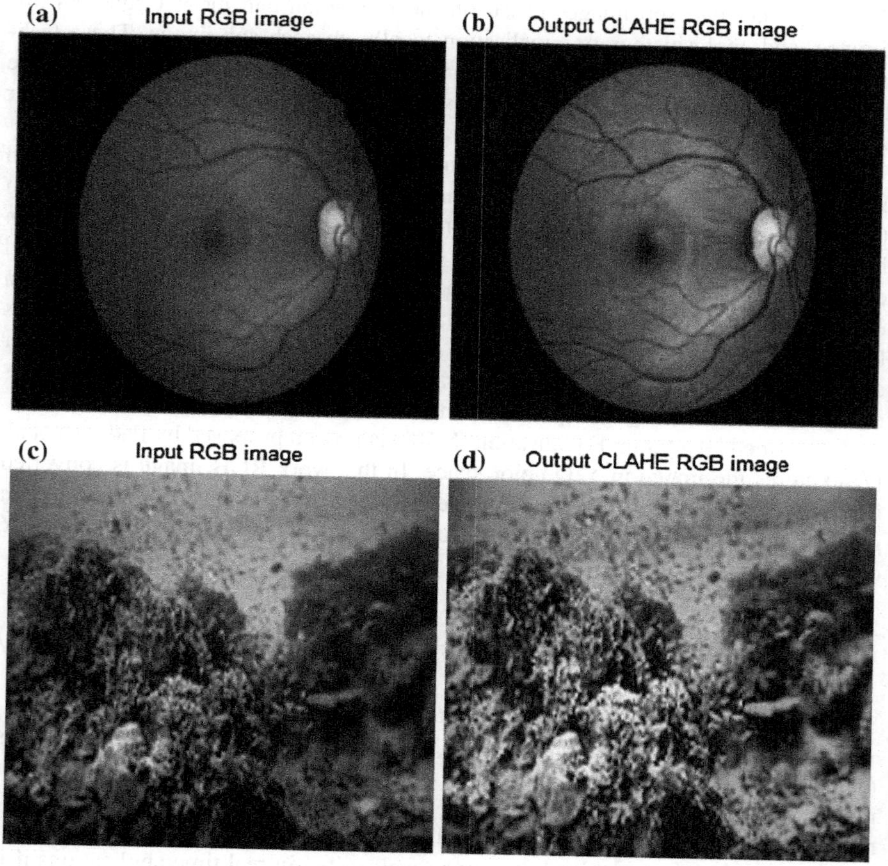

Fig. 2 Experimental results of contrast limited adaptive histogram equalisation. **a** Retinal fundus input image of size 256 × 256. **b** CLAHE output image with RMSE = 18.66 PSNR = 22.71 dB. **c** Under water input image of size 256 × 256. **d** CLAHE output image with RMSE = 35.85 PSNR = 17.04 dB

Adaptive histogram equalisation techniques such as Contrast Limited Adaptive Histogram Equalisation (CLAHE) [8] and Balanced Contrast Limited Adaptive Histogram Equalisation (B-CLAHE) [9] adjusts the local contrast of an image adaptively. Experimental test is conducted on few medical images and an underwater images of resolution 256 × 256, the result shows that adaptive HE techniques improves the local contrast of an image. In Fig. 2a, b the vascular network in a retinal fundus image is enhanced by CLAHE algorithm than the original retinal fundus image. From Fig. 2c, d it is found that AHE algorithms are also well suited in the field of oceanography for analysis of underwater images as the fish cloud is more clearly seen in the enhanced output image.

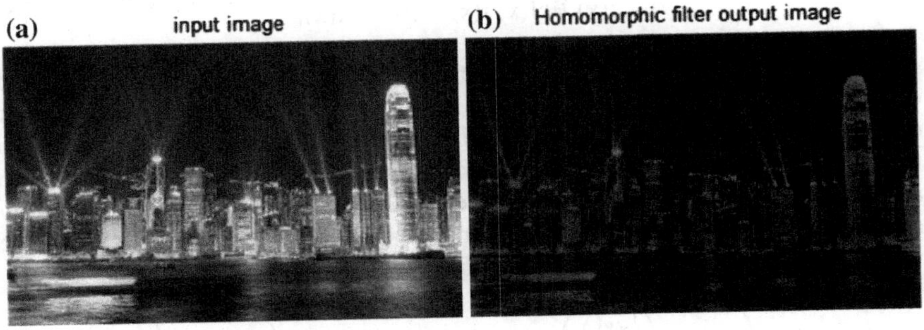

Fig. 3 Experimental results of homomorphic filter. **a** Input image of size 256 × 256. **b** Output image with RMSE = 0.25 PSNR = 60.07 dB

1.2 Homomorphic Filter

Homomorphic filter [10] is a frequency domain enhancement technique which improves the visual appearance of the image by compressing the brightness range of the image while simultaneously improving the contrast to much higher extent. Image is transformed to log domain first and then a linear filtering process is applied in frequency domain of the image so that the low frequency illumination component is filtered to reduce the noise effect and then inverse process is applied to obtain the enhanced image in spatial domain. Figure 3 shows that this technique is found to be most useful in reducing the unnatural effects created under heavy lighting and uneven illumination.

1.3 Dark Channel Prior

In most real time visual applications like traffic monitoring system and outdoor recognition system, the visual appearance of an image is affected due to diverse weather situations like haze or fog, smoke, rain and snow. In such kind of applications various procedures like dark channel prior, CLAHE, wiener filtering and bilateral filtering are used to remove the multifaceted visual effects on an image. A study on various digital image fog removal algorithms by Gangadeep [11] evaluates the effectiveness of dark channel prior, CLAHE, wiener filtering for the removal of fog. The study shows that dark channel prior technique performs efficiently than other methods. Xiaoqiang et al. [12] used dark channel prior method to remove the fog effects in a traffic monitoring system. In dark channel prior method the atmospheric attenuated image model is represented through Mccartney model in Eq. (2).

$$I(x) = J(x)\, t(x) + A[1 - t(x)] \tag{2}$$

I(x) is the observed image, J(x) is the original real image which got attenuated due to fog, A is the global atmospheric light, t is the transmission. Generally dark channel of an image is a color channel with low intensity in the non sky region of a fog free image. But the intensity of dark channel points becomes higher due to additive atmospheric light in a fog image. So the thickness of fog can be roughly approximated though the intensity of dark channel points. For a haze image J the value of global dark channel pixel is defined as given in Eq. (3)

$$\min_{c\in\{R,G,B\}}\left(\min_{Y\in I}\left(\frac{I_{c(Y)}}{A_c}\right)\right) = t_{global}(x)\min_{c\in\{R,G,B\}}\left(\min_{Y\in I}\left(\frac{J_{c(Y)}}{A_c}\right)\right) + 1 - t_{global}(x) \tag{3}$$

where $J_c(y)$ is the dark channel point of the haze free image with value approaches to 0, A_c is the atmospheric light of the dark channel. The global transmission is obtained using dark channel theory and McCartney model as given in Eq. (4)

$$t_{global}(x) = 1 - \omega(d_c/A_c) \tag{4}$$

Atmospheric light A is estimated according to dark channel points and substituting the value of A and t in Eq. (5), the scene radiance output image is obtained.

$$J(x) = \frac{I(x) - (1 - t_{global}(x))A'}{t_{global}(x)} \tag{5}$$

where J is the scene radiance (i:e) real image after fog removal, I is the observed image, t is global transmission, A is atmospheric light. Dark Channel prior is experimentally demonstrated on a natural outdoor scene covered by fog in Fig. 4.

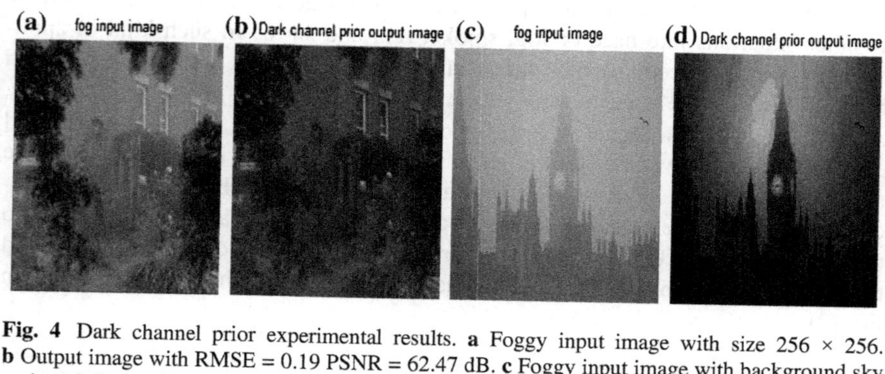

Fig. 4 Dark channel prior experimental results. **a** Foggy input image with size 256 × 256. **b** Output image with RMSE = 0.19 PSNR = 62.47 dB. **c** Foggy input image with background sky region. **d** Output image with RMSE = 0.36 PSNR = 56.96 dB

As dark Channel prior is based on accurate estimation of haze transmission using dark channel, it can effectively remove the haze by improving the local contrast of the output image. In Fig. 4d an unnecessary artifact is introduced as dark channel can accurately estimate the haze transmission only on images without uniform non-sky region.

2 Comparison of Experimental Results and Analysis

The effectiveness of different existing contrast enhancement techniques is evaluated for a real time fog affected image and their results are given in Fig. 5.

Comparative analysis of different contrast enhancement techniques on a non sky foggy image shows that dark channel prior method is found to give efficient contrast enhancement with reduced noise and excellent PSNR. Histogram equalisation based methods improve the overall brightness of a image but fail to reduce the fog effects in an image.

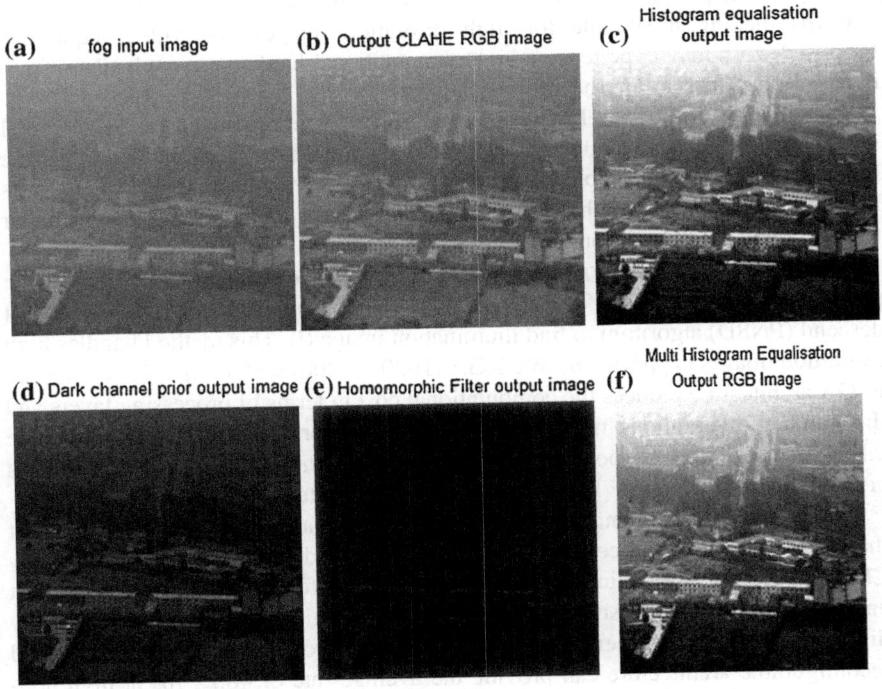

Fig. 5 Comparative analysis of different contrast enhancement techniques on foggy image. **a** Input image of size 256 × 256. **b–f** Output images. **b** CLAHE with RMSE = 10.93 PSNR = 27.36 dB. **c** HE with RMSE = 54.78 PSNR = 13.36. **d** Dark channel prior with RMSE = 0.3834 PSNR = 56 dB. **e** Homomorphic filter with RMSE = 0.42 PSNR = 55.67 dB. **f** Modified HE with RMSE = 144.88 PSNR = 4.91 dB

3 FPGA Implemented Contrast Enhancement Techniques

As Field Programmable Gate Array (FPGA) supports pipelining and parallelism in processing of images, use of FPGA to implement image enhancement techniques will reduce the processing time and the computational load. Our review work focuses on some of the image enhancement techniques that are implemented on FPGA. A comparison table that illustrates the FPGA devices used, device utilization in FPGA, the number of frames processed per second (fps) and the enhancement techniques used.

A novel FPGA architecture of adaptive color image enhancement proposed by Hanumantharaju et al. [13] uses a simple arithmetic mean filter on saturation (S) and value (V) component and preserves the hue (H) component to avoid color distortion. The experimental result shows that the image enhanced by this work is clearer and vivid compared to histogram equalisation with reduced PSNR.

A novel FPGA architecture for Gaussian based color image enhancement system for real time application proposed by Hanumantharaju et al. [14] provides better dynamic pixel range compression and color rendition effect with color constancy in an image. This proposed method process images of size 1600 × 1200 at the rate of 116 fps making it suitable for high resolution images and video sequences. Proposed work may experience a high computational hardware complexity since it is based on Gaussian approach.

Li et al. [15] proposed a FPGA structure to realize an optimized Gaussian kernel with address decoding and parallel scale convolution to implement multi scale Retinex for an outdoor application. The proposed technique outperforms with its Matlab implementation with a frame rate of 60 fps for monochrome images under fog and poor light conditions.

An FPGA implementation of real time retinex Video image enhancement using pipelining architecture by Tstutsui et al. [16] use projected normalised steepest descend (PNSD) algorithm to find illumination image (l). This method handles high resolution images supported by WUXGA (1920 × 1200) with 60 fps. This proposed FPGA architecture reduces the computational cost in terms of processing layers and the number of iterations making it more suitable for real time application. The effectiveness of this proposed work for enhancing endoscopic images with less computational complexity has been shown by Okuhata et al. [17].

A contrast enhancement algorithm with efficient hardware architecture to meet the needs of real time processing for high definition video applications proposed by Chen et al. [18] uses pipelined architecture of 5 stages for implementing the contrast enhancement algorithm using adaptive gamma correction and cumulative intensity distribution. The experiment results show that the proposed parameter controlled reconfigurable architecture can provide the average rate of 48.22 fps at high definition resolution 1920 × 1080 which means the proposed hardware architecture can run in real time.

An FPGA implementation of improved real time contrast enhancement technique proposed by Tang et al. [19] demands low computational complexity than

Table 1 Comparison of FPGA implemented color image enhancement algorithms

Name of researcher	Hanumantharaju et al. [12]	Hanumantharaju et al. [13]	Li et al. [14]	Tstutsui et al. [15]	Chen et al. [17]	Tang et al. [18]
Technique used	Arithmetic mean filter on S and V component	Gaussian smoothing on R, G, B component	Multiscale retinex	Retinex using PNSD for illuminance estimation	Gamma correction, cumulative intensity distribution	Adaptive HE based technique
FPGA device used	Vertex II XC2V2000-4ff896	Vertex II XC2VP40 7FF1148	XC4VSX55	Stratix III EP3SL150F1152C2	VirteX C5V110T	Cyclone II EP2C35F672C6
Maximum operating frequency	–	224.840 MHZ	–	–	66 MHZ	132.71 MHZ
Frames per second	–	116	–	60	48.22	–
Image size	256 × 256	1600 × 1200	1920 × 1200	1920 × 1200	1920 × 1080	640 × 480
Device utilization report	No. of slice flipflops—3128	No. of slice flipflops—28,899	No. of slice flipflops—12,724	Combinational ALUTs—4320/113,600	Gate count —11.5 K slices	LUTs—246
	Total no. of 4 input LUTS—7483	Total no. of 4 input LUTS—3195	Total no. of 4 input LUTS—18,186	Memory ALUTs—156/56,800	Memory—35.2 K slices	Registers—252
	No. of slices—3427	No. of slices—15,198	No. of FIFO16/RAMB16s—101	Dedicated logic registers—2611/113,600		Block RAMs—11,776 bits
	Total gate count—64,312	Total gate count—340,572				

improved histogram equalisation algorithms like CLAHE [6] for processing video images. As this work uses mean probability as the threshold value to modify the different gray levels of the image it may not ensure processing quality for any type of image.

Table 1 shows the comparison of FPGA implemented color image contrast enhancement techniques in terms of device utilization, enhancement techniques used, resolution of the image, maximum operating frequency and number of frames processed per second. It is clear that Gaussian based smoothing technique [13] can process more number of frames per second, but this method costs much computational load. Adaptive gamma correction and cumulative intensity distribution method [17] could process high resolution images of size 1920 × 1080 utilizing less computational resources compared to all the other methods with 48 fps. Retinex using PNSD for illuminance estimation algorithm by Li et al. [15] gives a clear vivid image for real time application by processing 60 fps.

4 Conclusion

Image enhancement is an essential preprocessing step in many real time image processing applications. Enhancement of Images is done by many approaches and choice of every approach depends on the type of images. Among all histogram equalisation techniques multi histogram equalisation techniques improves the contrast and brightness of the images. Adaptive histogram equalisation techniques improves the local contrast of the image making it well suited for enhancing real time medical images and underwater images. Homomorphic filtering works efficiently to reduce the excess lighting effect on images. Dark channel prior method gives excellent enhancement of fog images. Survey on FPGA implemented contrast enhancement techniques reveals the effectiveness of reconfigurable architectures for enhancing color images in less processing time to meet the real time application requirements.

References

1. Dr. Vipul Singh: Digital Image Processing with MATLAB and Lab view, Elsevier (2013).
2. Donald G., Bailey: Design for Embedded Image Processing on FPGAs, John Wiley & sons, August (2011).
3. Kim, Y. T.: Contrast enhancement using brightness preserving bi histogram equalisation, In: IEEE transactions on Consumer Electronics, vol. 43, no. 1, pp. 1–8, (1997).
4. Wang, E. Y., Chen, Q., Zhang, B.: Image enhancement based on equal area dualistic sub-image histogram Equalisation method, In: IEEE transactions on Consumer Electronics, vol. 45, no. 1, pp. 68–75, (1999).
5. Chen, S. D., Ramli, A. R.: Minimum mean brightness error bi - histogram equalisation in contrast enhancement, In: IEEE transactions on Consumer Electronics, vol. 49, no. 4, pp. 1310–1319, (1999).

6. Srikrishna, A., Srinivasa Rao, G., Sravya, M.: Contrast enhancement techniques using histogram equalisation methods on color images with poor lighting, In: International Journal of Computer science, Engineering and Applications, Vol. 3, No. 4, pp. 1–24, (2013).
7. Padmaja, P., Pavankumar, U.: An Approach to Color Image Enhancement using Modified Histogram, In: International Journal of Engineering Trends and Technology (IJETT), Vol. 4, pp. 4165–4172, (2013).
8. Zuiderveld, K.: Contrast Limited Adaptive Histogram Equalisation, In: Chapter Vl11.5, Graphics Gems IV. P.S. Heckbert (Eds.), Cambridge, MA, Academic Press, pp. 474–485, (1994).
9. Raheel Khan, Muhammad Talha, Ahmad S., Khattak, Muhammad Qasim: Realization of Balanced Contrast Limited Adaptive Histogram Equalisation (B-CLAHE) for Adaptive Dynamic Range Compression of Real Time Medical Images, In: IEEE proceedings of 10th IBCAST, pp. 117–121, (2013).
10. Delac, K., Grgic, M., Kos, T.: Sub-Image Homomorphic Filtering Technique for Improving Facial Identification under Difficult Illumination Conditions, In: International Conference on Systems, Signals and Image Processing, pp. 95–98, (2006).
11. Gangadeep singh: Evaluation of Various Digital Image Fog Removal Algorithms, In: International Journal of Advanced Research in Computer and Communication Engineering, Vol. 3, pp. 7536–7540, (2014).
12. Xiaoqiang Ji, Jiezhang Cheng, Jiaqi Bai, Tingting Zhang: Real time Enhancement of the Image Clarity for Traffic Video Monitoring System in Haze, In: 7th International congress on Image and Signal processing, pp. 11–15, (2014).
13. Hanumantharaju, M. C., Ravishankar, M., Rameshbabu, D. R., Ramachandran, S.: A Novel Reconfigurable Architecture for Enhancing Color Image Based on Adaptive Saturation Feedback, In: Information Technology and Mobile Communication in Computer and Information Science, Springer, vol.147, pp. 162–169, (2011).
14. Hanumantharaju, M. C., Ravishankar, M., Rameshbabu, D. R.: Design of Novel Algorithm and Architecture for Gaussian Based Color Image Enhancement System for Real Time Applications, In: Advance in computing, communication and control, Vol. 361, pp. 595–608, (2013).
15. Yuecheng Li, Hong Zhang, Yuhu You, Mingui Sun: A Multi-scale Retinex Implementation on FPGA for an Outdoor Application, In: 4th International Congress on Image and Signal Processing, IEEE, pp. 178–1792, (2011).
16. Hiroshi tsutsui, Hideyuki Nakamura, Ryoji Hashimoto, Hiroyuki Okuhata, Takao onoye: An FPGA implementation of real-time Retinex video image enhancement, In: World Automation Congress, TSI Press, (2010).
17. Hiroyuki Okuhata, Hajime Nakamura, shinsuke hara, Hiroshi Tsutsui, Takao Onoye: Application of the real time retinex image enhancement for endoscopic images, In: 35th annual international conference of the IEEE EMBS, Osaka, Japan, pp. 3407–3410, (2013).
18. Wen-Chieh Chen, Shih-chia huang, Trong-yen Lee: An efficient reconfigurable architecture design and implementation of image contrast enhancement algorithm, In: IEEE 14th International conference on high performance computing and communications, pp. 1741–1747, (2012).
19. Jieling Tang, Xiaoqing Du, Xiangkun Dong: Implementation of an improved real time contrast enhancement technique based on FPGA, In: Springer Engineering and Applications, pp. 380–386, (2012).

A Conceptual Model for Acquisition of Morphological Features of Highly Agglutinative Tamil Language Using Unsupervised Approach

Ananthi Sheshasaayee and V.R. Angela Deepa

Abstract Construction of powerful computer systems to understand the human languages or natural languages to capture information about various domains demands morphologically featured modeled architected appropriately in a core way. Morphological analysis is a crucial step that plays a predominant role in the field of natural language processing. It includes the study of structure, formation, functional units of the words, identification of morphemes to endeavor the formulation of the rules of the language. Since natural language processing applications like machine translation systems, speech recognition, information retrieval rely on large text data to analyze using linguistic expertise is not viable. To overcome this issue morphological analysis using unsupervised settings is incorporated. It is an alternative procedure that works independently to uncover the morphological structure of the languages. This paper gives a theoretical model to analysis morphologically the structure of the Tamil language in an unsupervised way.

Keywords Morphological segmentation · Clustering · Agglutinative · Unsupervised · Lexicon

1 Introduction

Morphology is an important phase in the science of languages, which narrates about the structure and the formation words [1]. The linguistics rules play a key role in the structural pattern of the natural language and the morphological rules helps in the

Ananthi Sheshasaayee (✉) · V.R. Angela Deepa
Department of Computer Science, Quaid E Millath Govt College for Women, Chennai 600002, India
e-mail: ananthi.research@gmail.com

V.R. Angela Deepa
e-mail: angelrajan.research@gmail.com

© Springer India 2016
S.C. Satapathy et al. (eds.), *Information Systems Design and Intelligent Applications*, Advances in Intelligent Systems and Computing 435,
DOI 10.1007/978-81-322-2757-1_49

formation of the lexicon or word in a sentence. Morphological analysis deals with the structural formation, functionality of words, and identification of morphemes (smallest meaningful units) and formulation rules that can model the natural language to the machine understandable mode. It is a deep analytic study of morphological features which eventually favors in understanding the morphological components of the natural languages. A computational tool that analyzes the word forms into roots and functional elements is termed as a morphological analyzer (MA). The transpose course of morphological analyzer is morphological generator (MG) in which a well formed word is generated with given root and functional elements. This task is difficult and of high demand since the availability of the text corpus consists of a large number of different word forms. Morphological complexity being the prime factor categories the natural languages based on the degree of grammatical and derivational relations [2]. The characteristic feature of Compounding in few languages attributes to combine different word forms into a single unit challenges the effectiveness of language systems. Indic Languages, especially Tamil language which is grouped under Dravidian type are highly agglutinative and complex in grammatical nature. It takes the traits of lexical morphology and inflectional morphology [3]. A motivation for the morphological learning by machine benefits different areas like machine translation, stemming, speech recognition etc. leads to various approaches like rule based-suffix based approach, finite state approach, semi-supervised and unsupervised learning. Rule based approach approaches for morphological languages requires high ratio of human effort with heavy package of linguistic input [4–7]. The unsupervised approaches requires an input in which the algorithmic models with the input data identify and examines the patterns by mapping them progressively to relate their correlation and facet in the data [8–10]. The combination of minimal linguistic inputs with the automated learning process of morphology is termed as semi-supervised approaches [11–13]. Finite state approaches rely on paradigm type of input with linguistic information which uses finite state machines [14, 15]. For Morphology acquisition of natural languages, various unsupervised approaches have been deployed. But for highly agglutinative languages like Tamil very less computational work has been done in this direction. This paper depicts a theoretical framework for acquisition of morphological features of highly agglutinative Tamil language using unsupervised settings. This framework portrays various phases that captures the intricate morphological structures in an unsupervised way. The organization of the paper is as follows; Sect. 2 is about the related works. Section 3 the nature of the language. Section 3 demonstrates the nature of Tamil languages. Section 4 gives an elaborate view on morphological analysis. The framework of this model followed by the discussions related to the state of the model is presented in Sects. 5 and 6. The summarized idea is elucidated as a conclusion.

2 State of Art

A diverse body of existing works has been focused on morphology acquisition through various approaches. The following is an assortment of morphological analyzers and generators developed for the Tamil languages. Finite state transducers called AMAG [16] developed by Dr. K. Soman et al. developed a Morphological Analyzer (MA) and morphological generator (MG) based on the rule based approach. This system incorporates nearly fifty thousand nouns with a mentionable list of adjectives followed by three thousand verbs. The competence of the system was based on a two-level morphological system arranged on the lexicon and orthographic rules. When compared to existing system this proposed AMAG predicted a remarkable results. Suffix stripping algorithms were used to develop the morphological generator(MG) [17] by Dr. K.P. Soman et al. It consist of proper nouns, compound words, and transitive/intransitive words. This system encompasses of two components; first one handles the stem part and the second handles the Morpholexical related information. The system demands the following data file to maintain the morpholexical information, table to handle the suffix, paradigm classification rules and stemming rules for efficient functioning. The improvised morphological analyzer cum generator (MA/MG) proposed by Parameswari were implemented on Apertium-Open Source toolkit [18, 19]. It consists of Finite state transducers algorithm for one-pass analysis followed by generators to hold the Word/Paradigm based database. The system showed better experimental results when compared to other analyzers developed by CALTS and AUKBC research centers. Morphological Analyzer (MG) for Tamil language proposed by Anand Kumar et al. is based on the supervised approach. It resolved the problems faced by the morphological analyzer through sequence labeling approach [20] which narrates the classification problem. It is merely a corpus based approach which consists of nearly 130,000 verb forms and 70,000 noun forms. It is deployed for training and testing in Support vector machine (SVM). The system was relatively efficient when compared to the existing approaches. The morphological generator (MG) [21] based on finite state automata were developed by Menaka et al. to focus on highly agglutinative languages like Tamil. It contains two separate systems for evaluating nouns and verbs. Accuracy of the results was based on the comparison of the expected outcomes. The morphological analyzer (MA) constructed by Dr. Ganesan [22] aimed to analyze the CIIL Corpus based on the phonological and morphophonemic rules as well as morphotactic constraints of Tamil to structure it in a better efficient way. Deivasundaram's Morphological analyzer morphological analyzer [22] was built in order to increase the language efficiency in Tamil Word Processor. It exploits the phonological and morphophonemic rules incorporating the morphotactic constraints in order to build an effective parser.

3 Nature of the Language

Tamil language is rich in morphological structure and complex in grammatical notes. It inflects to person, gender and invariably joins with auxiliaries to indicate mood, aspect, attitudes, etc. The noun form of the language takes up almost five hundred word forms inflecting based on the post positions. The verb form inflects with the auxiliaries taking up approximately two thousand word forms. To incorporate this language in the machine understandable mode the words are needed to be segmented at the root level to build a successful language application.

4 Morphological Analysis

Tamil being a morphological rich content, language and agglutinative in grammatical content requires deep analysis at the root level to represent the appropriate meaning of the word in the forms of morphemes and categories. Ideally, for Tamil language the inflections at the root level are post propositional, which takes up nearly few thousand forms of words. The morphemes which are derivational and compounding in grammatical nature added to the root of the lexicon/word undergo a change in the meaning and their respective classes. Such coherence is termed as lexical morphology. For inflectional morphology there is a change in the form of a word when the inflectional morphemes are added to the root word. To resolve this problem of morphological complexity morphological analysis tends to support the factor of decomposing the given tokens or word forms into a sequence of morphemes with their categories.

5 Proposal Model

The tasks of morphological acquisition in an unsupervised way [23] is broadly classified into three main components. The prime task involves the segmentation of lexicons in which words or lexicons are decomposed into stem words with their respective suffixes. Clustering of words invokes the joining of words into conflation sets (i.e.) as a set of morphological related words. Based on the clustered group new words are generated from the particular stem bag with the related suffix bags.

5.1 Morphological Segmentation

This comprises of breaking the words into their respective root words and suffixes followed by labeling to identify it. This plays an important role in the analysis of

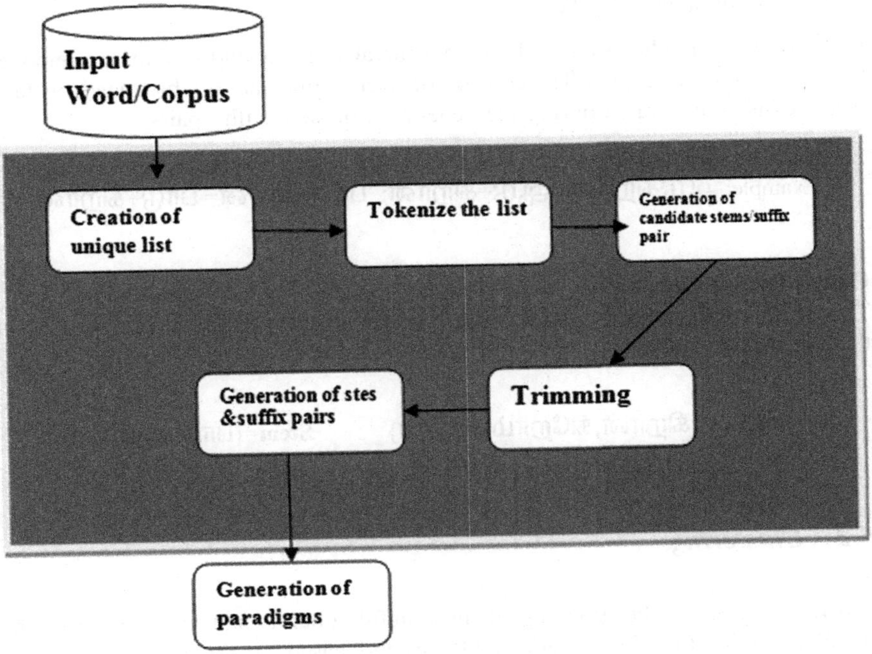

Fig. 1 Detail representation of proposed unsupervised morphological segmentation

morphological features. The main advantage of using an unsupervised setting for morphological segmentations is that the platforms for finding the morphemes as stems and suffixes do not depend on the linguistic details of labelling. Therefore the extracted analysis of results is straightforward without any hindrance.

Edifice of Morphological Segmentation

The proposed model (refer Fig. 1) aims to build families based on the derivational process in a language. The suffix pairs play a prominent chore in this construction.

Mining of Candidate Stems-Suffix Pairs

Highly agglutinative and complex languages like Tamil, the bound of morphemes are incorporated in the form of grammatical and derivational relations. The morphological segmentation algorithm separates each word types independently by classifying them as stems and suffixes.

Example: **Stem=** {பாடு, ஆடு,….}

Generation of Stem-Suffix Pairs

Based on the candidate stem-suffix pairs generated the invalid suffixes are removed by pruning the given pair. The concept of over segmentation which leads to false positives is concentrated during the generation of stem-suffix pairs.

Example: ஆடுகிறான்= ஆடு+கிறான்: பாடுகிறான்= பாடு+கிறான்

Generation of Paradigms
This is an enumerated list of all valid stems with the detailed information about their respective suffixes.

Suffix= { கிறான்,கிறோம்,கிறாள்} **Stem**={பாடு,ஆடு,....}

5.2 *Clustering*

Clustering refers to the grouping of meaningful subunits named as clusters. This predicted (refer Fig. 2) clusters highly distend based on the maximization and minimization of inter-cluster similarity. Thus the generated suffixes are clustered based on a simple criterion: any two suffixes belong to one cluster if and only if they form a pair rule. The characteristic form of Data distribution can be explained through the discovered clusters.

Optimum Clustering

The best clustering that cede the finest lexicon compression is assumed as optimum.

Creation of Conflation Sets

The best compressed lexicon during the clustering phase benefits the creation of conflation sets. Based on the optimum property of the clustering the conflation classes are constructed.

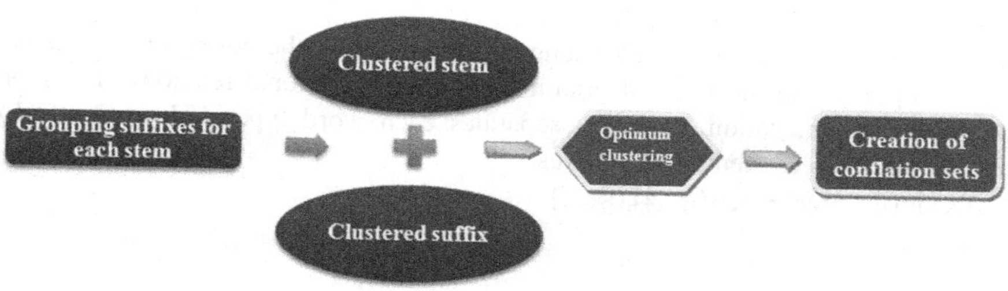

Fig. 2 Diagram to depict the mechanism of clustering

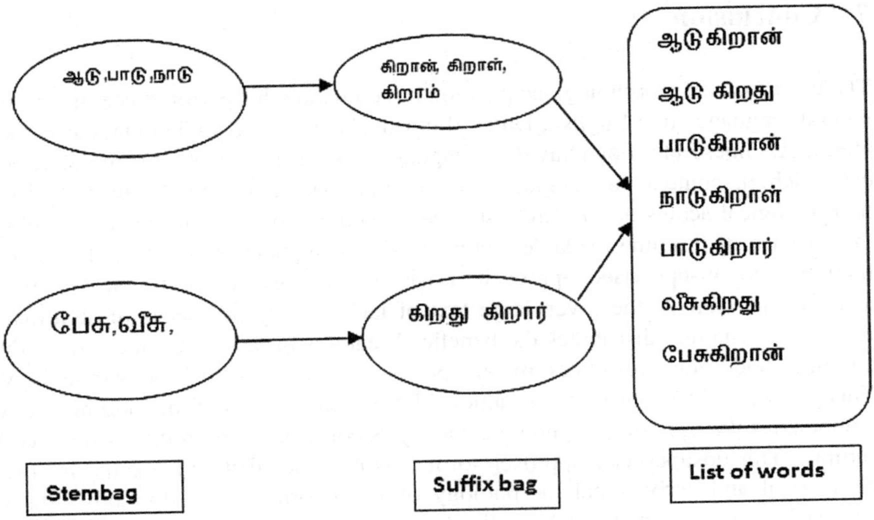

Fig. 3 Example to show the list of generated tamil words

5.3 *Word List Generation*

The generation of words is possible through the combination of the stems of the particular stem bag with respect to the corresponding suffix bag. Figure 3 illustrates the preparation of a new word list through permutations of stems and suffix pairs. Since the vocabulary size of the languages rapidly grows the generation of new words would definitely assist the language systems.

6 Discussions

Tamil languages are highly agglutinative and complex in nature. It incorporates bound of morphemes that acquires both inflectional and derivational morphology. This proposed unsupervised morphology acquisition can contribute an appreciative result in the lexical morphology. It identifies lexemes in all possible inflected forms lining up with an effective analysis of inflectional morphology. In case of Inflectional morphology where the root words undergoes a change when the suffixes are added in the post propositional, the featured unsupervised acquisition model tend to perplex because of the bound morphemes. Similarly, there is a possibility for scrutinizing some unreal lexicons.

7 Conclusion

The unsupervised morphology acquisition has embarked his performance in various natural languages like English, Dutch, Finnish, German and Indian languages like Hindi, Assamese etc. For Dravidian languages like Tamil which are morphologically rich in content and agglutinative in grammatical nature the unsupervised morphological acquisition is rarely studied. Resource poor languages like Tamil in the absence of machine readable dictionaries and language tools are highly benefited by the unsupervised approach. Basically, an unsupervised morphological analysis can analyze the given large text of language by no means of additional resources. It merely distributes the beneficial features of less economic power, the absence of additional resources, manual power and utility of different methodology. This paper provides a theoretical framework for identification of the morphological components through the morphological analysis constructed upon the unsupervised settings. This unsupervised approach for morphology acquisition is mainly focused on Lexical and derivational morphology of word forms. Thus the above stated approach is an attempt to identify the morphological features of Tamil language which can benefit the NLP applications.

References

1. Jayan, Jisha P., R. R. Rajeev, and S. Rajendran. "Morphological Analyser for Malayalam-A Comparison of Different Approaches." International journal of computer science and Information technologies (ICSIT), Volume 02, Issue-02, Page no(155–160), 2009.
2. Ananthi, S., and Angela Deepa. V.R., 2015. Morpheme Segmentation for Highly Agglutinative Tamil Language by Means of Unsupervised Learning. In Proceedings of IJCA on International Conference on Communication, Computing and Information Technology (Chennai, India 2015). ICCCMIT'14,1, 32–35.
3. Dr. A.G. Menon, S. Saravanan, R. Loganathan Dr. K. Soman,, "Amrita Morph Analyzer and Generator for Tamil: A Rule Based Approach", In proceeding of Tamil Internet Conference, Coimbatore, India. Page no(239–243), 2009.
4. Loftsson, H. 2008. Tagging Icelandic text: A linguistic rule-based approach. J. Nordic Journal of Linguistics 31(1).47–72.
5. Lovins, J. B. 1968. Development of a stemming algorithm. J. Mechanical Translation and Computer Linguistic.11(1/2), 22–31.
6. Paice, C.D. 1990. Another stemmer. In ACM SIGIR Forum, 24, 56–61.
7. Porter, M. F. 1980. An algorithm for suffix stripping. Program 14, 130–137.
8. Freitag, D. 2005. Morphology induction from term clusters. In Proceedings of the Ninth Conference on Computational Natural Language Learning (Ann Arbor, Michigan) CoNLL-'05. Association for Computational Linguistics, 128–135.
9. Goldsmith, J.A. 2001. Unsupervised learning of the morphology of a natural language. J. Computational Linguistics. 27(2), 153–198.
10. Hammarstrom, H. 2006. A naive theory of morphology and an algorithm for extraction. In Proceedings of the Eighth Meeting of the ACL Special Interest Group on Computational Phonology and Morphology (NewYork City, USA). Association for Computational Linguistics, 79–88.

11. Chan, E. 2008. Structures and Distributions in Morphology Learning. Ph.D Dissertation, University of Pennsylvania.
12. Dreyer, M. 2011. A non-parametric model for the discovery of inflectional paradigms from plain text using graphical models over strings. Ph.D Dissertation, Johns Hopkins University, Baltimore, Maryland.
13. Lindén, K. 2008. A probabilistic model for guessing base forms of new words by analogy. In Computational Linguistics and Intelligent Text Processing (Heidelberg, Berlin). Springer, 106–116.
14. Beesley, K. and Karttunen, L. 2003. Finite State Morphology. CSLI Publications, CA.
15. Koskenniemi, K. 1983. Two-level morphology: a general computational model for word-form recognition and production. Department of General Linguistics, University of Helsinki. Finland.
16. Dr. A.G. Menon, S. Saravanan, R. Loganathan Dr. K. Soman,, "Amrita Morph Analyzer and Generator for Tamil: A Rule Based Approach", In proceeding of Tamil Internet Conference, Coimbatore, India. Page no (239–243), 2009.
17. M. Anand Kumar, V. Dhanalakshmi and Dr. K P Soman, "A Novel Algorithm for Tamil Morphological Generator", 8th International Conference on Natural Language processing (ICON), IIT Kharagpur, December 8–11, 2010.
18. Parameshwari K, "An Implementation of APERTIUM Morphological Analyzer and Generator for Tamil", Language in India. 11:5 Special Volume: Problems of Parsing in Indian Languages, May 2011.
19. Parameswari K, "An Improvized Morphological Analyzer cum Generator for Tamil: A case of implementing the open source platform APERTIUM", Knowledge Sharing Event 1 – CIIL, Mysore, March 2010.
20. Anand Kumar M, Dhanalakshmi V, Soman K.P and Rajendran S, "A Sequence Labeling Approach to Morphological Analyzer for Tamil Language", (IJCSE) International Journal on Computer Science, International Journal of Computer Sciences and Engineering Volume 02, Issue No. 06, Page no(2201–2208), 2010.
21. Menaka, S., Vijay Sundar Ram, and Sobha Lalitha Devi. "Morphological Generator for Tamil." Proceedings of the Knowledge Sharing event on Morphological Analysers and Generators, LDC-IL, Mysore, India, Page no (82–96), March 22–23, 2010.
22. Antony, P. J., and K. P. Soman. "Computational morphology and natural language parsing for Indian languages: a literature survey."International journal Computer Science Engineering Technology" Volume-03, Issue-04, ISSN: 2229–3345, Page no(136–146), April 2012.
23. Moon, Taesun, Katrin Erk, and Jason Baldridge. "Unsupervised morphological segmentation and clustering with document boundaries." Proceedings of the 2009 Conference on Empirical Methods in Natural Language Processing: Volume 2-Volume 2. Association for Computational Linguistics, 2009.

A Fuzzy Approach for the Maintainability Assessment of Aspect Oriented Systems

Ananthi Sheshasaayee and Roby Jose

Abstract Maintainability of the software is well-thought-out as one of the vital quality that software should possess according to ISO standards. Software Maintainability Assessment (SMA) of aspect oriented software has been a focus of research for some time. Statistical and machine learning approaches have been used for assessing the maintainability of software. Fuzzy logic acts as an alternative approach to SMA of aspect oriented systems. Fuzzy logic has emerged as an important tool for use in a variety of applications that range from control system engineering to the design of automated intelligence systems. Fuzzy logic has the ability to deal with uncertainty and multivalued data and does not rely on historic data. This characteristic of data free model building enhances the prospect from using fuzzy logic for software metrics. The paper presents a fuzzy logic based algorithm for SMA of aspect oriented systems.

Keywords Aspect oriented maintenance metrics · Concern modularization approach of programming · Fuzzy algorithms · Fuzzy logic · Fuzzy inference systems · Maintainability model · Software quality estimation

1 Introduction

Design of a maintainable software system requires the developer to be aware of the phenomena that are observable while the system concerns are being evolved. One such observational phenomenon is static and dynamic measures of software. Since Aspect Oriented Software Development (AOSD) is one of relatively new branch, it

Ananthi Sheshasaayee (✉) · R. Jose
Department of Computer Science, Quaid E Millath Govt College for Women, Chennai 600002, India
e-mail: ananthi.research@gmail.com

R. Jose
e-mail: roby.research@gmail.com

© Springer India 2016
S.C. Satapathy et al. (eds.), *Information Systems Design and Intelligent Applications*, Advances in Intelligent Systems and Computing 435,
DOI 10.1007/978-81-322-2757-1_50

should tackle various challenges when compared to the precursor object oriented software development. Therefore, there is substantial interest for research in terms of the application of sophisticated techniques for software maintainability assessment.

This concern modularization approach of programming is a software development approach that overpowers the drawbacks of Object-Orientation (Programming) which offer appropriate constructs for sectionalizing crosscutting concerns [1, 2] which are difficult to be isolated from the remaining software artifacts. AOP sanctions the segmental recognition of crosscutting concerns–concerns wherein entire system modules handles its implementation. Concern modularizing approach focus on issues of modularity that are not treated by other existing approaches such as Object Oriented Programming and structured programming.

Common infomercial and internet applications are subjected to address "concerns" such as persistence, security, logging et al. The object oriented software development approach's programming abstractions namely objects and classes were proved impossible for designing concerns of a system in entirety. The new programming paradigm is a promising model to support concerns compartmentalization. The aspect oriented systems can be evolved from object oriented systems by altering the static OOP model to meet the new code that should be included to accomplish the further requirements. Thus it could be correctly stated that Aspect Oriented Programming supplements Object Oriented Programming but not replaces.

Maintainability is viewed a software trait that plays a vital role in software quality level. It is widely accepted that less effort in terms of cost can be achieved for maintenance cycle if the quality of the software is higher. Concern modularization approach of programming is extensively tapped software skill in academic set-ups [3]. The recognition of this new approach by the information technology enterprises may expand if studies, offering perfect information about its benefits and downsides are conducted. This target can be achieved by pointing out whether the sway by the newfangled scheme on the upkeep efforts of the software is beneficial or not.

The utilization of software metrics at the time of development controls the maintenance cost [4]. Studies [5] investigated in the realm of Object based technology had shown that upkeeps efforts are well foreseen by metrics.

Through this paper the authors give it a try for the formulation of a fuzzy approach for the maintainability assessment of aspect oriented systems. The paper structure is as follows. The next session briefs the fuzzy logic approach and aspect oriented software's basic terminology. Section 3 surveys the studies initiated in the maintainability assessment of software systems. Section 4 sketches the approach and the paper is concluded in Sect. 5.

2 Basic Lexis for Fuzzy Logic and Aspect Oriented Soft Wares

2.1 Fuzzy Logic

Fuzzy logic is a powerful tool that has the ability to deal with ambiguity and vagueness [6]. Fuzzy logic approach can be used to solve those problems which have complexity in understanding quantitatively. One advantage with this approach is that it is less dependent on historic data and is a viable approach to build a model with less data [7, 8].

The fuzzy system accepts unclear statements and inaccurate data by means of membership functions and provide conclusion. The membership function maps out units of a domain which is specified as universe of discourse Y for any number which is real in the range (0, 1). Formally $\tilde{A}: Y \rightarrow (0, 1)$. Hence, a fuzzy set is offered as a group of ordered pairs where the initial element is $y \in Y$, and the subsequent, $\mu_{\tilde{A}}(y)$, is the membership function of y in $\tilde{A}$, which maps y in the interval (0, 1), or, $\tilde{A} = (y, \mu_{\tilde{A}}(y)) \mid y \in Y\}$ [9]. Thus, a fuzzy set arises from the "enlargement" of a crisp set that creates the integration of traits of uncertainty. Fuzzification is this course of action. Defuzzification is the reverse process, that is, it is the transformation of a fuzzy set into single value [10].

In theory, a function of the form of $\tilde{A}: Y \rightarrow (0, 1)$ can be linked with a fuzzy set, determined by the perceptions and characteristics that is the requisite to be symbolized beside the situation in which the set is inserted. For each of these functions, a portrayal by a fuzzy number [10, 11] which is an arched and standardized fuzzy set defined in the real number range R, such that its membership function has the form $\mu_{\tilde{A}}: R \rightarrow (0, 1)$ is given.

Fuzzy logic could be comprehended to an extension to boolean logic that may be reflected as a simplification of multi-valued logic. By holding the capability of modeling the ambiguities of natural language through the theories of partial truth—that is truth-values falling someplace between totally right and totally wrong [12]—fuzzy logic manages these denominations over fuzzy sets in the interval (0, 1). Aforesaid points permit fuzzy logic to handle real-world objects that have imprecise limits.

By making use of fuzzy predicates (tall, weekend, high etc.), fuzzy quantifiers (much, little, almost none etc.), fuzzy truth-values (wholly false, more or less false) [13] and generalizing the significance of connectors and logical operators, fuzzy logic is perceived as denoting rough reasoning [14].

The fuzzy model gives mapping from input to output and the key means for doing this is a list of if-then statements called rules. All rules are contemplated in parallel and the order of the rules is rather insignificant.

To make it concise, fuzzy inference is the approach that interprets the crisp input values and based on a set of rules that act upon the input vectors a value is assigned to the output.

2.2 *Aspect Oriented Software Development*

The new programming architype termed as Aspect Oriented Software Development is an encouraging approach that permits the partition of concerns. According to [15] concern in reality is a section of the subject that is only a notional unit. These concerns are termed crosscutting concerns as they split crossways modularity of other concerns. Aspect Orientation has been suggested as an approach for bettering concern sectionalizing [15, 16] in the edifice of Object Oriented development of software. This style of programming utilizes aspects as new conceptual unit and make available structures for creating aspects and constituents at join point.

In Aspect-Oriented Programming (AOP) modularization is achieved by means of language abstractions that contributes to the modularization of crosscutting concerns (CCs). CCs are concerns which cannot be perfectly modeled to achieve modularity if traditional programming paradigms [16] are used. If not with apt programming development conceptions, crosscutting concerns become speckled and twisted with other fragments of the source code thereby affecting quality attributes like maintainability and reusability.

The programmers of concern handling languages are acceded to have the flexibility to plan and realize crosscutting concern detached from the core concerns. The Aspect Oriented Programming (AOP) compiler is provided with the facility to merge the decoupled concerns jointly in order to realize an exact software system. Even though, there is a thorough separation of concerns at the source-code level, the final publicized version delivers the full operationality anticipated by the users.

The basic concepts of AOP are illustrated with AspectJ language [2], considered to be an aspectation supported version for Java, letting the Java code to be compiled impeccably by the AspectJ compiler. The key notions in this language are: Aspect—a construction to epitomize a crosscutting concern; Point cut—an instruction used to catch join points of other concerns; Join Point—a locus of concern in some entity of the software growth through which two or more concerns may be constituted; JoinPoint model-A JoinPoint model offers the common frame of reference to permit the definition of the configuration of aspects; Advices —forms of actions to be completed when a join point is captured; and intertype declarations—the capability to include static declarations from the external side of the affected code.

3 Related Works

In literature it is possible to locate numerous models that have been proposed either for the prediction or for assessment of the software maintainability. Even though many academicians and researchers tried to offer several lexicons and techniques to appraise maintainability so that there may be enhancement of systems functionality and systems complexity, there exist the inconsistency and elusiveness in the lingo and ways occupied with assessment of maintainability of the software applications.

The subsequent paragraphs, look over the works initiated in the direction of maintainability of software systems.

In the study [17] Kvale et al. tried to show how the employability of AOP improved the maintainability of Commercial off the shelf (COTS)-based systems. The claim of the researchers in the study, is that maintenance task turned out to be trying when the source instructions for accessing the COTS libraries is disseminated through the glue code. They claimed the COTS based system to be easily maintainable one if the aspect oriented techniques were employed to construct the glue code. The study was conducted using the mail server developed using Java. By using the size based metrics, they showed that use of AO paradigm in the glue code betters the changeability aspect of studied system as the code that has to be altered is marginal in this cases.

The change impact in Aspect Oriented (AO) systems was studied by Kumar et al. [18]. The experimental analysis used AO systems which were refactored from their Object Oriented (OO) versions. For the experiment, the AO system used were refactored into 129 AO modules from 149 OO modules. A module in Object Oriented version denote the classes, and for Aspect Oriented version, a module suggests both classes and aspects. The study was based on some metrics and the metrics was collected using the tool developed. In the study, it was established that the effect of change is lesser in AOP systems when compared to the OO counterpart. Another notable point of the study was that if aspects are used to represent concerns that does not crosscut, then change impact for these modules (aspects) is higher. The study brought about that the maintainability of the Aspect Oriented system can be established on the changeability of the system.

Three elements were cogitated by Bandini et al. [19], namely; design convolution, maintenance efforts and programmer's skill to calculate the maintenance performance for object-oriented systems. Degree of interaction, dimensions of interface and operation argument intricacy were preferred to gauge design complexity. Perfective and corrective maintenance were the two types of maintenance assessed by the study.

Kajko et al. [20] pointed out some problems that were connected with the maintenance of the systems. The work suggested two maintainability models. One each for product and process. Product model, discuss the effect of ordinary attributes, variable aspects, maturity level and others on maintainablity whereas process aspects dealt with the usual assignments to be carried out to oversee maintainability during the development life—cycle of software.

The work [21] utilizes fuzzy model to compute the maintainability of a system. Four factors were obtained to be affecting maintainability. The study used fuzzy sets to categorize inputs while maintainability model based on the four factors of maintainability is categorized as very good, good, average, poor and very poor. The validation of the model was executed by software projects developed by engineering graduates. This attributed to one drawback for the proposal, that fuzzy model was not corroborated for real life complex projects.

In [22] an assortion of soft computing techniques were used by authors in the calculation of software maintenance work. The authors used OO metrics of design for the input to ANN, FIS and ANFIS as interpreters of maintenance effort. According to this study authors settle for Adaptive Neuro Fuzzy Inference System as the best one for the expectation of system maintenance effort. The study demonstrated using correlation analysis that the above discussed approaches are capable of predicting maintenance works associated with a system of software.

4 The Proposed Fuzzy Methodology

Maintainability of software is determined by both qualitative and quantitative data. Obtainable maintainability models amass data into orders of features with agreed wants. However, data applied to score the characteristics can be indeterminate or even entirely unknown. Therefore, it would be significant to evaluate sensitivity of the summed up result, i.e. the maintainability, with respect to the uncertainty and incompleteness of data. Fuzzy models serve as the best alternative to be adapted to evaluate maintainability [23] in view of these issues. There is the suggestion that the difficulties happening with the use of formal and linear regression modeling of software metrics for various purposes could be trounced by using fuzzy logic metrics and models [24].

Because of the reason that parameters in software metric models are either difficult to quantify (for example complexity), or are only known to a rough degree (such as system size), the use of fuzzy variables seems naturally agreeable. Since a fuzzy logic model can be initialized with expert rules, and given that the movement of membership functions and rules can be limited, it is possible that such a model will perform significantly better than alternatives such as regression and neural network models given lesser amounts of data.

As sketched out earlier, the paradigm that allows for the modularization of aspects is an emerging one and works initialized and contributed from the academic scenario for the valuation of maintainability seems to be ostensibly a smaller amount as compared to object oriented paradigm. If academic research comes out with maintainability models, then software industry norms and guidelines are definitely impacted. The model of maintainability which are instituted on metrics will act as a means for delivering norms of quality that has to be taken care of during the development of software product. The model also supports in business decision and budget planning by giving acumens on the software product and underlying process along with the effect on each other. On the other side, a model of maintainability makes available preliminaries for various methodologies that can revise and extend existing standards or develop new ones in the academic research.

4.1 The Pseudocode for the System

In [25] authors have identified some aspect oriented metrics that influences maintainability. The proposed approach for SMA of aspect oriented systems considers those metrics to quantify maintainability and is grouped into categories complexity, coupling, size and cohesion. These metrics takes three values low, medium and high in the fuzzy system and based on the rules gives five values for maintainability, viz low, very low, medium, high and very high. The fuzzy approach for SMA of aspect oriented systems is sketched below.

```
Pseudocode to AOFMM ()
// input: complexity, coupling, size and cohesion values
// output: Maintainability value based on metrics.
begin
1./*create the fuzzy rules using if-then statements*/
2./* Calculate membership value for premise parameter */
For i=1to n
    Output(i)=μAi(x)

where
```

$$\mu_{Ai}(x) = \exp\left(-\frac{(c_i - x)^2}{2\sigma_i^2}\right)$$

```
3./* Enthuse and normalize strength of rule */
For i=1 to n
{
Assign Σw=1
F(n)(A1,........An)=ΣwiBi
B1 to Bn is decreasingly ordered nature of fuzzy numbers
A1 to An so that Bi is the i-th largest element.
}
 4./* find consequence of the rule through the combination of outputs in step 2 and
step 3*/
 Calculate (a1x1+a2x2+.........+anxn+r)
     where a1, a2, ...an are consequent parameters
 5. /*Combine the output to an output distribution */
     Use fuzzy OR to combine the outputs of all fuzzy rules to
     obtain a single fuzzy output distribution
 6. /* Defuzzify the output */
     Use mean of maximum method to find a sharp value for
     Output
```

$$z = \sum_{j=1}^{m} (z_j / m)$$

```
end
```

The performance consideration of the proposed approach is based on the complexity of the fuzzy XNOR approach as compared to fuzzy OR and Fuzzy AND.

The above approach tries to build a maintainability model for aspect oriented systems. As mentioned earlier the resultant action of the implemented version of the above pseudocode gives the aspect based systems quality trait maintainability.

5 Conclusion

The intention of concern modularization approach of programming is that segmentalization may be achieved in software as long as configurations that enable the compartmentalization of programming aspects whose depiction using program code cannot be secluded using traditional approaches of programming. The engagement of object oriented methodology in such a case may give rise to code replication, thereby creating impediments in maintenance period. Concern modularization approach of programming is a worthwhile transformation. Since the paradigm is at its momentarily accumulating phase, the approach demands maturity at the maintenance metrics dimension. A fuzzy system for the assessment of maintainability of aspect systems find a sizeable place in the academic research scenario. A preliminary step towards this goal, is attempted in this paper. The approach is to be implemented with the fuzzy XNOR approach. Once this approach is implemented the system can be enhanced further for the automated assessment of aspect based systems quality.

References

1. Elrad, Tzilla, Robert E. Filman, and Atef Bader. "Aspect-oriented programming: Introduction. "Communications of the *ACM* 44.10 (2001): 29–32.
2. Kiczales, Gregor, et al. "An overview of AspectJ. "*ECOOP 2001—Object-Oriented Programming*. Springer Berlin Heidelberg, 2001. 327–354.
3. Eaddy, Marc, Alfred Aho, and Gail C. Murphy. "Identifying, assigning, and quantifying crosscutting concerns. "*Proceedings of the First International Workshop on Assessment of Contemporary Modularization Techniques. IEEE Computer Society*, 2007.
4. Bandi, R.K, Vaishnavi V.K., Turk, D.E.,. Predicting maintenance performance using object-oriented design complexity metrics. IEEE *Transaction on Software Engineering* 29 (1) 2003, 77–87.
5. Li, Wei, and Sallie Henry. "Object-oriented metrics that predict maintainability." Journal of systems and software 23.2 (1993): 111–122.
6. Sivanandam, S. N., Sai Sumathi, and S. N. Deepa. *Introduction to fuzzy logic using MATLAB*. Vol. 1. Berlin: Springer, 2007.
7. MacDonell, Stephen G., Andrew R. Gray, and James M. Calvert. "FULSOME: fuzzy logic for software metric practitioners and researchers." *Neural Information Processing, 1999. Proceedings. ICONIP'99. 6th International Conference on*. Vol. 1. IEEE, 1999.
8. Ryder, Jack. "Fuzzy modeling of software effort prediction." *Information Technology Conference, 1998. IEEE*. IEEE, 1998.

9. Zadeh, L. A., 1965, "Fuzzy Sets, Information and Control", vol. 8 (338–353).
10. Von Altrock, C., B. Krause, and H-J. Zimmerman. "Advanced fuzzy logic control of a model car in extreme situations." *Fuzzy Sets and Systems* 48.1 (1992): 41–52.
11. Klir, George, and Bo Yuan. *Fuzzy sets and fuzzy logic*. Vol. 4. New Jersey: Prentice Hall, 1995.
12. Kantrowitz, Mark, Erik Horskotte, and Cliff Joslyn. "Answers to frequently asked questions about fuzzy logic and fuzzy expert systems." *Comp. ai. fuzzy, November 1997. ftp. cs. cmu. edu:/user/ai/pubs/faqs/fuzzy/fuzzy. faq* (1997).
13. Dubois, Didier, and Henri Prade. "Random sets and fuzzy interval analysis." *Fuzzy Sets and Systems* 42.1 (1991): 87–101.
14. Grauel, A., and L. A. Ludwig. "Construction of differentiable membership functions." *Fuzzy sets and systems* 101.2 (1999): 219–225.
15. Tarr, Peri, et al. "N degrees of separation: multi-dimensional separation of concerns." *Proceedings of the 21st international conference on Software engineering*. ACM, 1999.
16. Kiczales, Gregor, et al. Aspect-oriented programming. *11th European Conference on Object Oriented Programming in: LNCS, vol. 1241, Springer Verlag*,1997, pp 220–242 1997.
17. Kvale, Axel Anders, Jingyue Li, and Reidar Conradi." A case study on building COTS-based system using aspect-oriented programming. "*Proceedings of the 2005 ACM symposium on Applied computing*. ACM, 2005.
18. Kumar, Avadhesh, Rajesh Kumar, and P. S. Grover. "An evaluation of maintainability of aspect-oriented systems: a practical approach." *International Journal of Computer Science and Security* 1.2 (2007): 1–9.
19. Bandini, S., Paoli, F. D., Manzoni, S., Mereghetti, P., 2002. A support system to COTS based software development for business services, Proceedings of the 14th International Conference on Software Engineering and Know ledge Engineering, Ischia, Italy, Vol. 27, pp: 307–314.
20. Kajko-Mattsson, Mira, et al. "A model of maintainability–suggestion for future research." (2006).
21. Aggarwal, K. K., et al. "Measurement of software maintainability using a fuzzy model." *Journal of Computer Sciences* 1.4 (2005): 538–542.
22. Kaur, Arvinder, Kamaldeep Kaur, and Ruchika Malhotra. "Soft computing approaches for prediction of software maintenance effort." *International Journal of Computer Applications* 1.16 (2010).
23. Canfora, Gerardo, Luigi Cerulo, and Luigi Troiano. "Can Fuzzy Mathematics enrich the Assessment of Software Maintainability?." *Software Audit and Metrics*. 2004.
24. Gray, Andrew, and Stephen MacDonell. "Applications of fuzzy logic to software metric models for development effort estimation." *Fuzzy Information Processing Society, 1997. NAFIPS'97., 1997 Annual Meeting of the North American*. IEEE, 1997.
25. Sheshasaayee Ananthi, and Roby Jose. "International Journal of Emerging Technologies in Computational and Applied Sciences (IJETCAS) www. iasir. net.", June 2014.

Split and Merge Multi-scale Retinex Enhancement of Magnetic Resonance Medical Images

Sreenivasa Setty, N.K. Srinath and M.C. Hanumantharaju

Abstract Image contrast enhancement and highlighting the prominent details based on edge preserving are the fundamental requirement of medical image enhancement research. This paper presents the Multi-Scale Retinex (MSR) enhancement of Magnetic Resonance (MR) medical images using split and merge technique. The main limitations of the retinex based image enhancement schemes is computational complexity, halo artifacts, gray world violation, over enhancement etc. There exist various extensions for retinex methods developed by over a dozens of researchers, but most of the techniques are computationally complex. This is due to the fact that the image details are improved at the cost of increased computational complexity. The proposed method is efficient in computation, since the original input image is splitted into a size of 8 × 8 and then gray level Fast Fourier Transform (FFT) based retinex algorithm is exploited to improve the quality of sub-image. Reconstructed image is produced by merging each of the enhanced versions to the original image resolution. This scheme is validated for MR medical images and the parameters of the retinex method is adjusted to improve the contrast, details and overall quality image. Experimental results presented confirms that the proposed method outperforms compared to existing methods.

S. Setty (✉)
Department of Information Science & Engineering,
Don-Bosco Institute of Technology, Bangalore, India
e-mail: sairamsst@gmail.com

N.K. Srinath
Department of Computer Science & Engineering,
R. V. College of Engineering, Bangalore, India
e-mail: srinathnk@rvce.edu.in

M.C. Hanumantharaju
Department of Electronics and Communication Engineering,
BMS Institute of Technology & Management, Bangalore, India
e-mail: mchanumantharaju@gmail.com

© Springer India 2016
S.C. Satapathy et al. (eds.), *Information Systems Design and Intelligent Applications*, Advances in Intelligent Systems and Computing 435,
DOI 10.1007/978-81-322-2757-1_51

Keywords Medical image enhancement · Split and merge · Multi-scale retinex · Histogram equalization

1 Introduction

Image enhancement is a pre-processing operation in digital image processing used for manipulation images to improve certain details in the image. For example, image can be denoised or brighten up, making it easier to identify significant region of interest in the image. The image captured by the camera under poor illumination conditions may be dark. Image enhancement operations performed on these type of images will increase the contrast or intensity of the image. With the increased use of digital imaging systems in medical diagnostics, processing of digital images becomes prominent in health care [1]. Medical imaging plays a leading role in modern diagnosis and is useful in assisting radiologist or surgeons to detect pathology or abnormal regions. Enhanced processing of medical images is good for display a perfect human tissues and organs, and supports in aided diagnosis.

Enhancement methods used in the field of medicine allows the doctor to observe interior portions of the body for flexible analysis. Further, enhancement schemes assists radiologists to make keyhole operations to reach the inner parts of the human body without really opening much of it. Computed Tomography (CT) scanning, Ultrasound and Magnetic Resonance Imaging (MRI) replaced X-ray imaging by allowing the clinicians to observe the body's indefinable third dimension. With the CT scanning, interior portion of the body can be displayed with simplicity and the unhealthy parts can be located without causing uncomfort or pain to the patient. MRI imaging captures signals from the body's magnetic particles spinning to its magnetic tune and with the assist of powerful computer, the scanner data are converted into revealing images of internal parts. Image enhancement schemes developed for studying aerial image data may be modified to analyse the outputs of medical imaging systems to get best advantage to analyse symptoms of the patients with ease. Although, medical imaging techniques such as Ultrasound, MRI, CT etc. provide detailed 3D images of internal organs, but further processing these images eases the decision of the pathologists. Quantifiable information, like organ size and shape, can be extracted from these images to support activities such as diagnosis of disease monitoring and surgical planning.

As per the statistics of American Society of Clinical Oncology (ASCO), in 2014, 22,850 adults (12,900 men and 9,950 women) in the US has been diagnosed with primary cancerous tumours of the brain and spinal cord. Also, it was predicted that 15,320 adults (8,940 men and 6,380 women) are going to die from this disease every year. Moreover, 4,300 children and teens have been diagnosed with a brain or central nervous system tumour every year. Surprisingly, more than half of these cases are the children aged less than 15 years. From these statistics, it is evident that the early detection, proper diagnosis and treatment place a substantial role in reducing the cancer cases.

The proposed split and merge retinex based MR image enhancement method is efficient in computation, improves contrast and offers greater detail enhancement for medical images. This framework may be prominently used by the clinician, researcher and practitioners etc., since the developed approach is faster and is capable of producing high quality enhancement. Further, the proposed scheme is highly suited for the implementation in programmable devices such as Field Programmable Gate Arrays (FPGAs) and greater throughputs can be obtained by adopting high degrees of pipelining and parallel processing.

Further, this paper is categorized into the following sections: The brief review of related work is described in Sect. 2. Section 3 elaborates the developed split and merge multi-scale retinex enhancement of magnetic resonance medical images. Section 4 describes results and discussions. Finally, the conclusion arrived for the proposed work is discussed in Sect. 5.

2 Review of Related Work

Wheeler et al. [2] presents the state-of-the-art methods of spinal cord imaging and applications. Hajar et al. [3] proposed contrast enhancement method for rapid detection of Micro-Calcification Clusters (MCCs) in mammogram images. The proposed system consist of four main steps including: (i) image scaling; (ii) breast region segmentation; (iii) noise cancellation using a filter, which is sensitive to MCCs; and (iv) contrast enhancement of mammograms using Contrast-Limited Adaptive Histogram Equalization (CLAHE) and wavelet transform. To evaluate this method, 120 clinical mammograms were used. To evaluate the performance of the image enhancement algorithm, Contrast Improvement Index (CII) was used. The enhancement method in this research achieves the highest CII in comparison with other methods presented in this study. The validity of the results was confirmed by an expert radiologist through visual inspection. Detection of MCCs significantly improved in contrast-enhanced mammograms. This method could assists radiologists to easily detect MCCs; it could also decrease the number of biopsies and reduce the frequency of clinical mis-diagnosis. Moreover, it could be useful prior to segmentation or classification stages.

Volker et al. [4] developed a hardware for the contrast limited adaptive histogram equalization. The application is intended for processing of image sequences from massive dynamic-range infrared cameras. The strategy divides the image into sub-images and then computes a transformation function on each of the sub-image and additionally interpolates each of the sub-image. The hardware implementation in this approach is aided by the modification of contrast limiting function. However, the error introduced with in the modification of the contrast limiting function is negligible. In this approach, design latency is reduced by performing the sequence of steps at the same time on the same frame and between the frames with the vertical blank pulse exploitation.

A frequency domain smoothing and sharpening method was proposed by Patel et al. [5]. The method is exploited to enhance mammogram images. This method ensures the benefit of enhancement and sharpening technique that aims to improve the abrupt changes in the image intensity. This scheme is generally applied to remove random noise in digital images. This approach combines the sharpening as well as smoothing techniques into the mammogram image enhancement domain. The parameter selection is invariant to the type of background tissues and severity of abnormality providing significant better results for denser mammographic images. This scheme is validated for breast X-ray mammograms. The simulation results presented reveals that the contrast enhanced images offer additional details. This assists the radiologists to observe and classify mammograms of breast cancer images.

Chanda et al. [6] proposed the method using multi-scale morphology for local contrast enhancement of Gray-scale Images. The same authors further enhanced their method and presented another method on nonlinear enhancement of extracted features from the images where features are selected based on shape, size and scale. Contrast enhancement proposed by Economopoulos et al. [7] using partitioned iterated function systems (PIFS) is difficult for mammogram image enhancement in the sense that it gives more irrelevant information.

3 Proposed Method

The enhancement of MR images is achieved by decomposing the image into sub-images. These sub-images are enhanced independently using a well known image enhancement scheme, namely, multiscale retinex. The retinex (retina and cerebral cortex) idea was first proposed by Edwin Land in 1970 as a model of lightness and color perception of human vision. There is no theoretical proof of retinex method, but experimentally validated. Currently, numerous retinex versions with wide range of modifications are available in the literature. The latest developments in the retinex includes, retinex image enhancement via a learned dictionary by Huibin et al. [8], the combination of MSR and bilateral filter by Wang et al. [9], fast retinex for color image enhancement by Analysa et al. [10], retinex based color image enhancement using Particle Swarm Optimization (PSO) by Shibudas et al. [11] etc. The main limitation of these schemes are the halo artifacts, over enhancement, increased computational complexity, and gray world violation etc. The proposed method overcomes limitations such as computational complexity and over enhancement by applying the algorithm into the sub-images instead of applying for the entire image. Reconstructed sub-images are merged into the size of original image to compare with the quantity of enhancement achieved.

The steps incorporated in the proposed method is summarized as follows:

Step 1: Read the medical image that need to be enhanced
Step 2: Separate the image into red, green and blue channels
Step 3: Divide image into the size of 8×8 pixels
Step 4: Apply gray-level 2D-FFT based multi-scale retinex algorithm
Step 5: Merge the enhanced sub-images to produce the composite image
Step 6: Apply adaptive histogram equalization to improve overall contrast of the medical image, adaptively
Step 7: Reconstruct the original image by combining three channels red, green and blue, respectively.
Step 8: Validate the proposed method by applying the image enhancement metrics such as Peak Signal to Noise Ratio (PSNR) (see Fig. 1).

Fig. 1 Proposed split and merge technique

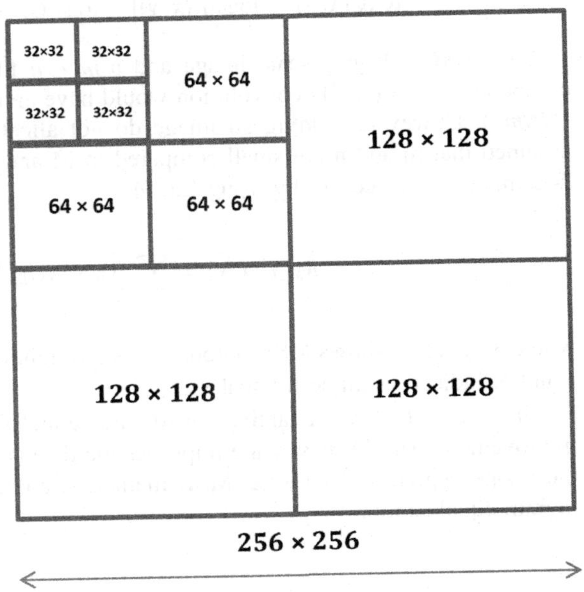

3.1 2D Fast Fourier Transform Based Multiscale Retinex Algorithm

The proposed split and merge based enhancement scheme utilizes the multiscale retinex algorithm applied to the splitted red, green and blue channels, respectively. The core part of in retinex based image enhancement scheme is the Gaussian surround function [12]. The general expression for the Gaussian surround function is given by Eq. (1)

$$G_n(x, y) = K_n \times e^{-\frac{(x^2+y^2)}{2\sigma^2}} \tag{1}$$

and K_n is given by the Eq. (2)

$$K_n = \frac{1}{\sum_{i=1}^{M} \sum_{j=1}^{N} e^{-\frac{x^2+y^2}{2\sigma^2}}} \tag{2}$$

where x and y indicates spatial co-ordinates, $M \times N$ indicates resolution of an image, n is preferred as 1, 2 and 3 since the three Gaussian scales are used for each splitted red, green and blue channels.

The SSR for the gray image is given by Eq. (3)

$$R_{SSR_i}(x, y) = \log_2[I_i(x, y)] - \log_2[G_n(x, y) \otimes I_i(x, y)] \tag{3}$$

For an $M \times N$ gray-scale image and a $m \times n$ filter, each pixel needs $O(mn)$ computations, so the 2D convolution would have an approximate complexity of $O(MNmn)$. However, padding an image do not affects this complexity. Also it is assumed that m and n are small compared to M and N. The MSR operation on a 2-D image is carried out by using Eq. (4)

$$R_{MSRi}(x, y) = \sum_{n=1}^{N} W_n \times R_{SSRni}(x, y) \tag{4}$$

where $R_{MSRi}(x, y)$ shows MSR output, W_n is a weighting factor which is assumed as $\frac{1}{3}$ and N indicates number of scales.

This step follows adaptive histogram equalization method for contrast improvement. The MSR scheme improves the details of the image, but contrast of the image degrades. Therefore, MSR method is contended with the adaptive histogram equalization.

4 Experimental Results and Comparative Study

The developed method is simulated using Matlab (Ver. R2014a) software and is validated with the local database spinal cord images. The split and merge based retinex method is compared with the conventional histogram equalization. Figure 2a shows the original image of resolution 1024 × 1024 pixels. The same image is enhanced using histogram equalization proposed by Ali et al. [13] and is

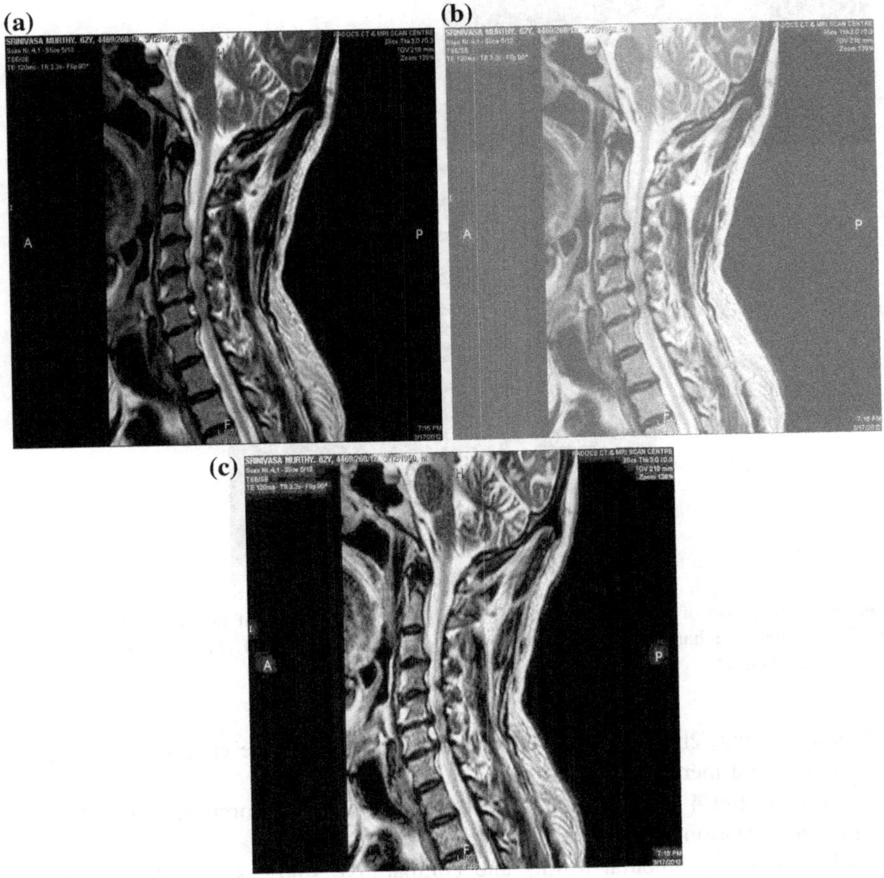

Fig. 2 Comparison of reconstructed MR images. **a** Original image of resolution 1024 × 1024 pixels. **b** Image enhanced based on histogram equalization of Ref. [13], PSNR: 31.2 dB. **c** Proposed method

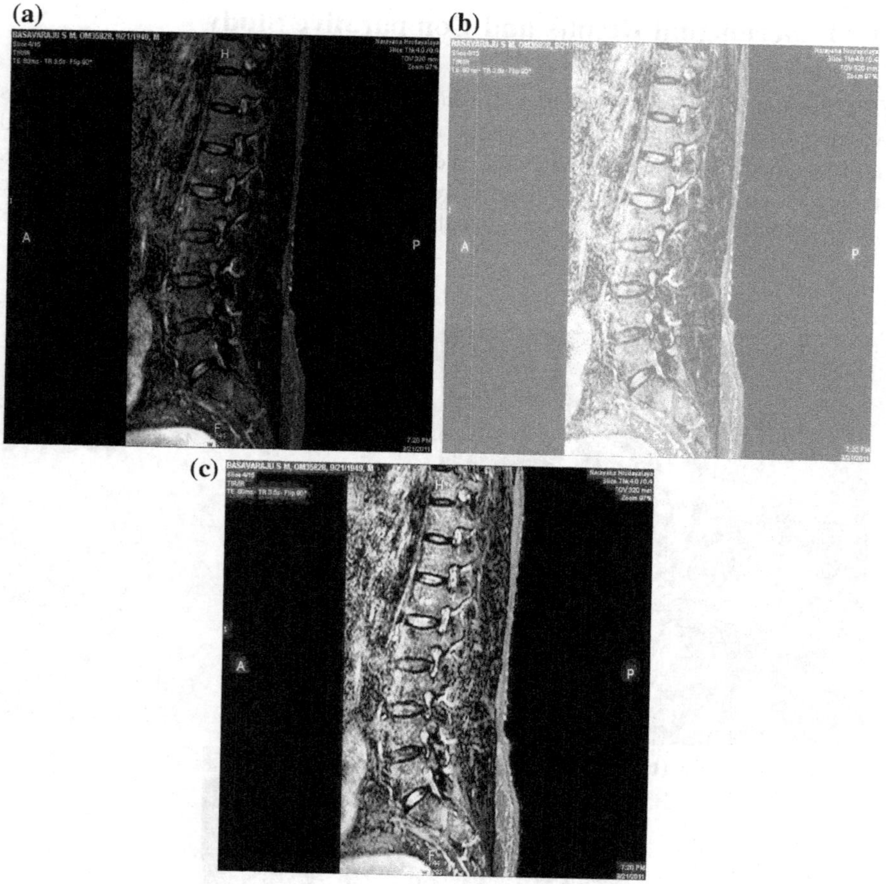

Fig. 3 Comparison of reconstructed MR images. **a** Original image of resolution 1024 × 1024 pixels. **b** Image enhanced based on histogram equalization of Ref. [13], PSNR: 31.2 dB. **c** Proposed method

presented in Fig. 2b. Figure 2c shows the spinal cord image enhanced using proposed split and merge technique.

Figures 3 and 4 shows second and third set of results compared with the histogram equalization, respectively.

Additional experimental results and comparative study is provided in Fig. 5. First column of Fig. 5 shows the original image. Histogram equalized images are

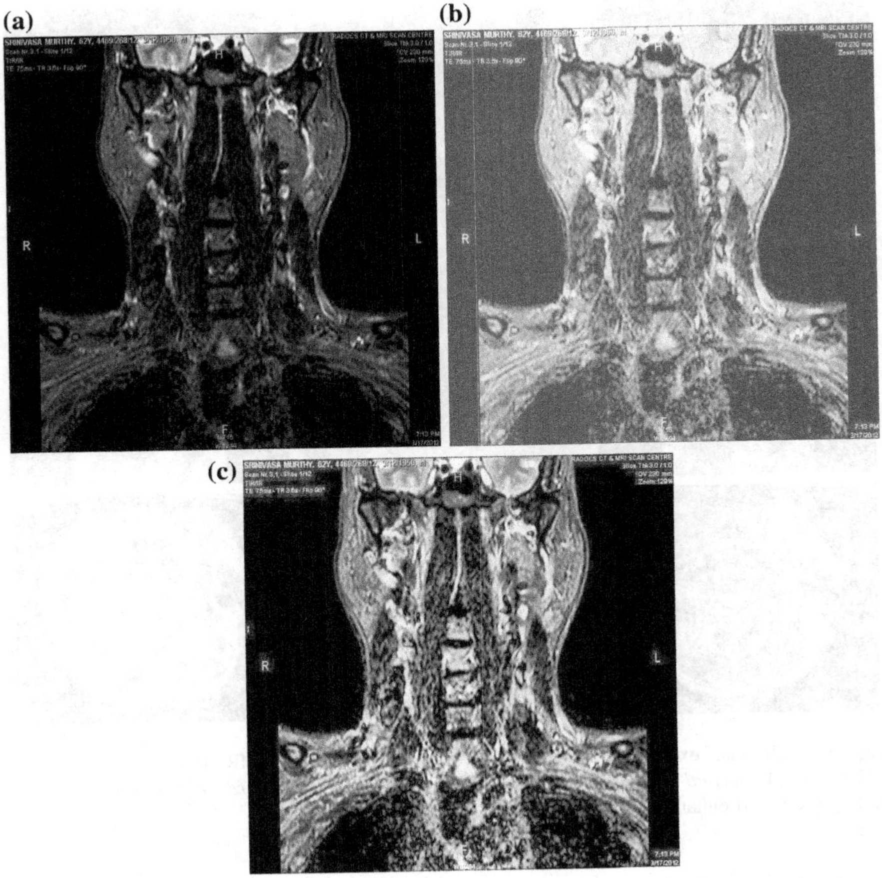

Fig. 4 Comparison of reconstructed MR images. **a** Original image of resolution 1024 × 1024 pixels. **b** Image enhanced based on histogram equalization of Ref. [13], PSNR: 31.2 dB. **c** Proposed method

presented in the second column of Fig. 5. The last column of Fig. 5 presents the proposed split and merge based enhancement technique. The experimental results described confirms that the proposed method is better compared to other existing methods.

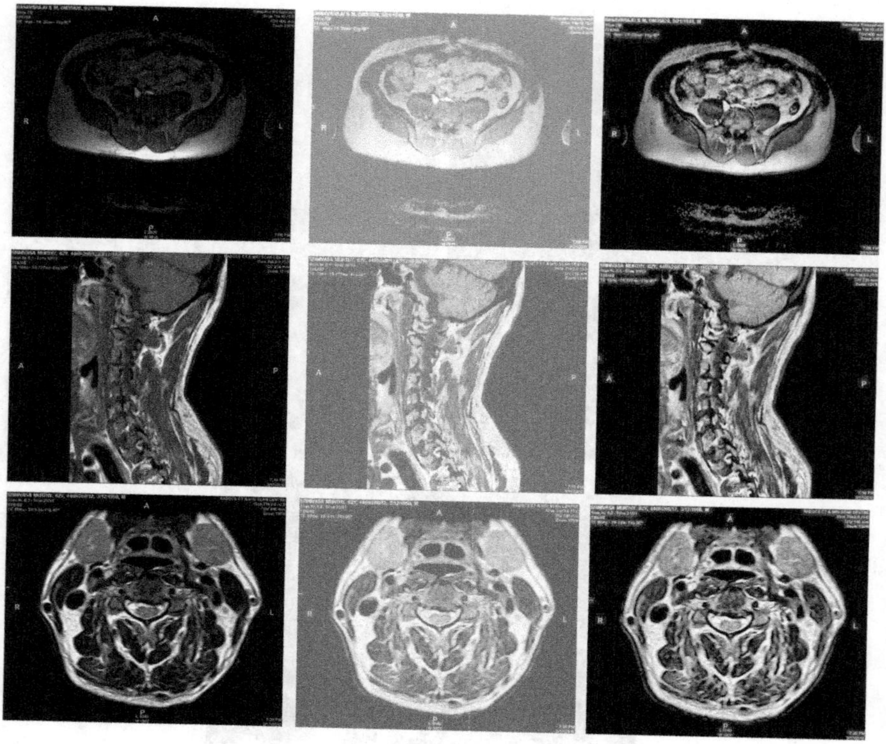

Fig. 5 Additional experimental results. *First column* Original MR images of resolution 1024 × 1024. *Second column* Histogram equalized images, and *Third column* Proposed split and merge based enhanced images

5 Conclusion

The Multi-Scale Retinex (MSR) enhancement of Magnetic Resonance (MR) medical images using split and merge technique is proposed in this paper. This paper overcomes the limitations of retinex method exhibited by other researchers such as computational complexity and improvement of image details. Detail enhancement of the MR medical images is achieved by splitting each of the channels of composite image into the level of 8 × 8 pixels. The gray level retinex algorithm is applied on this sub-image is capable of producing the results in real-time. This scheme is validated by the spinal cord MR images from the local database. It is clear from the experimental results presented that the images enhanced by the proposed method are brilliant, vivid and excellent.

References

1. Thomas Martin Deserno.: Biomedical Image Processing. Springer Science & Business Media, Berlin Heidelberg (2011).
2. Wheeler-Kingshott, C. A., et al. "The current state-of-the-art of spinal cord imaging: applications." Neuroimage 84 (2014): 1082–1093.
3. Moradmand, Hajar, et al.: Contrast Enhancement of Mammograms for Rapid Detection of Microcalcification Clusters. Iranian Journal of Medical Physics 11.2.3 (2014): 260–269 (2014).
4. Schatz, Volker. "Low-latency histogram equalization for infrared image sequences: a hardware implementation." Journal of real-time image processing 8.2 (2013): 193–206, (2013).
5. Patel, Vishnukumar K., Syed Uvaid, and A. C. Suthar. "Mammogram of breast cancer detection based using image enhancement algorithm." Int. J. Emerg. Technol. Adv. Eng 2.8 (2012): 143–147. (2012).
6. Zuiderveld, Karel. "Contrast limited adaptive histogram equalization." Graphics gems IV. Academic Press Professional, Inc., 1994.
7. Economopoulos, T. L., Pantelis A. Asvestas, and George K. Matsopoulos. "Contrast enhancement of images using partitioned iterated function systems." Image and vision computing 28.1 (2010): 45–54.
8. Chang, Huibin, et al. "Retinex image enhancement via a learned dictionary." Optical Engineering 54.1 (2015): 013107.
9. Ji, Tingting, and Guoyu Wang. "An approach to underwater image enhancement based on image structural decomposition." Journal of Ocean University of China 14.2 (2015): 255–260.
10. Gonzales, Analysa M., and Artyom M. Grigoryan. "Fast Retinex for color image enhancement: methods and algorithms." IS&T/SPIE Electronic Imaging. International Society for Optics and Photonics, 2015.
11. Subhashdas, Shibudas Kattakkalil, et al. "Color image enhancement based on particle swarm optimization with Gaussian mixture." IS&T/SPIE Electronic Imaging. International Society for Optics and Photonics, 2015.
12. Jobson, Daniel J., Zia-ur Rahman, and Glenn Woodell.: Properties and performance of a center/surround retinex. IEEE Transactions on Image Processing, Vol. 6, Issue 3, pp. 451–462, (1997).
13. Ali, Qadar Muhammad, Zhaowen Yan, and Hua Li. "Iterative Thresholded Bi-Histogram Equalization for Medical Image Enhancement." International Journal of Computer Applications 114.8 (2015).

Review on Secured Medical Image Processing

B. Santhosh and K. Viswanath

Abstract Medical image processing techniques requires continuous improve quality of services in health care industry. In the real world huge amount of information has to be processed and transmitted in digital form. Before transmission the image has to be compressed to save the bandwidth. This is achieved by alternate coefficient representation of image/videos in a different domain. Processing of images in transform domain takes comparable less computation by avoiding inverse and re-transforms operations. The fundamental behind the transform domain processing is to convert the spatial domain operations to its equivalent transform domain. This paper describes the analysis in the field of medical image processing.

Keywords Medical imaging · Compression · Spatial transform · Wavelet transform

1 Introduction

The medical image processing techniques contributes an important aspect in diagnosis of diseases, therapy of doctor, teaching and research. Plenty of modern imaging processing techniques likes Magnetic Resonance Imaging (MRI), ultrasound, Computerized Tomography (CT) and Radiography etc. provides different information to health industry. But the use of information by manual analysis is difficult. With the advancement in the field of computer, imaging technology,

B. Santhosh (✉)
Department of Telecommunication Engineering, Dayananda Sagar College of Engineering, Bangalore, India
e-mail: santhoshmehtre@gmail.com

K. Viswanath
Department of Telecommunication Engineering, Siddaganga Institute of Technology, Tumkur, India
e-mail: kviitkgp@gmail.com

© Springer India 2016
S.C. Satapathy et al. (eds.), *Information Systems Design and Intelligent Applications*, Advances in Intelligent Systems and Computing 435,
DOI 10.1007/978-81-322-2757-1_52

modern processing tools and advanced image processing techniques are become the part of medical field. This increased the ability of understanding, there by the level of diagnosis gradually improved and given a great contribution by using the processing and analysis of different medical images. Spatial domain approach require more computational load and they produce distortions in the processed images.

Everyday enormous amount of information is processed and then stored for a data base and transmitted in digital form. Compressing the data for memory management prior to transmission is of significant practical and commercial concern. Compression is the technique which saves the bandwidth before the transmission of digital image/video. This is achieved by alternate coefficient representation of image/videos in a different domain. Processing of images in the transform domain saves computation by avoiding re-transform operations. The principle behind the domain processing is to convert the spatial domain operations to its equivalent in the transform domain.

2 Related Work

Literature on medical image processing has been reviewed and discussed in this section based on general medical image processing research works.

2.1 *Protection and Authentication*

Protection and authentication of medical images one of the major issues in the management of medical data base. The data has to be secured from non-authorized updating or the quality degradation of information on the medical images.

Pal et al. [1] proposed a scheme where biomedical image watermarking performed by multiple copies of the same data in the concealed image make its hidden using bit replacement in horizontal and vertical resolution approximation image components of wavelet domain. This paper uses a method for recovering the secret information from the damaged copies due to non-authorized modification of data.

Discrete wavelet transform (DWT) segments an image into lower resolution approximation image as well as horizontal, vertical and diagonal components. In this method the technique is adopted to hide several sets of the same information logo into the cover image. If the data is lost, the information is retrieved from the cover images and the secret information can be retrieved using the bit majority algorithm which is almost similar to the original data.

Xin et al. [2] proposed a protection and authentication technology for images. This paper proposed a digital watermarking algorithm which is mainly based on integer wavelet transform to support confidentiality, authentication and integrity of digital medical images. Here, the matrix norm quantization and watermarks are impacted into low, medium and high-frequency sub-bands of image's integer wavelet domain. A blind-testing watermarking algorithm is developed, outcome of this algorithm proved that this achieves robustness and sensitivity.

The data images which are transformed by integer wavelet transform can avoid rounding error which occurs in floating point. High frequency coefficients are more exposed to sensitivity. Medium frequency coefficients lie between them, watermarks can be separately impacted into low, medium and high frequency sub-bands of images inter wavelet domain. This assures both robustness and sensitivity, to locate the tempered area, block the sub-bands and calculate the spectral norm and embedded watermarks through norm quantification. Security can be improved by scramble the watermark by chaotic sequences.

The block diagram approach of this work shown in Fig. 1.

2.2 *Medical Image Processing Techniques Available in the Transform Domain*

Tian et al. [3] proposed that the wavelet transform and inverse transform algorithm can be applied on the Medical images. This describes study of various applications in medical image, such as ECG and EEG signal processing, compression, Image reinforcing and edge detection, medical image registration etc.

Wang et al. [4] proposed a survey on algorithm based on correlation and contourlet transform on medical images. In this paper a review on edge detection, reinforcing, denoising and image compression is discussed. compared to 2-D wavelet Transform c is better. Figure 2 denotes that contourlet transform has exceptional performance on medical image compression (SNR = 23.35 dB when compression ratio is 52.5).

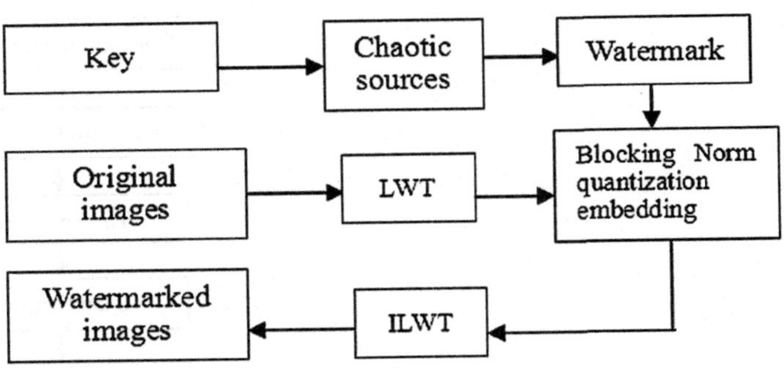

Fig. 1 The block diagram approach

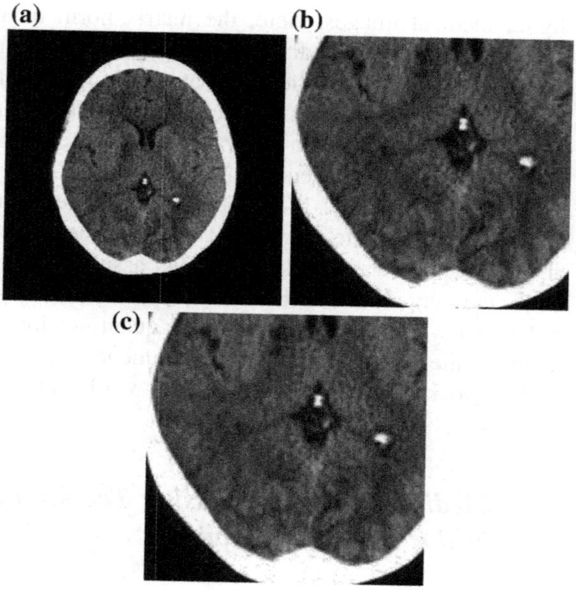

Fig. 2 **a** Original cerebral CT image. **b** Reconstructed image with wavelet transform (using 6554 coefs; SNR = 22.76 dB). **c** Reconstructed image with contourlet transform (using 6554 coefs; SNR = 23.35 dB)

Li et al. [5] proposed a novel algorithm to segment a lumbar vertebra in CT scan images in contourlet domain. Implementation of algorithm is as follows:

1. Contourlet transform is applied on the input images, take the approximation image Aj and all sub bands Wi j at scale j, direction i.
2. Step 1: Select M-most significant coefficients in all W, and let the other W = 0;

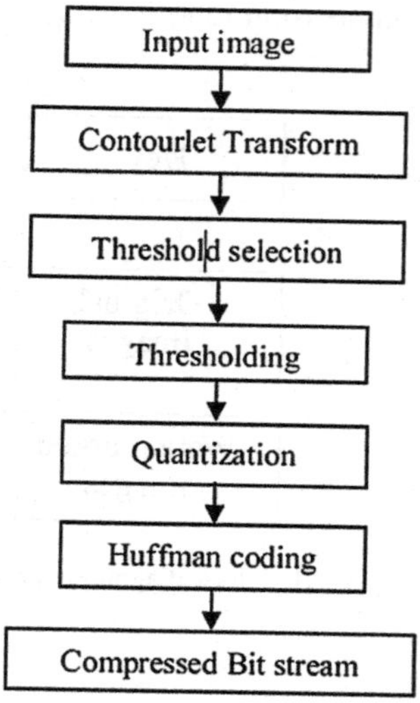

Fig. 3 Proposed compression method block diagram

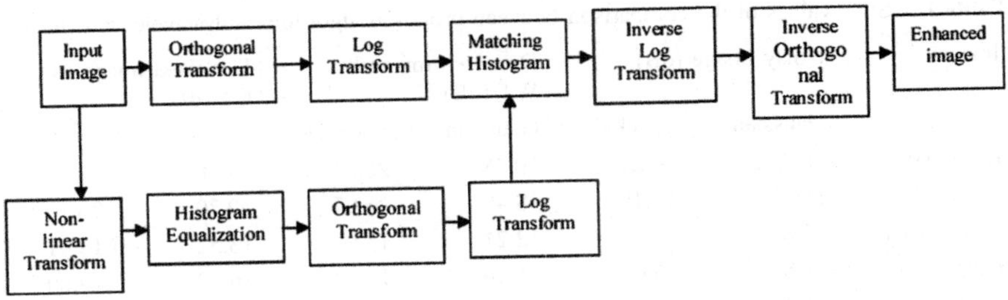

Fig. 4 Block diagram approach

Step 2: Aj = 0; Flag = CM (Wi, j, 2), denote the larger class as 1, the other as 0; then W′i, j = Flag.* Wi, j. Here, CM is C-Means clustering algorithm.

3. Reconstructed image is obtained from these modified coefficient

Hashemi-Berenjabad et al. [6] proposed a method of threshold based lossy compression of ultrasound images using contourlet transform, in this method image decomposition is obtained by using this contourlet transform, and then applying a new thresholding process on these coefficients before quantization, and by using these thresholding algorithm speckle noise can also be reduced and can be used real time image transmission systems.

The proposed block diagram shown in Fig. 3.

Hossain et al. [7] proposed a novel technique to enhance the medical images of different modalities based on integrating non linear and logarithmic transform coefficient histogram matching. A New robust method of enhancement is discussed which improves the visual resolution and disclose the dark shadow of medical images. This paper also discusses the global dynamic range correction and local contrast enhancement which improves the dark shadows by keeping the tonality of the desired images consistent. Evaluation tool is used to evaluate the performance contrast and to measure the enhancement (Fig. 4).

Hassin AlAsadi et al. [8] proposed contourlet transform based method. This paper discussed a contourlet transform based denoising algorithm for medical images. The Contourlet transform has the advantages of multistate and time-frequency-localization properties of wavelets with high degree of directionality. For verifying the denoising performance of the Contourlet transform, two kinds of noise are added into samples; Gaussian noise and speckle noise. Soft thresholding value for the Contourlet coefficients of noisy image is computed. The proposed algorithm is compared with the results of wavelet transform and the proposed algorithm has achieved acceptable results compared with those achieved by wavelet transform.

Table 1 shows the comparison results with WT.

Table 1 PSNR values of the comparison between proposed algorithm and wavelet transform

Image	Noisy image (dB)		Denoised image by WT (dB)		Denoised image by CT (dB)	
	Gaussian	Speckle	Gaussian	Speckle	Gaussian	Speckle
Brain MRI	16.85	21.27	20.78	26.31	22.10	28.44
CT scan	14.57	17.19	18.49	23.97	19.79	26.10
Tunor MRI	7.82	12.36	12.21	15.98	13.52	18.01
Ultrasound	11.85	15.07	15.29	17.57	16.60	19.60
X-ray	16.35	18.30	19.34	20.29	20.66	22.37

3 Discussion

Author	Year	Technique	Advantages
Da-Zeng Tian	2004	Wavelet transform	• Reasonable computing time
Jung-Wang	2008	Counterlet transform	• Medical image analysis • Feature extraction • Image smoothing • Image enhancement • Image reconstruction
Foisal Hossain	2010	Medical image enhancement based on non-linear technique and logarithmic transform	• Improves the visual quality
Liu Xin	2010	Wavelet transform	• Protection • Authentication
Qiao Li	2010	Counterlet transform	• Accuracy is more
Hashemi	2011	Counterlet transform	• Less computation time • Online image transmission
Pal, K	2012	Wavelet transform	• Protection • Authentication
Abbas H.	2015	Counterlet transform	• Supports the denoising better than WT

4 Conclusion

In this paper the survey study of medical image processing is done. The analysis of medical image processing and its applications in healthcare industry are described in this paper. The advanced Medical image processing techniques and algorithms are reviewed.

Acknowledgments I am really grateful to my research supervisor Dr. K Viswanath, Professor Department Of Telecommunication Engineering, Siddaganga Institute of Technology, Tumkur as well as Department Telecommunication Engineering, Dayananda sagar college of engineering for all kinds of support and encouragement to carry out this research work.

References

1. Pal, K., Ghosh, G., and Bhattacharya, M., "Biomedical image watermarking for content protection using multiple copies of information and bit majority algorithm in wavelet domain," *IEEE Students Conference on Electrical, Electronics and Computer Science (SCEECS)*, pp. 1–6, 1–2 Mar 2012.
2. Liu Xin, Lv Xiaoqi, and Luo Qiang, "Protect Digital Medical Images Based on Matrix Norm quantization of Digital Watermarking Algorithm," *International Conference on Bioinformatics and Biomedical Engineering (ICBBE)*, pp. 1–4, 18–20 Jun 2010.
3. Da-Zeng Tian, Ming-Hu Ha," Applications Of Wavelet Transform In Medical Image Processing" Proceedings of the Third International Conference on Machine Learning and Cybernetics, Shanghai, 26–29 August 2004.
4. Jun Wang and Yan Kang. (2008). Study on Medical Image Processing Algorithm based on Contourlet Transform and Correlation Theory. IEEE. pp. 233–238.
5. Qiao Li and Haiyun Li, "A Novel Algorithm Based on Contourlet Transform for Medical Image Segmentation," *2010 4th International Conference on Bioinformatics and Biomedical Engineering (ICBBE)*, pp. 1–3, 18–20 Jun 2010.
6. Hashemi-Berenjabad, S., Mahloojifar, A and Akhavan, A., "Threshold based lossy compression of medical ultrasound images using contourlet transform," 18th Iranian Conference of Biomedical Engineering (ICBME), pp. 191–194, 14–16 Dec 2011.
7. Hossain, M.F., Alsharif, M.R. and Yamashita, K., "Medical image enhancement based on nonlinear technique and logarithmic transform coefficient histogram matching," *IEEE/ICME International Conference on Complex Medical Engineering (CME)*, pp. 58–62, 13–15 Jul 2010.
8. Abbas H. Hassin AlAsadi "Contourlet Transform Based Method For Medical Image Denoising" International Journal of Image Processing (IJIP), Volume (9) Issue (1): 2015.

Identification of Stages of Malignant Tumor in MRM Images Using Level Set Algorithm and Textural Analysis

M. Varalatchoumy and M. Ravishankar

Abstract An efficient Computer Aided Detection and Classification system has been developed to detect, classify and identify the stages of malignant tumor. Preprocessing stage that involves image enhancement and removal of noises has been carried out using histogram equalization and morphological operators. As a novel initiative Level set method has been used for segmentation. High accuracy at low computation time has been attained using level set method. Feature extraction stage involves extracting both wavelet and textural features. Wavelet analysis was best suited for the task as it aided in analyzing the images at various resolution levels. Initially a neural network is trained using the wavelet features to classify the tumor as normal, benign or malignant tumor. Few textural features are extracted from detected malignant tumors. Out of all textural features, energy values are used to train another neural network which is used to identify the stage of the malignant tumor. The performance of the system has been tested on 24 patient's data obtained from a hospital. An overall accuracy of 97 % has been achieved for MIAS database images and 85 % for patient's dataset.

Keywords Stages of malignant tumor · Histogram equalization · Morphological operators · Level set · Wavelet and textural features

M. Varalatchoumy (✉)
Department of Information Science and Engineering, B.N.M. Institute of Technology, Banashankari II Stage, Bangalore, India
e-mail: Kvl186@gmail.com

M. Varalatchoumy
Dayananda Sagar College of Engineering, Bangalore, India

M. Ravishankar
Vidya Vikas Institute of Engineering and Technology, Mysore, India
e-mail: ravishankarmcn@gmail.com

© Springer India 2016
S.C. Satapathy et al. (eds.), *Information Systems Design and Intelligent Applications*, Advances in Intelligent Systems and Computing 435,
DOI 10.1007/978-81-322-2757-1_53

1 Introduction

Breast cancer [1] is usually defined as, abnormal growth of malignant cells in the breast. The cancer growth is considered to be very crucial, because, undetected tumors get spread all over the body at a rapid rate. In order to improve the survival rates of women affected by breast cancer, in both developed and developing countries, early detection of breast cancer seems to play a vital role. This paper proposes a Computer Aided Diagnostic System (CAD) [1] to Detect and Classify the Breast Tumors. Moreover the system has been mainly developed to identify the stages of malignant tumors [2]. It aids in providing the radiologists a second opinion regarding the confirmation towards malignancy and also gives an idea about the extent to which the tumor growth has occurred. The system is highly promising for breast cancer patients, as it reduces the stress and strain involved in undergoing biopsy and also aids in early treatment thereby reducing the mortality rates.

Mammography [3] has proved to be the most effective technique [3], till date, in order to detect breast cancer. Images of Mammography Image Analysis Society (MIAS) database [3] was used for detecting the tumors and its stages. The developed CAD system was tested on a hospital's patient's data. The system was trained on the features extracted from the MIAS database images and was tested on the real patients mammographic images obtained from the hospital.

In the broad sense the CAD system developed is said to comprise of three major modules: Segmentation of tumors, classification and identification of stage of malignant tumor. Initially the MRM images have to be preprocessed before performing segmentation. The preprocessing stage can be further divided into Image enhancement and removal of noises in the image. A most efficient and simple method, histogram equalization, has been used for image enhancement. Morphological operators have been used for denoising the image. A preprocessed image always provides better segmentation accuracy. Level set algorithm has been used for segmentation. Extensive survey was carried out and it was identified that level set algorithm has been used in the detection of brain tumors and other tasks. Due to its efficient performance with minimal computation time this method has been used in the proposed system.

Prior to classification, the features of the image have to be extracted in order to train the neural network for classifying the images. Wavelet features has been used for classifying the tumor as normal, benign or malignant and textural features of malignant features are used to detect the stage of the tumor.

The aim of the research work is to develop an efficient CAD system for detecting the stages of malignant tumor. Overall system performance of 97 % has been attained for MIAS database images and 85 % for patients data. The related work on various modules is described as follows:

Wadhwani et al. [3] has presented a research on 68 proven breast tumor samples. The research work composed of analyzing and classifying the tumor into benign and malignant. Artificial Neural Network was used for classification. GLCM and

Intensity based features are compared. The author had concluded that GLCM method performs better than the other method.

Singh et al. [4] has proposed a computer aided breast cancer detection system that classifies, scores and grades tumors in order to aid pathologists in classifying tumors. Artificial neural network has been used to classify the tumors and has been proved to be a promising tool for classifying breast tumors.

Ananth et al. [5] had proposed level set geodesic active contour for segmenting medical images for diagnosing diseases. Many cluster objects were segregated based on previous knowledge on disease contours. Diseased region in segmented method was identified using fuzzification. Geodesic active contour method for level set algorithm involves parametric evaluations which proved to reduced the efficiency and increase the computation time.

Asad et al. [6] had proposed various feature set that can be used for analyzing breast tissues. Kohnan neural network had been used to train the features. Only few images were selected from the mini MIAS database for study. Eighty percent classification rate was achieved using the new feature set and classifier.

Elsawy et al. [7] had proposed band-limited histogram equalization for detecting the histogram of mammograms. Wavelet-Based contrast enhancement method was used in combination with this method to improve the overall performance.

2 Proposed System

The aim of the proposed system is to detect the tumors in MRM images and patient's data. The detected tumors are classified into normal, benign or malignant tumor. The tumor identified as malignant tumor is categorized based on the stages as Stage I or Stage II tumor. Figures 1 and 2 depicts the various steps involved in detection and classification of tumor and identification of stages of malignant tumor. Figure 1 shows the Computer Aided Diagnosis system proposed for MIAS database images and Fig. 2 shows the system proposed for 24 patient's data collected from a hospital.

The various modules of the CAD system for both MIAS database and patient's data is explained as follows:

2.1 *Module I-Preprocessing*

Preprocessing module is comprised of two phases, image enhancement phase and denoising. Due to difficulty involved in the interpretation of MIAS database images and hospital's patients data, preprocessing module is highly necessary to improve the quality of the MRM images.

Image enhancement phase deals with enhancing the image appearance by accentuating the various image features. A simple and efficient technique called

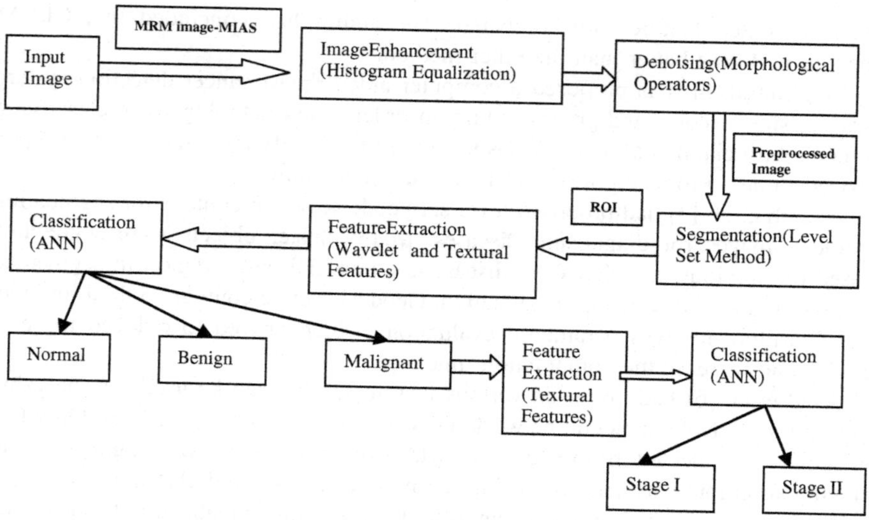

Fig. 1 Proposed CAD system for MIAS database images

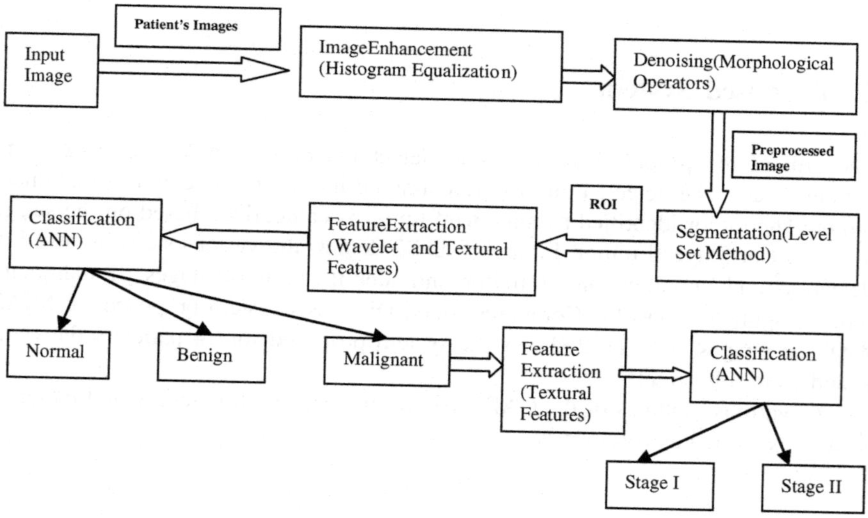

Fig. 2 Proposed CAD system for patient's images

Histogram Equalization is used for image enhancement. Through this technique an uniform histogram [8] of an image is obtained by gray level redistribution. This technique replaces every pixel of an image by integral of the histogram [8] of that particular pixel in the image. This method aids in better distribution of intensities on the histogram.

Denoising phase deals with removal of various artifacts in the image by the application of morphological operators. The various artifacts in the image are removed using the morphological open [2] and close [2] operation. These operations are followed by image reconstruction.

2.2 Module II-Segmentation

Level set algorithm [5] has been used for the first time for segmentation of tumors from mammograms. Through various surveys it has been identified that level set algorithm has proved to be effective in different applications like detection of brain tumor, detection of oil spills, motion tracking, color, texture and motion integration [5].

A level set is defined as a function F1 which is real valued and consists of 'm' real variables. The function is described as,

$$L_d(F1) = \{(y1, \ldots\ldots, ym) | F1(y1, \ldots\ldots, ym) = d\} \tag{1}$$

In the above equation, the level set is a set in which the function F1 works on the provided constant value d. The level set function works as a curve, when the variable count is equated to two. Level set algorithm performs numerical analysis of various surfaces and region of varying topology. The efficiency of level set algorithm lies in its capability of computing numerical values on surfaces by using a fixed Cartesian grid [5], without the need for parameterization.

2.3 Module III-Feature Extraction

Feature extraction is mainly used for describing a large dataset accurately with minimal resources. Usually more variables are involved in analyzing a large dataset, which tends to increase the memory usage as well as the computation time. In this paper a combination of wavelet and textural features are used on the segmented region. This module describes the implementation of Two-Dimensional Wavelet Transform in identifying the wavelet features followed by Texture Feature analysis that aids in detection of malignant tumors as well as identification of stages of malignant tumors.

Feature extraction is performed using wavelet transform approach and the multi resolution approach. The powerful mathematical theory underlying the wavelet methods aids in its application for medical imaging. Wavelets play a very crucial role in diagnosing diseases.

A Discrete wavelet transform is defined as a type of wavelet transform that contains wavelets that are sampled discretely. Temporal resolution serves to be the major advantage of discrete wavelet transform when compared to other transforms.

Using discrete wavelet transform, frequency information and location information can be captured.

The Two Dimensional Discrete Wavelet Transform (2D-DWT) is considered to be a transformation that is linear, operating on an input vector. The 2D-DWT is considered to be a DWT in which the wavelet function is applied for the rows in the images and then to the columns. The length of the input vector should be of an integer value raised to the power of two. The DWT transforms the input vector into a vector of numerically different value. DWT [9] is computed through a series of filtering operations which is followed by factor 2 subsampling [9]. DWT leads to two outputs namely, the approximate coefficients and detailed coefficients. The transform uses the approximate coefficients for next level scaling and the detailed coefficients are considered to be the transform output.

DWT aids in representing a function in multiple scales. Hence the function used is analyzed at different resolution levels. Using 2D-DWT the image is down sampled to an approximate size. After the application of DWT the textural features of the down sampled image region is determined. Figure 3 shows the application of DWT for MRM images and on patients data.

The textural features of the sub sampled images are identified by formulating the Gray Level Co-occurrence Matrix (GLCM). Specifically five textural features, namely, correlation, energy, entropy, contrast and homogeneity are calculated. The energy values calculated is used in identifying the stages of the malignant tumor.

In analyzing the texture using statistical approach, the textural features [10] are calculated using statistical distribution [10] of observed combinations of intensities [10] at specific locations. GLCM is mainly used to extract the second order statistical features [10]. A GLCM is basically a matrix in which the row and column count is similar to the gray levels in the image. Each and every element in the matrix denoted as $Q(m, n|a, b)$ gives information about the relative frequency with which the two pixels m and n occur within a specified neighborhood. Parameters a, b specifies the pixel distance. The matrix element $Q(m, n|l, \Phi)$ give the second order

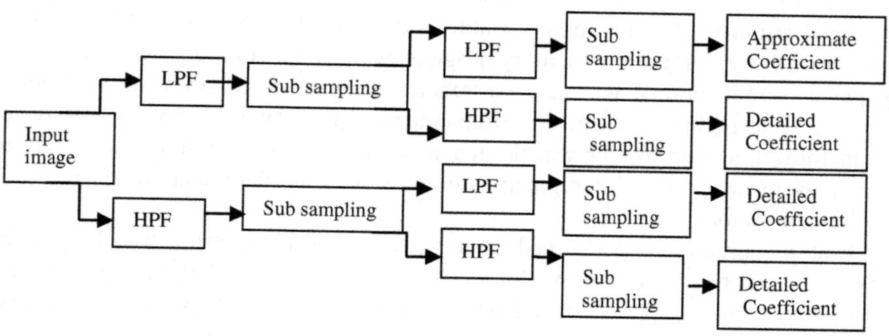

Fig. 3 Various steps involved in DWT operation on MRM images

statistical probability values [10] pertaining to the gray level changes in pixel values m and n, separated by a distance ‘l’ and at a specific angle Φ. The large amount of information stored in the matrix reduces the performance of the GLCM as the size of the sample increases. Hence, GLCM is applied to sub sampled region of the segmented image in which the gray levels are reduced.

2.4 Module IV-Classification and Stage Detection

A neural network is basically a connection of various input, output units. Each connection has a specific weight linked to it. Neural Network training mainly involves adjusting the weights, to enable the neural network to identify the appropriate class.

Feed-forward back propagation [4] neural network is used for classification. Two neural networks are used in the proposed system. The first neural network is trained using the wavelet features as input, in order to classify the segmented tumor as normal, benign or malignant tumor. The second neural network is trained using the energy values of the textural features, in order to classify the malignant tumor as belonging to stage I or stage II.

The classification module can be considered as a combination of training and testing phase. During training phase the Neural Network is trained using the features of the images, whose details are available. This trained algorithm is saved to be used in the classification stage.

In the proposed system whenever an input image is considered, either from MIAS database or from patient’s data, the trained network is used to simulate it and the data is used for further testing [4].

The accuracy of the classification of the tumor as well as the detection of stages solely depends on the training efficiency. The proposed system has achieved good classification accuracy on 24 patient’s data, due to the efficient wavelet and textural feature training of the neural networks.

3 Experimental Results and Discussion

This section discusses the output attained in various modules of the CAD system. As discussed earlier the first module is preprocessing, which performs image enhancement and denoising. Figure 4 depicts the sample output of this module for MIAS database images. Histogram equalization has proved to be more efficient for enhancing the image. The efficiency of this method has aided in achieving the best segmentation output.

Enhanced image serves to be the input for denoising phase of preprocessing. Morphological operators were used to remove the pectoral muscle and the artifacts

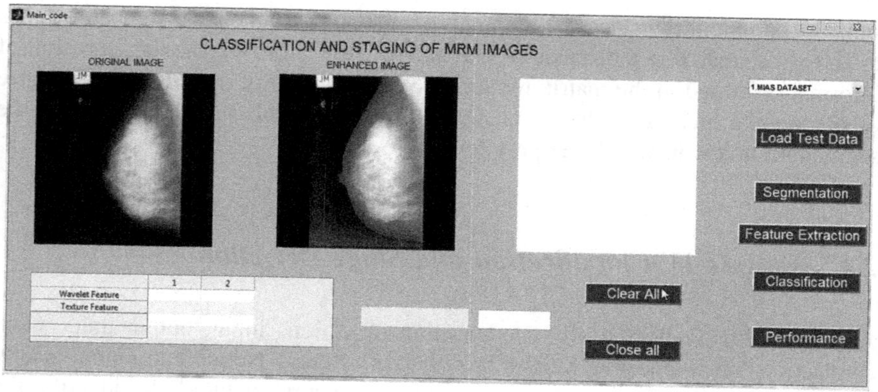

Fig. 4 Sample output for preprocessing of MIAS database image

in the image. The output of this phase, namely, the completely preprocessed image is given as input for the segmentation module. Level set algorithm used for segmentation has achieved accurate segmentation result at less computation time and cost, when compared to all other methods. Figure 5 depicts the sample out for denoising and segmentation of MIAS database images. The textural features of the detected malignant tumors are used to train the neural network to detect the stage of the tumor.

The next module, Feature extraction module deals with extracting the wavelet and textural features. The wavelet features extracted using the discrete wavelet transform is used for training the Feed-forward back propagation neural network to classify the segmented tumor as normal, benign or malignant tumor. Figures 6, 7 and 8 gives the sample output for feature extraction, classification and identification of stage of malignant tumor.

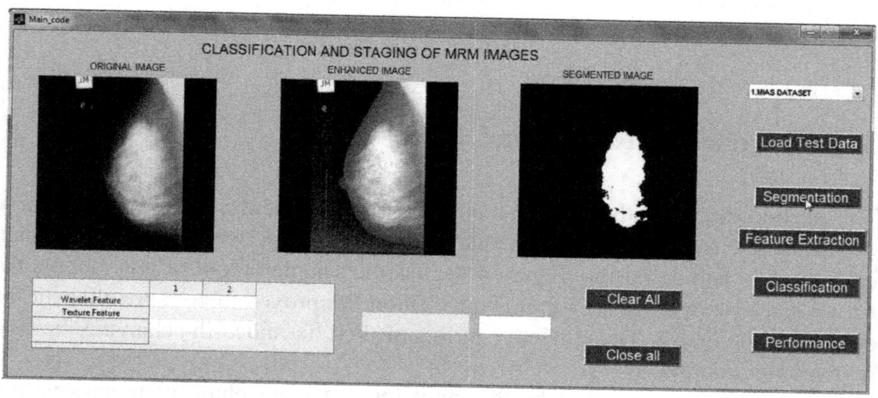

Fig. 5 Sample output for image denoising and segmentation of MIAS database image

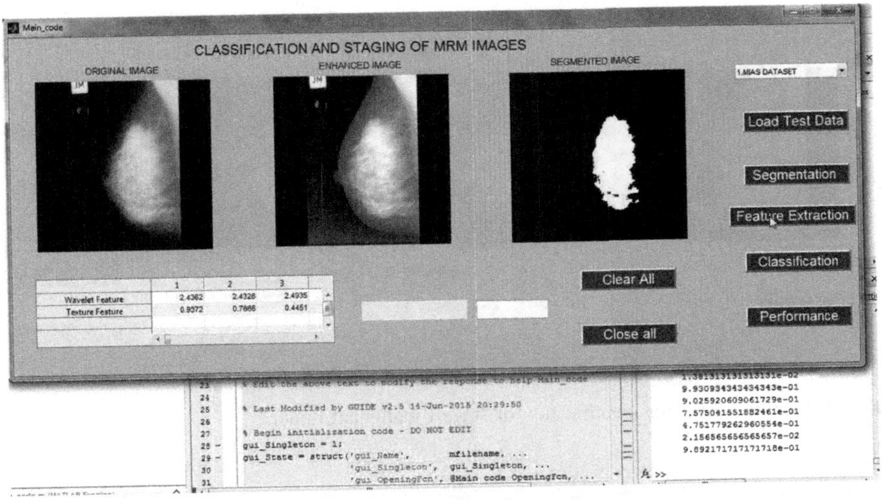

Fig. 6 Sample output for feature extraction of MIAS database images

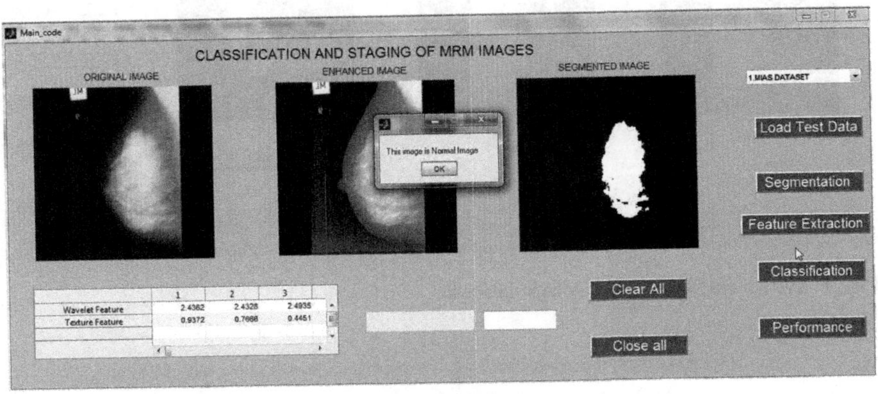

Fig. 7 Sample output for classification of tumor

Figure 9 depicts the overall system performance for MIAS database images. The system has achieved an overall accuracy of 97 % in detecting the stages of malignant tumor.

Figures 10, 11, 12, 13 and 14 depict the output of various modules of CAD system when hospital patient's data was used as input. An overall accuracy of 85 % was achieved for patient's data.

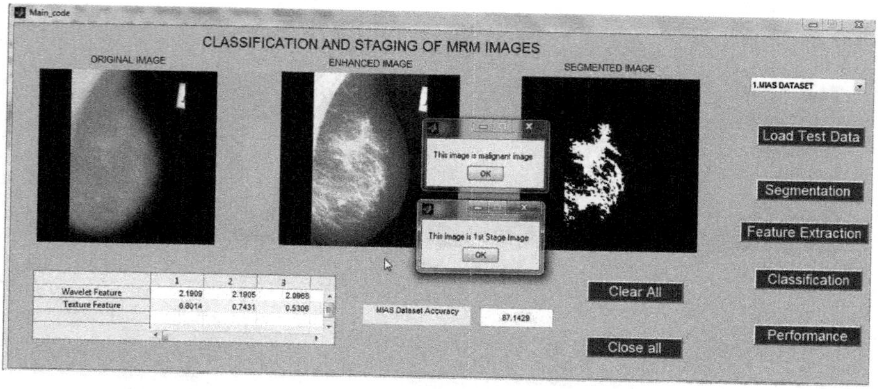

Fig. 8 Sample output for identification of stage I malignant tumor

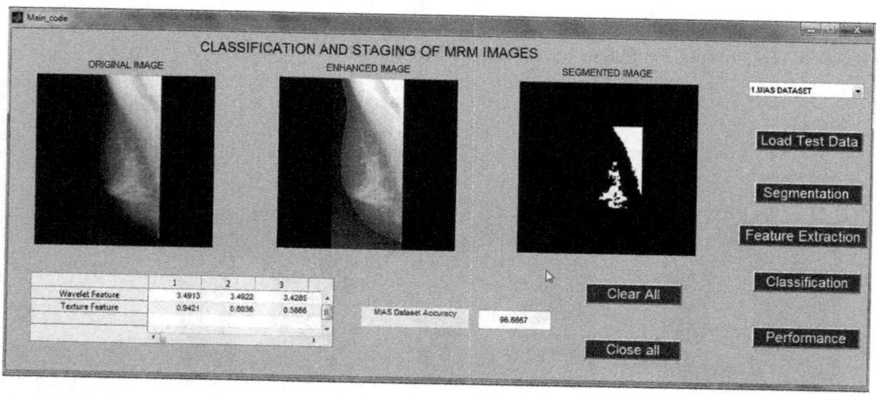

Fig. 9 System performance for MIAS database

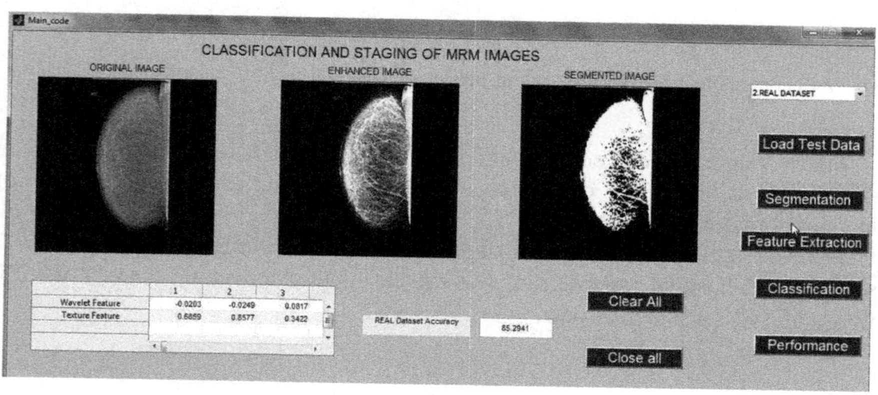

Fig. 10 Sample output of preprocessing and segmentation of hospital patient's MRM

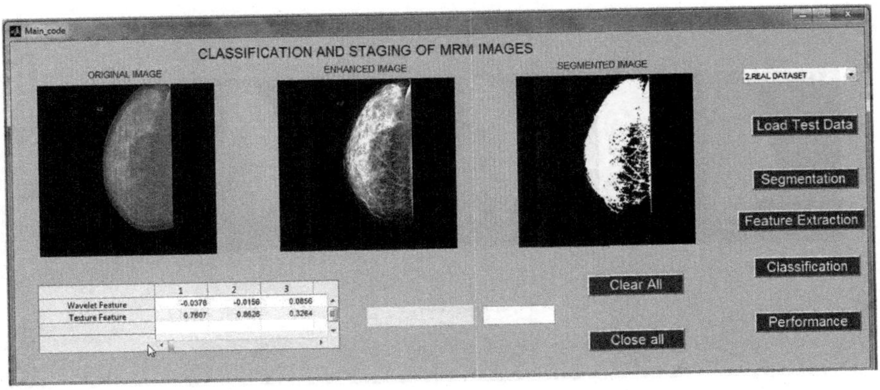

Fig. 11 Sample output of feature extraction of hospital patient's MRM

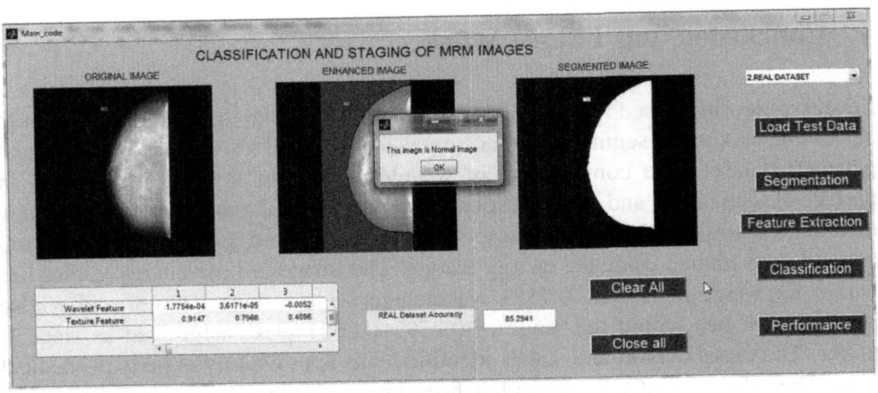

Fig. 12 Sample output of classification of hospital patient's MRM

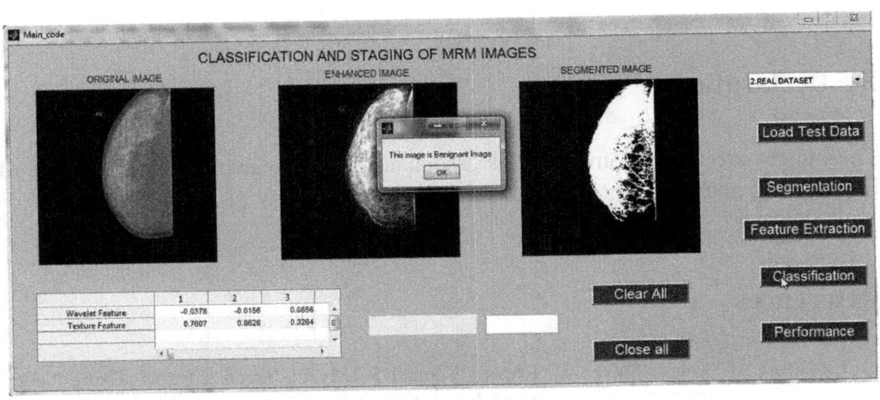

Fig. 13 Sample output of classification (benign tumor)

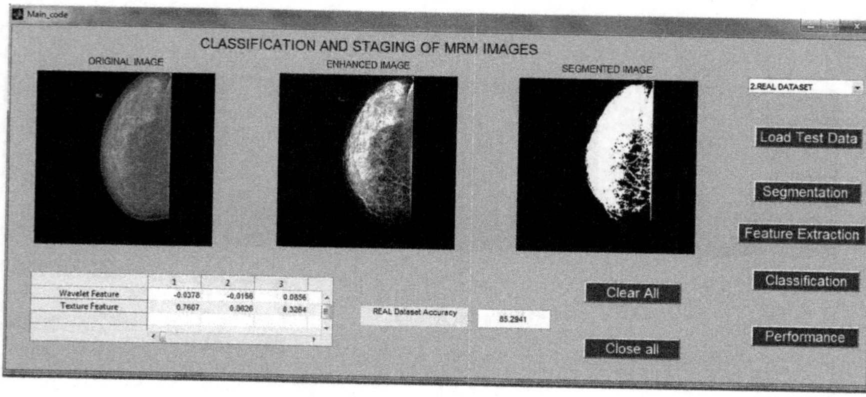

Fig. 14 System performance for patient's data

4 Conclusion and Future Work

Level set algorithm used for image segmentation has proved to be highly efficient in achieving maximum segmentation accuracy for both MIAS database images and patient's dataset. The combination of wavelet and textural feature has aided in perfect classification and identification of stages of malignant tumor. Only few textural features of malignant images were extracted and specifically the neural network was trained using the energy feature. The images whose energy values lies within the range of 0.7082–0.7581 was categorized as stage I tumor and whose energy value lies within the range of 0.7597–0.8627 was categorized as stage II tumor. The classification and stages identified was approved by experts in medical field. The efficiency of the system has been proved by the maximum accuracy, of 97 %, that was attained at very low computational cost and time. In future the work can be extended to 3D and 4D images.

5 Declaration

The images are taken from hospitals associated with Dayananda Sagar College of Engineering and the authors would take the responsibility of the images.

References

1. Catherine M. Kocur, Steven K. Rogers, Lemuel R. Myers, Thomas Burns: Using Neural Networks to Select Wavelet Features for Breast Cancer Diagnosis. IEEE Engineering in Medicine and Biology (1996), 0739-5175.

2. M Egmont-Petersen, D. de Ridder, H. Handels: Image Processing with neural networks-a review. Pattern Recognition 35(2002) 2279–2301.
3. Sulochana Wadhwani: Classification of Breast Cancer Detection Using Artificial Neural Networks. Current Research in Engineering, Science and Technology (CREST) Journal, 2013.
4. Shekar Singh: Breast Cancer Detection and Classification Using Neural Network. International Journal of Advances Engineering Sciences and Technologies, 2009.
5. K.R. Ananth: A Geodesic Active Contour Level Set Method for Image Segmentation. I. J. Image Graphics and Signal Processing, 2012.
6. Muhammad Asad: Early Stage Breast Cancer Detection through Mammographic Feature Analysis. IEEE, 2011.
7. Nabila Elsawy: Band-Limited Histogram Equalization for Mammograms Contrast Enhancement. Cairo International Biomedical Engineering Conference, Egypt, IEEE, 2012.
8. Fahd Mohessen, Mohiy Hadhoud, Kamel Mostafa, Khalid Amin: A new Image Segmentation Method based on Particle Swarm optimization. The International Arab Journal of Information and Technology, Vol. 9(2012).
9. Nizar Ben Hamad: Wavelets Investigation for Computer Aided Detection of Microcalcification in Breast Cancer. IEEE Transactions, 2009.
10. P. Mohanaiah: Image Texture Feature Extraction Using GLCM Approach. International Journal of Scientific and Research Publications, Volume 3, Issue 5, May 2013.

Reviews Based Mobile Application Development Model (RBMAD)

Manish Kumar Thakur and K.S. Prasanna Kumar

Abstract Mobile development is the area of heavy activities in the software industry. The exponential growth in the users of smart phones has opened the doors of opportunities for e-commerce, m-commerce, banking, health and almost all walks of business. There has been a shift in software engineering models to device-centric models. The open-source technologies like Android and large number of handset manufactures under the umbrella of OHA (Open Handset Alliance) has changed the way software industry looks at mobile application development. Industry suffers from lack of a model and hence depends on best practices. This paper presents a model that has an application developed by one of the best practices and then review-based modelling is applied to enhance the overall development of the mobile application. This model assumes more importance as additional worries such as code-efficiency interaction with device resources, low-reusability and lack of portability are common in mobile application.

Keywords RBMAD · App knowledge base · App core domain · Logical framework · App context

1 Introduction

Mobile applications are the primary link between the company and the end-users. It is difficult to hold customers without an effective, efficient, secure, and timely app. Customers these days exhibit zero tolerance towards inefficient apps and hence

M.K. Thakur (✉)
Department of MCA, Acharya Institute of Technology,
Bangalore, India
e-mail: manishkthakur@acharya.ac.in

K.S. Prasanna Kumar
PM Inno Lab, Bangalore, India
e-mail: drprasannaks@gmail.com

© Springer India 2016
S.C. Satapathy et al. (eds.), *Information Systems Design and Intelligent Applications*, Advances in Intelligent Systems and Computing 435,
DOI 10.1007/978-81-322-2757-1_54

system feedbacks are to be considered seriously if the app needs to survive in the market and intern have positive impact on revenue.

Software Engineering for developing mobile apps has grown to the extent of catering to almost all the business requirements and finds no boundaries as seen in the real world as we find its growth today. The process has to ensure the development of protected, superior mobile apps which incorporates the quickly changing need of the users. There are a number of programming environments for mobile applications development available with the focus mainly for small single developers and medium-sized application. There is no formal approach for mobile application development is available which insight all the concepts and aspects of mobile applications development.

All mobile systems are open systems which respond to users' reviews. The model is decomposed into various sub-processes as a result of which the model fulfills all functions incorporating needed changes.

Inefficient apps leads to client lose.

- Scaling of app from vertical to horizontal and vice-versa needs to be corrected in many apps because they do not follow these types of models and intend it's a loss for client.
- Reviews not addressed properly so, not up to the client needs.

As in [1], there are a total of 3.8 million mobile apps on May-2015 in Amazon App Store, Window Phone World, Google Play, Apple App Store and BlackBerry these are based on various uses for communications, games, multimedia, productivity, travel, and utility purposes. Mobile computing, cloud computing, actionable analytics, in-memory computing, personalization and gamification are newest technologies that have been enabled by achievements in mobile technology. All of them have significant impact on mobile applications design, development, and utilization.

Business Models for App Development

- Free, But with Ads (In-App Advertising)
- Freemium (Gated Feature)
- Paid App (Cost Money to Download)
- In-App Purchases (selling Physical/Virtual Goods)
- Paywall (Subscriptions)
- Sponsorship (Incentivized Advertising)

We find various full-fledged mobile app development environments already existing, available to developers these days. iOS Dev center headed by Apple provides Xcode package. This Xcode package has an efficient iPhone emulator for the taste of app, interface builder and complete application development environment for all its products [2]. Google provides Android Development KIT along with a good high graded documentation for developers, tools and plugins for easy integration [3]. Microsoft provides Microsoft's Visual Studio environment exclusively for the purpose of mobile app development [4].

These are the powerful tools and frameworks based on core programming knowledge which simplify the task of development and implementation of a mobile application. However, these technologies for mobile application development mainly focused on the finished application not the common steps followed by the individual developer. These technologies do not provide any model for the mobile application development but, mainly to provide an app developer with good design to develop the required app at ease and in time. These technologies are the next versions of tools allowing the developers to concentrate on requirements and modules rather than the complexity of architecture.

Nevertheless, the growing world of mobile application in recent years and complexity addition to the simple mobile application, when more complex applications are considered software engineering becomes a necessity in order to produce efficient and secure applications.

It is not fare expect classical software engineering models to provide this level of efficient technique as the mobile application is a latest inclusion into our computer echo system. This paper is focused to provide a software engineering model for mobile application development which is based on the classical and new areas for the mobile application development.

1.1 Related Work

In Wasserman [5], the development practices for mobile applications are

1. Compare to other conventional projects mobile app development needs less or most of the time one or two developers who will build relatively less source code and will also be doing most of the jobs in the life cycle of any software development method.
2. We find a large difference between convention computer programs and the applications that are built for mobile and held specific devices.
3. All mobile developers along with experience build own best practices as there exists no development method.
4. In comparison to any software development methods these developers deal with fewer parameters and analytics during development.

In [6], we find that the services market of mobile app has breaded a number of business models over the number of years. The latest trend being development for specific devices rather than a generic development. The concentration is on the service and content rather than generic application.

In [7], it is easy to get carried away to understand that there exists a formal method for development of mobile applications by seeing the large number of releases and the areas these mobile applications are reaching. But in reality there is lot of gap between the best practices and the software development skills that could enhance the security and over all working due to development method that is involved.

In [8], the paper proposes a MDE approach for Android applications development. This method views the development in the form of intent, data and service relaying on the standard UML. This methods provides a platform for utilization of UML and sequence diagrams.

In [9], the paper proposed a model-driven approach for automated generation of mobile applications for multiple platforms. They use a carefully identified subset of UML that fits our purpose of mobile application modelling. There focus is on the modelling of business logic layer for the mobile application and provide a controller in the generated code to bridge the user interface with it.

In [10], the paper reviewed mobile devices and platforms differences, different cross-platforms approaches and highlighted software engineering issues during the development of multi-platform application and raised some questions for each phase of SDLC which may help the stakeholders to improve the application quality.

2 Reviews Based Mobile App Development

2.1 Introduction to the Model

The framework models uncertainty into development of software-intensive systems to create enhanced possibilities for adaptation of reviews provided by various real-time users of that software environment. Mobile application developments requires considerable knowledge provided by software engineers and others to help explain the problem domain. The quickly changing various environment needs and use expectations can be adjusted in the mobile application by the reviews provided by several users. The primitives of this hierarchal structure can be rules, frames, semantic networks, concept maps, logical expressions, behavior after release and reviews to overcome the application bottleneck due to fast changing environments. Knowledge Based Mobile Application Development Model will also follow the footsteps of an adaptive system based on reviews. This will features:

1. App will respond to changing operational context, environments, and system characteristics.
2. App will achieve greater flexibility, efficiency, and resiliency.
3. App will become more robust, energy-efficient, and customizable

This model for mobile app development is based on the efficiency of the model which is indicated by:

1. **Reviews by the real-time users**, provide this the flexibility to be more effective and efficient app.
2. **Context-Aware System**—a universal course of mobile system development that can sense and adapt their behavior accordingly.

3. The model is more efficient, because it tries to address the need arises due to *functional and non-functional requirements*.
4. The steps/stages can be *modified based on the environmental change that leads* to a more efficient app.

2.2 Explanation of the Model

This model for mobile app development is based on the core development knowledge and the knowledge gathered real-time user's reviews. The model updates its functionalities based on the knowledge gained from reviews of the users. The Mobile Application Development requires

1. The knowledge of problem domain.
2. A possible solution that could be provided by software engineers and other core domain developers.
3. The knowledge of mobile app based on functional and non-functional requirements.
4. Release of the app.
5. The identification of users who will use the app in various environment and provide reviews for the overall app functionalities.
6. Required changes based on the reviews and various other development factors.

In the broad changing and diverse world of mobile app development, we cannot adopt a model which is fixed and has no room for any changes. We need a model which can be extended based on the requirements and fulfils the need of users. Adaptive system based on reviews of user can be a good practice to this, if properly managed.

This provides a formal way for mobile app development and enhancement of the app based on early technical, logical, and behavioral knowledge influenced by reviews of real app users. The whole idea of this model is to use an app knowledge-base for development and enhancement of the app for the future. This provides a framework for the app development in current and future scenario based on each and every aspects of mobile app development.

The model works on the principle that collects a lot of information and analyses it to provide knowledge of any problem and start rectifying it based on technical feasibility. The most important aspects of this model is app knowledge base from the past and available technologies to solve it in the present drawbacks.

As in Fig. 1, the App Knowledge Base is divided into three important parts:

1. App Knowledge Corpus—focuses on the domain concepts and relationships
2. App Operation's Knowledge—represents particular and general factual knowledge
3. App Knowledge Re-structuring—uses the adaptive probabilities to represent degree of belief in uncertain knowledge.

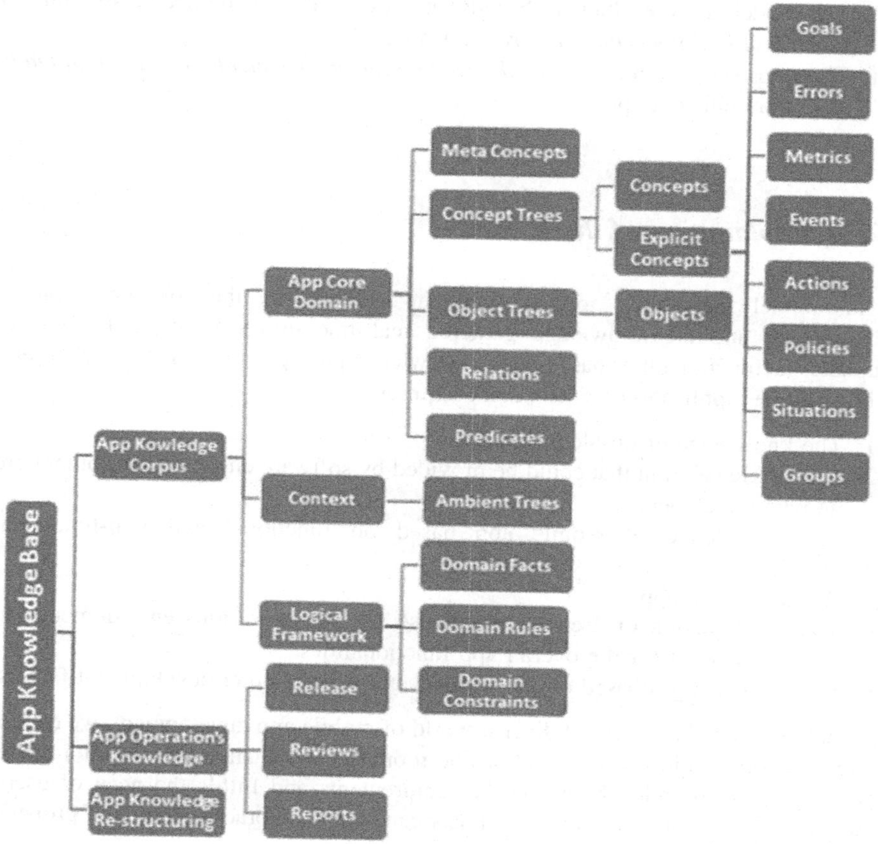

Fig. 1 Block-representation of RDMAD

App Core Domain

1. **The Core App Developer** will do development and all the steps in this stage intend to support the development of mobile app.
2. **Meta Concept**—The basic concepts to understand how to develop. This step will explain the other steps needed for developing a successful app.
3. **Concept-Tree**
   - Requirement Understanding (functional/non-functional)
   - Concept that pre-defined/pre-requisites

4. **Explicit Concepts**
   - Non-functional factors keeping environmental factors in mind needed to be considered for the effective app development.
   - May change based on the app development context/scenario/environment.
   - Broader view of app-execution needed for consideration
   - Real-world scenarios addressed, because open-system development

The App incremental development is a complex process, often based on

1. Quickly changing software world and need of the users
2. Reasoning for drawbacks based on probability and statistics.

Ambient Trees

- A universal course of mobile systems that can understand their environment, and adjust their performance as required.
- Properties of mobile devices.
- Applied more to the smart phone users which links to the changing environmental factors.
- Algorithm in the form of decision tree to decide what to do in the physical changing environment.
- Tree-structure for better understanding and considering the different aspects of physical changes, which helps in the successful development of the mobile app.

The model supports logical and statistical perceptive in its very core. This is comprehensive yet multiplier model, where app development information can be presented at different stages of concept. These concepts are assembled by both hierarchical and functional patterns.

App Knowledge Base incorporates concepts organized through concept trees, object trees, relations, and predicates. Each concepts is specified with particular properties and functionalities and is hierarchically linked to other concepts.

The four important stages of this model:

1. Initial Core Activity—It is done by the core experts of app development. It defines functional as well as non-functional attribute of the problem domain and also provides a working solution based on the various requirements and observations.
2. App Behavior Definition—This stage focuses on release of app for the end-user and collection of review. Review report for the further enhancement of the mobile apps is developed.
3. App knowledge Structuring—This stage encapsulates the identified domain entities, situations, and behaviour into Mobile App core constructions, i.e., perceptions, facts, objects, relations and rules.
4. App Building—Based on the above three activities the required activity is planned for the next version/release of the mobile app.

App Knowledge Restructuring

- Based on the changes asked in the App Knowledge Operator step.
- It will be done based on the fundamental steps followed in the case of the App Knowledge Corpus step.
- All the necessary stages and steps will be followed to make the necessary changes available for the app.

Control Releases

- Releases will be based on the reviews by the user and the reports based on the reviews.
- Release will follow time-line and only allowed when there is greater need for restructuring sought in the reviews.
- Releases will be adaptive in nature and always follow the review and report generated.

3 Conclusion

The paper attempts to collectively capture all the important aspects of incremental development of mobile app during its complete life-time. Building of knowledge-base for each app, its utilization in the development of app is the key factor highlighted. Since, there exists many tools and techniques for instantaneous development of apps, developers these days are more concentrating on the version and release rather than initial development of mobile apps. The paper provides an insight on all the factors that influences the development of the effect of them on final release.

Acknowledgments The authors express their deep sense of appreciation for **PM Inno labs**, Bangalore for the support extended to them during all phases of this model development.

References

1. Statista, the Statistic Portal http://www.statista.com/statistics/276623/number-of-apps-available-in-leading-app-stores/.
2. Apple Developer Connection. http://developer.apple.com/iphone/index.action.
3. Eclipse web site. http://eclipse.org.
4. Windows Phone developer site http://developer.windowsphone.com/windows-phone-7-series/.
5. Agrawal, S. and A.I. Wasserman, "Mobile Application Development: A Developer Survey", (2010).

6. Allan Hammershøj, Antonio Sapuppo, Reza Tadayoni, "Challenges for Mobile Application Development", (2010).
7. Anthony I. Wasserman, "Software Engineering Issues for Mobile Application Development", (2010).
8. Abilio G. Parada, Lisane B. de Brisolara, "A model driven approach for Android applications development", (2012).
9. Muhammad Usman, Muhammad Zohaib Iqbal, Muhammad Uzair Khan, "A Model-driven Approach to Generate Mobile Applications for Multiple Platforms", (2014).
10. Hammoudeh S. Alamri, Balsam A. Mustafa, "Software Engineering Challenges in Multi-Platform Mobile Application Development" (2014).

Three Phase Security System for Vehicles Using Face Recognition on Distributed Systems

J. Rajeshwari, K. Karibasappa and M.T. Gopalakrishna

Abstract There is a continuing problem of automobile theft which is a greater challenge that needs to be solved. Traditional security system needs many sensors and it is costlier. Modern security system needs to be implemented which uses biometric verification technology in the vehicles. In the proposed method the GSM controlled by the Renesas Microcontroller is used to find the location of the vehicle. The vehicle ignition will on only when the person accessing the vehicle authorises with the three phase security system. An SMS alert is sent to the authorised person if an unauthorised person accesses the vehicle who fails in the three phase security system. In this paper, Haar features are used to detect the face and Adaboost classifier is used to combine all weak classifiers into strong classifiers for deciding whether the captured image from the video is face or not. Principal Component Analysis method is used to recognize the detected face in the video. The proposed system provides an inexpensive security system for the automobiles using Face Detection and Recognition technology. This is an efficient method to authenticate the person in the vehicle security system.

Keywords Face detection · Face recognition · GSM · Renesas microcontroller

J. Rajeshwari (✉)
Dayananda Sagar College of Engineering, Bangalore, India
e-mail: raji_jkl@yahoo.co.in

K. Karibasappa
Oxford College of Engineering, Bangalore, India
e-mail: K_karibasappa@hotmail.com

M.T. Gopalakrishna
K.S School of Engineering and Management, Bangalore, India
e-mail: gopalmtm@gmail.com

© Springer India 2016
S.C. Satapathy et al. (eds.), *Information Systems Design and Intelligent Applications*, Advances in Intelligent Systems and Computing 435,
DOI 10.1007/978-81-322-2757-1_55

1 Introduction

During past few years vehicle thefts have been increased and advance technology is needed to prevent vehicle thefts. The traditional means of verifying the driver is a non-biometric method such as licence card, ID card or Passport. In the recent technology biometric methods are used since it is more trustworthy than the non-biometric methods. Biometric methods are characterised into two forms physiological and behavioural. Physiological forms are correlated to shape of the body and behavioural forms are correlated to the behaviour of the person. Different Biometric methods are Voice, Signature, Retinal, Iris, Fingerprint, and Face Recognition technology. Different biometric methods have its own advantages and disadvantages. Face Recognition is more efficient compared to other biometric methods since it requires no physical interaction and non-interfering with the application.

Seshasayee and Manikandan [1] proposed method to track the position of the vehicle and SMS is sent to the vehicle owner when someone as theft the vehicle. As security of the vehicle is raising its importance Padmapriya and Kala James [2] proposed method, where car door will open only for the authorised users using Face Detection and Recognition method. Saranya et al. [3] proposed method for controlling the car by the owner of the vehicle when unauthorised face of the driver is detected and sent via SMS through GPS module. Patil and Khanchandani [4] proposed a system which disables the automobile when it is lost. Meeravali et al. [5] proposed to know the location of the car and control of the speed of the vehicle when unauthorised person access the car. Nandakumar et al. [6] proposed vehicle security system using Viola-Jones Face Detection technique and LDA as the Face Recognition technique. After detecting an unauthorised face an MMS is sent to the owner of the vehicle through the operating GSM modem. Shankar Kartik et al. [7] proposed security system where authorised person is not allowed to access the door.

This paper deals with the advance anti-theft security system for the vehicles. Three phase security is implemented to gain the access to the vehicle. When a person access the vehicle, camera installed in the vehicle captures the drivers image and compares with stored dataset consisting of authorized person. If unauthorized person is identified then GSM module present in the system sends an SMS alert to the owner. Only authorised users permits the driver to trigger the ignition process thereby turning on the engine of the vehicle. Haar features are used for Face Detection and PCA algorithm is used for Recognition.

2 Proposed Method

This paper describes the anti-theft security for vehicles under three-phase process using Face Detection and Recognition system. The Proposed block diagram is shown in Fig. 1. In the proposed method three phase security is provided for the user to drive the vehicle. If there is an unauthorised access then the SMS is sent to the owner.

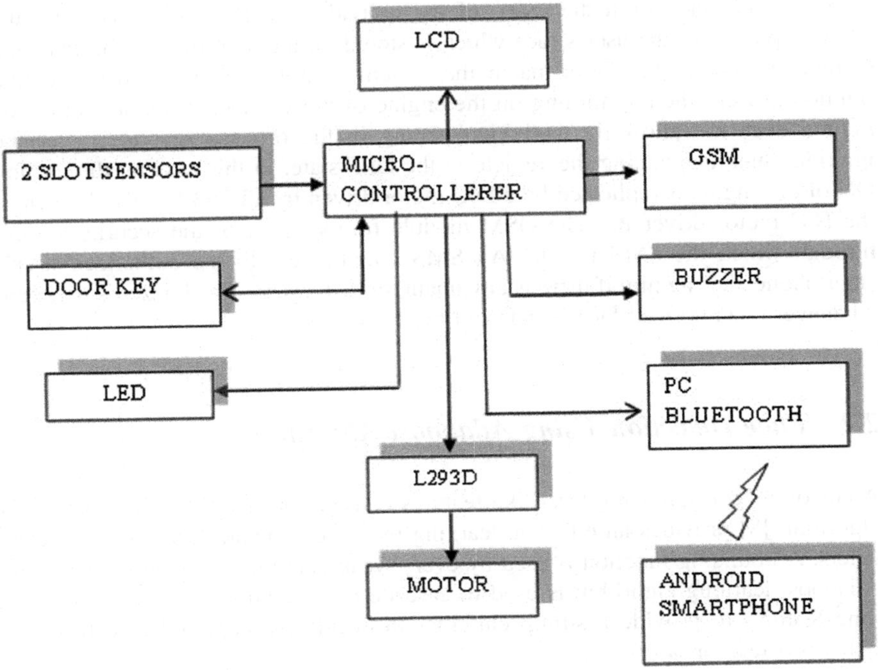

Fig. 1 Smart security system

2.1 Smart Security System

It consists of a microcontroller and a Personal Computer. This unit is placed within an automobile. There are three phases wherein the security system is implemented.

- Door Key
- Bluetooth password
- Face recognition of the driver and SMS alert

In the first phase two slot sensors is used to replicate the point where the key is actually placed to begin the ignition process hence turning on the engine. When the key is placed in the first slot sensor it indicates the car is turned on but the ignition does not take place. When the key enters the second slot the ignition takes place and the engine is powered up.

In the second phase, an Android Smartphone is used to gain access to the automobile. Once the user of the automobile sends the password via an application, Personal Computer placed within the automobile having a Bluetooth capability will receive a file from the smartphone. On receiving this file, the Personal Computer will open the camera placed in the system.

The third phase of the security is activated once the camera is switched on. The face of the individual sitting in the driver's seat will be captured. The captured picture of the face from the video of the individual will be compared with the existing picture of the user's face which is stored on the hard disk of the Personal Computer. If both the faces match the system permits the driver to trigger the ignition process thereby turning on the engine of the vehicle. If this test fails, the system does not allow the individual sitting in the driver's seat to trigger the ignition, thus maintaining the vehicle in the OFF state. In this system, the turning ON of the engine is replicated by a D.C motor driven by a L293 IC, also known as the D.C motor driver IC. The GSM module forms a part of the security system through which the SMS is sent. An SMS will be sent to the authorized user's smartphone alerting him if there is any unauthorized access. The buzzer is triggered whenever unauthorized face is detected by the system.

2.2 *Face Detection Using Adaboost Algorithm*

Adaboost Algorithm using Haar like features is used for Face Detection. Adaboost algorithm [8] provides an effective learning by the combination of the weak classifiers. A bounding function is used by every weak classifier on one of the feature. Adaboost learning algorithm is used to cascade only small set of weak classifiers (one feature) to provide a strong classifier to decide the captured face from the video is a face or not.

Haar features and integral image: Haar features is used for detecting the face region. Each haar features consists of two or three rectangles. The two rectangle haar feature is calculated by subtracting the sum of the pixels of the white region with the sum of the pixels of the of the black region. The three rectangle haar like features is considered by subtracting the sum of the two outside region with the inside region. The four rectangle haar feature is calculated from the diagonal pairs of regions. Face region is identified by searching the haar features at each stage. In Fig. 2a, b shows the two rectangle haar feature, (c) and (d) shows three rectangle haar feature and (e) shows four rectangle haar feature.

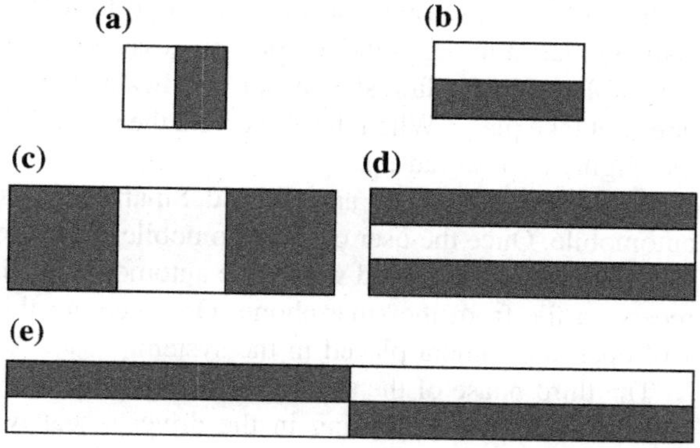

Fig. 2 Different types of haar like features

The difference value obtained is used as feature value. The area of the region is found using the integral image. The integral image is calculated by adding all the pixels to the above and to the left at the location (x, y). Integral image is defined as follows

$$\mathrm{ii}(\mathrm{x}, \mathrm{y}) = \sum_{x' \leq x,\, y' \leq y} i(x', y') \tag{1}$$

ii is the integral image of the image i. Number of haar features generated may be too large and use of all these features may be expensive in time so Adaboost learning algorithm is used which selects only small set of features. Best features are selected from the combination of the weak classifiers which forms a strong classifier.

Adaboost is a method of training a boosted classifier. A boosted classifier is a classifier of the form shown in Eq. 2. F(t) is taken as an object x as an input and returns the result indicating the class of the object. Weak classifier output identifies the predicted object class of t-layer classifier which will be positive if the samples belong to positive class otherwise belong to negative class.

$$\sum_{t=1}^{T} Fx(t) \tag{2}$$

2.3 *Face Recognition Using PCA*

PCA [9] is used for obtaining the small number of uncorrelated components in the form of the principal components from the correlated components. PCA is not only used for feature extraction but also to reduce the number of dimensions without any great loss of information. It is used for finding the directions that maximize the variance in the dataset. The entire dataset is projected into different subspace to find the maximum variance for the data. Principal components are obtained by calculating the eigenvectors and eigenvalues of the covariance matrix. Sort the eigenvector based on Eigen value from highest to the lowest. First principal component is the eigenvector with highest eigenvalue and second principal component is the eigenvector with the second highest eigenvalue and so on. From the given dataset of n-dimensions, only p-dimensions are selected on the basis of the first p eigenvectors which are called as eigenface. Each face is represented as the combination of the eigenfaces.

Specific steps and equations:

1. Let N bet the set of training images with m × n dimensions.
2. Calculate the covariance matrix

$$C = 1/N \sum_{i=1}^{n} (x_i - \bar{x})(y_i - \bar{y}) \quad (3)$$

3. Eigenvalues and eigenvectors are calculated using

$$AX = \lambda X \quad (4)$$

   where X is the eigenvector and λ is the eigenvalue.
4. Eigenvectors are ordered based on the eigenvalue from high to low. The eigenvectors are the eigenface.
5. Each of the images in the training set are projected onto the eigenface.
6. The test images are projected onto the same eigenface.
7. The test image is matched with image present in training set which is having the closet match using the similarity measures technique.

3 Result Analysis

The proposed system is implemented in Microsoft Visual Studio 2008 on dot net framework. SMS sending method is done in Cube Suit+. Figure 2 shows the sequence of messages displayed on the LED screen. The system includes some hardware components like Microcontroller, GSM module, buzzer, and L293 motor driver, DC motor, 2 slot sensors, LED and a LCD screen, toggle button. When the system starts, all the components of the system will be initialized. Then the messages "CAR DOOR MONITOR" and "DOOR CLOSED" will be displayed on the LCD screen. When the door key is open the message "DOOR OPENED" will be displayed on the LCD otherwise it will wait for input. Then the system will wait for the input from the Personal Computer, which is used for Face Recognition. Once the input is received from the Personal Computer, if the input is "F" then the message "FACE RECOGNISED" will be displayed on LCD and the car ignition turns on. If the input from Personal Computer is "D" then the message "ALERT SMS SENT" will be displayed on the LCD. Then the controller will tell the GSM module to send a SMS message to the car owner and the buzzer beeps. There will be no ignition if the SMS alert is sent to the car owner. If the driver has left the key in the car by mistake, after the ignition process then also an alert SMS is sent to the owner and buzzer beeps once again. Figure 3 shows Android Login and Sending Password File, Fig. 4 shows selecting the Android Device, Fig. 5 indicates Snapshot of the model, Fig. 6 shows the result if authorized person is accessing the vehicle and Fig. 7 shows outputs on the LED screen.

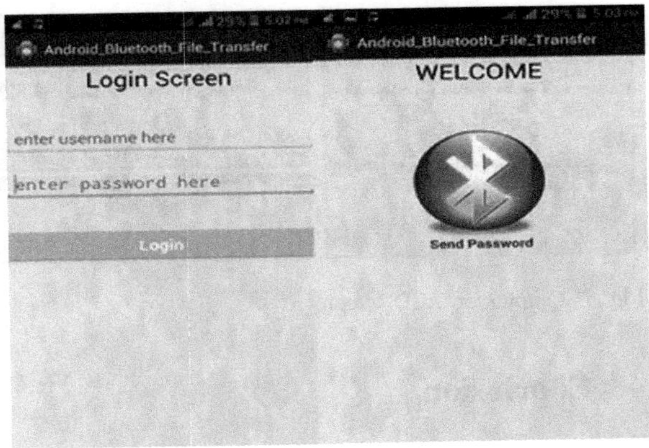

Fig. 3 Android login and sending password file

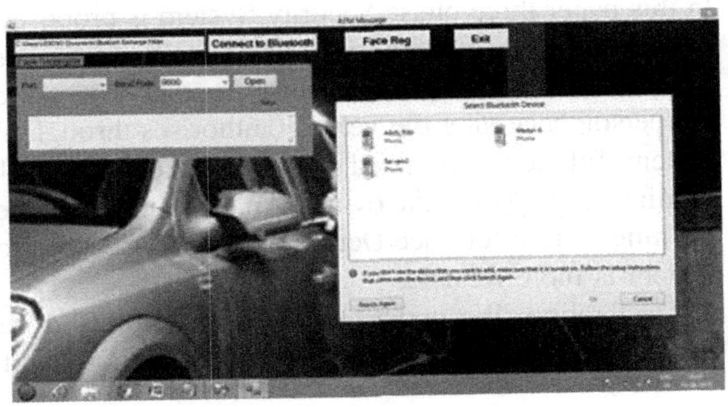

Fig. 4 Selecting the android device

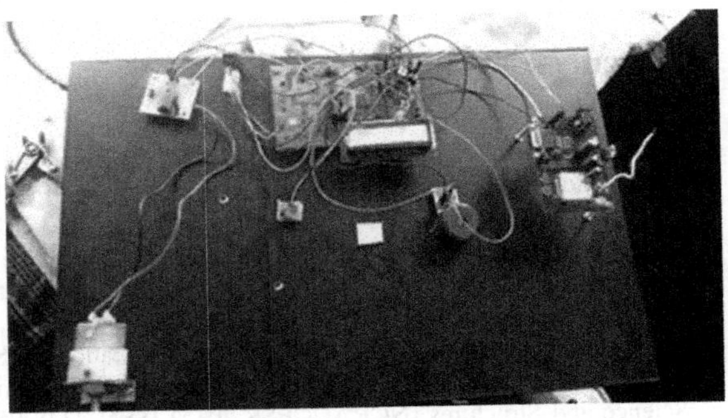

Fig. 5 Snapshot of the model

Fig. 6 Face Recognition

Fig. 7 Outputs on LED screen

4 Conclusion

In this paper three phase Security System is proposed for vehicle users to decrease the theft from the unauthorized persons. It is more cost-effective and efficient when compared to the traditional security system. The vehicle ignition will on only when the person accessing the vehicle authorises through all the three phases of security system. When unauthorised person operates the vehicle, GSM has been used for the sending alert SMS to the owner. An efficient technique Haar features with Adaboost learning is used for Face Detection and PCA is used for Face Recognition hence the system is more reliable.

Declaration: We undertake the responsibility of use of figures/persons etc. and the figure/person in the paper are of the authors photograph.

References

1. V. Balajee Seshasayee and E. Manikandan,: Automobile Security System Based on Face Recognition Structure Using GSM Network, Advance in Electronic and Electric Engineering. ISSN 2231-1297, Volume 3, Number 6 (2013), pp. 733–738.
2. S. Padmapriya and Esther Annlin Kala James,: Real Time Smart Car Lock Security System Using Face Detection and Recognition, 2012 International Conference on Computer Communication and Informatics (ICCCI-2012), Jan. 10–12, 2012, Coimbatore, INDIA.
3. Saranya V, Sabitha Tamilanjani V," Face Identification: In Smart Car Security System In Real Time, National Conference on Research Advances in Communication, Computation, Electrical Science and Structures (NCRACCESS-2015) ISSN 2348-8549.
4. Vishal P. Patil, Dr. K.B. Khanchandani,: International Journal of Engineering Science and Innovative Technology (IJESIT) Volume 2, Issue 1, January 2013 487 Design and Implementation of Automotive Security System using ARM Processor.
5. Shaik Meeravali, A. Vamshidhar Reddy, S. Sudharshan Reddy,: An Inexpensive Security Authentication System Based on a Novel Face-Recognition Structure, International Journal of Engineering Trends and Technology (IJETT) Volume 4 Issue 9 Sep 2013.
6. C. Nandakumar, G. Muralidaran and N. Tharani,: Real Time Vehicle Security System through Face Recognition, International Review of Applied Engineering nal.org Page 4094 ISSN 2248-9967 Volume 4, Number 4 (2014), pp. 371–378.
7. J. Shankar Kartik, K. Ram Kumar and V.S. Srimadhavan: Security System with Face Recognition, SMS Alert And Embedded Network Video Monitoring Terminal, International

Journal of Security, Privacy and Trust Management (IJSPTM) Vol. 2, No. 5, October 2013 doi:10.5121/ijsptm.2013.2502 9.
8. Yue Ming, Qiuqi Ruan, Senior Member, IEEE, 2010, : Face Stereo Matching and Disparity Calculation in Binocular Vision System, 2nd International Conference on Industrial and Information Systems, 281–284.
9. Kevin W. Bowyer, Kyong Chang, Patrick Flynn, 2006, : A survey of approaches and challenges in 3D and multi-modal 3D + 2D face recognition, Computer Vision and Image Understanding (101), pp. 1–15.

SLA Based Utility Analysis for Improving QoS in Cloud Computing

Ananthi Sheshasaayee and T.A. Swetha Margaret

Abstract A service rented or leased from cloud resource providers follows a systematic procedure on working with the resource and returning them in the same accordance. In relevant terms it can be stated like policies that a user need to adhere in order to utilize the resources. These polices are clearly declared on an agreement to define the service level policies to the cloud users. This agreement stances or acts as a legal document between the user and the resource provider. The most important role of an SLA treaty is to provide quality assured service to its users as stated on the agreement. Quality Agreement of negotiation among the contributors helps in defining the Quality of Service necessities of critical resource based progressions. Though, the negotiation process for users is a momentous job predominantly when there are frequent SaaS providers in the Cloud souk. Consequently, this paper proposes a novel briefing on negotiation agenda where a SaaS broker is employed as a resource provider for the customers to achieve the required service efficiently when negotiating with multiple providers. Negotiation framework simplifies intelligent mutual negotiating of SLAs between a SaaS agent and multiple providers to achieve different objectives for different participants. To capitalize on revenue and mend customer's contentment levels for the broker, the paper also proposes the design of strategies based counter generation techniques.

Keywords Cloud utility computing · Market-oriented slant · SaaS platform broker and agents · SLA treaty · QoS parameters

Ananthi Sheshasaayee (✉) · T.A. Swetha Margaret
Department of Computer Science, Quaid E Millath Government College for Women, Chennai 600002, India
e-mail: ananthi.research@gmail.com

T.A. Swetha Margaret
e-mail: swethaaugustes.research@gmail.com

© Springer India 2016
S.C. Satapathy et al. (eds.), *Information Systems Design and Intelligent Applications*, Advances in Intelligent Systems and Computing 435,
DOI 10.1007/978-81-322-2757-1_56

1 Introduction

Cloud based slant of computing is a refined evolution of technology and this is instantly obtainable computing resources is exposed and treated as a service paradigm. Cloud as a computing resources is generally accessible as pay-as-you-go model and hence have become striking to cost sentient customers. Apart, from IT and education sectors cloud computing also betterments on ecological provisions. This resource growing trend is divesting the previous in-house service systems when associated to cloud systems. On the arbitrary end business and enterprise processes enhance this terminology in various work tactics. Resources can be allocated on basis of short rage to the customer or the user and this is considered to be the most prominent artifice in allocating the resources. The Service level agreement is an authorized document between provider and the end-user that defines the QoS, which is accomplished through negotiation process. Negotiation processes on Cloud are vital because contributing terms are self-governing entities with diverse intents and quality parameters. By utilizing the negotiation process, users in the cloud souk place their given prospect to maximize their return on investment (RoI). Design of Service oriented architecture (SoA), assist the users in providing dependable service and becomes an important phase. Although systematic procedure is available to estimate user resources their demands over resource requisition varies significantly. This contract can be laid off as per the terms and conditions mentioned on the agreement [1, 2] (Fig. 1).

Resource organization is done by load balancing and scalability where virtualization serves to be the major feature of cloud utility computing. Agreement grounded service assurance is a legal document between cloud service provider (cloud server side) and cloud resource users [1, 2]. The intact resources are hired using cloud and utility computing on the other hand in cloud terminology the organization that handles the cloud resources has less knowledge about source of the services or the type of service that the cloud functions. Fixed time interval under negotiation mechanism makes an agreement between the provider and consumer, which is completely taken care by the agent.

2 Basic Understanding SLA

The service level treaty is defined as: "An explicit statement of expectations and obligations that exist in a business relationship between two administrations i.e., service provider and end-user (customer)" [1, 2]. Since service treaty have been used from 1980s in a range of domain, most of the retained terminology are circumstantial and vary from domain to domain [3, 4].

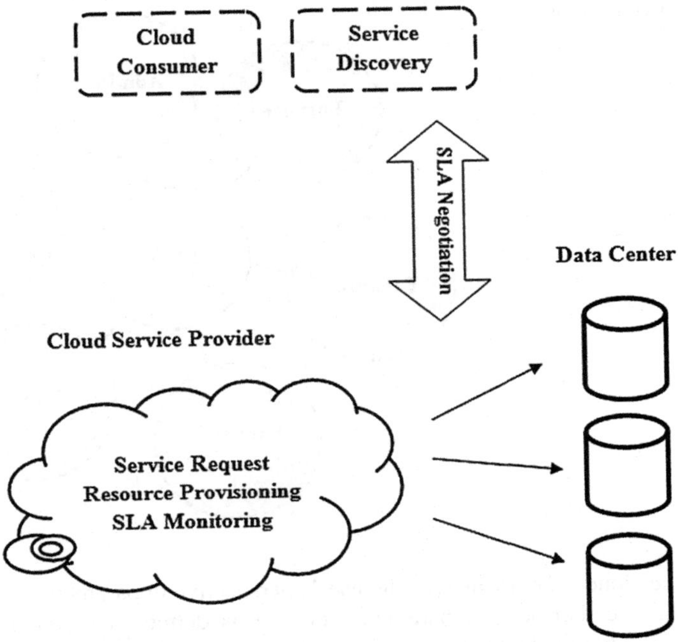

Fig. 1 SLA negotiation between cloud user and service provider

2.1 Components Requirements of SLA

The treaty delineates the capability of handling the requirements for the cloud users from the direct cloud resource providers. The components involved in the treaty enhance the services that are leased from the providers and therefore give concrete customer satisfaction to the users. The components include (Fig. 2).

2.2 Life-Cycle of Parametric View on Service Treaty (Quality SLA Retort)

Any life cycle for a development involves different approaches and stages to uniquely determine the capability of the functioning process. These stages are furthermost spilt to simplified steps to make its understanding precise. In terms of cloud terminology, service that is bargained or leased undergoes a strict security filtering and only after this procedural cycle the user is given access permission to user the cloud resource that is rented from the resource provider. Maintaining a balance between the standard of quality provided and utilized is an important aspect in quality exploration. In cloud resources a six step methodology for SLA life-cycle is

Fig. 2 SLA components in cloud utility

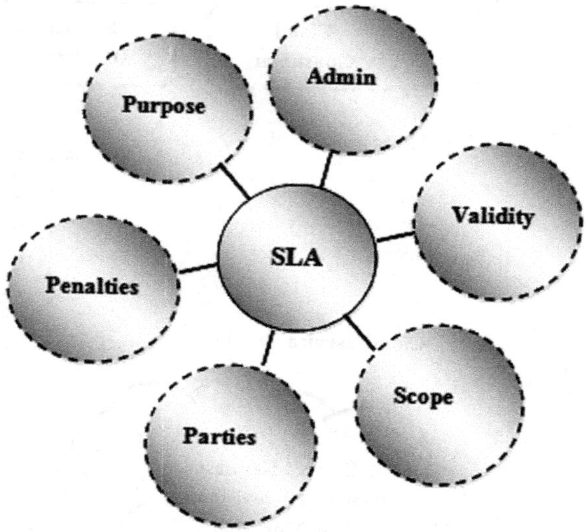

given by the cloud users to analyze the user's perspective in parametric view [5, 6]. A diagrammatic represent of parametric process is defines the important stages involved in the life-cycle. Once a resource is been leased, the provider takes at most possibility of analyzing the procurement of the resource state. Creating the procurement of resources is the initial phase in quality degree [7]. This should yield favorable results to the provider and henceforth gives an positive result of proceeding to the next phase of operating the cloud resources to its desired and right full users (availing the resources).

Figure 3, explains the phases and steps that cloud service resources procures in terms distributing the resources to its users. Once the resources are leased a batch sealing will be done and then the resource are accessed. Prior to this systematic approach, the phases are enhanced with order oriented steps. This involves finding the right resource provider. As per the user's needs and the resource momentum, refinement is made and then it is established made available for mainspring (monitor). All these are a kind of reframing and enhancing, to assure the user with the right resource and procurement.

3 SLA in Cloud Environment

The quality assurance treaty or the service level treaty is a legal manuscript which bears quality parameters which is necessary to be analysed by both the cloud resource user and the cloud service provider. This treaty contains the components that satisfy the user and the provider in terms of resource leasing and renting. Quality ascertainment needs to be supported in the negotiation activities for which

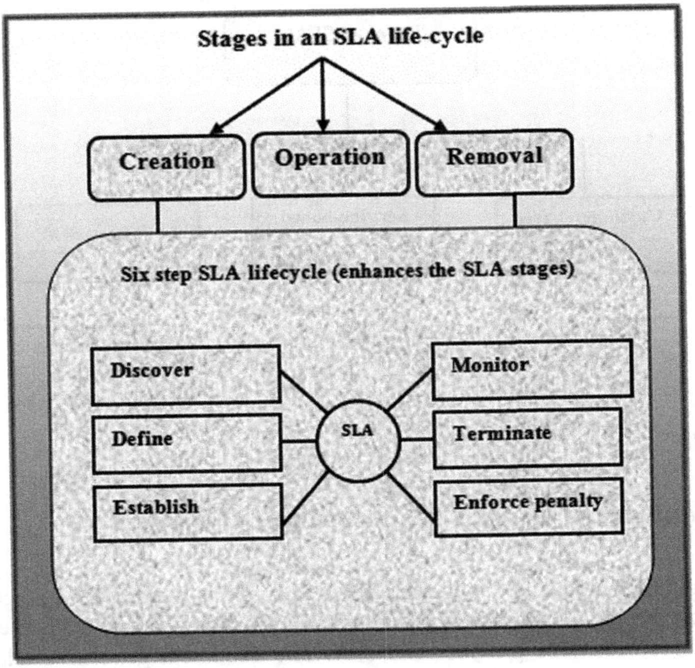

Fig. 3 A parametric quality view in SLA life-cycle

QoS parameters are considered to be part of the same legal treaty. This systematic procedural activity is performed by a cloud broker; he acts as a conduit between the resource provider and user [8, 9]. Monitoring the treaty violations, in resource provisioning is an important role with resource utilization. This increases the resource utilization in aspects of cloud users from the resource provider. Different types of SLA's that are bounded in cloud resources are:

3.1 Categories of SLA's in Cloud Environment

The resources hired from the cloud boundaries are given to the user only after a refined filtering based security approaches. Hence treaty based SLA is altered with five important orders of categories and implemented to its resource effort points with the same source of order for effective utilization of services and oriented approach towards customer beneficial satisfaction (Fig. 4).

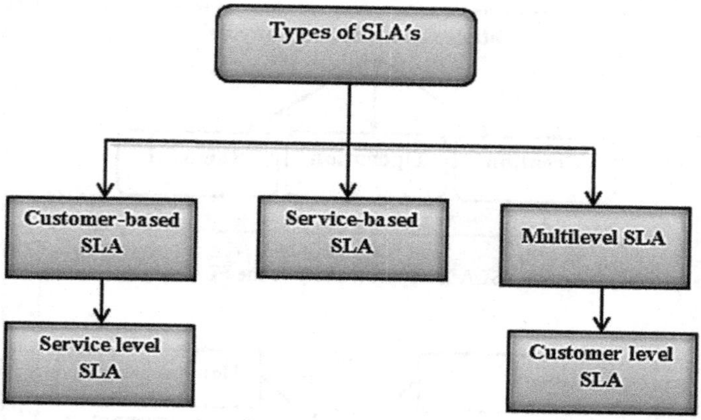

Fig. 4 Types of SLA's orders in cloud

3.2 Contents in an SLA Manuscript of Document

On a communal note the service treaty of SLA binds and stores collective data combined with the occurrence of cloud resources which is approved and enhanced between cloud resource provider and user. The document governs a relationship that can analyze the resources pooled from the cloud service provider to the source which it requires to be utilized. The document states the necessary content that need to be followed by the user at the time of leasing the service. This acts the same with the provider, as they are responsible for ensuring quality protocols to the user. Customer satisfaction is an important parameter in terms of cloud resource sharing and pooling. This template also monitors the usage from the user's perspective which is also referred as client oriented approach. On-demand access and service approaches in cloud are two important quality metrics that enhance the service level treaty to cloud users. Dynamically provisioning the resources from cloud to its users is a challenging task and predominant one. In such scenarios the parameter ascertained in the SLA document acts as a parametric board to evaluate the client requirements. Some of the significant parameters in the document are (Fig. 5):

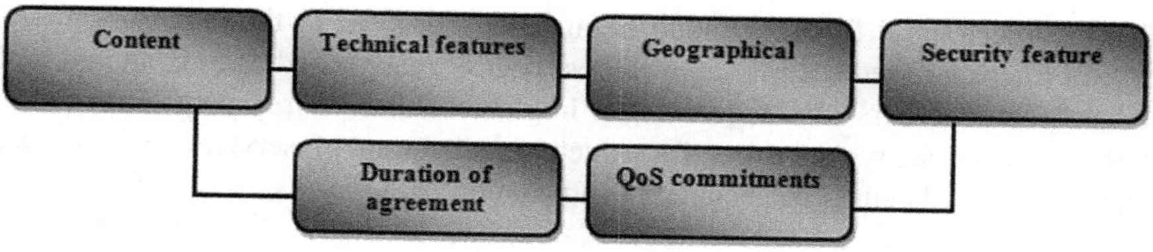

Fig. 5 Contents in the legal manuscript for cloud services

These parametric values, which are mentioned in the legal document of quality assurance, are responsible for maintaining a healthy relationship accompanied with quality measures between the cloud resource provider and its users [10–12].

4 An Architectural Approach for SLA Violation—DeSVi

In this segment we would learn all conceivable considerations of SLA negotiation process, its design and framework for service level treaty Violation. DeSVi—is a design for detecting service level treaty violations through resource monitored access (watch access) on CC (cloud computing) infrastructures. All the necessary resources that are allocated and organized by the DeSVi design component is deployed and analyzed through a virtualized environment (this purely works on the basis of user requirements) [13]. Low level resources are available for cloud and this is monitored by the service level treaty s and in core cases these resources are also defined by treaty s. With the existing service level objectives, violations on SLA are grounded to the knowledge based databases and this kind of implementation will reduce the rate of violations on cloud. These DB's of knowledge base are instigated with techniques like reasoning case based which is defined on the basis of system acquaintance. According to the design, end users or the customers will fall on the top most layer of the system and these users are liable for service provisioning. Emulations (AEF) are available which helps or lubricates pooling resources of cloud from virtual machines (VM's) or virtual environment. Application deployment is also done via these frameworks which measures the metrics on cloud environment [14, 15].

4.1 Importance of Standardizing SLA's in Multi-cloud Environment

Some of the network access to cloud is provided through same domain addresses and these resources can be accessed via service provider's acceptance. This practice is very common on terms of enterprise providers [16]. These providers deliver dedicated services on private cloud which works on all the three platforms of the cloud services. Mathematical combinations can be used to distinguish the relationships between the requesting source and the ancillary providers. Normally the requesting source is always the user of the resource or the customer who demands the service. Resources that are requested from the provider by the user involves a rigorous data transfer which often reflects into load balancing concepts and cloud acts as a substitute in transferring these high priority data. Service parameter that works with SLA's proactively manages the Quality of Service on multiple resource targets which maintains the lag offs on SLA management [17, 18]. The demanding

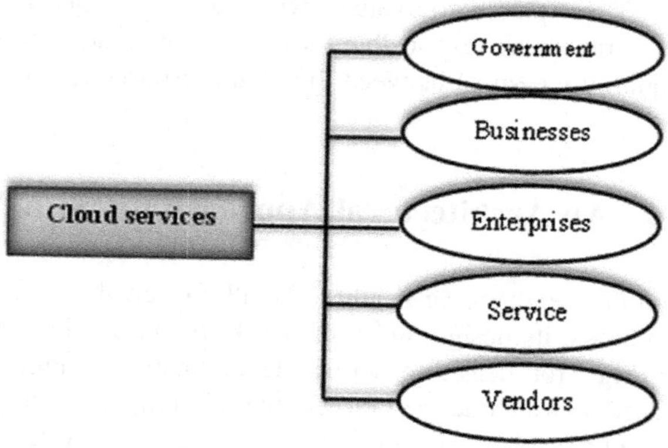

Fig. 6 Cloud services in different fields

service provider will always differ from the ancillary service provider or vendor. Their ailments are not standardized with SLA parameters or delineate in different ways to meet the performance. Service level treaty is a legal word of call on a typed format between the users (customers) and the service provider of the resources [18]. Value chain is also included in these documents for the services obtained in partnerships of enterprises. Some of the implied service sectors are [19] (Fig. 6).

5 Conclusion

In the fraternity of Cloud computing terminology, service level treaty is a legal call on method which provides terms and conditions between the user demands (Requisition) and service provider of the resources. This in turn maintains legitimate results on quality parameters of the resource that are provided to the customers. Negotiation polices becomes an essential tool for all the user of cloud service wherein prioritizing the objectives of service level treaty is vital (Service providers initiative). This paper emphases and discusses on technical strategies decision making heuristics for cloud resources. A step forward the study also focuses on major understanding of service level treaty and its stages of working with cloud resources.

References

1. Sheshasaayee, Ananthi, and Swetha Margaret TA. "A Structured Analysis on Lexicons For Open Source Cloud Technologies", International Journal of Engineering Sciences & Research Technology (2015).
2. Dash, S. B., et al. "Service Level Treaty Assurance in Cloud Computing: A Trust Issue." SB Dash et al./(IJCSIT) International Journal of Computer Science and Information Technologies 5.3 (2014).

3. Kertész, Attila, Gabor Kecskemeti, and Ivona Brandic. "An interoperable and self-adaptive approach for SLA-based service virtualization in heterogeneous Cloud environments." Future Generation Computer Systems 32 (2014): 54–68.
4. Kyriazis, Dimosthenis, et al. "A real-time service oriented infrastructure." Journal on Computing (JoC) 1.2 (2014).
5. Wu, Linlin, et al. "Automated SLA negotiation framework for cloud computing." Cluster, Cloud and Grid Computing (CCGrid), 2013 13th IEEE/ACM International Symposium on. IEEE, 2013.
6. Ariya, T. K., Christophor Paul, and S. Karthik. "Cloud Service Negotiation Techniques." computer 1.8 (2012).
7. Goudarzi, Hadi, Mohammad Ghasemazar, and Massoud Pedram. "Sla-based optimization of power and migration cost in cloud computing." Cluster, Cloud and Grid Computing (CCGrid), 2012 12th IEEE/ACM International Symposium on. IEEE, 2012.
8. Wu, Linlin, and Rajkumar Buyya. "Service level treaty (sla) in utility computing systems." IGI Global (2012).
9. Wu, Linlin, Saurabh Kumar Garg, and Rajkumar Buyya. "SLA-based admission control for a Software-as-a-Service provider in Cloud computing environments." Journal of Computer and System Sciences 78.5 (2012): 1280–1299.
10. Xiao, LI Qiao Zheng. "Research survey of cloud computing." Computer Science 4 (2011): 008.
11. Subashini, Subashini, and V. Kavitha. "A survey on security issues in service delivery models of cloud computing." Journal of network and computer applications 34.1 (2011): 1–11.
12. Zhang, Qi, Lu Cheng, and Raouf Boutaba. "Cloud computing: state-of-the-art and research challenges." Journal of internet services and applications 1.1 (2010): 7–18.
13. Zhang, Jian-Xun, Zhi-Min Gu, and Chao Zheng. "Survey of research progress on cloud computing." Appl Res Comput 27.2 (2010): 429–433.
14. Emeakaroha, Vincent C., et al. "DeSVi: an architecture for detecting SLA violations in cloud computing infrastructures." Proceedings of the 2nd International ICST conference on Cloud computing (CloudComp'10). 2010.
15. Dillon, Tharam, Chen Wu, and Elizabeth Chang. "Cloud computing: issues and challenges." Advanced Information Networking and Applications (AINA), 2010 24th IEEE International Conference on. IEEE, 2010.
16. Buyya, Rajkumar, et al. "Cloud computing and emerging IT platforms: Vision, hype, and reality for delivering computing as the 5th utility." Future Generation computer systems 25.6 (2009): 599–616.
17. Rimal, Bhaskar Prasad, Eunmi Choi, and Ian Lumb. "A taxonomy and survey of cloud computing systems." INC, IMS and IDC, 2009. NCM'09. Fifth International Joint Conference on. IEEE, 2009.
18. Buyya, Rajkumar, Chee Shin Yeo, and Srikumar Venugopal. "Market-oriented cloud computing: Vision, hype, and reality for delivering it services as computing utilities." High Performance Computing and Communications, 2008. HPCC'08. 10th IEEE International Conference on. IEEE, 2008.
19. Youseff, Lamia, Maria Butrico, and Dilma Da Silva. "Toward a unified ontology of cloud computing." Grid Computing Environments Workshop, 2008. GCE'08. IEEE, 2008.

A Descriptive Study on Resource Provisioning Approaches in Cloud Computing Environment

Ananthi Sheshasaayee and R. Megala

Abstract Cloud computing has become a promising technology in many organizations. A huge amount of applications is accessed through Cloud at anytime and anywhere. Hence provisioning the resources at the right time is a challenging task in Cloud computing environment. The Cloud consumers utilize resources using Virtual machines based on a "Pay-as-you-go" basis. For this, the two types of resource provisioning plans were offered by the Cloud providers, namely On-Demand plan and Reservation plan. In common, the reservation scheme has low cost than On-Demand plan. Cloud computing environment provides different types of resource provisioning approaches for minimizing total cost. The good resource provisioning approach should avoid disintegration of resources, lack of resources, disputation of resources, over provisioning and under provisioning. This paper mainly focuses on giving an overview of Cloud computing, resource provisioning and descriptive analysis of various resource provisioning algorithms and techniques.

Keywords Cloud computing · Resource provisioning · On-demand · Reservation plan · Virtual machine

1 Introduction

The most increasing technologies in the advanced world are Cloud computing, which delivering a Network form of utilities for permitting clients to access the broad range of services based on a 'Pay-per-use' basis [1]. In the Cloud, the enormous quantity of computational resources is spread over the data centers of

Ananthi Sheshasaayee (✉) · R. Megala
Department of Computer Science, Quaid E Millath Government College for Women, Chennai 600002, India
e-mail: ananthi.research@gmail.com

R. Megala
e-mail: megala.research@gmail.com

© Springer India 2016
S.C. Satapathy et al. (eds.), *Information Systems Design and Intelligent Applications*, Advances in Intelligent Systems and Computing 435,
DOI 10.1007/978-81-322-2757-1_57

Virtual, in which accessing has made through web browsers. The collection of computational resources is available on the Internet, accessing by the consumers depending upon their requirements. In Infrastructure-as-a-Service (Iaas), consumer desires the accessing Cloud resources using virtualization technologies [2]. The On-Demand plans and Reservation plans are fundamental plan for provisioning the resource which can be recommended by Cloud Service Provider (CSP). The plan of On-Demanding is around one year plan and is charged as 'Pay-What-You-Use' method. The plan Of Reservation is to pay the cost as a one time and this resource will be used by the consumer and is roughly a 3 year plan. By comparing these two plans cost wise, the Reservation plan is low cost compared to the On-Demand plan [3]. Thus the main aim is to reduce the amount for clients who leasing the resources from the Cloud Provider and the provider's perspective to capitalize on profit by efficiently allocating the resources [4].

The paper contains the following parts: Sect. 2 discusses about cloud computing overview. Section 3 discusses resource provisioning. Section 4 discusses various approaches for resource provisioning. Section 5 gives conclusions and future workplace.

2 An Overview of Cloud Computing

According to the Bohn et al. [5], Cloud utility computing is regarded as a framework, ever-present, suitable and provisioning the resources with minimum cost for Client and maximizing the profit for Providers [6]. Cloud computing has an elastic feature where scale-out and scale-in resource provisioning can be implemented rapidly in order to attain resources available, and it can be obtained at any time by the Cloud subscriber.

2.1 Cloud Deployment Models

Figure 1 shows the three types of Deployment model of Cloud [1].

2.1.1 Private Cloud

The Private Cloud is a utility for a collection of people in an organization, which is largely employed in the corporate sectors where a Cloud infrastructure is preserved and it used for the inner functioning of the organization also accessing has made by the people with their access privileges, the Private Cloud is measured to be safer than Public Cloud because Private Cloud is accessed inside and without access rights, access from external is not provided.

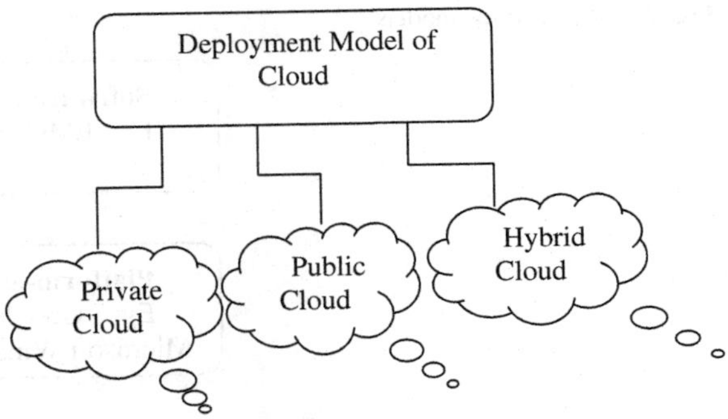

Fig. 1 Deployment model of cloud

2.1.2 Public Cloud

Public cloud is obtainable by all the members who required resources as a service below a convention depend on a pay-as-use model. The most benefit of Public Cloud is it to be used by any person who was necessitated in a scalable model.

2.1.3 Hybrid Cloud

The Hybrid Cloud is growing Popular and reason of its popularity is the exercise of individual and Public Cloud infrastructure as needed. The means of Hybrid Cloud is Combination of both Private and Public Cloud which is more efficient of cost and benefit.

2.2 Services of Cloud

The services of Cloud are classified as follows [1] (Fig. 2).

3 Resource Provisioning

Resource provisioning is a procedure for processing consumers work and warehousing the data by supplying required resources. Service Provider can supply two types of resource provisioning schemes, specifically On-demand and reservation plan. The two problems will occur in resource provisioning namely under provisioning and over provisioning. These problems are overcome by various algorithms [2].

Fig. 2 Cloud service models

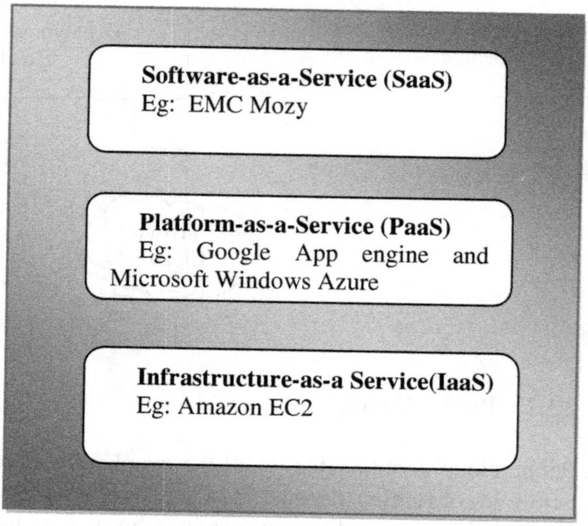

3.1 Types of Resource Provisioning

The resource provisioning can be sorted as follows [4].

(a) *Static Provisioning* In the Static provisioning, the applications have convened and usually fixed demands/workloads. By an advance provisioning, the customer contracts with the provider for services and the provider arranges the suitable resources prior to the commencement of service. The customer is charged a flat fee or is payable on a monthly basis.

(b) *Dynamic Provisioning* In the dynamic provisioning, the demand by applications may change or vary; this technique has been recommended whereby Virtual Machines may be transferred spontaneously to compute nodes within the Cloud. By the dynamic provisioning, the provider allots many resources as they are preferred and eliminates unused resources. The customer is payable on a "pay-per-use basis". When dynamic provisioning is used to construct a hybrid Cloud, it is meant that "Cloud bursting".

(c) *User Self-provisioning* In the User Self provisioning (also called as "Cloud self-service"), the customer buys resources from the Cloud provider through Web services, developing a client invoice and paying for resources with a recognition card. The provider's resources are used for customer uses within hours, if not minutes.

4 Various Approaches for Resource Provisioning

The following Table 1 shows the various existing approaches of resource provisioning.

Table 1 The various approaches of resource provisioning

S. no	Survey	Techniques	Description	Limitations
1	Chen et al. [7, 8]	An approach based on modeling	It is efficient for web content	This attack was not desirable to every cloud consumer
2	Kusic et al. [7, 9]	Guided redundant submission (GRS) technique	Earthly resource availability can be secured by chances of joint failure	It doesn't handle techniques for cutting the cost resource
3	Kusic et al. [7, 10]	LLC dynamic approach	It is a virtual choice for implementing self-organizing actions of provisioning technologies	No functionality for reducing the operating cost and running time
4	Peng-yeng yin et al. [7, 11]	Algorithm for particle swarm optimization	It is efficient for finding the solution of the optimization task with low time consistent	It is applicable for large scale system and also fast pace
5	Juve et al. [7, 12]	Multi-level scheduling technique	To keep off or minimize job delays and facilitates resources by client-level managers	It doesn't provide an optimal result of the whole resource provisioning problem. And doesn't meet on security issues
6	Chaisiri et al. [3, 7]	Algorithm for optimal virtual machine placement (OVMP)	It is accompanied with prior reservation and allocate virtual machines to other utility services	Integer programming is the only way to get the optimal solution for this problem
7	Chaisiri et al. [3, 13]	Server provisioning algorithm for virtualization	The cost of the on-demand and reservation plan has to be reduced. The finest number of held in reserve Virtual servers and the fine amount of tender server-hours also be figured out	Valuable chances for proffering spot cases, low price and demand variations
8	Dawoud et al. [13, 14]	Security model for IaaS	This model used to develop security in every layer of the delivery model of IaaS	To develop privacy and reliability of virtual machines and to take more control isolation environment of IaaS
9	Tsakalozos et al. [13, 15].	Microeconomics inspired approach	This system monitors financial capability of the user depending upon the	The time has to be reduced to reach the highest profit position

(continued)

Table 1 (continued)

S. no	Survey	Techniques	Description	Limitations
			number of virtual machines allowed to unique user and also monitors the amount of resources used and processing time of user applications	
10	Zou et al. [13, 16]	Genetic algorithm	It has been planned and staked out to compute virtual machine to other head and the resources allocated to every virtual machine, thus as to optimize resource consumptions	When the node increases the optimized state of the virtual machines doesn't provided by this system
11	Zou et al. [16]	Cloud reconfiguration algorithm	It is made to transfer from the current state of the cloud to the better one processed by the genetic algorithm	After the reconfiguration of the system, the allocated resources to the virtual machines in the node would go over the physical resource capacity and switching time causes more expenses
12	Sharma et al. [17]	Algorithm for dynamic provisioning	It requires 2 ways: (i) use the algorithm for resource allocation, and (ii) check the way for reducing the infrastructure cost using this provision algorithm	It is depending upon the queuing theory model for provisioning capability evaluation in the cloud
13	Chaisiri et al. [3, 7]	Algorithm for optimal cloud resource provisioning (OCRP)	This algorithm helps to reduce the charge of the resource allocating within the specific time period	Applicable only for SaaS users and Saas providers
14	Meera et al. [18]	Robust cloud resource provisioning (RCRP) algorithm	The main objectives of this algorithm are to carry down the optimization problems into several minor problems which can be figured out independently and comparatively	It depends on only reservation plan
15	Subramanian et al. [19]	Federated architecture	The resource provisioning using federated cloud is the operation of determining best position plans for virtual machines and reconstruct the virtual machines according to others in the surroundings	It doesn't satisfy personalized SLA based on application requirements

5 Conclusion and Future Work

Resource provisioning has become a challenging duty in the Cloud utility computing. This analysis paper discussed about dynamic provisioning algorithms, Cloud deployment based resource provisioning techniques and optimized resource provisioning algorithms. In this paper, the Table 1 described each and every technique and their virtues and faults. These techniques are used to improve reaction time, minimize cost, maximize profit, public display and save energy. The major involvement of this detailed study is to analyze and given the best algorithms to minimize the amount for provisioning the resource in the Cloud utility system. This will helpful to the researchers to explore their new minds to solve resource provisioning problems. The future workplace will be developing an algorithm which satisfies the low price and high profit with low energy.

References

1. Girase, Sagar, et al. "Review on: Resource Provisioning in Cloud Computing Environment." *International Journal of Science and Research (IJSR)* 2.11 (2013).
2. Menaga. G and S. Subasree, *"Development of Optimized Resource Provisioning On-Demand Security Architecture for Secured Storage Services in Cloud Computing", International Journal Engineering Science and Innovative Technology (IJESIT)*, 2013.
3. Chaisiri, Sivadon, Bu-Sung Lee, and Dusit Niyato. *"Optimization of resource provisioning cost in cloud computing."* Services Computing, IEEE Transactions on 5.2 (2012): 164–177.
4. Nagesh, Bhavani B. "Resource Provisioning Techniques in Cloud Computing Environment-A Survey." *IJRCCT* 3.3 (2014): 395–401.
5. Bohn, Robert B., et al. "NIST cloud computing reference architecture."*Services (SERVICES), 2011 IEEE World Congress on*. IEEE, 2011.
6. Armbrust, Michael, et al. "M.: Above the clouds: a Berkeley view of cloud computing." (2009).
7. Karthi, M., and S. Nachiyappan. "Survey on Resources Provisioning in Cloud Systems for Cost Benefits".
8. Chen, Jin, Gokul Soundararajan, and Cristiana Amza. "Autonomic provisioning of backend databases in dynamic content web servers." *Autonomic Computing, 2006. ICAC'06. IEEE International Conference on*. IEEE, 2006.
9. Kusic, Dara, and Nagarajan Kandasamy. "Risk-aware limited lookahead control for dynamic resource provisioning in enterprise computing systems." *Cluster Computing* 10.4 (2007): 395–408.
10. Yin, Peng-Yeng, et al. "A hybrid particle swarm optimization algorithm for optimal task assignment in distributed systems." *Computer Standards & Interfaces* 28.4 (2006): 441–450.
11. Juve, Gideon, and Ewa Deelman. "Resource provisioning options for large-scale scientific workflows." *eScience, 2008. eScience'08. IEEE Fourth International Conference on*. IEEE, 2008.
12. Chaisiri, Sivadon, Bu-Sung Lee, and Dusit Niyato. "Optimal virtual machine placement across multiple cloud providers." *Services Computing Conference, 2009. APSCC 2009. IEEE Asia-Pacific*. IEEE, 2009.

13. M. Uthaya Banu et al, "A Survey on Resource Provisioning in Cloud", *Int. Journal of Engineering Research and Applications, Vol. 4, Issue 2 (Version 5), February 2014, pp. 30–35.*
14. Dawoud, Wesam, Ibrahim Takouna, and Christoph Meinel. "Infrastructure as a service security: Challenges and solutions." *Informatics and Systems (INFOS), 2010 The 7th International Conference on.* IEEE, 2010.
15. K. Tsakalozos, H. Kllapi, E. Sitaridi, M. Roussopoulous, D. Paparas, and A. Delis, "Flexible Use of Cloud Resources through Profit Maximization and Price Discrimination", *in Proc of the 27th IEEE International Conference on Data Engineering (ICDE 2011)*, April 2011, pp. 75–86.
16. He, Ligang, et al. "Optimizing resource consumptions in clouds." *Grid Computing (GRID), 2011 12th IEEE/ACM International Conference on.* IEEE, 2011.
17. Sharma, Upendra, et al. "A cost-aware elasticity provisioning system for the cloud." *Distributed Computing Systems (ICDCS), 2011 31st International Conference on.* IEEE, 2011.
18. Meera et al.," Effective Management of Resource Provisioning Cost in Cloud Computing ", *International Journal of Advanced Research in Computer Science and Software Engineering 3 (3)*, March-2013, pp. 75–78.
19. Subramanian, Thiruselvan, and Nickolas Savarimuthu. "A Study on Optimized Resource Provisioning in Federated Cloud." *arXiv preprint* arXiv:1503.03579 (2015).

Analyzing the Performance of a Software and IT Growth with Special Reference to India

Molla Ramizur Rahman

Abstract Effective management is essential in the software industries to produce quality software. The use of software has enabled software industries to manage effectively with reduced human effort. This paper aims to investigate the scenario of software utility in the Indian industry through a case study. This investigation has enabled to prepare a basic flowchart model for on-demand software. The paper also aims to study the number of errors made by a programmer for different projects having different number of program lines. It also identifies various parameters to rate and analyze the performance of software.

Keywords Software defects · Software performance · Software scenario · Software classification

1 Introduction

Software is an essential tool used in almost every industries of twenty-first century. Due to the advancement of technology and to match with the fast moving world, software has gained its importance. The use of software for commercial purpose started when the first business machine which was developed by IBM. At this early period of time, the business machine or the computer was limited to certain high class industries. But as the time rolled on affordability of computer came within the reach of less capital industries and slowly in the hands of personals. This led to an increase in development of various types of software. This is very evident from the rise of various information technological companies. Initially computers were used only in developed countries and so most of the software developing companies were limited to such nations but presently after the rapid growth in use in the

M.R. Rahman (✉)
Prin. L. N. Welingkar Institute of Management Development and Research, Mumbai, India
e-mail: ramizurscience@yahoo.com

© Springer India 2016
S.C. Satapathy et al. (eds.), *Information Systems Design and Intelligent Applications*, Advances in Intelligent Systems and Computing 435,
DOI 10.1007/978-81-322-2757-1_58

developing countries presently, such software companies have become a huge source of employment.

Software is generally developed to replace the existing manual work or to simplify the work which was initially performed by some existing software. The author [1] in his work gave an enhanced approach for computing Least Common Multiple (LCM) which can be efficiently used for programming microcontrollers. Software is also used for research purposes to carry out different types of analysis. Author [2] in his paper developed a code in MATLAB software to analyze the concept of friction. Various software companies develop a wide variety of software depending on the requirements of their clients. Clients generally may be any government or private entity which traditionally made use of manual practices. After globalization these companies used software for their day to day operations. Rationale to embed software in their operational field may be attributed to achieve customer satisfaction, to capture fast growing markets, to provide a better service, to maintain a better internal management or to maintain their reputation.

The above mentioned reasons made software companies to develop software and vested a huge responsibility in doing so. Hence, software can be visualized as common and on-demand software as shown in Fig. 1. Software development can be for use by some specific sector or common use, which may be available in the market on payment or sometimes free and follows one to many matrix as shown in Fig. 2. Examples of such common software available in the market are generally gaming software, graphical software etc. which are widely used and much needed in daily applications in society. This type of software is generally developed by analyzing the market needs with a proper procedural market survey. Other type of software may be on-demand software which is developed by the software vendor, when a client approaches the software company to develop software products according to their specifications as required for their work. This software is not available in the market as in case of the aforementioned software and is customized built only for a single specific client which follows one to one matrix as shown in Fig. 3. For example, xyz bank which is client here requires on-demand software which is developed by a software unit called as vendor.

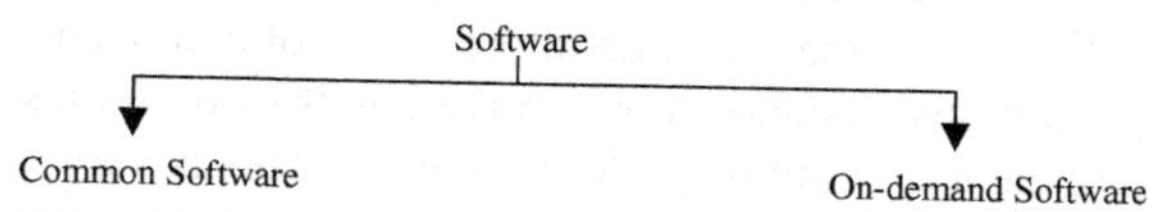

Fig. 1 Shows the classification of software

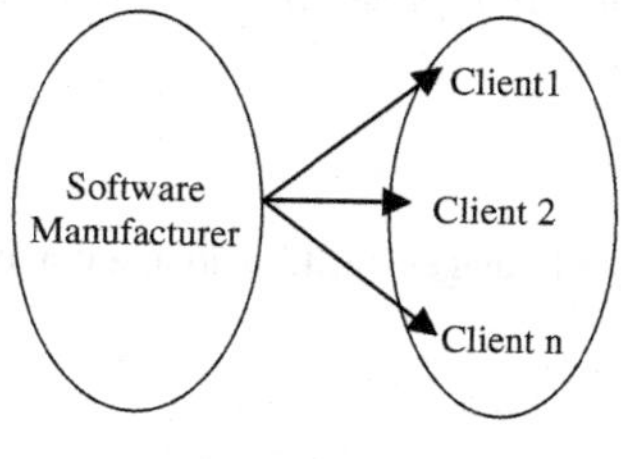

Fig. 2 Shows the matrix model for common software

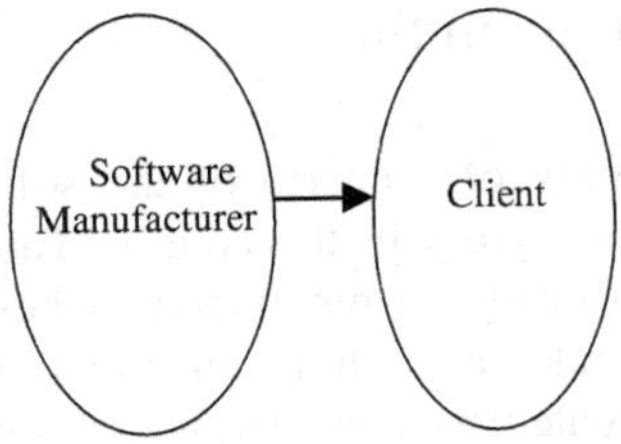

Fig. 3 Shows the matrix model for on-demand software

Thus, software developing centers must provide software which is free of defects. Thus, various strategies are followed to develop qualitative software which undergoes different stages in order to make their products free from defect. Defect management plays an important role in this process having various defect management techniques.

2 Literature Survey

Authors of [3] put forth a case study on defect patterns after appropriate investigations and provided observational inference for developing quality software. Authors in [4] investigated the post production defects in order to help the software developing team for the reduction of defects during the process of software development. Further, authors of [5] in their work provide post production defect patterns and the probability of such defect occurrence with complexity and effects of development hours.

The authors of [6] explained use of different software development models for various purposes and compared the models over certain important parameters. The authors [7] identified factors causing defects which when focused will enable us to reduce the magnitude of such defects leading to quality software. The authors [8] through a case study described defect management techniques and highlighted pre-production defects and its effects on making a project successful.

In paper [9], authors have put forward one pass data mining algorithm which is efficient for software defect prediction over the parameters such as time and performance. The authors [10] in their paper made a detailed analysis over genetic fuzzy data mining methods where they presented few methods along with their applications with respect to software defect analysis. Challenges with solutions over such methods are also mentioned.

In paper [11], the work focuses software defects from five projects and classified such defects, understanding the cause for such defects and developed preventive ideas which can be implemented in such projects to reduce defects.

3 Software Scenario in India

This section provides a study of scenario of the software industry involved in various important industrial sectors in the county. Table 1 indicate the usage of software in different departments as mentioned in below sectors and the utility performance of software is rated as high, average and low. For each of the sectors, utility performance is evaluated by considering the software usage during input, processing and output phases. Table 2 indicates one more analysis where programming errors are considered with various programming languages.

From the table it is clear that overall utility performance of software in industry is average. High as well as low software utility sectors are also observed.

The above sample as shown in Table 2 is for five different projects using and not using object oriented programming concept. It is clear that errors made by the programmer during the first attempt of programming are a function of program lines and increases with the increase in program lines. It is also evident that errors made is more in the case of programming language which use object concept compared to non-object concept for the same program lines.

Table 1 Depicts the utility performance of software in various sectors in India

Sector	High	Average	Low
Railways	◆		
Healthcare		◆	
Electrical		◆	
Academics		◆	
IT	◆		
Agro industry			◆
Financial services		◆	
Marketing			◆

Table 2 Depicts the errors performed by a programmer for different projects with different program lines with and without OOP's concept

Project	Program lines	No. of errors in the first attempt of programming using OOPs concept	No. of errors in the first attempt of programming not using OOPs concept
P1	15	1	0
P2	25	2	1
P3	40	3	1
P4	65	5	2
P5	80	6	3

4 Parameters to Determine the Performance of a Software

(1) Client's Satisfaction: It is one of the major parameter to determine the performance of software. To obtain this, all the specification as given by the client to the company must be fulfilled. Hence, a proper understanding of the industry for which the software is prepared is required and thus client interaction clarifies the specifications. A proper initial plan must be prepared and the functions to be carried out during the developmental process of the software are to be noted. A proper groundwork will definitely make the possibility to reduce the errors in the first attempt of programming. Study should also be made on the other industries of similar type. Client's satisfaction can be obtained by a feedback from the client. Software developer should develop effective software than already existing one which is either used by the client or by a similar industry of the client's type.

(2) Time Consumption: Time is one of the major factors which determine not only the rating of the software developed but also the reputation of the software industry. There should be required resources available to produce the software on time. Before taking the software order in hand, a proper time duration required in order to develop the software should be given to the client. Internal time limits should be allocated to various departments involved in development process of the software. Parallel work is recommended wherever possible to save time. A less time for providing quality software is always desired and hence increases the rating of the software.

(3) Hiding of Codes: Codes must be hidden in such a way that it remains inaccessible by any outsiders. Codes basically are the program parts which are generally hidden. It should be well protected. Minimum usage of code to produce quality software is always entertained. Codes should be flexible enough so that future change if required by client is easily done.

(4) Cost: Cost is one of the primary factors to determine the rating of the software. Software with minimum cost is always desirable.

(5) User-Friendly: The software must be user-friendly so that with adequate training to the client, the software becomes easy to operate. The software company must also ensure that the software should not undergo any complexity during its operations. It should be such that an operator with little or no knowledge of programming can operate.

5 Flowchart for a General On-Demand Software

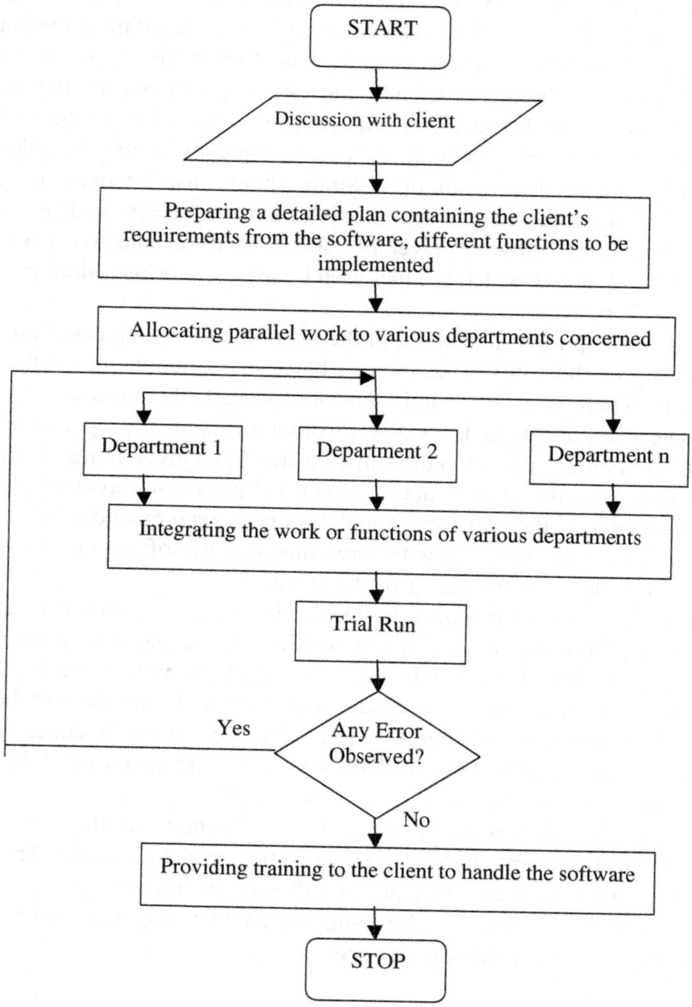

Above illustrated flowchart which is formulated based on the study made in the software market, can be followed to develop on-demand software. It initially starts with discussion with the client regarding the software specifications. Then the next stage is the most important stage which includes framing a detailed plan about how to develop in work and proceed. It may contain various departments which would get involved, assigning work to various such departments, framing time limits to such departments, explaining them about their various functions. According to the plan, allocations of the work to various departments are made. Upon the completion of work within specified time, integration work of the various functions is performed and complete software is developed. A trial run of the software is made and any error or defects if identified is send to the department concerned for rectification. Once the trial run becomes successful, the software is handed over after providing training to the client.

6 Results

The case study is performed on different industrial sectors to understand the scenario of the software industry. Investigation also formulated several parameters to be kept in mind to analyze the performance of software. The study also analyzed the number of errors made during the first attempt of programming with and without the concept of object for five different projects. After a detailed study, a flowchart is drawn which includes development steps to provide an on-demand software.

References

1. Molla Ramizur Rahman, "Enhanced Approach of Computing Least Common Multiple (LCM)" Dayananda Sagar International Journal (DSIJ), ISSN: 2347-1603, Vol. 1, Issue 1, pp. 65–69.
2. Molla Ramizur Rahman, "Software and Graphical Approach for Understanding Friction on a Body", Proceedings of the 3rd International Conference on Frontiers of Intelligent Computing: Theory and Applications (FICTA) 2014, ISSN:2194-5357, Vol. 2, pp. 637–644.
3. Bhagavant Deshpande, Jawahar J. Rao and V. Suma, "Comprehension of Defect Pattern at Code Construction Phase during Software Development Process", Proceedings of the 3rd International Conference on Frontiers of Intelligent Computing: Theory and Applications (FICTA) 2014, ISSN: 2194-5357, Vol. 2, pp. 659–666.
4. Divakar Harekal and Suma V, "Implication of Post Production Defects in Software Industries", International Journal of Computer Applications, ISSN:0975 – 8887, Vol. 109 – No. 17, January 2015, pp. 20–23.
5. Divakar Harekal, Jawahar J. Rao and V. Suma, "Pattern Analysis of Post Production Defects in Software Industry", Proceedings of the 3rd International Conference on Frontiers of Intelligent Computing: Theory and Applications (FICTA) 2014, ISSN: 2194-5357, Vol. 2, pp. 667–671.
6. Shikha Maheshwari and Dinesh Ch. Jain "A Comparative Analysis of Different types of Models in Software Development Life Cycle", International Journal of Advanced Research in Computer Science and Software Engineering, ISSN: 2277 128X, Vol. 2, Issue 5, May 2012, pp. 285–290.
7. Mrinal Singh Rawat and Sanjay Kumar Dubey, "Software Defect Prediction Models for Quality Improvement: A Literature Study", International Journal of Computer Science Issues, ISSN: 1694-0814, Vol. 9, Issue 5, No 2, September 2012, pp. 288–296.
8. Bhagavant Deshpande and Suma V, "Significance of Effective Defect Management Strategies to Reduce Pre Production Defects", International Journal of Innovative Research in Computer

and Communication Engineering, ISSN: 2320-9801, Vol. 2, Special Issue 5, October 2014, pp. 124–129.
9. K.B.S Sastry, B.V. Subba Rao and K.V. Sambasiva Rao, "Software Defect Prediction from Historical Data", International Journal of Advanced Research in Computer Science and Software Engineering, ISSN: 2277 128X, Vol. 3, Issue 8, August 2013, pp. 370–373.
10. V. Ramaswamy, T. P. Pushphavathi and V. Suma, "Position Paper: Defect Prediction Approaches for Software Projects Using Genetic Fuzzy Data Mining", ICT and Critical Infrastructure: Proceedings of the 48th Annual Convention of Computer Society of India- Vol. II, ISBN: 978-3-319-03094-4, pp. 313–320.
11. Sakthi Kumaresh and R Baskaran, "Defect Analysis and Prevention for Software Process Quality Improvement", International Journal of Computer Applications, ISSN: 0975 – 8887, Vol. 8, No. 7, October 2010, pp. 42–47.

Stock Price Forecasting Using ANN Method

Thangjam Ravichandra and Chintureena Thingom

Abstract Ability to predict stock price direction accurately is essential for investors to maximize their wealth. Neural networks, as a highly effective data mining method, have been used in many different complex pattern recognition problems including stock market prediction. But the ongoing way of using neural networks for a dynamic and volatile behavior of stock markets has not resulted in more efficient and correct values. In this research paper, we propose methods to provide more accurately by hidden layer data processing and decision tree methods for stock market prediction for the case of volatile markets. We also compare and determine our proposed method against three layer feed forward neural network for the accuracy of market direction. From the analysis, we prove that with our way of application of neural networks, the accuracy of prediction is improved.

Keywords Stock price · Index · Investor · Neural networks · Data mining · Stock market prediction · Decision tree · Volatile market

1 Introduction

Financial market is the most complex and difficult concept to understand the trend and the pattern of the equity share price fluctuation in the system. There can be various factors and reason which can lay foundation to understand the stock price

T. Ravichandra (✉)
Department of Professional Studies, Christ University, Bangalore, India
e-mail: ravichandran.singh@christuniversity.in

C. Thingom
Centre for Digital Innovation, Christ University, Bangalore, India
e-mail: chintureena.thingom@christuniversity.in

© Springer India 2016
S.C. Satapathy et al. (eds.), *Information Systems Design and Intelligent Applications*, Advances in Intelligent Systems and Computing 435,
DOI 10.1007/978-81-322-2757-1_59

fluctuation. Therefore, predictions of stock market price and its direction are quite a difficult task to perform. To explain with some proof, various data mining methods have been tried for prediction of the fluctuation in the financial market. Most of the research studies have been found focusing on the technique to forecast the possibility of the stock price accurately. However, there are various type of investors which are based on risk appetite and as such they adopt different trading strategies; hence, the model based on minimizing the standard error between the input and output variable may not be suitable for them. Moreover, accurate prediction of stock index is crucial for investor to make impactful market trading strategies to maximize the financial wealth of the investor. To put forward few important and vital factors that cause the fluctuation of financial index/market directions are as the following: socio economics status, political dilemmas, expectation of buyer's, expectation from the supply side other dynamic unexpected events. For prediction problem, the neural network can be used as there are many special features. The statistical linear regression is a model (test) based but, the technique of neutral networks is a system which have a self-adjusting techniques or methods which are based on experiment data, and it can solve the issue by itself. Also, the neural networks can be used for many difficult and complex situations in which determining the data dependence is really difficult. The next vital point is that in Neural Networks, generalization and conclusive ability are well equipped which means that after the training work, they can recognize the unknown new patterns even if they are not included in training set. Often the recognition or detection of pattern and prediction of future events are based on the historical data set (training set), the application of networks would be very helpful. Third point indeed is that the neutral networks have been also included as function approximations. It is also said that MLP neural network can approximate any multi stage continuous function that enables us to identify and learn any dependence relationship between the output and input variable of the system. Also for volatile stock market, volatility and fluctuation is less compared to stable market. In this work, we propose a new mechanism in processing of data and using the neutral with decision trees to get more approximation of stock prices.

2 Literature Survey

In the year 1964, Mr. Hu expressed that the idea of using neural networks for predicting fluctuation problem was first was used for weather forecasting [1]. The possibility of not having any method for identifying and learning for multi-layer networks made it highly impossible to determine and predict complex problems but during the decade of 80s there was an introduction of back propagation algorithm which was used for training an MPL neutral network. Werbos [2] have conducted a study in 1988 about this technique which concluded that the method requires to train a neural network and also proves that proposed neural networks are effective than regression methods. In the recent years, many researchers have been trying to conduct research experiments to predict the stock market changes and fluctuation. Mr. Kimoto was one of the first efforts made in which they used neutral networks to predict the market index of Tokyo stock market [3]. The participants who have done research on neutral network application continued to show up that all participants were winners of the prediction contest in Santafa institute who have all neutral networks [4]. Mizuno and his co-authors have performed the experimental research with various neutral networks to forecast the equity share price fluctuation of many stocks in Tokyo stock market. The research outcome was able to predict with 63 % precision [5].

3 Problem Definition

Market price of the equity shares is very difficult to predict and it is based on historical prices of stock but it is an important tool for short term investors for achieving maximum profits. MLP neural networks have been the existing method for price predictions. However, MLP Model does not always give accurate prediction in case of volatile markets. In our research work, we are introducing a new way of collaborating both neural networks and decision tree to forecast the stock market price more accurately than MLP.

4 Research Work

Our current method is based on application of both neural network and decision tress for prediction of stock market price of a stock based on its previous day prices. The architecture of our solution is given below (Fig. 1).

The stock market price for last 1 month is taken as input. But the data in the form of day versus price is not suitable for processing for neural networks. So we take the previous 4 day's stock market value as input and the fifth day equity share price as output and create the hypothesis testing datasets. The neural network is the three feed forward neural network with 4 inputs and 1 output. The 4 days previous price is given as input and fifth day price is taken as output and the training is done on the neural networks. Like previous works we are not focused on the output neurons in our work, we are more focused on the hidden layer neurons in our work. Based on

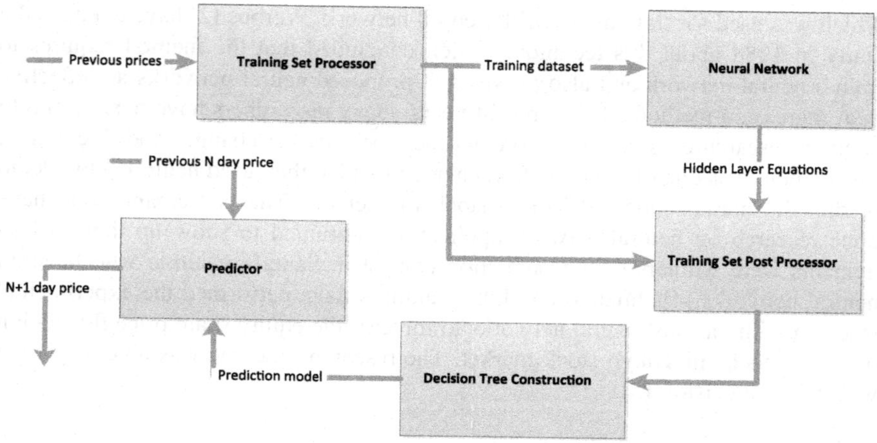

Fig. 1 Architecture for the proposed system

Fig. 2 Structural workflow of 2 input and 2 hidden layer

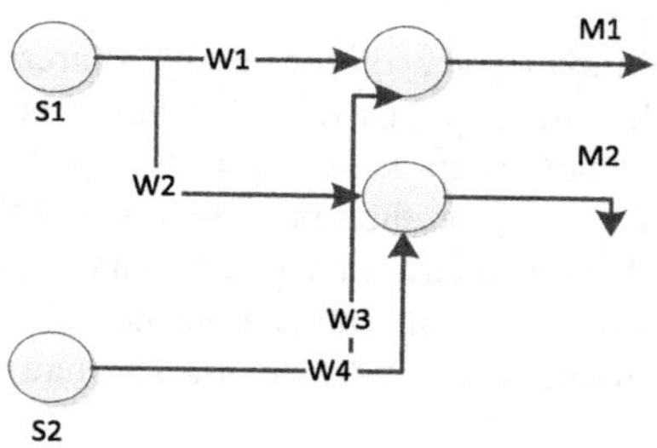

the weight between the input layer to middle layer neurons hidden layer equations are constructed. An example is given below for 2 input and 2 hidden layer neuron (Fig. 2).

$$M1 = F(S1 * W1 + S3 * W3)$$
$$M2 = F(S1 * W2 + S2 * W4)$$

Using the hidden layer equations the training data set is processed from the form of S1, S2 to M1, M2. By applying this processing on the training dataset linear and nonlinear relations in the training dataset can be represented. From this processed training dataset, C-4.5 decision tree algorithm is invoked to learn the decision tree model for prediction. The decision tree model will be able to predict any fifth day price by using previous four day prices.

5 Performance Analysis

Parameters	Values
No. of stocks used for evaluation	3
Training dataset	30 consecutive days stocks price of 3 stocks
No. of neural network	3 neural network
No. of layers in neural network	3 layers
No. of neurons in input layers	4 neurons
No. of neurons in middle layer	9 neurons
No. of neurons in output layer	1 neuron
Neural training error	0.001
Learning method	Back propagation neural network

With these parameters, we measured the accuracy of prediction of 5 stocks for next 10 days and plotted the error in accuracy in prediction for those 10 days. We compared our proposed algorithm MLP neural network of 3 layers (Figs. 3, 4 and 5).

From the results we see that the prediction error is less in case of our proposed solution compared to 3 layer neural network.

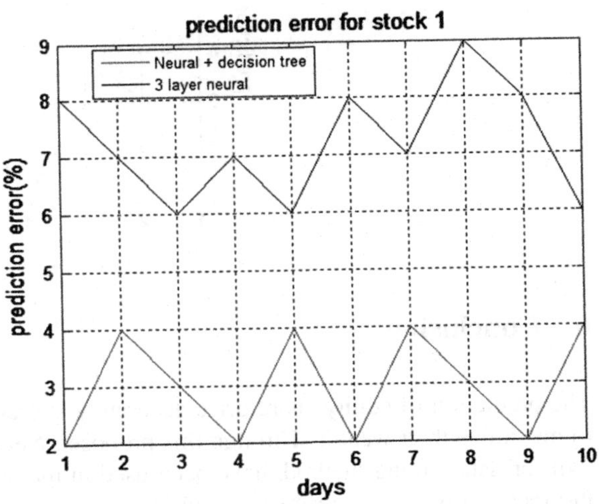

Fig. 3 Prediction error for stock 1

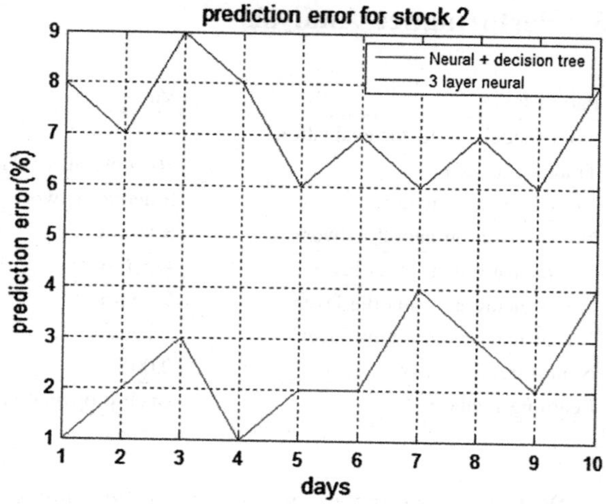

Fig. 4 Prediction error for stock 2

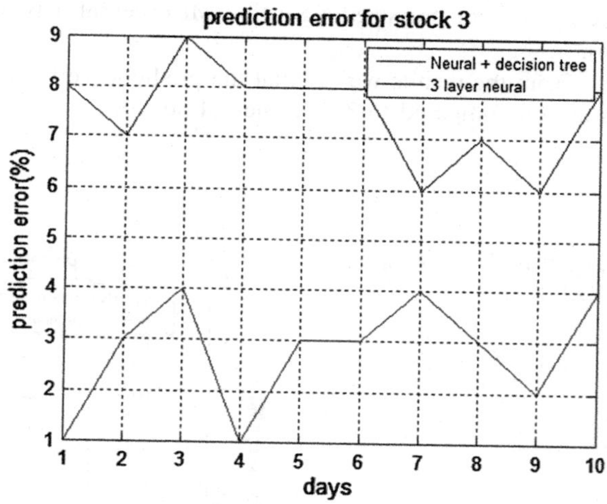

Fig. 5 Prediction error for stock 3

6 Conclusion

The prediction of equity share price accurately is crucial for the financial investors to maximize their wealth. Through this proposed Neural Networks as an intelligent tool for data mining method, have been used in many different complex pattern for the recognition of the problems such as stock market prediction. In our research work, we are introducing this new method for combining neural network and decision trees to forecast the stock market price of the volatile stocks and to provide outcome with reliability and accuracy through the support of both neural networks and decision tree than MLP. With the extensive experimental evaluation, we have proved that our proposed method is able to predict the stock prices with very good accuracy.

References

1. Hu, J. C. *Application of the Adaline System to Weather Forecasting*. Stanford, CA. (June 1964).
2. Werbos, P. J. *Neural Networks, Vol 1*, Generalization of back propagation with application to a recurrent gas market model. 339–356. (1988).
3. T Kimoto, K. A. *International Joint Conference on Neural Networks*, Stock market prediction system. (pp. 01–06). (1990).
4. A S Weigend, N. G. Time Series Prediction: Forecasting the Future and Understanding the Past. *Addison – Wesley*. (1993).
5. H Mizuno, M. K. Application of Neural Network to Technical Analysis of Stock Market Prediction. *Vol 7, Issue No. 3, Studies in Informatics and Control*, 111–120. (1998).

Recognition of Handwritten English Text Using Energy Minimisation

Kanchan Keisham and Sunanda Dixit

Abstract In handwritten character recognition one of the most challenging task is segmentation. This is mainly due to different challenges like skewness of textlines, overlapping characters, connected components etc. This paper proposes a character recognition method of handwritten English documents. The textlines are segmented based on information energy that is calculated for every pixel in the scanned document and the characters are recognized using Artificial Neural Network (ANN). The recognition has an accuracy of almost 92 %. The proposed method can also be further improved to work on other languages as well as increase the accuracy.

Keywords Line segmentation · Word segmentation · Character segmentation · Information energy · Neural networks · Recognition etc.

1 Introduction

Text recognition plays an important role in document image processing. The segmentation of machine printed documents are performed more effectively than handwritten documents. This is mainly because in machine printed documents the text are straight, uniform alignment, uniform spacing between the characters and stable alignment etc. However in handwritten documents many challenges occur due to the following reasons: (1) Multi-orientations (2) skewness of the textlines (3) overlapping of characters (4) connected components etc. The segmentation methods for machine printed and handwritten documents are different. Some of the methods for machine printed are as follows. [1] Ping uses an approach to segment

K. Keisham (✉) · S. Dixit
Dayananda Sagar College of Engineering, Bangalore, India
e-mail: kanchankeisham@gmail.com

S. Dixit
e-mail: sunanda.bms@gmail.com

© Springer India 2016
S.C. Satapathy et al. (eds.), *Information Systems Design and Intelligent Applications*, Advances in Intelligent Systems and Computing 435,
DOI 10.1007/978-81-322-2757-1_60

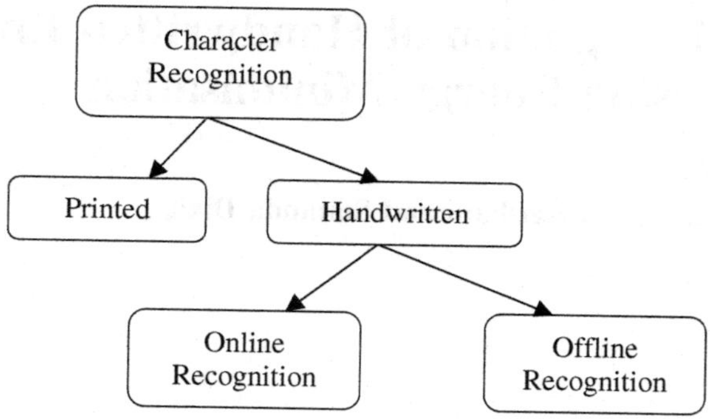

Fig. 1 Block diagram for character recognition

printed text using Markov Random fields (MRF). This method separates noise, printed and handwritten text. Shetty uses an approach to segment and label by using Conditional random fields (CRF). One advantage of this method over MRF is that it does not assume the conditional independence of the data. Handwritten character recognition can be either offline or online. In offline character recognition the text are written in paper or document are automatically converted into letter codes which are then used within the computer. In online character recognition it involves the automatic conversion of text which are written on special digitizer or PDA. Offline character recognition is more difficult since different people have different writing styles. Offline handwritten character recognition involves segmentation of textlines, words and characters which are then recognized. Many methods have been proposed for text, line and word segmentation. Studies have shown that text and word segmentation have achieved very high accuracy. This paper proposes an English document handwritten character recognition using information energy that are used to formulate an energy map. The energy map is then used to construct seams to separate the text lines [2]. The seam passes through minimum energy levels. The characters are then segmented and recognized using feed forward artificial neural network [3]. The rest of the paper is organized as follows. Section 2 gives related work. The proposed method is explained in Sect. 3. Section 4 gives the experimental result and the conclusion is given in section.

The general diagram for types of character recognition is given in Fig. 1.

2 Related Work

This section gives a summary of works that has been done for line, word and character segmentation and recognition. [4] performs text line extraction using Hough transform. This method also involves post processing in order to extract the lines which the Hough transform fails to do so. [5] uses Minimum spanning tree (MST) to segment the lines. In this method the characters of the same textline will be present on the same subtree. [6] proposes a method based on Active Contour

(Snakes). In this method the snakes are adapted over the ridges of the textline. The ridges passes through the middle of the textlines. [7] uses Mumford-Shah Model by Du. It proposes a segmentation algorithm know as Mumford-Shah model. It is script independent and it achieves segmentation by minimizing the MS energy function. Morphing is also used to remove overlaps between neighboring text lines. It also connects broken lines. The result does not depend on the no. of evolution steps involved. [8] Text line segmentation of handwritten document using constraint seam carving. It proposes a constraint seam carving in order to tolerate multi-skewed text lines. This method extracts text lines by constraining the energy that is passed along the connected component of the same text lines. It achieves an accuracy of 98.4 %. It is tested on the Greek, English and Indian document image. [9] uses a MLP based classifier in order to recognize the segmented words. It is a holistic approach as it recognizes the word from its overall shape. [10] uses a method know as Transformation based error driven learning. This algorithm generates a set of rules and scores are evaluated. This method applies the rules to different sentences repeatedly until the scores remain unchanged. [11] uses pivot based search algorithm to recognize the words. This method select some words from the training set as pivot words. The remaining words are then compared with the pivot words and similarity is computed. [12] Local gradient histogram features for word spotting in unconstrained handwritten document by Rodriguez. It proposes two different word spotting systems-Hidden Markov Model and Dynamic time wrapping. In the window this method a sliding window moves over the word image from left to right. A histogram of orientations is calculated in each cell by subdividing the window into cells at each position. Uses Convex Hull to determine the initial segmentation points and SVM classifier is used to classify the segmented characters. [13] uses mutilple Hidden Markov Model (HMM). Multiple HMM are used since different characters can be written in different writing styles. [14] uses horizontal, vertical and radial histogram to segment and recognize the characters. [15] Space scale technique for word segmentation by Manmatha. It proposes a method that is based on the analyses of 'blobs' that is present on the scale representation of an image. In this technique the pages are first segmented into words and then a list containing instances of the same word is created. It achieves an average accuracy of 87 % [16]. Segmentation of off-line cursive Handwriting using Linear Programming by Berlin Yanikoglu. It proposes a segmentation technique Thai is based on a cost function at each point on the baseline. This cost function is calculated by the weighted sum of four feature values at that point. It achieves an accuracy of 97 % with 750 words.

3 Proposed Method

Our proposed method computes line segmentation based on information energy of the pixels [2] and segmented characters are recognized based on feed forward Artificial Neural Network (ANN). In this case the image is first binarized. The binarized image undergoes line segmentation based on the information energy of the pixels. Word and Character segmentation is also performed by plotting vertical histograms and then finding local minima. The segmented characters are then recognized using Artificial Neural Networks. The proposed method is implemented on English Handwritten sentences.

Our schematic diagram for the proposed method is given as follows.

In a document image every pixel has certain level of information present in it. The blank spaces between the text lines represent low information pixels. If high energy pixels are removed it leads to information loss. The idea is to generate an energy map based on the information content of each pixel. From the computed energy map the location of the text lines can be found out.

The energy value of each pixel is computed as follows []:

$$E(1, m) = \left|\frac{I(1, m+1) - I(1, m-1)}{2}\right| + \left|\frac{I(1+1, m) - I(1-1, m)}{2}\right| \quad (1)$$

The energy map between two lines is given by

$$E_a = E(K, :) \in R^{d*e} \quad \text{where } K = \{i \ldots m\} \quad (2)$$

which gives the set of co ordinates between the start and end of energy block map. Here $d = m - i + 1$.

Now based on the energy map a seam is generated by dynamic programming [17]. The image block is transverse and the minimum energy Nb is computed.

The energy Nb is computed backwards. Now the optimal seam is finally generated which segments the text line.

After the lines are extracted and skew and slant are corrected the next step is word segmentation. This paper uses vertical histogram to identify the segmentation points. Sometimes the gaps between the characters may be larger than the gaps between word. To segment the words in such cases K-means clustering is used with k = 2.

First vertical histogram is constructed by using the following equation:

$$S_{pt} = \begin{cases} 1 & V_h(k) = 0 \quad V_h(k+1) > 0 \\ 0 & \text{otherwise} \end{cases} \quad (3)$$

Here $S_{pt}(k)$ refers to the starting point of all the words.

The end point of the words is given as follows:

$$St_{Pt}(k) = \begin{cases} 1 & (V_h(k+1) == 0 \; \&\& \; V_h(k) > 0) \\ 0 & \text{otherwise} \end{cases} \quad (4)$$

Based on the start and end points the words are segmented.

Now vertical histogram of the words are computed to find the initial segmentation points. After the characters are segmented its features are extracted using

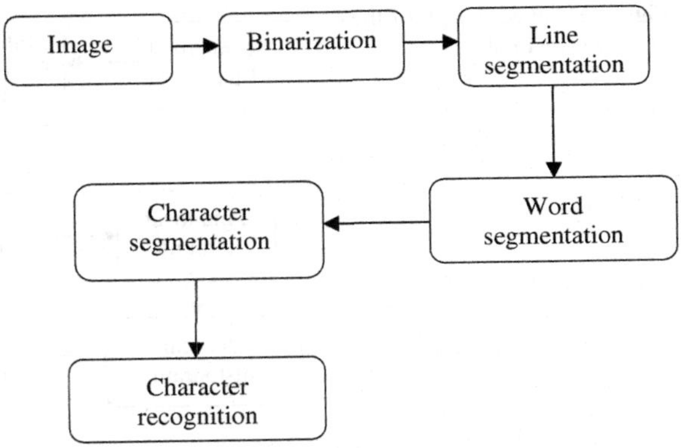

Fig. 2 Block diagram for proposed method

diagonal based feature extraction method and feed forward artificial neural network (ANN) is used for classification and recognition.

Our detailed process is given in the Fig. 2.

The diagonal based feature extraction is given below:

i. Each character of size 90*90 pixel is taken.
ii. Divide each character into 54 zones of 10*10 pixel
iii. Each zone has 19 diagonals. Now features are extracted by moving along the diagonals.
iv. Thus 19 subfeatures are extracted from each zone.
v. These 19 subfeatures are eaveraged to form a single feature and it is then placed in the corresponding zone.
vi. In case of empty diagonals the feature corresponds to zero
vii. Thus 54 features are extracted for each character. Also (9 + 6) features are also extracted for rows and columns zones. Thus totally (54 + 16) features are extracted.

Feed forward Artificial Neural Network (ANN) is then trained for classification and recognition. The neural network has two hidden layers of 30-100-100-30 to perform classification. The output layer gives the identified character. The feature vector is denoted by V where V = (d1, d2…dn) where d denotes feature and n is the number of zones for each character. Here no. of neurons in the outer layer is given by the total no of characters. Input layer neurons by the feature vector 'n' length. However trial and error method is used to determine the hidden layer neurons. It is as shown in the figure

The network training parameters are:

- Input node: 35
- Hidden node: 100
- Output node: 62 (52 alphabets, 0–9 numerals)
- Training algorithm: Gradient propagation with adaptive learning rat
- Perform function: Sum square error

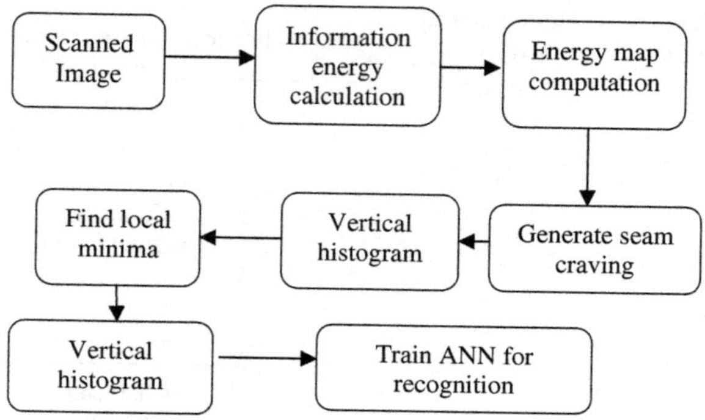

Fig. 3 Detailed design of the proposed method

- Training goal achieved: 0.000001
- Training epochs: 5000
- Training momentum constant: 0.9

Output node: 62 (52 alphabets, 0–9 numerals).

- Training algorithm: Gradient propagation with adaptive learning rate
- Perform function: Sum square error
- Training goal achieved: 0.000001
- Training epochs: 5000
- Training momentum constant: 0.9

Our detailed design is given in Fig. 3.

4 Experimental Results

The neural network is trained with 52 English characters capital and small. It is also trained using numbers ranging from 0 to 9. The input consist of handwritten sentences which in first segmented with respect to lines, words and characters which is as shown in the figure Finally it is then recognized using the trained neural network.

The segmentation accuracy achieved can be tabulated in Table 1.

Table 1 Segmentation accuracy

Segmentation	Accuracy (%)
(1) Line segmentation	95
(2) Word segmentation	94
(3) Character segmentation	94

The output is as shown below:

(a) Line Segmentation

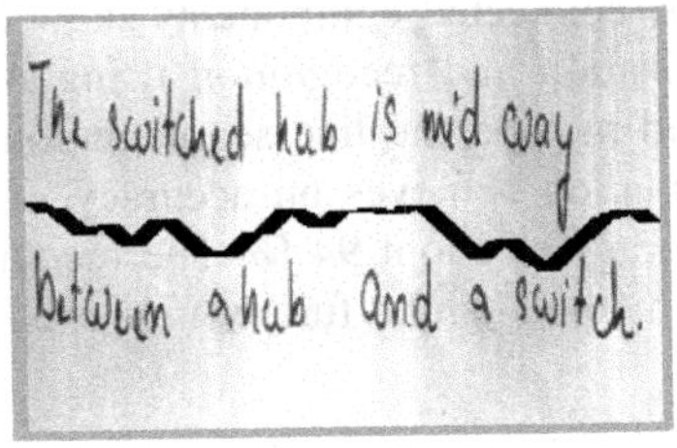

(b) Word Segmentation

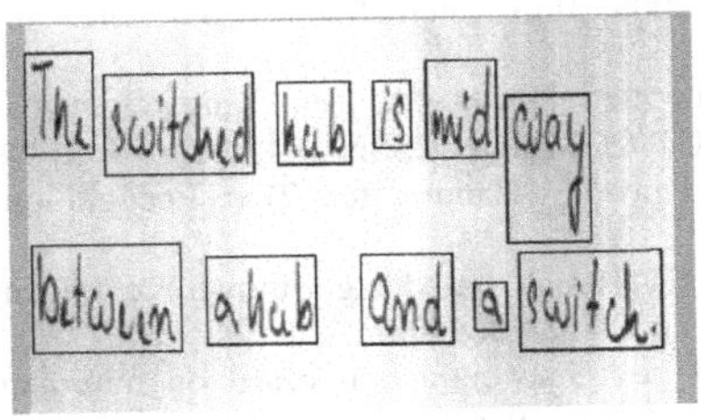

(c) Character Segmentation

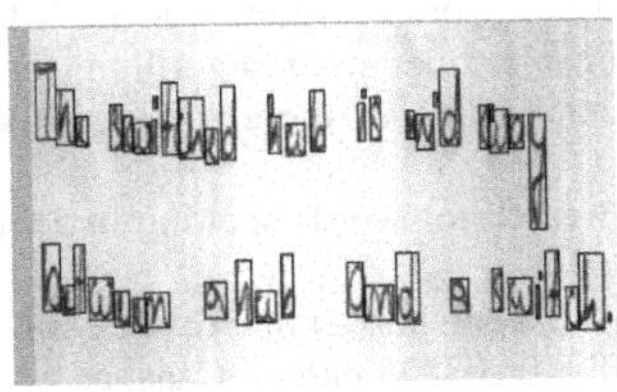

(d) Character Recognition

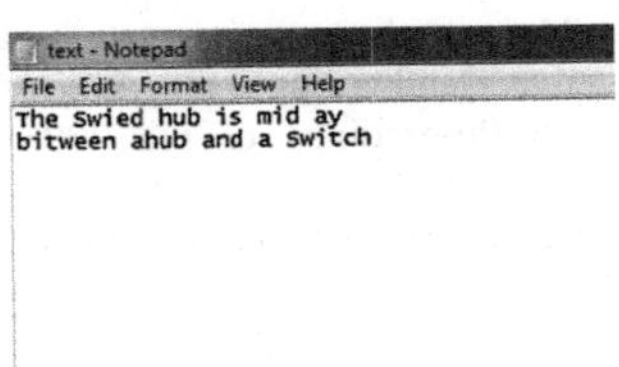

5 Conclusion

Handwritten character recognition is a difficult task. The main reasons are due to overlapping of characters, connected components skew etc. The above method is based on Energy minimization and recognition using Artificial Neural Network. Using the above proposed method the line segmentation achieves an accuracy of 95 %. The word segmentation achieves an accuracy of almost 94 %. Character segmentation has an accuracy of about 94 %. The recognition has an accuracy of almost 92 %. The above method can be further improved to increase the accuracy.

References

1. Xujun Peng, "Markov Random Field Based Text Identification from Annotated Machine Printed Documents" 2009 10th International Conference on Document Analysis and Recognition.
2. C.A. Boiangiu "Handwritten Documents Text Line Segmentation based On Information Energy, INT J COMPUT COMMUN, ISSN 1841-9836.
3. Neeta Nain Subhash Panwar, "Handwritten Text Recognition System Based on Neural Network".
4. L. Likformann-Sulem "A Hough based algorithm for extracting text lines in handwritten documents", International Conference, 95.
5. Fei Yin, "Handwritten Text line extraction based on minimum spanning tree clustering" International Conference Of wavelet Analysis and Pattern Recognition, 2–4 Nov, 07.
6. Bukhari S.S "Script Independent Handwritten text line segmentation using Active contour", ICDAR, 09.
7. Xiaojun Du, "Text line segmentation using Mumford-shah model", ICFHR, 2008.
8. Xi Zhang "Text line segmentation using Constrains seam carving", ICFHR, 2014.
9. Ankush Acharyya, "Handwritten word recognition using MLP based classifier" IJCSI, 2013.
10. David D. Palmer, "A trainable rule-based algorithm for word segmentation", ACL'98 Proceedings of the 35th annual ruling of the association for computational Linguistics and eight conference.
11. Laszlo Czuni. "Pivot-based search for words spotting in achieve documents", ICEPAF, 15th Aug, 14.
12. Jose A. Rodriguez, "Local gradient histogram features for word spotting in unconstrained handwritten documents", ICFHR, 08, Montreal, Canada.
13. Rajib Lochan Das, "HMM based Offline Handwritten Writer Independent English Character Recognition using Global and Local Feature Extraction".
14. E. Kavallieratou, "Handwritten Character Recognition based on Structural Characteristics".
15. Manmatha R, "A space scale approach for automatically segmenting words from historical documents", IEEE trans, aug 05.
16. Puolo Rodrigo Cavalin, "An implicit segmentation-based method for recognition Segmentation of off-line cursive Handwriting using Linear Programming n of handwritten string of characters", SAC'06 Proceedings of the 2006 ACM symposium of Applied computing.
17. PP Roy, "Muti oriented and Multi sized touching character segmentation using Dynamic Programming", ICDAR, 09.

A Novel Codification Technique for Tacit Knowledge in Software Industry Using Datamining Techniques

Jeffrey Bakthakumar, Murali Manohar and R.J. Anandhi

Abstract Tacit knowledge is an important resource which comes from experience and insight, and is not in any pre-recorded form. But, it has a strong contribution to the success of decision making procedure. This paper focuses on the summarization of various efforts of codification methodologies to convert tacit to explicit knowledge of an IT company, where the later plays a key role in decision making. This paper also tries to bring out the lacuna or technical gaps of various methodologies and propose a novel method to capture the tacit technical knowledge, so that the technical knowledge transfer in case of staff relocation does not affect the growth of the small scale IT industry. The challenge in software development life cycle is clearly captured by product management software tools like JIRA. We are using text mining techniques on these reports to gather all the tacit knowledge, of technical person assigned to the project. Added to that, from our scrum report, we will extract consolidate generic knowledge. These two, will ensure that we capture most of the needed tacit knowledge of the project coding or maintenance phase. The mining results from the data are promising and future work is to include sentimental analysis.

Keywords Tacit knowledge · Text mining · Software life cycle · JIRA and scrum reports · Knowledge codification

1 Introduction

Information is associated with all the intellectual assets from the human resources of a company and its proper management helps firms to turn this into intelligence values. The goal of knowledge management (KM) is to enhance the skills and expertise of the human resources and assist in decision making process. KM is basically achieved by collaboration and connection of people's skillset. The capability to find an expert who can provide valuable suggestions and that too

J. Bakthakumar (✉) · M. Manohar · R.J. Anandhi
Bangalore, India
e-mail: jeffbk@hotmail.com

© Springer India 2016

S.C. Satapathy et al. (eds.), *Information Systems Design and Intelligent Applications*, Advances in Intelligent Systems and Computing 435,
DOI 10.1007/978-81-322-2757-1_61

within a specific timeframe, for solving a problem is critical in business success. Knowledge management (KM) avoids the wastage of valuable resources of companies from doing repetitive tasks, which results in is wastage of talent, and in turn reduces the turnover of the staff. The preliminary KM technologies emerge from document-centric approaches and hence directly relates to the KM process steps like classifying, storing and retrieval of knowledge. When explicit knowledge is available as data and written documents, the tacit form of knowledge is inherent and implicit. The latter is available as deed, processes and sentiments, acquired over a quite long period of the project phase.

This gives rise to two kinds of knowledge management, explicit knowledge (EK) and tacit knowledge (TK). Explicit knowledge (EK) basically can be codified, recorded, or actualized into some form which is easy to process and comprehend. The explicit knowledge management process includes using clear policies, goals and strategies and can be maintained in the form of papers and reports which can be easily codified. In a software company the explicit knowledge can be spread using various repositories like periodicals, journals, audio-recordings, WebPages, and portals. As stated, tacit knowledge is derived from experience and insight, using intuition. Tacit knowledge is difficult to codify by any normal means of knowledge capture methods. For instance, stating to someone that Bangalore is an IT hub or silicon city of India is obvious knowledge. But, the capability to master an art, express a poetic feeling [1] or operate a complex electronic equipment requires mostly implicit knowledge, and experts in that field also find it difficult to document. Though this implicit knowledge looks like very elementary, it is not widely understood nor can be expressed.

The optimal utilization of the employees knowledge and capacities, are essential for small-scale business success. The small scale companies are noted for their fast and streamlined decision making process as it has less bureaucracy and lower fixed costs. Due to these processes, the response time is very fast to the requirements of firsthand clienteles. Hence when it comes to product development, these features with their advantages make them well suited to incorporate novelty.

It is very crucial to maintain the tacit knowledge in small scale IT Company due to the limited number of employees and also the attrition rate is more. The Tacit Knowledge derived from experience and insight, are subjective, cognitive and experiential learning; and also highly personalized and difficult to formalize, hence also referred as sticky knowledge. As the concept of tacit knowledge refers to individual, it has to be acquired without formal symbols or words.

The strategic way to acquire TK is mostly through familiarity and intuition. With support of practice, it is feasible for human resources to communicate their thoughts [2]. Tacit knowledge can be differentiated from explicit knowledge [3] mainly in three metrics; first metric is codifiability and the mechanism of transferring knowledge, second factor lies in the method for the acquisition and accumulation and the third differentiation is the possibility of combination. EK can be acquired through logical deduction whereas tacit needs to be acquired mostly by real-world practice.

Codification is a process of converting tacit knowledge into explicit knowledge. The characteristics of this knowledge can only be transmitted via training or added through personal understanding.

2 Literature Survey

Schindler and Eppler [4] gives various ways in which one can extract the know—hows from projects and then use them in practice. Process-based techniques focus on practical methods to capture project lessons. Documentation-based approaches are templates for project learning. This paper brings out the variances between hypothetical views and the practical management. This paper deliberates on problems in probing, like the deficiency in enthusiasm to study from faults along with the inappropriate use of project management handbooks.

The technical gap identified in this work is that the managing the project intuitions require substantial progress in terms of the organization, process, and use of acquired knowledge.

Jabar et al. [5] not only focuses on the fact that knowledge and experiences as a significant benefit to a company, but also brings out the challenges in gathering, possessing, distributing and salvaging the information. Evaluating tacit knowledge is a hard task as the employee's knowledge is elusive. It is also complex due to staff attrition in the firm or when they are reassigned on different work. Even though KEPSNet contribution reveals techniques of management of acquired information, acceptance of skilled resources knowledge and recommendation of the professional for distribution of knowledge needs to be considered.

Prencipe and Tell [6], in their survey of articles on categorization of knowledge are concerned mainly with the financial aspects. They have given minimal importance to the fundamental methods used in learning or knowledge acquisition. The paper differentiates the basic learning ways into three categories, such as collection of the know—how knowledge, storing the information in acceptable formats and then to classify the same. The authors also have suggested a structure to analyze the capabilities to learn in project-based companies.

The main gap in this work is that they have short sight in concentration of process and the authors also have given lesser weightage to the codification outcomes. It is stated to be necessary that the relationship between efficiency in learning and outcome of the learning process in terms of company performance statistics is to be proven.

Bugajska [7] in his paper has done a detailed study on successful transfer of knowledge between two companies which are important for subcontracting. It is needed even within an enterprise as sometimes assignments needs to shifted or completed faster. The author focuses on the group of individuals who are responsible for transfer, basically the team which possesses the know—how skills and end user. He proposes a framework which lists all the knowledge transfer initiatives that has to be followed by any subcontracting company. This framework in turn also supports utilities to sustain the process such as profiler and catalogue, which helps in the process knowledge transition.

From the study there are some areas that we believe needs further research. Complete automation may enhance supplementary value to the catalogue and the profiler. Patterns identified during transfer can bring out more value to the instruments and makes them reusable in firms, especially when they have an idea of subcontracting in future.

Keith et al. [8] explores the way a new project team learns to resolve project issues and also highlights the importance to share the acquired knowledge in that process. The main challenge in such situation is that so acquired knowledge is implicit and people find it difficult to explain, to record or to publish. This is one of the main problems faced by team leaders to inspire their team members to learn. Normal practice in some firms is to have reviews after the project completion, to establish the lessons learned from the project along with recording. Their research shows that most of the formal documents do not convey the significant learning from development teams. The only way out of this situation is that the managers have to inspire the individuals to learn and conduct the reviews so as to create and capture the tacit knowledge.

In analyzing the paper we find that the development encompasses solving many practical problems faced in the project phase. The experience so generated, by teams is substantial and more valuable. The problem solutions are unique and it is important to codify them with much ease. Due to the complexity involved in this process, there is a need for formulizing a new codification method and it becomes inevitable.

Prieto-Diaz [9] in his paper explores the understanding in developing, executing and installing the reusability of knowledge in library. The primary aim is on organizing various components involved in software, for reuse, using classification techniques. It also discusses some success stories involved in increasing the reuse of technical components. Initial discussion is regarding the technology developed. Further discussion is on how it was conveyed. The inferences from the study are numerous as it ascertains the possibility of reuse of library information, apart from the feasibility of making it transferable; it also explores the affirmative impact on the economic benefits to the company, thereby justifying its implementation.

We find that the study has also given guidelines needed to support research in reuse of technology such as domain analysis and related activities, integration of tools for supporting such applications.

Storey [10] in her paper describes a hypothetically designed tool which is used to pool resources amongst development team members who do not work in a synchronous manner. The aim was to create and use a simple source tool which will provide annotation, which can be latter used to enhance mobility within code, thereby helping in achieving synchronization. This will in turn capture knowledge pertinent to a development of the software. The authors have tried out a novel design by combining outputs of geographical navigation and tag log details from social bookmarking products. The most needed support for establishing synchronization and message transfer is hence forth achieved between team members.

This paper helps us to visualize the application and benefits of usage of tag categorizations and terminologies. This will also increase consciousness of development team on tasks beyond the borders of the development environment.

Midha [11] in his work tries to explore if the positive and negative experiences actually help us to learn and along with that he also tries to find if we eagerly share such lesson with our counterparts. The paper also questions if most of the project lessons learnt is used effectively in bigger firms.

A methodical approach for gathering and translating knowledge acquired in a process, such as collecting lessons learnt during the execution of different events, labeling in an organized data warehouse, transfer of such acquired information apart from applying them lessons are the first few preprocessing steps discussed in this work. The second part of the approach involves in integrating such knowledge into processes and commissioning the acquired information.

Kwasnik [1] studies the underlying connection that exists in between knowledge and cataloging. The systems used in catalog or classifying schemes give a mirror image of various factors and also their relations with each other. This is exactly what happens when domain knowledge is classified and arranged using ontology. Various techniques that were discussed include faceted analysis, paradigms, trees and hierarchies. Various parameters such as their ability to replicate and recreate new knowledge are considered during this analysis.

Renowned researcher Ikujiro Nonaka has suggested the SECI model, which deliberates upon socialization, externalization, combination and internalization. Being considered as one of the most widely cited theories, it shows the growing processes of collaboration between explicit and tacit knowledge.

However, classifications or cataloging approaches make it possible to manipulate the knowledge for exploring the information; they can be inflexible and very fragile, and the success is proportional to the amount of new knowledge in it. It is important to comprehend the properties of different approaches, whose strengths and pitfalls will help us to maximize their usefulness. There is a possibility that new techniques and approaches can be used in cataloging by considering innovative ways of exploring for pattern strings, bringing out the correlation between them, and representing them in new data structures along with novel presentation ways.

3 Methodologies of Tacit Knowledge Codification

3.1 Existing Methodologies of Tacit Knowledge Codification

Knowledge management is a multitier active process used for creation of derived or fresh knowledge, followed by sharing the same. The first stage involves in finding the source followed by extraction, storing and circulation of the knowledge [12]. The formulation of new knowledge can be done basically through either sharing in between tacit knowledge or between tacit to explicit knowledge [13]. Normally in an IT firm, the key persons who will be utilizing the codified tacit knowledge could be managers, and team members. The various ways to identify and codify this inbuilt knowledge includes subject expert interviews, knowledge obtained from listening and by observing. One other method which is explored is to extract similar knowledge from scrum meeting minutes and project management documents from tools.

3.2 Proposed Methodology for Tacit Knowledge Codification

An abstract, yet essential process in knowing how the experts work and how their expertise can be captured revolves around how a knowledge developer captures the expert's views and experiences. It also involves in converting that expertise into a coded program. The main challenge in the whole process is mainly because even the expert might not be aware of the latent knowledge in him or her. The next bottleneck is the need for voluntary and eager involvement of the employee, in the firm's quest for his tacit knowledge, showing no reluctance in sharing the same. Our main problem definition includes in how to effectively organize software project lessons learnt and devise some practices to convert those tacit to explicit knowledge for usage in maintaining and developing future software projects.

We have found that the systematic way of gathering tacit knowledge reduces the challenges in later codification process. We have taken the JIRA reports and retrospective learning of selected participants and focused in extracting the opinions using text, mining techniques using the simple yet effective queries like what was done rightly (positive measures in sentiment analysis), what went wrong (negative measures in sentiment analysis) and the critical bottlenecks where we could improve. JIRA is project management software which helps in tracking different tickets and problems which harms the product functionality, along with product maintenance features and requirement of customized new features.

We have created a unique lesson-learnt questionnaire which primarily covers lesson learnt, root cause for the problem, response for critical issues, selection criteria for proposing such responses. We have tried to extract knowledge about a specific case from these sources and then consolidate into the Knowledge base.

4 Inferences and Results

One of the popular Product Management Tool used by many small-scale IT industries is JIRA. The major information available in an issue ticket of JIRA report are type of the bug, status, priority, resolution, components, labels, environment and Description of the bug fix. The data used for this analysis is taken from a Bangalore based healthcare MNC company's log data of their project management software tool.

We have used text mining tools to extract information from the description log of the bug fixes. The description log available in one of the sample data obtained from the Bangalore based IT firm states that: "When a build has no root cause (possible when firing the job programmatically), the code at service. BuildCauseRetriever #build Cause Environment Variables produces a StringIndexOutOfBoundsException because the substring method is applied to an empty string". Using our parser we were able to extract four variables, which are listed in the parsing report in Fig. 1.

In Fig. 1, the environmental variable (C1) cites when a particular bug has occurred and what are the environmental parameters at the time of incident. The location variable (C2) stores the parsed output of where in the code the error has occurred and the cause variable (C3) holds the parsed string of the end cause of the error, and action variable (C4) explains what the programmer has done to fix the bug. The initial approach in our methodology uses only C3 and C4 and builds frequent pattern string and from that association rules are derived. There are two places where information can be extracted; one is Bug Write-up Procedure and second one is the step where issues are resolved and closed. There are some assumptions we have used before extracting information from the logs. First one is that the reporter of the ticket needs to write the description correctly using common nomenclature and without ambiguous wordings; this will enable to frame the environment variable, C1, correctly. The success of the knowledge codification and reuse of the same depends on it. The next requirement is that the solution part also needs to have clarity and be generic. This will enable the correctness of the action taken variable, C4. This will also benefit any newly recruited team member, for whom training is not needed to associate himself, in the ongoing projects. Normally, the bug report contains sequence of stages that is needed for reproducing the error, actual response from the system due to the error, correct results in case of absence of the bug, date and platform details and additional debugging information. The standard method for resolving the issue, expects the programmer to specify the code change clearly in the description section of the "Resolve Issue" tab of the JIRA product. This information can be utilized for the analysis to extract the knowledge. The cause and remedy has been captured as frequent patterns.

C1: Environment variable When a build has no root cause(possible when firing the job programmatically)	**C2 : Location Variable** service.BuildCauseRetriever#buildCauseEnvironme ntVariables
C3: Cause Variable Produces a StringIndexOutOfBoundsException	**C4: Action taken variable** the substring method is applied to an empty string

Fig. 1 Parsed output from JIRA issue ticket

5 Conclusion

Significant research is in progress in the field of Tacit Knowledge management, but very few researchers are keen to extract knowledge from either logs or predefined templates, questionnaires available or from recorded sections of any project management tool log books. This paper explores the codification of tacit knowledge from one such project management tool, named JIRA. The KM implementation from tacit to explicit is very critical in medium size IT Companies, who have basically accumulated a lot of log data and also starves form knowledge deficiency from staff attrition. This paper pursues to take account of several other areas for consideration when trying to implement the codification process using text mining approach. The future work planned in this area is the usage of opinion mining and sentiment analysis which might throw a deep understanding in capturing the hidden tacit knowledge.

References

1. Kwasnik, Barbara H. "The role of classification in knowledge represantation and discovery." Library trends 48.1 (2000): 22–47.
2. Lam, Alice. "Tacit knowledge, organizational learning and societal institutions: an integrated framework." Organization studies 21.3 (2000): 487–513.
3. Polanyi, Michael. Personal knowledge: Towards a post-critical philosophy. University of Chicago Press, 2012.
4. Schindler, Martin, and Martin J. Eppler. "Harvesting project knowledge: a review of project learning methods and success factors." International journal of project management 21.3 (2003): 219–228.
5. Jabar, Marzanah A., Fatimah Sidi, and Mohd H. Selamat. "Tacit knowledge codification." Journal of Computer Science 6.10 (2010): 1170.
6. Prencipe, Andrea, and Fredrik Tell. "Inter-project learning: processes and outcomes of knowledge codification in project-based firms." Research policy 30.9 (2001): 1373–1394.
7. Bugajska, Malgorzata. "Piloting knowledge transfer in IT/IS outsourcing relationship towards sustainable knowledge transfer process: Learnings from swiss financial institution." AMCIS 2007 Proceedings (2007): 177.
8. Goffin, Keith, et al. "Managing lessons learned and tacit knowledge in new product development." Research-Technology Management 53.4 (2010): 39–51.
9. Prieto-Diaz, Ruben. "Implementing faceted classification for software reuse." Communications of the ACM 34.5 (1991): 88–97.
10. Storey, Margaret-Anne, et al. "Shared waypoints and social tagging to support collaboration in software development." Proceedings of the 2006 20th anniversary conference on Computer supported cooperative work. ACM, 2006.
11. Midha, A. "How to incorporate "lessons learned" for sustained process improvements." NDIA CMMI Technology Conference. 2005.
12. Smith, Elizabeth A. "The role of tacit and explicit knowledge in the workplace." Journal of knowledge Management 5.4 (2001): 311–321.
13. Parsaye, Kamran, and Mark Chignell. "Expert systems for experts." New York, Chichester: Wiley, 1988 1 (1988).

An Analysis on the Effect of Malicious Nodes on the Performance of LAR Protocol in MANETs

R. Suma, B.G. Premasudha and V. Ravi Ram

Abstract Mobile Ad Hoc Networks (MANETs) are more affected by various security problems because of the inherent characteristics such as dynamic network topology, unavailability of fixed infrastructure, lack of centralized control and high mobility of the nodes. One of the major security problems in MANETs is nodes' misbehavior. The misbehaving nodes can advertise themselves of having a short route to the destination to transmit the data. Also, the misbehaving nodes participating in the route may stop forwarding the data at some point of time resulting in loss of packets. The main objective of our work is to analyze the effect of misbehaving nodes over the performance of Location Aided Routing (LAR) in MANETs. The work analyzes the performance of LAR protocol with the parameters such as throughput, packet delivery ratio, average delay and routing overhead using NS2 simulator. The simulation results clearly show that the routing performance of LAR decreases in the presence of misbehaving nodes. Thus there is a need for introducing authentication mechanism for secured data transmission using LAR protocol.

Keywords MANET · Malicious node · LAR · Performance analysis

R. Suma (✉) · V. Ravi Ram
Department of MCA, SSIT, Tumkur, Karnataka, India
e-mail: sumaraviram@gmail.com

V. Ravi Ram
e-mail: raviramv@gmail.com

B.G. Premasudha
Department of MCA, SIT, Tumkur, Karnataka, India
e-mail: bgpremasudha@gmail.com

© Springer India 2016
S.C. Satapathy et al. (eds.), *Information Systems Design and Intelligent Applications*, Advances in Intelligent Systems and Computing 435,
DOI 10.1007/978-81-322-2757-1_62

1 Introduction

Mobile Ad Hoc Network (MANET) is a self-configurable and infrastructure-less network having collection of several mobile nodes [1]. A node in MANET can act both as a host and router simultaneously. The real time applications of MANET such as military battlefield, emergency rescue operations and disaster recovery demands co-operation among the participating nodes. To enable communication among nodes inside the transmission range, it is required that the nodes in MANETs must forward packets on the basis of mutual understanding [2].

Due to high mobility of nodes in MANETs, finding out the correct destination is challenging and important. Thus, an efficient routing protocol has to be used to maintain the routes in MANETs [3].

One of the main problems in MANETs is node misbehavior [4]. The nodes can advertise themselves of having a shortest route to the destination to transmit the data or at some point of time they may stop forwarding the data to avoid further transmission leading to packet drop [5, 6].

The objective of this paper is to analyze the effect of node misbehavior in MANETs using LAR protocol. The effect of node misbehavior in LAR protocol is evaluated on the performance parameters such as throughput, packet delivery ratio, average delay and routing overhead using NS2 simulator.

The route calculation procedure of LAR is illustrated in Sect. 2, the performance and the simulation parameters are defined in Sect. 3, the results obtained through simulation and the corresponding result analysis is presented in Sect. 4 and finally Sect. 5 concludes our work.

2 Location Aided Routing (LAR)

The routing protocols in MANETs are classified as Flat routing, Hierarchical routing, Geographical routing, Power aware routing and multicast routing [7, 8]. The hosts in Geographical routing protocols are aware of their geographic position [9]. An important feature of Geographic routing protocols is that they avoid the entire network to be searched for locating the destinations [10]. If the geographical coordinates are known, the control overhead is reduced by sending the control and data packets in the direction of the destination. A disadvantage of Geographical routing protocols is that all the nodes must be aware of their geographical coordinates. For the effective functioning of location-based routing, any update on routing shall take place quicker than that of the node mobility. This is because of the fact that the locations of the nodes change quickly in a MANETs [11].

There are two methods for selecting geographical locations in mobile ad hoc networks. One is by making use of the geographic coordinates obtained through Global Positioning System (GPS) and the other by using reference points in some fixed coordinate system.

Location-Aided Routing (LAR) is based on restricted directional flooding [12] and uses partial flooding for the purpose of path identification. Thus LAR utilizes location data to improvise the route discovery process. With the help of the position information, an expected zone is located relative to the source node. Similarly, a request zone is identified as a collection of nodes that are intended to forward the packets pertaining to route discovery. Typically a request zone encloses the expected zone and there are two schemes of LAR that are used to identify the request zone.

In Scheme 1 of LAR, a node forwards the route discovery packets if and only if it is within the selected rectangular geographic region. In Scheme 2 of LAR, a message will be forwarded by a node (source/intermediate) to others if they are found closer to the destination than itself. Thus, a node after receiving the route request message will retransmit that message if it is close to the destination than the previous hop and if not so it drops the message.

In the process of finding the shortest path in the network, multiple nodes are identified to transmit the route request packet and each such node saves its IP address in the header of the route request packet. This enables the tracking of the route followed by the route request message.

LAR considerably reduces the routing overhead as it uses location information to restrict the flooding to a limited region [11]. With the use of GPS, a desired route can be obtained with limited number of messages pertaining to the discovery of route [13].

The scheme 1 of LAR identifies the request zone by using the expected location of the destination. A smallest rectangle that encompasses the source node and the

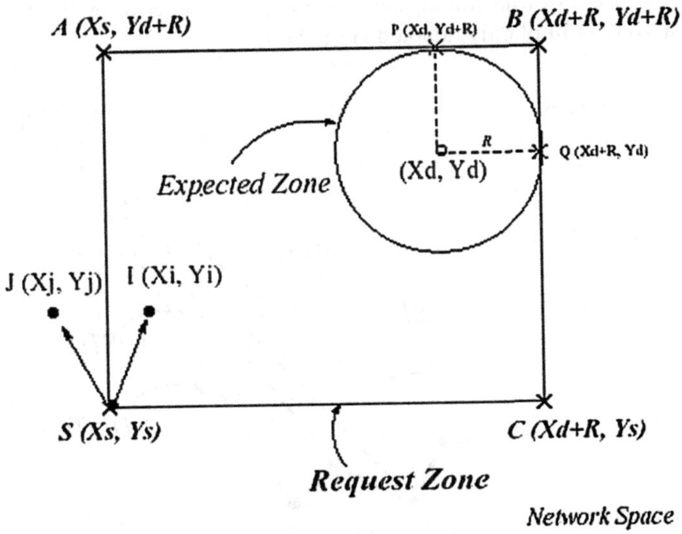

Fig. 1 Source node outside the expected zone [12]

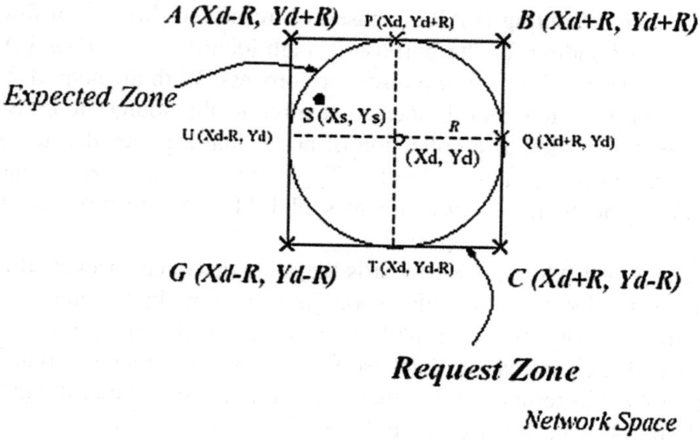

Fig. 2 Source node within the expected zone [12]

expected zone of the destination is considered as the request zone whose sides are parallel to the X and Y axes. During the route discovery process, the four coordinates of the rectangle are included in the route request message. The coordinates of the request zone help the intermediate nodes in taking the forwarding decision and the intermediate nodes are not authorized to alter these coordinates (Figs. 1 and 2).

The scheme 2 of LAR identifies the request zone by considering the distance from the former geographic position of the destination. Any intermediate node N forwards route request if it is nearer to or not much far from the previous position of the destination than the node M which had transmitted the request packet to N. Thus in the process of transmitting the route request packet through several nodes, the request zone is implicitly adapted (Fig. 3).

Fig. 3 Scheme 2 of LAR [12]

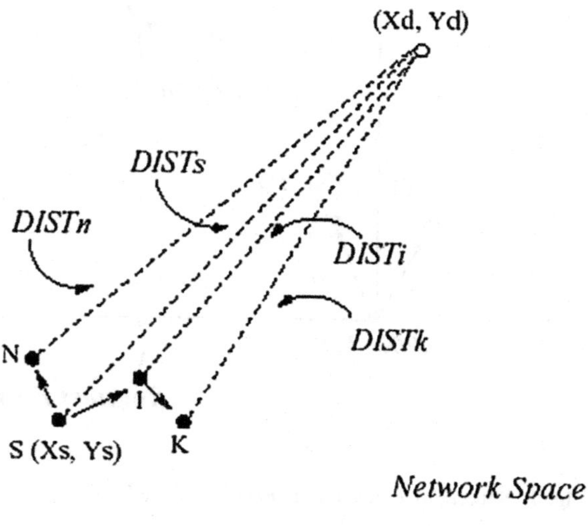

Table 1 Simulation parameters

Protocol verified	LAR
Number of nodes	50
Area size	1000×1000 m^2
MAC	802.11
Radio range	250 m
Simulation time	200 s
Traffic source	Constant bit rate (CBR)
Packet size	512 bytes
Mobility model	Random way point
Max. and min. speed	5, 10, 15, 20, 25, 30 m/s
Initial energy	100 J
Simulator	NS 2.34
Performance Parameters	Throughput, average delay, packet delivery ratio and routing overhead

LAR adopts restricted flooding as the nodes in the request zone are only expected to carry forward the route request. Due to timeout concept LAR does not ensure identification of a path comprising of only hosts belonging to a specified request zone. Thus there is always a tradeoff between the latency in route discovery and the underlying message overhead.

3 Simulation Tool and Parameter Setup

NS 2.34 network simulator is used to investigate the influence of misbehaving nodes over the performance of LAR protocol in MANTEs. The malicious nodes are introduced in the network scenario by creating nodes without proper header information and considering these nodes in the route identified from source node to destination node. The effect of the misbehavior of these malicious nodes on the performance of LAR protocol is analyzed using parameters such as throughput, packet delivery ratio, average delay, and routing overhead. The parameters considered during the simulation are listed in Table 1.

4 Simulation Results and Analysis

The effect of nodes' misbehavior on the performance of LAR protocol is represented on the basis of the performance parameters such as throughput, average delay, packet delivery ratio and routing overhead. Throughput signifies the percentage of frames that are transmitted successfully per unit amount of time. Delay is considered as the interval between the time of arrival of the frame at the transmitter's MAC layer and the time when the transmitter confirms the successful reception of the frame by the receiver. This not only includes the propagation time but also the time consumed due to buffering, queuing and retransmitting of data packets. The data packet delivery ratio is termed as the ratio between the number of packets generated at the source and the number of packets received by the destination.

With varied number of nodes and a simulation time of 200 s, a comparison of the throughputs obtained using LAR protocol with and without malicious nodes is given in Table 2. The geographical representation of the same is represented in Fig. 4 and it can be seen that there is a decrease in throughput using LAR-with malicious nodes for the entire range of nodes as compared to LAR without

Table 2 Throughput comparison

No. of nodes	Throughput (kbps)	
	LAR	LAR with malicious nodes
10	64	59
20	49	42
30	27	24

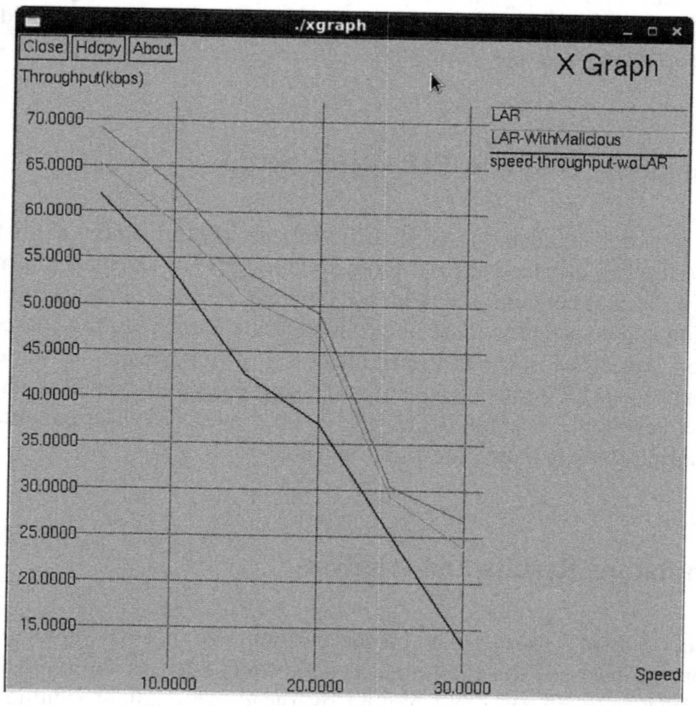

Fig. 4 Throughput comparison of LAR with and without malicious nodes X-axis: number of nodes, Y-axis: throughput (kbps)

malicious nodes. The difference in throughputs is between a minimum of 3 kbps and a maximum of 7 kbps for different number of nodes.

With varied number of nodes and a simulation time of 200 s, a comparison of the Average Delay produced using LAR protocol with and without malicious nodes is given in Table 3. The geographical representation of the same is represented in Fig. 5 and it can be seen that there is an increase in the delay by using LAR-with malicious nodes for the entire range of nodes as compared to LAR without

Table 3 Average delay comparison

No. of nodes	Average delay(s) x 10^{-3}	
	LAR	LAR with malicious node
5	20	25
10	40	45
15	160	179
20	110	140
25	170	220
30	120	139

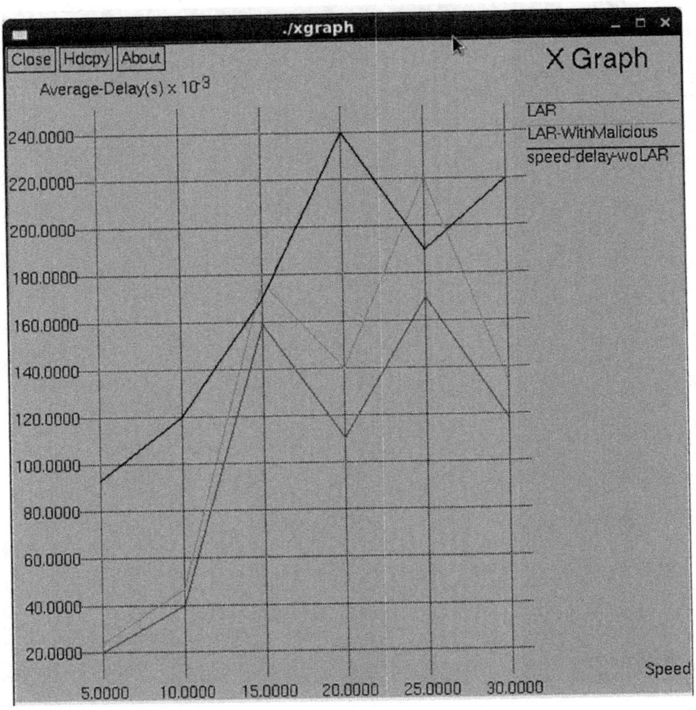

Fig. 5 Average delay comparison of LAR with and without malicious nodes X-axis: number of nodes, Y-axis: average delay(s) × 10^{-3}

malicious nodes. The difference in delay is between a minimum of 5 s and a maximum of 50 s for different number of nodes.

With varied number of nodes and a simulation time of 200 s, a comparison of the Packet Delivery Ratio obtained using LAR protocol with and without malicious nodes is given in Table 4. The geographical representation of the same is represented in Fig. 6 and it can be seen that there is a decrease in Packet Delivery Ratio by using LAR-with malicious nodes for the entire range of nodes as compared to

Table 4 Packet delivery ratio comparison

No. of nodes	Packet delivery ratio (in %)	
	LAR	LAR with malicious node
5	100	98
10	93.5	90.5
15	91.5	89
20	87	83.5
25	83	78.5
30	81	77

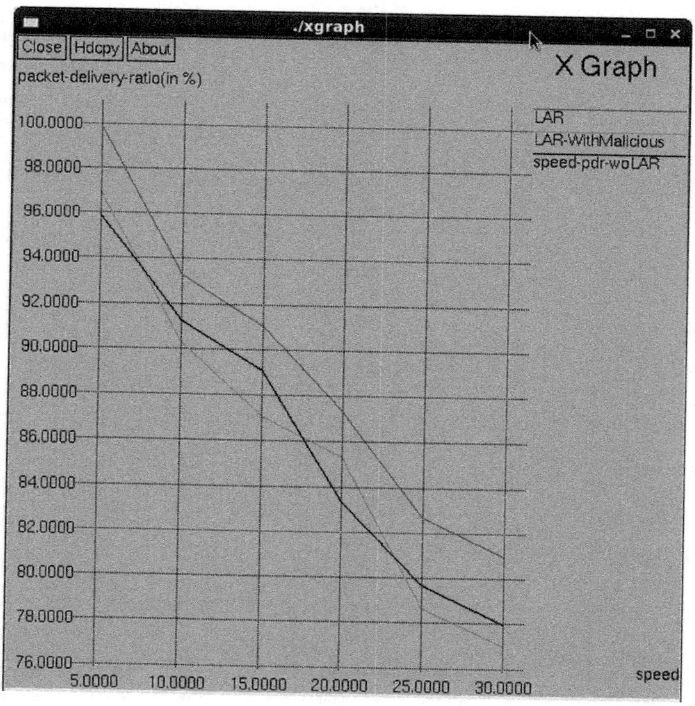

Fig. 6 Comparison of packet delivery ratio of LAR with and without malicious nodes X-axis: number of nodes, Y-axis: packet delivery ratio (in %)

LAR without malicious nodes. The difference in Packet Delivery Ratio is between a minimum of 2 % and a maximum of 4.5 % for different number of nodes.

With varied number of nodes and a simulation time of 200 s, a comparison of the Routing overhead produced using LAR protocol with and without malicious nodes is given in Table 5. The geographical representation of the same is represented in Fig. 7 and it can be seen that there is an increase in Routing overhead by using

Table 5 Routing overhead comparison

No. of nodes	Routing overhead	
	LAR	LAR with malicious node
5	1.2	3.2
10	1.4	4.3
15	1.49	6.4
20	1.6	7.5
25	1.8	8.6
30	1.9	9.8

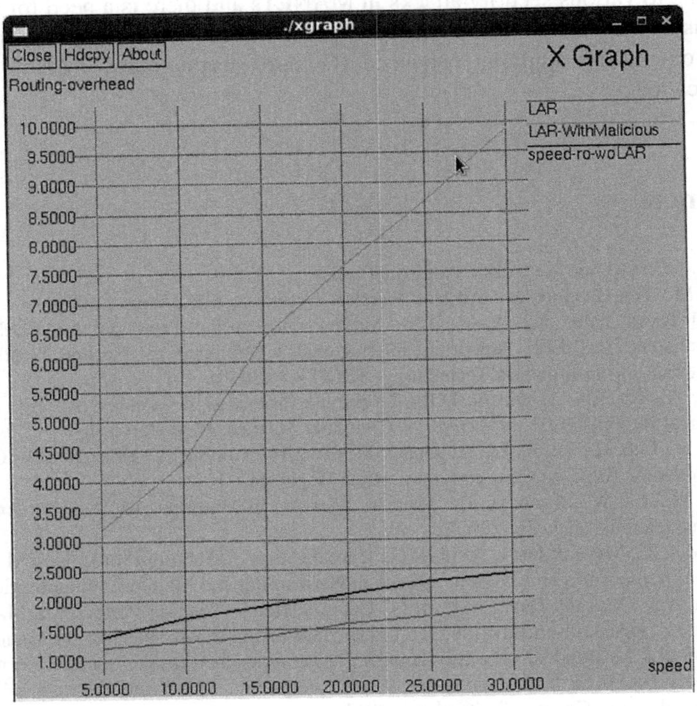

Fig. 7 Comparison of routing overhead of LAR with and without malicious nodes X-axis: number of nodes, Y-axis: routing overhead

LAR-with malicious nodes for the entire range of nodes as compared to LAR without malicious nodes. The difference in Routing overhead is between a minimum of 0.2 and a maximum of 7.9 for different number of nodes.

5 Conclusion

MANETs suffer from various security problems because of high mobility of nodes. One of the main security issues in MANETs is node misbehavior. In this work, we have implemented LAR protocol with malicious nodes and analyzed the effect of malicious nodes on the performance of LAR protocol. From the simulation results it is observed that the throughput got decreased by a minimum of 3 kbps and a maximum of 7 kbps, the average delay got increased by a minimum of 5 s and a maximum of 50 s, the packet delivery ratio got decreased by a minimum of 2 % and a maximum of 4.5 % and finally the routing overhead got increased by a minimum of 0.2 and a maximum of 7.9. These statistics clearly convey the adverse effect of misbehaving nodes on the performance of LAR protocol. We conclude that LAR is vulnerable to various security attacks in MANETs and there is a need for defensive mechanisms to overcome these attacks. Thus in our future work, we will overcome these problems by applying enhanced ID-based cryptography method for node authentication.

References

1. Perkins, C.: Ad hoc Networks. Addison-Wesley.
2. Ilyas, M.: The Handbook of Ad Hoc Wireless Networks. CRC Press (2003).
3. Kamal Kant, Lalit, K., Awasthi: A Survey on Routing Protocols for MANETs and Comparisons, In: IEEE supported International Conference on Recent Trends in Soft Computing and Information Technology, TSCIT-09 (2010).
4. Perrig, A., Hu, Y.C., Johnson, D.B.: Wormhole Protection in Wireless Ad Hoc Networks. Technical Report TR 01-384, Dept. of Computer Science, Rice University.
5. Yang, H., Luo, H., Ye, F., Lu, S., Zhang, L.: Security in mobile Ad Hoc networks Challenges and solutions. IEEE wireless communication Magazine.
6. Deng, H., Li, W., Agrawal, D.: Routing Security in Wireless Ad Hoc Networks. IEEE Communications Magazine (2002).
7. Verma, A.K., Mayank Dave, Joshi, R.C.: Classification of Routing Protocols in MANET. In: National Symposium on Emerging Trends in Networking and Mobile Communication (2003).
8. Verma, A.K., Mayank Dave, Joshi, R.C.: Classification of Routing Protocols in MANET. In: National Symposium on Emerging Trends in Networking & Mobile Communication (2003).
9. Blazevic, L., Le Boudec, J.Y., Giordano, S.: A location-based routing method for mobile Ad Hoc networks. IEEE Transactions on Mobile Computing, Broadband Wireless LAN Group, STMicroelectron, Geneva, Switzerland (2011).
10. Atekeh Maghsoudlou, Marc St-Hilaire, Thomas Kunz: A Survey on Geographic Routing Protocols for Mobile Ad hoc Networks. Technical Report, Systems and Computer Engineering, Carleton University (2011).

11. Toh, C.K.: Ad Hoc Mobile Wireless Networks: Protocols and Systems. Prentice Hall, Englewood Cliff, NJ.
12. Young Bae Ko, Nitin Vaidya H.: Location-Aided Routing (LAR) in Mobile Ad Hoc Networks. Department of Computer Science, Texas A&M University College Station, TX 77843- 3112, youngbae, vaidya@cs.tamu.edu.
13. Suma, R., Premasudha, B.G.: Location Based Routing Protocols in MANETs. In: National Conference on Trends in Computer Engineering and Technologies. T John College, Bangalore, India (2012).

Qualitative Performance Analysis of Punjabi and Hindi Websites of Academic Domain: A Case Study

Rupinder Pal Kaur and Vishal Goyal

Abstract The dependency on websites has increased manifold. More and more websites are being developed in local languages also. However, most users feel that the websites developed in their local languages are not reliable and updated. So, the quality of the websites of Academic institutes which are in local languages have been performed. There are 49 academic institutes in India whose websites are in local languages. Using stratified sampling technique, the sample of websites that are selected for case study are 2 (66.6 %) of Punjabi and 20 (40.8 %) of Hindi. The testing has been performed on the selected websites by implementing a web quality model According to the testing, 12 (54.5 %) websites' score is less than 50 %, 7 (31.8 %) websites' score is between 50 and 60 % while only 3 (13.6 %) websites' score is more than 60 %.

Keywords Academic domain · Punjabi and Hindi websites · Qualitative performance analysis · Testing of websites

1 Introduction

Day by day most of the information is available on websites. India being the third country India in the world where large number of people depend on Internet [1]. The dependence on websites has increased manifold. However, most users feel that the websites developed in their local languages are not reliable and updated. So, the qualitative performance analysis whose websites are in local languages of academic institutes has been done in this research.

R.P. Kaur (✉)
Sri Guru Gobind Singh College, Chandigarh, India
e-mail: rsandhu_18@yahoo.com

V. Goyal
Punjabi University, Patiala, India
e-mail: vishal.pup@gmail.com

© Springer India 2016

S.C. Satapathy et al. (eds.), *Information Systems Design and Intelligent Applications*, Advances in Intelligent Systems and Computing 435,
DOI 10.1007/978-81-322-2757-1_63

The structure of the paper is as under. Following the introduction are objectives, and research methodology implicated in Sects. 2 and 3 respectively. Concerned literature is in Sect. 4. Section 5 outlines the strategy for the data collection and discusses about the results and analysis. Finally, Sect. 6 concludes the paper that the future work may be undertaken.

2 Objectives

The main objective of the research is to evaluate the quality of websites in local languages of academic institutes. The main objective of the research can be divided into:

- To propose a web quality model to measure the external quality of websites in local languages of academic institutes.
- To prepare the set of websites those are in local languages of academic institutes.
- To select a sample of websites for testing.
- To implement the developed web quality model on the sample of websites.

3 Research Methodology

- The ISO/IEC 9126 [2] is the quality model for Software Engineering that is also being used by the researchers of the Web Engineering. But, recently, ISO/IEC 9126 has been amended by ISO 25010 therefore ISO 25010 [3] acts as a framework for developing a web quality model (Punjabi and Hindi Website Quality Tester-PH.WQT). The PH.WQT consists of two parts first includes the global attributes that are discussed in [4] and the other part that includes the academic domain dependent attributes is described in [5] to measure the external quality of websites in local languages of academic institutes. The model [4] includes more than 100 attributes that have been defined and a metric is chosen for each indicator. In the domain dependent quality model [5] 41 attributes of academic institutes have been proposed on the basis of their characteristics mentioned in ISO 25010 guidelines. There are 11 attributes that have been proposed by other researchers and 30 new have been proposed.
- An exhaustive and exploratory research method is used for preparing the set of websites in local languages of academic institutes with citations from references [6, 7].
- Using Stratified Sampling technique the sample of websites has been selected for case study.
- The web quality model (PH.WQT) discussed above is implemented on the sample of websites.

4 Related Work

Olsina et al. [8] include results of a case study about academic institutes. Six internationally well-known academic sites from four different continents are selected to carry out the case study. More than a 100 attributes for the academic sites are outlined. Zihou Zhou in his MPhil thesis [9] proposes the website quality metrics and means to measure the website quality factors. A case study is performed to evaluate a website: www.dmu.ac.uk which is the official website of De Montfort University. Finally, a validation process is applied by using the questionnaires to verify the web evaluation tool. Ten universities websites were selected in this evaluation process, i.e. Cambridge, St Andrews, College London, Warwick, Exeter, York, Leicester, Birmingham, Manchester, and De Montfort. Priyanka et al. [10] propose usability metrics for websites of Academic domain.

5 Data Collection and Results

A web quality model (PH.WQT—Punjabi and Hindi Website Quality Tester) has been developed and discussed in [4, 5] to measure the external quality of websites in local languages of academic institutes. The characteristics of quality depend on the type of website. Hence, the web quality model has been divided into two parts: global (domain independent) and domain dependent. The global web quality model (domain independent) developed [4] for websites in local languages is represented by hierarchical two-level tree structure that consists of five top-level characteristics i.e. functional suitability, operability, reliability, security and performance efficiency. The model is divided into two parts first includes the attributes that are to be tested manually and the other part includes the attributes that can be automated. Five open source testing tools have been selected for automated testing. The model consists of more than 100 attributes that have been defined and a metric is chosen for each indicator.

Figure 1 shows the template where Attribute Name specifies the name proposed for the attribute. Characteristic specifies the name of the characteristic that is mentioned in ISO 25010. Attribute definition describe the details of the attribute.

Attribute Name :
Characteristic :
Attribute Definition :
Proposed Evaluation Aspects :
Algorithm Used :
Scale: ; **Scale Type:** ; **Reference:** .

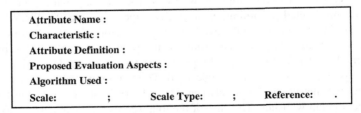

Fig. 1 Template for global attributes (manual testing)

Proposed Evaluation Aspects express the recommended results. Algorithm Used depicts the proposed measurement value; the proposed value lies between 0 and 1. Scale indicates whether the metric of the attribute is numeric or numerical.

Scale type signifies the scale of the metric whether nominal, ordinal, interval or ratio. Reference portrays whether the attribute has been referred from any research paper or not.

The quality model can be used to measure the external quality and to evaluate and compare the quality of for websites in local languages. In domain dependent quality model [5] 41 attributes of academic institutes have been suggested applying the characteristics mentioned in ISO 25010 guidelines.

Further the set of websites (Appendix 1) has been prepared for academic domain. Some assumptions and observations are made while preparing the set of websites.

Assumptions

Following assumptions are made while preparing the set of websites:

- Only websites in Hindi and Punjabi for the Indian Universities are included. The research centers' or other academic domain websites (like UGC, CBSE board) are not included as the quality characteristics like functionality, usability differs.
- The set of websites have been prepared from 1st September, 2011 to 30th November, 2011 keeping into consideration the deadlines for various phases of the research.

Observations

Following observations are made while preparing the set of websites in above mentioned time frame:

- There are 449 universities in India. Websites of forty nine universities are developed in Hindi and four in Punjabi as well as in English. Out of these 7 websites of universities did not use Unicode encoding

Results

Stratified sampling technique is used to select the sample of websites for testing. Table 1 show the exact number and percentage of sample considered domain wise for the academic domain.

Data is collected for the sample of websites. The manual testing is done and the values are interpreted according to the metrics mentioned in the PH.WQT (web quality model) developed and saved in the Microsoft Excel accordingly. The attribute values are interpreted according to the proposed metrics in the proposed quality models. Table 2 shows a glimpse of the testing of academic domain attributes as discussed and shown in Table 2 [5]. The maximum value of each attribute is 1.0 and the metric chosen is between 0.0 and 1.0. Table 2 shows the testing of academic domain websites. The sum of the total scores has been calculated and the

Table 1 Distribution of sites included in the study (domain wise)

Doman name	Frequency	Tested (in %)
.ac	1	4.55
.ac.in	12	54.55
.edu	2	9.09
.ernet.in	1	4.55
.in	2	9.09
.nic.in	2	9.09
.org	2	9.09
Total	22	100

Table 2 Glimpse of testing

S. no.	Attribute name	AIIMS	Delhi University	Punjabi University	IIT Bombay
1	Student hostel information	0.50	1.00	0.00	1.00
2	Hostel number images	0.50	0.75	0.00	1.00
3	Research details	0.50	1.00	1.00	1.00
4	News and press releases	0.00	1.00	1.00	1.00
5	Placement companies visited	0.00	0.50	1.00	1.00

percentage has been considered (Appendix 2) consequently. Being the domain independent attributes in large number thus, their weightage is more.

The results of testing are shown in Appendix 2. As, per the score, the website of Indian Institute of Management, Ahmedabad (69.13 %) scores highest followed by the website of Indian Institute of Technology, Bombay (67.12 %) in Hindi language and Punjabi University (53.03 %) scores highest among Punjabi language.

6 Conclusion and Future Work

Data collected can support maintenance of websites. Finally, the framework can feed design guidelines and can be embedded in web authoring tools. Hence, the quality of web sites developed in local languages can be improved. PH.WQT can be generalized for other languages also. The research is valuable to researchers and practitioners interested in designing, implementing and managing websites developed in other regional languages also. Analysis and comparisons among web sites developed in other regional languages could be performed in a consistent way.

To sum up, it may be said, working on this project was a very challenging task as the websites and their URL' s got changed with the time taken for the research therefore the quality of the particular website mentioned pertains to the day of testing only. Notwithstanding this change, it has to be mentioned that the websites should be seriously administered by the designers and owners.

Appendix 1

List of Punjabi and Hindi Websites

S. no.	Name of the institute	URL
1	Panjabi University	http://punjabiuniversity.ac.in
2	Baba Farid University of Health Sciences	http://bfuhs.ac.in
3	Panjab Agricultural University	http://www.pau.edu
Hindi websites		
1	Central University of Karnataka	http://www.cuk.ac.in
2	Central University of Jharkhand, Ranchi	http://www.cuj.ac.in
3	Indian Institute of Information Technology	http://www.iiita.ac.in
4	Indian Institute of Science	http://www.iisc.ernet.in
5	Indira Gandhi National Tribal University	http://igntu.nic.in
6	National Institute of Mental Health and Neuro Sciences	http://www.nimhans.kar.nic.in
7	National Institute of Technology, Hamirpur	http://nith.ac.in
8	Indian Institute of Technology, Patna	http://www.iitp.ac.in
9	University of Hyderabad	http://www.uohyd.ernet.in
10	All India Institute of Medical Sciences	http://www.aiims.edu
11	Central university, Orissa	http://cuorissa.org
12	Central University of Bihar	http://www.cub.ac.in
13	Central University of Gujarat	http://www.cug.ac.in
14	Central University of Haryana	http://www.cuharyana.org
15	Central University of Rajasthan	http://www.curaj.ac.in
16	Delhi University	http://www.du.ac.in
17	Dr. B.R. Ambedkar, NIT Jalandhar	http://www.nitj.ac.in
18	Guru Ghasidas Vishwavidayalaya, Bilaspur	http://www.ggu.ac.in
19	IIM Ahmedabad	http://www.iimahd.ernet.in
20	IIM Kashipur	http://www.iimkashipur.ac.in
21	IIM Bangalore	http://www.iimb.ernet.in
22	IIT Bombay	http://www.iitb.ac.in
23	IIT Kanpur	http://www.iitk.ac.in

(continued)

(continued)

S. no.	Name of the institute	URL
24	IIT Kharagpur	http://www.iitkgp.ac.in
25	IIT Roorkee	http://www.iitr.ac.in
26	Indian Agricultural Research Institute	http://www.iari.res.in
27	Indian Council of forestry research and Education, Dehradun	http://hindi.icfre.org
28	Indian Institute of Foreign Trade	http://www.iift.edu
29	Indian Institute of Technology, Ropar	http://www.iitrpr.ac.in
30	Institute of Microbial Technology	http://www.imtech.res.in
31	Jagad Guru Ramanandacharya Rajasthan Sanskrit University	http://www.jrrsanskrituniversity.ac.in/
32	Jawahar lal Nehru Centre for Advanced scientific Research	http://www.jncasr.ac.in
33	Jawahar lal University	http://www.jnu.ac.in
34	Jawaharlal Institute of Postgraduate Medical Education & Research	http://www.jipmer.edu/
35	Lakshmibai National Institute of Physical Education	http://lnipe.nic.in
36	Mahatma Gandhi Antarrashtriya Hindi Vishwavidyalaya	http://www.hindivishwa.org
37	Maulana Azad National Institute of Technology, Bhopal	http://hindi.manit.ac.in
38	National Dairy Research Institute	http://www.ndri.res.in
39	National Institute of Fashion Technology	http://www.nift.ac.in
40	Indian Institutes of Science Education and Research	http://www.iisermohali.ac.in
41	NIT Rourkela	http://nitrourkela-hindi.weebly.com
42	PDPM Indian Institute of Information Technology, Design & Manufacturing	http://www.iiitdmj.ac.in
43	Ramakrishna Mission Vivekananda University	http://www.rkmvu.ac.in
44	Rashtriya Sanskrit Sansthan	http://www.sanskrit.nic.in
45	School of Planning and Architecture, New Delhi	http://www.spa.ac.in
46	Shri Lal Bahadur Shastri Rashtriya Sanskrit Vidyapeetha	http://www.slbsrsv.ac.in
47	Sikkim University, Tadong	http://www.sikkimuniversity.in
48	University of Solapur	http://su.digitaluniversity.ac
49	Visvesvaraya National Institute of Technology	http://vnit.ac.in

Appendix 2

PH.WQT: Results (Academic Domain)

Name of the academic institute	Website link	Language	Date of testing	Domain independent (%)	Domain dependent (%)	Aggregate (%)
Punjabi University	http://www.punjabiuniversity.ac.in	English and Punjabi	26.02.14	45.05	68.29	53.03
Baba Farid University	http://bfuhs.ac.in/	English and Punjabi	27.04.14	37.50	12.50	28.73
Indian Institute of Management, Ahmedabad	http://www.iimahd.ernet.in	English and Hindi	13.02.14	61.68	86.25	69.13
Indian Institute of Technology, Bombay	https://www.iitb.ac.in	English and Hindi	17.02.14	55.00	96.25	67.12
Guru Ghasidas Vishwavidayalaya, Bilaspur	http://www.ggu.ac.in	English and Hindi	12.04.14	57.80	73.17	62.50
Delhi University	http://www.du.ac.in	English and Hindi	25.02.14	56.51	67.68	59.85
Maulana Azad National Institute of Technology, Bhopal	http://hindi.manit.ac.in	English and Hindi	29.03.14	50.26	67.50	55.37
Jawahar lal Nehru Centre for Advanced scientific Research	http://www.jncasr.ac.in	English and Hindi	26.03.14	55.59	46.25	52.80
Lakshmibai National Institute of Physical Education	http://lnipe.nic.in	English and Hindi	29.03.14	53.42	45.00	50.74
IIM Lucknow	http://www.iiml.ac.in	English and Hindi	10.04.14	55.73	37.50	50.37
Central University Haryana	http://www.cuharyana.org	English and Hindi	12.04.14	56.05	31.71	48.71
Sikkim University	http://www.sikkimuniversity.in	English and Hindi	10.04.14	48.37	48.78	48.68
IIM. Shillong	http://www.iimshillong.in	English and Hindi	10.04.14	45.36	35.00	42.34
All India Institute of Medical Sciences	http://www.aiims.edu	English and Hindi	02.01.14	47.53	21.25	39.50
Indian Institute of Foreign Trade	http://www.iift.edu	English and Hindi	25.03.14	46.05	22.50	39.07
National Institute of Technology (Jalandhar)	http://www.nitj.ac.in	English and Hindi	23.03.14	42.47	20.00	35.71
Central University, Gujarat	http://www.cug.ac.in	English and Hindi	30.03.14	48.61	6.25	35.58
Jawahar lal University	http://www.jnu.ac.in	English and Hindi	28.03.14	44.29	10.37	33.83
Rashtriya Sanskrit Sansthan	http://www.sanskrit.nic.in	English, Hindi and Sanskrit.	15.04.14	57.40	56.10	57.01
Shri Lal Bahadur Shastri Rashtriya	http://www.slbsrsv.ac.in	English, Hindi and Sanskrit	11.04.14	45.10	57.32	48.73

(continued)

(continued)

Name of the academic institute	Website link	Language	Date of testing	Domain independent (%)	Domain dependent (%)	Aggregate (%)
Sanskrit Vidyapeetha						
Solapur University	http://su.digitaluniversity.ac	English, Hindi and Marathi	10.04.14	54.22	4.88	37.90
Mahatma Gandhi Antarrashtriya Hindi Vishwavidyalaya	http://www.hindivishwa.org	Hindi only	29.03.14	44.12	30.00	39.68

References

1. http://www.internetworldstats.com/stats.htm (Accessed on 10th January, 2013).
2. ISO/IEC 9126: Information technology-Software Product Evaluation-Software Quality Characteristics And Metrics, International Organization for Standardization (ISO/IEC), Geneva, Switzerland, (2001).
3. ISO/IEC CD 25010 Software engineering–Software product Quality Requirements and Evaluation (SQuaRE)–Quality model and guide. (2009).
4. Rupinder Kaur, Vishal Goyal and Gagandeep Kaur, Web Quality Model for Websites Developed in Punjabi and Hindi, The International Journal of Soft Computing and Software Engineering [JSCSE], Vol. 3, No. 3, Special Issue: The Proceeding of International Conference on Soft Computing and Software Engineering 2013 [SCSE'13], San Francisco State University, CA, U.S.A., March 2013 pp. 557–563, doi: 10.7321/jscse.v3.n3.84. e-ISSN: 2251-7545. (2013).
5. Rupinder Kaur and Vishal Goyal, Empirically Validating a Web Quality Model for Academic Websites Developed in Punjabi and Hindi: A Case Study, International Conference on Innovations in Engineering and Technology, International Institute of Engineers (IEE), Bangkok, Thailand. ISBN: 978-93-82242-60-4, (2013).
6. http://www.4icu.org/in/indian-universities.htm. (Accessed on 20th November, 2011).
7. http://en.wikipedia.org/wiki/List_of_universities_in_India (Accessed on 16th November, 2011).
8. Luis Olsino, D Godoy, G.J. Lafuente and Gustavo Rossi, Assessing the Quality of Academic Websites: a Case Study, In New Review of Hypermedia and Multimedia(NRHM) Journal, Taylor Graham Publishers, UK/USA, ISSN 1361-4568, Vol. 5, pp. 81–103. (1999).
9. Zihou Zhou, Evaluating Websites Using a Practical Quality Model, MPhil Thesis, Software Technology Research Laboratory, De Montfort University, (2009).
10. Priyanka Tripathi, Manju Pandey and Divya Bharti, Towards the Identification of Usability Metrics for Academic Web-Sites, Proceedings IEEE Volume 2, pp. 394–397. (2010).

A Robust, Privacy Preserving Secret Data Concealing Embedding Technique into Encrypted Video Stream

B. Santhosh and N.M. Meghana

Abstract With tremendously increasing demand for multimedia communication, it is essential to safeguard the video data from various attacks during its transmission over internet. Video encryption ensures privacy and security of the video content. For the purpose of authentication or to identify tampering if any, secret information is concealed in these encrypted videos. This paper puts forward an efficient and robust methodology for embedding the secret information directly into encrypted videos, which guarantees the confidentiality of the video content. The input video is first compressed using the popular H.264/AVC compression technique and by analyzing the properties of H.264 encoding technique, the three portions containing the sensitive data are encrypted. A standard RC4 stream cipher has been employed for encrypting the codewords of Motion vector differences, Intra prediction modes and Residual coefficients. The secret data can then be hidden by the data hider using a unique Codeword substitution technique without being aware of the video contents. Thus the technique ensures confidentiality of the video content. The hidden data can then be directly extracted by the intended authenticated receiver even without decrypting the video sequence. To validate the feasibility and the performance of the proposed work, the result metrics PSNR, SSIM and VQM have been estimated.

Keywords Video encryption · Authentication · Tampering · H.264/AVC · Codeword substitution

B. Santhosh (✉) · N.M. Meghana
Department of Telecommunication Engineering, DSCE,
Bangalore, India
e-mail: santhoshmehtre@gmail.com

N.M. Meghana
e-mail: megha.meharwade@gmail.com

© Springer India 2016
S.C. Satapathy et al. (eds.), *Information Systems Design and Intelligent Applications*, Advances in Intelligent Systems and Computing 435,
DOI 10.1007/978-81-322-2757-1_64

1 Introduction

With the tremendous usage of internet, digital media such as images, audio and video suffer various attacks from intruders during transmission [1]. To bypass those attacks, it is customary to conceal the secret data in a cover medium, the technique being popularly known as steganography. If the data to be hidden is large, then choosing image or text as the cover medium is definitely not feasible due to several hindrances encountered during transmission. Thus, video based steganography proves to be ideal for hiding considerably huge amount of data. Since video comprises of large number of frames, the attacker faces difficulty in finding out as to which frame comprises the secret data and also due quick display of video frames.

With video being the most popular multimedia data, prior to transmission over internet, these videos are encrypted for safeguarding the content and to avoid unwanted interception [1]. Thus video encryption comes into picture. Pertaining to these discussions, there is a strong need for a robust technique which can hide data directly in encrypted video streams. Especially in cloud computing applications or medical surveillance videos, it is sometimes necessary to maintain the privacy of the video content even from the database manager or data hider. Thus, the paper discusses such a technique of concealing the secret data directly into encrypted video streams, thereby ensuring the privacy of the video content. Since video comprises of large amount of data and occupies a lot of bandwidth during transmission, it is necessary to compress the video data. H.264/AVC, which is the most popular video encoding technique for transmission of videos over internet, has been employed for this purpose [2].

2 Proposed Scheme

The methodology is divided into 3 sections:

A. Selective Encryption of H.264/AVC compressed bit stream
B. Data concealing
C. Data extraction

Figure 1a depicts the block diagram at the sender side.

The input video is compressed using H.264/AVC codec. The sender selectively encrypts the sensitive portions of the video using a standard RC4 stream cipher. Symmetric key encryption is employed thus producing an encrypted video. The data embedder will now hide the secret information into the encoded, encrypted video by incorporating the unique codeword substitution technique. Since the data hider conceals the secret information into the video without knowing the content of the video, this method assures privacy of the video content and is most suitable for cloud applications. In the view of making the hidden accessible only to the authenticated receiver, the secret data is secured with a data hiding key. The data

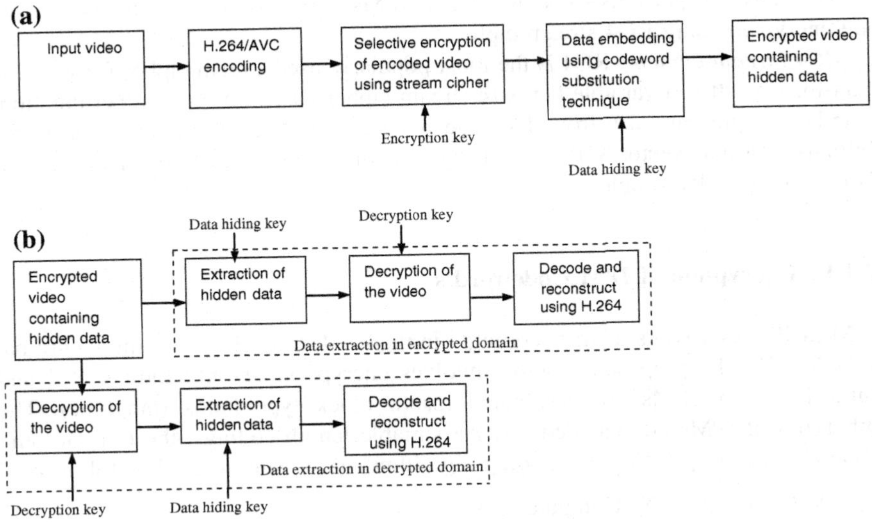

Fig. 1 **a** Block diagram at the sender's side. **b** Block diagram at the receiver side

cannot be retrieved unless the data hiding key of the receiver matches with that at the sender. Hence, the final output at the sender side is an encrypted H.264 video stream comprising the concealed data.

Figure 1b shows the block diagram at the receiver side.

At the receiver side, upon providing the right data hiding key, the authenticated receiver can extract the concealed data directly from the encrypted video. This provision especially helps when the data is of utmost importance and that the video can later be decrypted. Data can also be retrieved from decrypted domain. The decrypted video is then subjected to H.264 decoding and is reconstructed using filters to give the original input video.

2.1 *Selective Encryption of H.264/AVC Compressed Bit Stream*

Encryption of compressed video bit stream is a challenging task. Encryption of the entire bit stream works out to be expensive and time consuming. Hence selective encryption [3] has been adopted in the proposed method. The encoded bit streams are analysed and only selective bits are encrypted keeping considerable portion of the video as it is. Still the video appears to be greatly scrambled one cannot identify the content without decrypting it. Also, since H.264 codec has been used, care must be taken to accomplish format compliance, encryption of all bits results in enormous values disturbing the compliance of the H.264 bit stream.

The sensitive portions—codeword's of IPMs, MVDs and residual coefficients are encrypted using RC4 stream cipher.

RC4 Stream cipher—RC4 is the most popularly used stream cipher for wireless transmission. It is a variable key size stream cipher based on random permutation. The key length can vary from 1 to 256 bytes i.e., 8 to 2048 bits and is used for initialising a state vector S. RC4 works fastest compared to other stream ciphers and is computationally simple.

2.1.1 Encryption of IPM Codeword's

IPM facilitates 4 types of Intra coding—Intra_4 × 4, Intra_16 × 16, Intra_chroma, I_PCM [4]. The proposed scheme involves encryption of only Intra_4 × 4 and Intra_16 × 16. IPMs are designated macro block type fields (mb_type). The mb_types of IPMs are encoded using a variable length coding—the Exp-Golomb code [5]. The Exp-Golomb encoding algorithm can be summarised as follows:

(i) Let mb_type = X. Compute X + 1.
(ii) Represent X + 1 in binary.
(iii) Find the number of digits required to represent X + 1 in binary and denote it as N.
(iv) Append N − 1 zeroes to binary representation of X + 1. Thus the IPM codeword is framed. Table 1 shows the IPM encoding in Intra_16 × 16.

Encryption must be carried out on IPM code words such that the encrypted bit stream is strictly a H.264 format compliant. It is also desired to maintain the same codeword length even after encryption. It can be inferred from the table that the CBP is same for every four rows and every two lines have the same codeword length. For example mb_type 3 and 4 have the same codeword length. A careful observation yields that the last bit contains the sensitive information. Hence the last bit of every IPM codeword is XOR-ed with standard RC4 stream cipher so that the video appears unintelligible.

Table 1 IPM encoding (Intra_16 × 16)

mb_type	Intra_16 × 16 prediction mode	Chroma CBP	Luma CBP	IPM codeword
1	0	0	0	010
2	1	0	0	011
3	2	0	0	00100
4	3	0	0	00101
5	0	1	0	00110
..	..	…	….	…..

Table 2 MVD encoding using extended Exp Golomb coding

Motion vector differences	Code_num	MVD codeword using Exp-Golomb coding
0	0	1
1	1	010
−1	2	011
2	3	00100
−2	4	00101

Example: IPM codewords	010	011	00100	00101
Stream cipher	1	0	0	1
Encrypted codewords	011	011	00100	00100

2.1.2 MVD Encryption

IPM encryption alone will not prove to be secure enough since it is necessary to encrypt both texture and motion information. Motion vector is 2-D vector used in Inter prediction. Motion vector difference is obtained by performing motion vector prediction on motion vectors.

The Exp-Golomb coding is extended to encode MVDs.

Each MVD is expressed in terms of code_num which is an integer.

Table 2 shows encoding of MVDs.

The last bit of MVD codeword is encrypted with stream cipher. Though the XOR operation results in sign change of the codeword, the codeword length remains the same and is fully format compliant with H.264.

2.1.3 Encryption of Residual Coefficients

For ensuring better scrambling, the residual coefficients of both I and P frames are also encrypted.

The residual coefficients are encoded using Context Adaptive Variable Length Coding (CAVLC). CAVLC operates on every macroblock. Each macroblock (16 × 16) is subdivided into 4 × 4 submacro blocks. The coefficients of each submacro block are scanned in a zig zag manner (Fig. 2).

CAVLC encoding flow is as follows:

The format of CAVLC codeword is:

{Coefficient_token, Sign of trailing 1s, Levels, Total_zeros, Run_before}

Fig. 2 Steps in CAVLC encoding

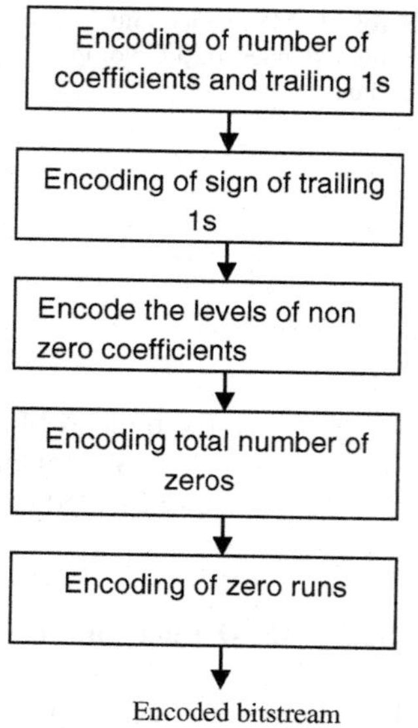

With the view of maintaining H.264 bit stream compliancy, there is no change done to the Cofficient_token, Run_before and Total_zeros during encryption of CAVLC codewords. Only the codeword's of Sign of trailing 1s and Levels are involved in encryption. The codeword of Sign of trailing 1s is XORed with the stream cipher.

Levels are non zero coefficients which are scanned in zigzag order during encoding. The Levels are encoded with reference to several VLC lookup tables. The codeword of Levels relies on suffix length.

Table 3 depicts the encoding of Levels based on suffix length.

The last bit of the Levels' codeword is encrypted with RC4 stream cipher [6]. This, though causes a change in the sign of Levels, but still maintains the same codeword length and is absolutely H.264 format compliant.

2.2 Data Concealing

The data bits are concealed only in the encrypted code words of Levels having suffix length greater than 1, since code words having suffix length less than 1 are of unequal length. A unique codeword substitution technique is adopted which must not modify the sign bit of the levels since the sign bit of the levels is encrypted and any modification on the sign during data embedding might result in insecure encryption.

Table 3 CAVLC codewords for levels

Suffix length	Level (>0)	Codeword	Level (<0)	Codeword
0	1	1	−1	01
	2	001	−2	0001
	3	00001	−3	000001
	4	0000001	−4	00000001
1	1	10	−1	11
	2	010	−2	011
	3	0010	−3	0011
	4	00010	−4	00011
	5	000010	−5	000011
	6	0000010	−6	0000011
	7	00000010	−7	00000011
	8	000000010	−8	000000011
2	1	100	−1	101
	2	110	−2	111
	3	0100	−3	0101
	4	0110	−4	0111
	5	00100	−5	00101
	6	00110	−6	00111
	7	000100	−7	000101
	8	000110	−8	000111
	9	0000100	−9	0000101
	10	0000110	−10	0000111
	11	00000100	−11	00000101
	12	00000110	−12	00000111
	13	000000100	−13	000000101
	14	000000110	−14	000000111

The necessary constraints to be met by the embedding technique:

(i) After the codeword substitution, the resultant codeword must still be a H.264 format compliant, so that it can be easily decoded at the receiver.
(ii) The resultant codeword after data concealing must have the same codeword length so that there is no increase in bit rate.
(iii) There can be only minimum degradation in video quality caused by codeword substitution.

The embedding is done only into the Levels cod words of P-frames. This is because, I frames being the most prominent frames during video transmission, error if any in I frame propagates to the subsequent frames and distorts the video content.

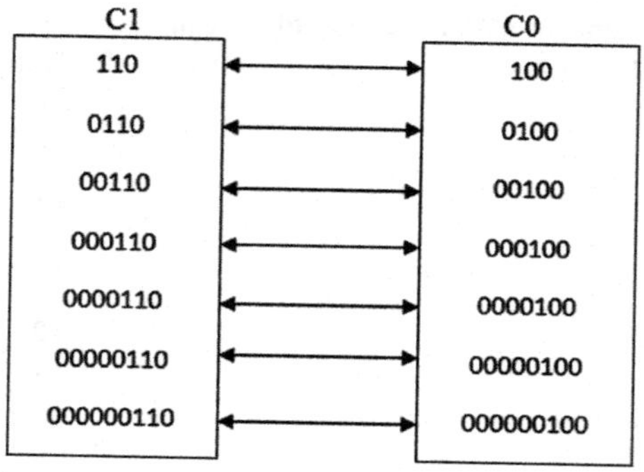

Fig. 3 Codeword mapping between code spaces C0 and C1 for suffix length = 2 and level > 0

The codeword replacement algorithm is as follows:

Step 1 Convert the data to be concealed into a stream of binary digits.
Step 2 The code words of Levels having suffix length greater than 1 are divided into two code spaces—C0 and C1 alternately.
Because, the levels code words of suffix length 0 and 1 have unequal lengths.
Step 3 Frame the mapping between code spaces C0 and C1 as shown in Fig. 3.

2.3 *Data Extraction*

At the receiver side, data can be extracted either in encrypted domain or in decrypted domain.

Data extraction in encrypted domain is useful especially when the data embedded by the database manager has to be manipulated. In this case, the database manager need not decrypt the video for the data manipulation; instead data can be directly extracted in encrypted domain. This feature is extremely useful in medical and surveillance fields. Also, in order to ensure that the data is available only to the authenticated receiver, we provide with a data hiding key, without which the receiver cannot extract the data. For extraction of data in encrypted domain, the steps are as follows:

Step 1 The encrypted bit stream is parsed to detect the codeword of Levels.
Step 2 If the codeword belongs to code space C0, the extracted data bit is 0;
If the codeword belongs to code space C1, the extracted data bit is 1.

For extracting the data in decrypted domain:

Step 1 The code words of Levels are detected.
Step 2 The encrypted video is decrypted using the same stream cipher with the appropriate decryption key which matches with the one employed for encryption. Symmetric encryption is employed.
Step 3 The rest of the procedure is similar to data extraction in encrypted domain.

3 Results

The work was implemented in MATLAB 2013 version software. The scheme was tested on different video sequences contained in different video containers (.avi, .wmv, .mp4). The secret data included both text and image data. The code was tested on nearly 60–70 frames for embedding text data and image data.

The data is hidden only in P-frames, frame rate being 30 frames/second, frame width = 320, frame height = 240.

The RC4 stream cipher proved to resist attacks, and the encrypted video was scrambled efficiently as shown in Fig. 4b thus providing perceptual security. Encryption is nothing but adding noise. The decoded video after data extraction suffered minimal degradation as shown in Fig. 4d. Figure 4a–d depict the results.

VQM is another video quality metric which is roped in the perceptual effects caused due to video impairments including blur effects, noise, jerks, color distortion etc. It is basically a subjective video quality assessment. VQM is calculated taking into account, the input original video sequence and the processed video. Also features from processed video are extracted and compared with the features of original video. Lower the VQM better is the perceptual quality of the processed video.

Table 4 shows the result metrics values computed for different values of Quantization Parameter. (QP), where QP indicates the amount of compression.

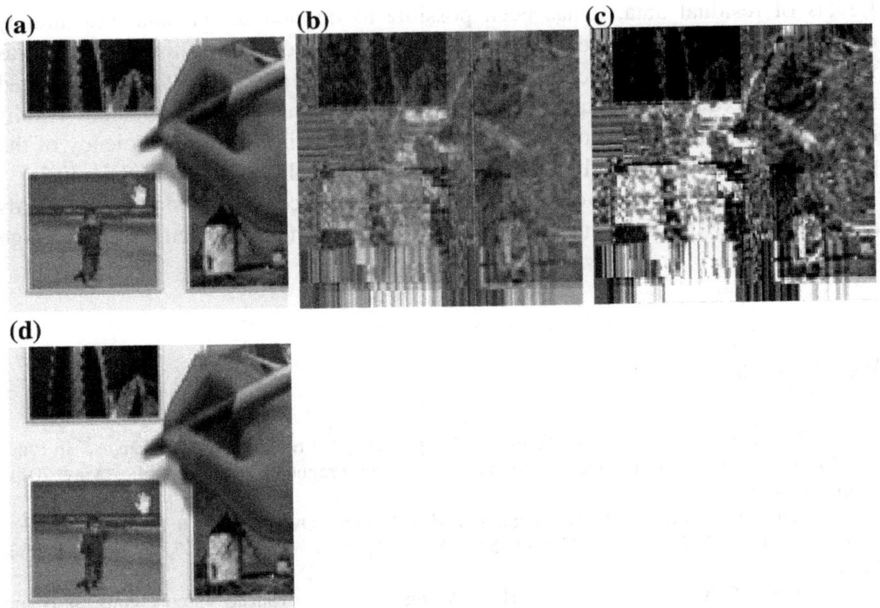

Fig. 4 **a** Input video frame. **b** Selectively encrypted video frame. **c** Encrypted frame containing hidden data. **d** Decoded frame

Table 4 Results

Quantization parameter (QP)	Video sequence	Compression ratio (CR)	MSE	PSNR (dB)	SSIM	VQM
24	.avi	5.4311	6.0967	40.4151	0.98314	0.9996
	.mp4	10.1335	7.1279	39.7018	0.98653	0.9984
	.wmv	13.4762	3.8088	40.5653	0.97807	1.0003
25	.avi	5.694	6.7206	40.1179	0.97998	1.0007
	.mp4	10.6108	8.7872	38.8322	0.98504	1.0006
	.wmv	14.3669	4.2794	40.0724	0.97591	0.9998
26	.avi	6.201	7.8228	39.4548	0.9775	0.9994
	.mp4	12.0954	9.4101	38.5532	0.98294	0.9988
	.wmv	16.2814	4.8715	39.5167	0.97397	0.9997

4 Conclusion

Due to increase in illegal intruder interference in multimedia communication during the transmission over the channel, there is a necessity for concealing the secret information into the encrypted video directly for preserving the privacy of such video data especially in cloud management. An attempt has been put forward to satisfy this objective by encrypting the IPM code words, MVD code words and the code words of the residual data. Using the codeword replacement technique for the code words of Levels of residual data, it has been possible to implant secret data into already encrypted video, thereby not leaking the actual video content. This technique was found to preserve the confidentiality of the video content and allows the authenticated receiver to extract the data in the encrypted domain. The experimental results involving the computation of PSNR, SSIM and VQM implied the efficiency of the proposed method wherein the degradation in video quality was found to be little.

A future enhancement would be to try out the replacement of MVD code words also to conceal the data bits, which would provide more capacity to hide large amount of data.

References

1. W. J. Lu, A. Varna, and M. Wu, "Secure video processing: Problems and challenges," in Proc. IEEE Int. Conf. Acoust., Speech, Signal Processing, Prague, Czech Republic, May 2011, pp. 5856–5859.
2. T. Wiegand, G. J. Sullivan, G. Bjontegaard, and A. Luthra, "Overview of the H.264/AVC video coding standard," IEEE Trans. Circuits Syst. Video Technol., vol. 13, no. 7, pp. 560–576, Jul. 2003.
3. S. G. Lian, Z. X. Liu, Z. Ren, and H. L. Wang, "Secure advanced video coding based on selective encryption algorithms," IEEE Trans. Consumer Electron., vol. 52, no. 2, pp. 621–629, May 2006.

4. D. W. Xu, R. D. Wang, and J. C. Wang, "Prediction mode modulated data-hiding algorithm for H.264/AVC," *J. Real-Time Image Process., vol. 7, no. 4, pp. 205–214, 2012.*
5. D. W. Xu and R. D. Wang, "Watermarking in H.264/AVC compressed domain using Exp-Golomb code words mapping," *Opt. Eng., vol. 50, no. 9, p. 097402, 2011.*
6. Z. Shahid, M. Chaumont, and W. Puech, "Fast protection of H.264/AVC by selective encryption of CAVLC and CABAC for I and P frames," IEEE Trans. Circuits Syst. Video Technol., vol. 21, no. 5, pp. 565–576, May 2011.

Trusted Execution Environment for Data Protection in Cloud

Podili V.S. Srinivas, Ch. Pravallika and K. Srujan Raju

Abstract Cloud Computing has become a major part of all organizations throughout the world because of the services offered such as Iaas, Saas, Paas and wide availability of these services. In spite of the benefits of cloud services they must consider how the security and privacy aspects are ensured, as users are often store some sensitive information with cloud storage providers which may be un trusted. In this paper, we are going to discuss about a novel approach where the Protection to data as included as a new service which will reduce the per application development cost to provide a secure environment and also provides secure access to data stored in public clouds.

Keywords Data protection in cloud · Cloud services · Cloud data protection

1 Introduction

The new concept of Cloud computing pretends a major change in how we store information and run applications. Instead of running programs and data on an individual desktop computer, everything is hosted in the cloud which a huge group

P.V.S. Srinivas (✉)
Department of Computer Science and Engineering, Gokaraju Rangaraju Institute of Engineering and Technology, Bachupally, Hyderabad 500090, India
e-mail: pvs.srinivas@griet.ac.in

Ch. Pravallika
Department of Computer Science and Engineering, CVR College of Engineering, Vastunagar, Mangalpally, Ibrahimpatnam (M) 501510, Telangana, India
e-mail: pravallika@cvr.ac.in

K. Srujan Raju
Department of Computer Science and Engineering, CMR Technical Campus, Kandlakoya (V), Ranga Reddy (D) 501401, Telangana, India
e-mail: India.ksrujanraju@gmail.com

© Springer India 2016
S.C. Satapathy et al. (eds.), *Information Systems Design and Intelligent Applications*, Advances in Intelligent Systems and Computing 435,
DOI 10.1007/978-81-322-2757-1_65

of computers and servers are connected together for accessing via the Internet. Cloud computing providing various services which are attracting the small and medium companies to use cloud computing as a resource for deploying their applications [1].

1.1 Cloud Architecture

The Major Services provided by cloud platform include Platform as a Service (PaaS) which provides platform on which application can build, test and deploy. Which will allow the users to pay based on the time they use the platform. Second, Infrastructure as a Service (IaaS), provides the service in the form of infrastructure which will reduce the cost for users to build and maintain the infrastructure. Thirdly, Software as a Service (SaaS) which provides required software online that you can use without installing any software on your own PC [2]. Cloud computing also includes deployment models which represents a specific type of cloud environment that can be selected by users based on their requirements. The Diagram shows various services and deployment models available in the cloud. This paper maintly we are discussing about how CSP can achieve the trustworthiness of users by providing more secure data storage approaches as a service (Fig. 1).

1.2 Security in Cloud

In Cloud Computing environment, it allows the users to lease the resources from a single service provider and store the data on its corresponding storage server. This type of services will decreases the costs of software and hard ware, but it may introduce a new security breach or violation because of the availability data over

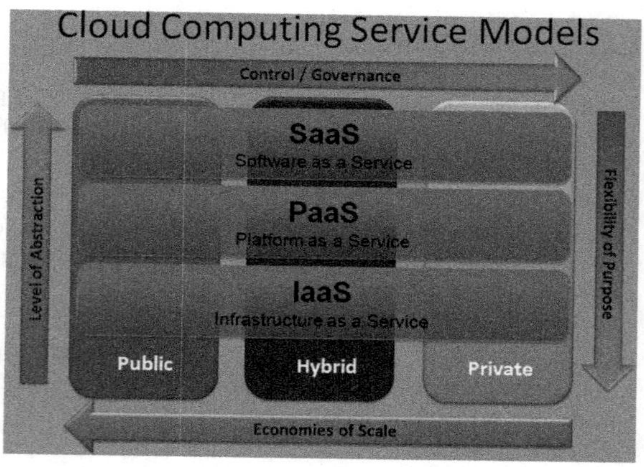

Fig. 1 Cloud computing services and deployment models

storage may disclose information malicious users inside and for privileged users [3]. Even if we follow encryption strategy to encrypt the data by using various Key algorithms before deploying it on to the cloud may create some unauthorized disclosure to managing authority and administrators of Data providers [4]. Because these people are going to maintain the keys information and encryption algorithm information [5].

The below are three sensitive issues that comes as result of processing the data remotely [6].

- Personal data is transmitted to cloud server,
- Sending the processed information back to client from cloud server
- Storing the client sensitive information on the cloud server which is not owned by him

All the above three issues of cloud computing are severely prone to security breach that makes the research to focus on providing security features to the data stored on cloud data server [7]. Several security and privacy challenges need to be taken care for effective implementation of data security approach in cloud which is going to be used in many applications.

This paper mainly focuses on the issues related to the providing data security in cloud computing. As data and information will be shared with a third party, cloud computing users want to avoid an untrusted cloud provider while Protecting their private and important information, such as credit card details, Bank details from attackers or malicious insiders. In the Proposed Approach we are going to provide integrity for the data uploaded by users as well as secure access to the user data.

2 Architecture of the Proposed System

Trustworthiness is one of the major aspects of users for storing their data on cloud. If the cloud develops an approach for providing the trustworthiness to the users then it becomes more popular. Data protection as a services enforces various access control policies on data through application confinement and information verification. It offers it through robust log in facilities as well as well audit records. DPaaS is more beneficial for small companies where they don't have much internal security experts.

The proposed approach can be used to improve the efficiency of cloud by giving better data protection to user data which will make the cloud become more popular by providing trustworthy cloud services among users. DPaas uses features from both the formal encryption strategies full Disk Encryption (FDE), fully homomorphic encryption (FHE). It will Places the encryption mechanism on sharable items like FHE and at the same time it will maintain the performance of FDE by using symmetric encryption techniques. We are using one of the best encryption technique AES (Advanced Encryption Standard) for symmetric encryption of data and forming data caspsules. Here, we can apply methods where it will enable searching without decrypting the data discussed by Majid et al. [8]. We can also maintain Access Control Matrix where it is going contain capability list and users list for each and every data item that is going to be included in the cloud data server. This architecture will provide a common platform for all the users by moving all access control polices and key management principles into middle tier (Fig. 2).

In the above architecture we have a trusted platform module which acts as medium for crypto processing where it is used to encrypt each of the data bit before it is stored on the cloud storage servers. DPaaS system combines the features of logging, access control, key management aa a whole to make it easier to verify the flow of information. Trusted Platform Module (TPM) is a secure crypto processor that can store cryptographic keys that protect information. The TPM specification is the work of the Trusted Computing Group for providing a secure environment for sharing the data. A Secure Data Capsule is a data bit encrypted and has some security policies for accessing it. We are going to use the Access control and Debugging information for giving access to these secure capsules. Secure Execution Environment provides a greater performance because we are using data marshalling techniques. Third party auditor is used to monitor the activities that are performed on shared data items. Administrator will give the permission to auditor to view all the user data without any key to verify the integrity of data. On request a client get the information about the operations and any modifications performed on his data capsules. Auditor works on behalf of a client to identify the unauthorized access by logging the information about who accessed and what are the operations performed on a data capsule.

By using the ACL (Access Control List) one can restrict the access to the data items which can be modified only by the creator of that list. DPaaS supports access

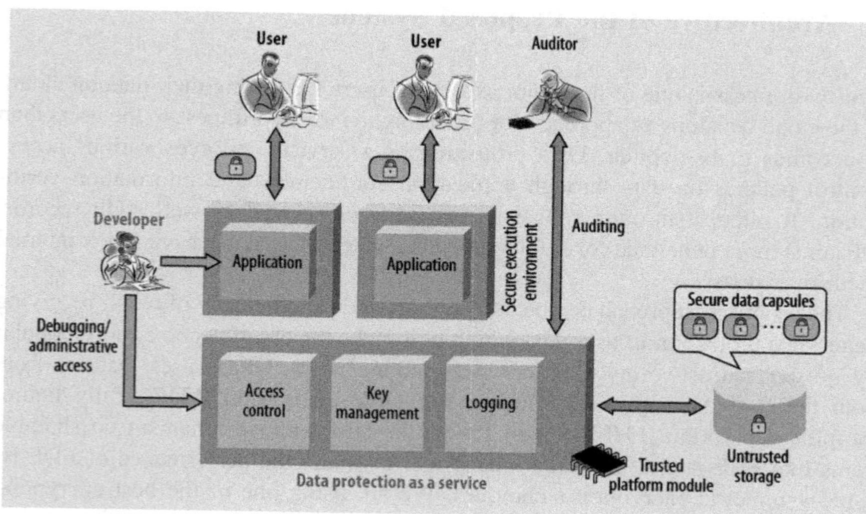

Fig. 2 Architecture of DPaaS system [1]

control list on secure data capsules. The key to enforcing those ACLs is to control the I/O channels available to the SEEs. To confine data, the platform decrypts the SDC's data only when the security policies are met. The deciphered text may be passed to the client directly or to a SEE. DPaaS can guarantee the integrity of the data by applying cryptographic authentication of the data in storage and by auditing the application code at runtime in this approach benefits both developers and users.

3 Conclusion

As we have discussed above if the user wants to share his sensitive information on the cloud, providing the security to that is very much essential. As vast availability of cloud services many users and applications are going to use the services provided by cloud. In this perspective this system will provide a secure execution environment in a cloud by using logging, key management and access control in a secure way using AES encryption techniques and also by maintaing ACL for protecting data shared on cloud which is going to be benefited for many users and cloud providers for providing trustworthiness to customers.

4 Results and Discussion

In the Proposed approach we mainly divided the functionality among 3 i.e., Admin, Users, Auditor. Where User can upload his data on to the cloud which will be further saved in a encrypted form which can be seen only by the authorized users with the key. Admin is allowed to view and maintain the information about files uploaded by the users. He can monitor the changes made by other users and auditor. The admin is not capable of viewing the data that is presented in the file uploaded by users. Where Auditor can view the data and he can maintain log information about files. If any changes done on the file can be sent to the user on request. The Auditor has a special privilege to see the data and maintain the integrity of it.

The following shows the data flow diagrams of admin, auditor and user.

4.1 Data Flow Diagrams

See Figs. 3, 4, 5, 6, 7 and 8.

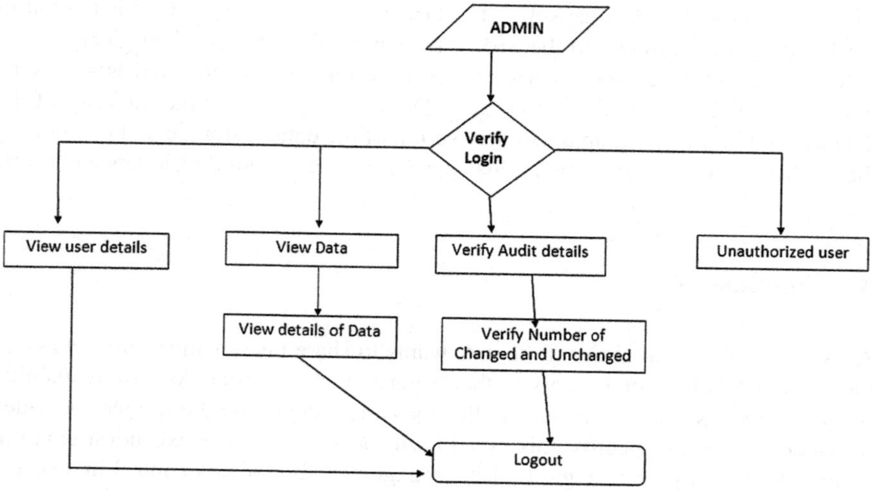

Fig. 3 Flow of activities by admin

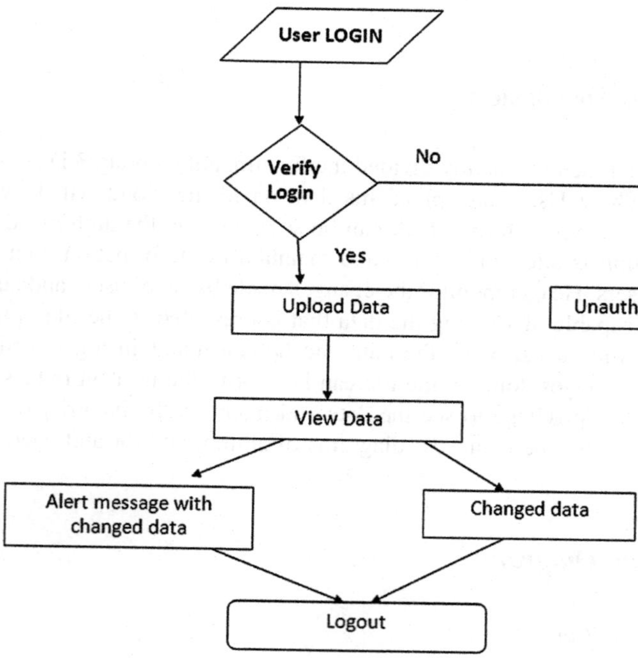

Fig. 4 Flow of activities by user

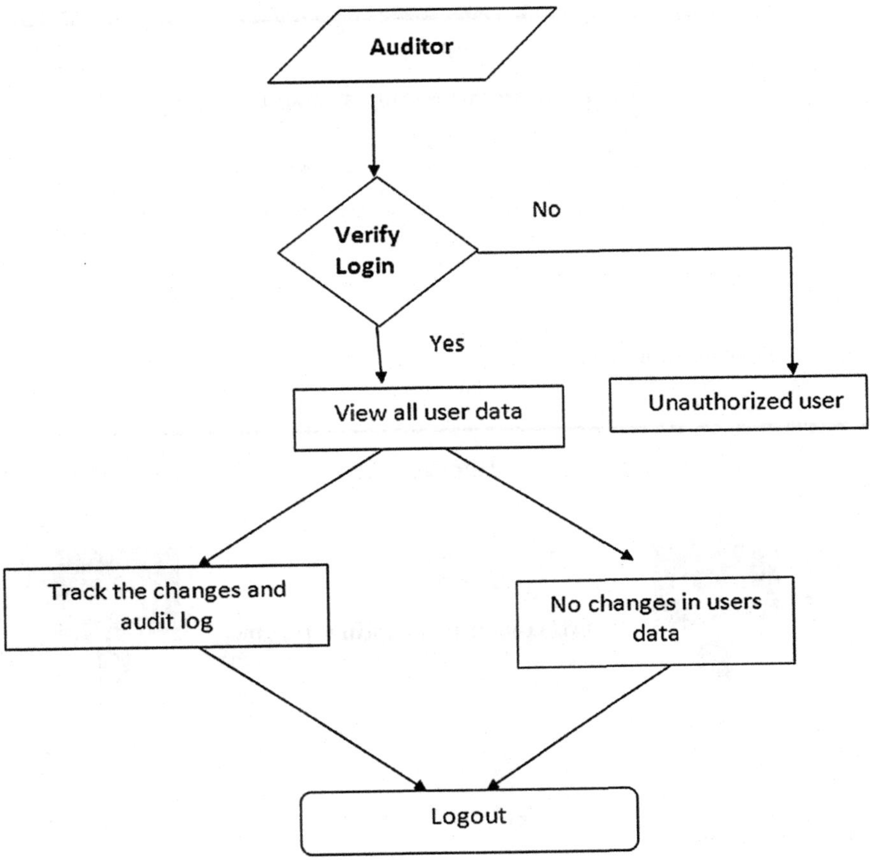

Fig. 5 Flow of activities by auditor

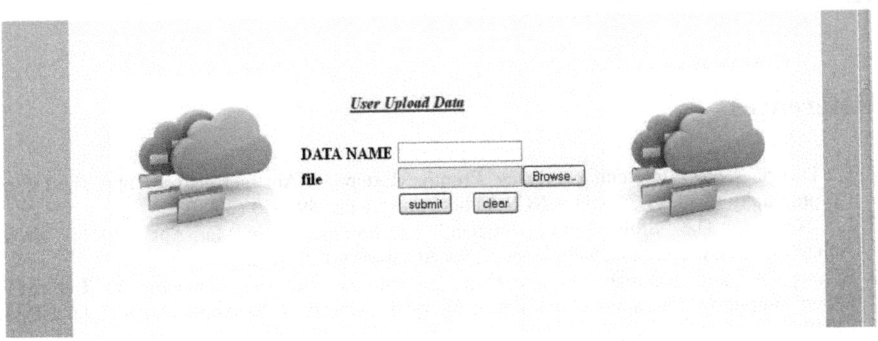

Fig. 6 User uploading the data

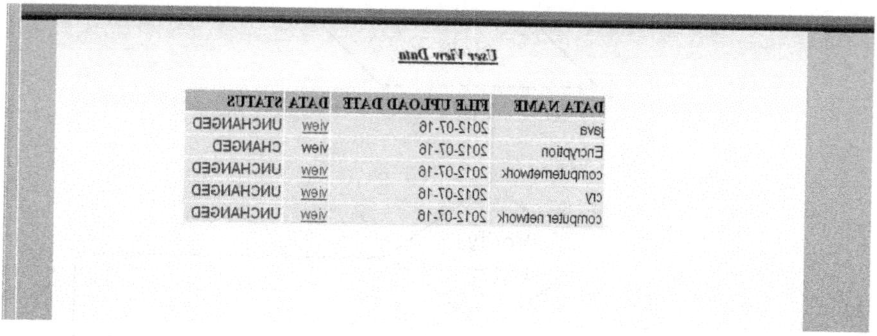

User View Data
DATA NAME
FILE UPLOAD DATE
DATA STATUS
java 2012-07-16 view UNCHANGED
Encryption 2012-07-16 view CHANGED
computernetwork 2012-07-16 view UNCHANGED
cry 2012-07-16 view UNCHANGED
computer network 2012-07-16 view UNCHANGED

Fig. 7 Viewing the status of files

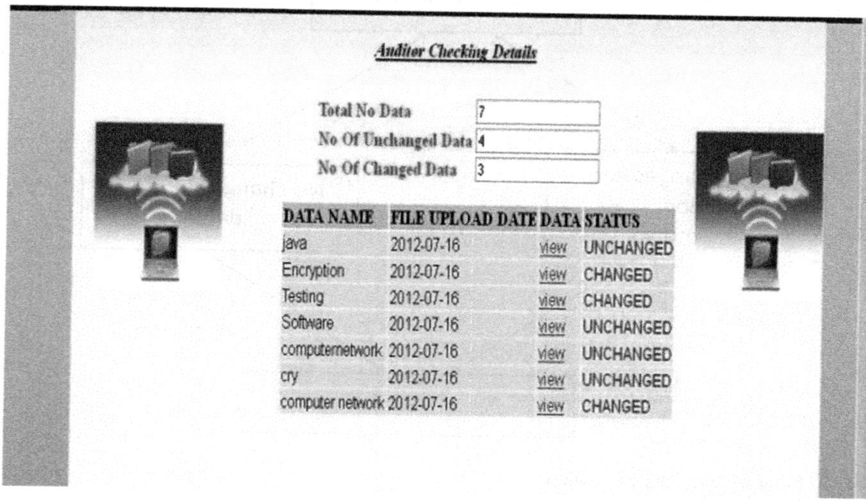

Auditor Checking Details
Total No Data 7
No Of Unchanged Data 4
No Of Changed Data 3
DATA NAME
FILE UPLOAD DATE
DATA STATUS
java 2012-07-16 view UNCHANGED
Encryption 2012-07-16 view CHANGED
Testing 2012-07-16 view CHANGED
Software 2012-07-16 view UNCHANGED
computernetwork 2012-07-16 view UNCHANGED
cry 2012-07-16 view UNCHANGED
computer network 2012-07-16 view CHANGED

Fig. 8 Auditor verifying the data

References

1. C. Dwork, "The Differential Privacy Frontier Extended Abstract," Proc. 6th Theory of Cryptography Conf. (TCC 09), LNCS 5444, Springer pp. 496–50 (2009).
2. E. Naone, "The Slow-Motion Internet," Technology Rev.,Mar./Apr. 2011; www.technologyreview.com/files/54902/GoogleSpeed_charts.pdf (2011).
3. Bisong, A. and Rahman, S. An Overview of the Security Concerns in Enterprise CloudComputing. International Journal of Network Security & Its Applications, 3(1), 30–45. (2010).
4. F. B. Shaikh, S. Haider, "Security Threats in Cloud Computing", 6th International Conference onInternet Technology and Secured Transactions, Abu Dhabi, United Arab Emirates (2011).

5. H. Takabi, J.B.D. Joshi, G.J. Ahn, "Security and Privacy Challenges in Cloud Computing Environments", IEEE Security & Privacy, Vol: 8, No:6, pp 24–31, 201 (2007).
6. W.A. Jansen, "NIST, Cloud Hooks: Security and Privacy Issues in Cloud Computing", Proceedings of the 44th Hawaii International Conference on System Sciences, Koloa, HI, 4–7 (2011).
7. Pvssrinivas et all, "Deploying an Application on the Cloud" IJACSA Vol 2, No 5, 2011, pp. 119–125. (2011).
8. Majid Nateghizad, Majid Bakhtiari, Mohd Aizaini Maarof "Secure Searchable Based AsymmetricEncryption in Cloud Computing", Int. J. Advance. Soft Comput. Appl., Vol. 6, No. 1, March 2014 (2014).

Erratum to: Information Systems Design and Intelligent Applications

Suresh Chandra Satapathy, Jyotsna Kumar Mandal, Siba K. Udgata and Vikrant Bhateja

Erratum to: S.C. Satapathy et al. (eds.), *Information Systems Design and Intelligent Applications*, Advances in Intelligent Systems and Computing 435, DOI 10.1007/978-81-322-2757-1_32, 10.1007/978-81-322-2757-1_33

The book was inadvertently published with an incorrect author's name as Shivalik Mahapatra in Chaps. "A New Block Least Mean Square Algorithm for Improved Active Noise Cancellation" and "An Improved Feedback Filtered-X NLMS Algorithm for Noise Cancellation", whereas it should be Shibalik Mohapatra in both the chapters. The book and the chapters are updated for the same.

The updated original online version for this chapter can be found at
DOI 10.1007/978-81-322-2757-1_32
DOI 10.1007/978-81-322-2757-1_33

S.C. Satapathy (✉)
Department of Computer Science and Engineering,
Anil Neerukonda Institute of Technology and Sciences, Visakhapatnam, India
e-mail: sureshsatapathy@ieee.org

J.K. Mandal
Kalyani University, Nadia, West Bengal, India
e-mail: jkm.cse@gmail.com

S.K. Udgata
University of Hyderabad, Hyderabad, India
e-mail: udgatacs@uohyd.ernet.in

V. Bhateja
Department of Electronics and Communication Engineering, Shri Ramswaroop Memorial Group of Professional Colleges, Lucknow, Uttar Pradesh, India
e-mail: bhateja.vikrant@ieee.org

© Springer India 2016
S.C. Satapathy et al. (eds.), *Information Systems Design and Intelligent Applications*, Advances in Intelligent Systems and Computing 435,
DOI 10.1007/978-81-322-2757-1_66

Erratum to: Information Systems Design and Intelligent Applications

Suresh Chandra Satapathy, Jyotsna Kumar Mandal, Siba K. Udgata and Vikrant Bhateja

Erratum to:
S.C. Satapathy et al. (eds.), *Information Systems Design and Intelligent Applications*, Advances in Intelligent Systems and Computing 435, DOI 10.1007/978-81-322-2757-1_32, 10.1007/978-81-322-2757-1_33

The book was inadvertently published with an incorrect author's name as [illegible] for chapters "A New Block Least Mean Square Algorithm for Improved Active Noise Cancellation" and "An [illegible] Cancellation", whereas it should be [illegible] in both the chapters. The book and the chapters are updated for the same.

The updated original online version for this chapter can be found at:
DOI 10.1007/978-81-322-2757-1_32
DOI 10.1007/978-81-322-2757-1_33

S.C. Satapathy (✉)
Department of Computer Science and Engineering
Anil Neerukonda Institute of Technology and Sciences, Visakhapatnam, India
e-mail: sureshsatapathy@ieee.org

J.K. Mandal
Kalyani University, Nadia, West Bengal, India
e-mail: jkm.cse@gmail.com

S.K. Udgata
University of Hyderabad, Hyderabad, India
e-mail: udgatacs@uohyd.ernet.in

V. Bhateja
Department of Electronics and Communication Engineering, Shri Ramswaroop Memorial Group of Professional Colleges (SRMGPC), Lucknow, Uttar Pradesh, India
e-mail: bhateja.vikrant@gmail.com

© Springer India 2016
S.C. Satapathy et al. (eds.), *Information Systems Design and Intelligent Applications*, Advances in Intelligent Systems and Computing 435,
DOI 10.1007/978-81-322-2757-1_69

Author Index

© Springer India 2016
S.C. Satapathy et al. (eds.), *Information Systems Design and Intelligent Applications*, Advances in Intelligent Systems and Computing 435,
DOI 10.1007/978-81-322-2757-1

Printed in the United States
By Bookmasters